Living the Lunar Calendar

Living the Lunar Calendar

edited by

Jonathan Ben-Dov, Wayne Horowitz
and John M. Steele

Oxbow Books
Oxford and Oakville

Published by
Oxbow Books, Oxford, UK

ISBN 978-1-84217-481-4

This book is available direct from:

Oxbow Books, Oxford, UK
(Phone: 01865-241249; Fax: 01865-794449)

and

The David Brown Book Company
PO Box 511, Oakville, CT 06779, USA
(Phone: 860-945-9329; Fax: 860-945-9468)

or from our website

www.oxbowbooks.com

A CIP record for this book is available from the British library

Library of Congress Cataloging-in-Publication Data

Living the lunar calendar / edited by Jonathan Ben-Dov, Wayne Horowitz and John M. Steele.
p. cm.
Includes bibliographical references.
ISBN 978-1-84217-481-4
1. Calendars--History--To 1500. 2. Calendar, Assyro-Babylonian 3. Calendar, Greek. 4. Calendar, Egyptian. 5. Calendar, Jewish. I. Ben-Dov, Jonathan. II. Horowitz, Wayne, 1957- III. Steele, John M.
CE25.L59 2012
529'.3--dc23
2012008148

Printed and bound in Great Britain by
Information Press, Eynsham

Contents

Foreword

In the winter of 2010, during the full moon of the Hebrew month of Shevat (*Tu Bishvat*), the Jewish New Year for Trees, the Bible Lands Museum Jerusalem proudly hosted the "Living the Lunar Calendar" conference. The conference brought together scholars from around the world to discuss a wide range of issues relating to the tracking of time, and the calendar in diverse cultures, ancient and modern, ranging from those of the Ancient Near East and Biblical Israel, Qumran and early Christianity, and broadening our scope at the Museum to include classical and medieval Europe, the native peoples of the Americas, and China and Japan. Conference participants, including speakers, and members of the Bible Lands Museum Jerusalem community, had the opportunity to share a common experience of discovery, both on site at the Museum and on a visit to the Dead Sea and the archaeological site at Qumran. Over the three days of the meeting our participants, experts and laymen alike, examined familiar and unfamiliar civilizations and text-traditions that all shared the common human experience of marking the passage of time by means of the universe around us, in particular the movements of the Moon, the Sun, and the Stars.

The Bible Lands Museum is a universal centre for people of all faiths dedicated to the study, understanding and appreciation of the cultures of the Ancient Near East. As Israel's largest museum dedicated solely to the ancient history of this region, the Museum reflects the rich city within which it exists: Jerusalem today remains deeply influenced by the lunar cycles of the Jewish, Islamic, and Christian liturgical calendars. Dictating the ebb and flow of our calendar year and the intertwining of religious celebrations and worship, the Museum is a natural home for this type of program, given its ongoing commitment to academic excellence, and to facilitating public participation in high level educational programming. It is in answer to both these goals that we at the Bible Lands Museum Jerusalem are pleased to be able to once again share in the "Living the Lunar Calendar" experience, this time through the written pages of this volume.

I would like to express my gratitude to the many outstanding participants in this conference and publication, and in particular to our partners in coordinating and organizing the conference, Prof. Jonathan Ben-Dov of Haifa University, Prof. Wayne Horowitz and Prof. Shalom Paul of the Hebrew University of Jerusalem, Prof. John Steele of Brown University and Bible Lands Museum Jerusalem Curator, Dr. Filip Vukosavović.

We are deeply indebted to the many supporters who made this conference possible, including Mr. Henry Zemel of the Caeno Foundation, without whose initial enthusiasm this entire conference may not have been organized. In addition we would also like to thank The Bible Department at the University of Haifa, The Orion Center for the Study of the Dead Sea Scrolls and Associated Literature at the Hebrew University of Jerusalem, Qumran National Park, Israel Nature and Parks Authority and The Haifa Forum for the History of Science for their support and assistance.

In closing, I would like to thank Carolyn Budow Ben-David and the staff of the Bible Lands Museum Jerusalem for all of the hard work that ensured a most successful conference from start to finish, and to John Steele for helping to preserve the research and knowledge that is shared in this publication for future study and greater understanding in this fascinating and timeless field.

Amanda Weiss
Director
Bible Lands Museum Jerusalem

Introduction

The opening scene of Aristophanes' *Clouds* famously shows how the protagonist, Strepsiades, fears the nearing arrival of the new moon, because on that day he would need to pay his debts. But Strepsiades is also well aware that the declaration of the new moon in Greek cities could be announced either *kata selene*—i.e. according to the actual sighting of the moon—or *kata archon*, i.e. according to the present needs of the ruler. For example, while a declaration of an intercalary month could be expected on astronomical grounds, a given ruler could choose to postpone the extra month if he needed the new year's tax payments urgently. Another anecdote related to the new moon is found in the Babylonian Talmud: the Jewish sage R. Ḥiyya, upon viewing the new moon, throws clods of earth at it because it destroys the pre-calculated scheme assigned to it by the rabbinic court (b.Rosh Hashana 25:1, cf. the Palestinian parallel story in y.Rosh Hashana 12:2, which mentions pebbles of stone instead of clods of earth, due to the different geology of the region).

These anecdotes from classical Greek comedy and rabbinic literature illustrate the main themes of the present volume. Calendars have often been subject to a qualitative study, surveying administrative texts as well as scientific treatises and ritual prescriptions, in order to ascertain the calendar or calendars practiced under the respective society, and more work remains to be done in understanding the technicalities of historical calendars (some examples are included in the present collection). But the articles collected here represent not only the mathematical dimension of calendar reckoning—they also consider the effect of the great forces of ancient history on the calendar: politics, identity, social cohesiveness, cultural hybridity, and ultimately the basic questions of human civilization, namely how does mankind enforce order on the endless flow of natural phenomena. Time is one of the most basic categories—if not the most basic one—in the rational matrix by which human beings create meaning in the world. This was acknowledged on the philosophical level by Immanuel Kant, Henri Bergson and Martin Heidegger, to name only a few. It has been applied to the study of historical and contemporary calendars by M. P. Nilsson in his groundbreaking *Primitive Time-Reckoning* (1920) and by E. Zerubavel in his *The Seven Day Circle* (1985). Some recent prominent studies which have integrated the study of the calendar into the fabric of culture and society include Jürgen Rüpke's *Kalender und Öffentlichkeit* (1995), Sylvie Anne Goldberg's *La Clepsydre* (2000), and Anthony Aveni's *Empires of Time* (1989). The papers in this volume too consider questions of how calendars and the management of time are embedded within the political, social, religious and other aspects of historical cultures, based upon careful study of primary source material.

Scientific Indeterminacy and Political Intervention

The most profound conflict of the lunar calendar is its indeterminacy, which comes to the fore both in the irregular length of months—29 or 30 days—and in the periodic need to intercalate the year, inserting an additional month in order to align the year with the march of the seasons. While the latter problem can be managed through the use of cycles of intercalations (for example the 19-year Metonic cycle), accurate calculation of whether the new moon will be visible on a given day is not possible. A variety of factors such as weather conditions and the acuity of the eye of the observer hamper the modeling of the moon's visibility even today. Human societies have thus needed to address these indeterminate situations: When should commodities for the festival be arranged if it might be postponed in the last minute? How are taxes and interest to be calculated along an indeterminate number of days? The article by Patrizia Marzillo addresses a related problem in the form of the Greek 'Old and New Day', a term which appears both in Aristophanes' *Clouds* and in the Archaic piece *Works and Days* by Hesiod. This ambiguous marker of time raised much speculation in post-classical times, when calendar reckoning stabilized under the later Roman Empire. Marzillo discusses the allegorical meaning attached to this term in scholia from Late Antiquity. Using Pythagorean number manipulation and other cultural templates connected with the moon and the sun, the Neo-Platonist philosopher Proclus highlighted the idea of rest connoted with the day of conjunction.

The indeterminacy of the calendar creates conspicuous lacunae in the routine conduct of society. These lacunae, in turn, call for an intensive involvement of political institutions and other organs of power, who stand in the breach by undertaking the duty—and the privilege—to make effective calendrical decisions. In some cultures time reckoning was associated so closely with kingship that it functioned not only at the level of ritual (the New Moon, the New Year), but also on the level of myth, anchoring the regulation of time in the metaphysical image of kingship. The article by John Steele traces such a phenomenon in Chinese tradition—the emperor's role in maintaining cosmic harmony and possessing the mandate of heaven—and compares it with the apparently more mundane administrative and cultic reasons for the regulation of the calendar in Mesopotamia.

It is curious to note that the question of authority over the calendar arose not only in societies which practiced *ad hoc* observations, but also in China and Mesopotamia where calendars based on calculations were in use. The Jewish tradition is a prominent example of a culture that gave a central role to political considerations in the calendrical procedure. Since this cultural tradition transmits more information on the policy of calendar-making than any other ancient tradition, it is discussed in the present volume quite extensively. Sacha Stern dedicates his article to the political context of the procedure for observing the new moon, as described in the Mishnah and later rabbinic sources. Stern stresses the character of this procedure as a judicial, even forensic, mode of activity. In his opinion, the judicial flavour derives from the procedures of the *boule*, the city council in Palestinian cities during the Roman period. Unlike earlier authors, Stern does not see the civil authority of the rabbinic court as an established fact, but rather submits that the rabbis had to compete with other organs of power. Hence the judicial character of the rabbinic procedure, which was aimed to compete with the force of the city council, otherwise the most natural regulator of the calendar. Stern assigns little place to the role of ideological factors in the calendrical realm.

Robert Hannah discusses the importance of the calendar in another aspect of national identity: the organization of the ancient Greek Olympic Games which brought together people from across the Greek city states. In an attempt to reconstruct when the games were held—and, crucially, how people in different parts of the Greek world knew when the games were to take place—Hannah discusses the conflicting testimonies found in a handful of scholia which must be correlated with the most obvious time constraint: the games must be held at a season in which fresh olive leaves are available for making the victors' wreaths! Hannah suggests a novel solution, which should be taken in account by ancient historians who study this emblematic expression of pan-Hellenism.

Schematic vs. Observational Rulings

During a relatively short interval in the Jewish tradition, a schematic calendar of 364 days was practiced by apocalyptic circles in Hellenistic Palestine. The most famous attestation of this calendar appears in the Dead Sea Scrolls from Qumran, and in the earlier books of 1 Enoch and Jubilees. This tradition arose out of priestly circles that promulgated the generative role of the week and other heptad time units in the construction of the calendar. This tendency more or less disappeared with the destruction of the Jerusalem Temple in 70 CE, with more mainstream Jewish traditions left to find their way between the solar Julian calendar of the empire and the more well-rooted luni-solar tradition of the Levant. Jonathan Ben-Dov discusses how the 364-day calendrical tradition in the Dead Sea Scrolls interacts with the luni-solar calendar. This is done on the basis of comparison with some Ptolemaic Egyptian texts, which were similarly required to synchronize their schematic (365-day) year with the (Macedonian) lunar calendar. The result is that, while the role of the moon is not altogether ignored in the Qumran texts, it is diminished with regard to the 364-day framework. Here, the role of the Moon is less pronounced than in the Egyptian synchronistic calendars. The reason for this must be ideological on the part of the sectarian authors.

Ron Feldman gives a different view of the distinction between the luni-solar and the 364-day calendars in the early Jewish tradition. Using a method from cultural studies he distinguishes frameworks of Wild Time from those of Tame Time, and assigns the different Jewish perspectives to these distinctive concepts. To him, the various Jewish positions are an outcome of varying ideological attitudes to the encounter between mankind and nature. Finally, Lawrence Schiffman supplies a comprehensive survey of the contradicting trends—observation and calculation—in Israelite and Jewish calendars, from the Hebrew Bible to the Middle Ages.

Every historical text which aims to predict astronomical phenomena must depend on a scheme of some sort. The paper by Michael Gorodetsky sheds light on a fascinating period in history in which the schemes available to astronomers were not particularly accurate, but in which the genealogy of astronomical schemes is more valuable than their actual content. As part of his survey of medieval Russian manuscripts, Gorodetsky studies the Russian tables of Kirillo-Belozersky, which record computed lunar phases and eclipses for a period of 19 years in the 14th century. Submitting these tables to a thorough statistical analysis using modern ephemerides, Gorodetsky is able to trace the origins of the tables to the city of Belgrade, thus shedding welcome light on the otherwise unknown discipline of astronomy in this part of the southern Slavic region in the late Middle Ages.

Tradition and the 'Other'

Calendars are in many ways the 'applied science' branch of astronomy, with this feature of the calendar naturally being more pronounced in calculation-based systems rather than in those based upon observation. Thus, an intercalation method usually reflects the achievements of astronomical science at the time of its inception. What should be done, however, when the underlying system is out of sync with the real passage of the seasons, due to solar and lunar anomalies and other elements that had not been considered at the time of the institution of the system? This is a significant challenge for traditional societies because calendrical schemes tend to acquire more prestige than the mere instruments of calculation that they are. Associated with prominent kings and patriarchs, they also serve as the basis for a whole set of civil and ritual statutes that are not easily amended. The range of sources discussed in Schiffman's article ends before this problem arose with the traditional Jewish calendar, as it is presently encountered. Contemporary Jewish authorities fail to face the upcoming crisis, which has already led to the occasional celebration of Passover on a later date than is theoretically permitted. The substantial article by Susan Tsumura gives some food for thought in this matter, as it describes the gradual recognition of the same problem in China and Japan in the 15th – 16th centuries CE. Not only was the Metonic cycle discarded in favour of more accurate systems, the whole set of rituals dependent upon it had to be gradually revised, to the dismay of calendar traditionalists. Tsumura draws a long history of compromises reached in the Chinese and Japanese states, including the current one which will lead to a calendrical crisis in Japan in 2033, and thus requires urgent action on the part of the authorities. In terms of social history this challenge is interesting because, now at the dawn of the 21st century, the functioning organs of traditional societies no longer retain their original modes of operation. The ever-turning wheel of history impacts on many areas of human experience; the calendar is one realm where this transition is easily detected and pointed out to the public, while on the other hand the regulation of the calendar has far-reaching implications for the most basic aspects of human existence.

The adoption of the Julian year—and its later Gregorian modification—by Christian authorities made life relatively easy for Christians in modern history. However, the solidification of Christianity's foundational calendars and rituals was not at all smooth. This long stabilization process of Christian identity was part of what is commonly called 'the Parting of the Ways' with Judaism. Once again we encounter how historical topics, hotly debated by theologians and philologists, can be more easily tracked by their ramifications in the realm of the calendar. The most notorious problem of early Christianity in this regard was the mode of fixing the date of Easter. Christians were not satisfied with fixing this festival according to the Jewish date of the Passover, and instead sought ways to fix the date of Easter by means of an independent Christian method. This debate produced many conflicting mechanisms, and lead to enormous advances in theory and practice. A glimpse into the materialization of this problem is supplied here by the article by Mark Dickens and Nicholas Sims-Williams (with contributions by T. A. Carlson and C. Reck). This article supplies a publication and commentary of a dozen or so previously unpublished Christian calendrical fragments from the environs of Turfan in North West China from the 9th to the 12th centuries. Written in Syriac and Sogdian (some are Sogdian in Syriac script), these fragments reflect the vicissitudes of the calendar in this distant branch of the Eastern Christian oikoumene. The church in this region preoccupied itself—like many other Christian institutions elsewhere—with the timing of Lent and Easter. The special point in these texts is the synchronization of three different calendar systems—Syrian, Sogdian, and the Chinese 12-year animal cycle – each using different sets of month names.

The Easter computus in another corner of the Christian world appears in the essay by Daniel Mc Carthy. Continuing his earlier work on Anatolius' *De ratione paschali*, Mc Carthy examines the principles adopted in this 4th century CE treatise in order to harmonize the Julian calendar with the luni-solar Jewish year. Having been preserved in Latin translation and spread mainly in the West, this book preserves a curious list of the Hebrew month names and provides an easily rhymed rule-of-thumb for the calculation of the day in the lunar month equivalent to a given Julian date. This rhyme is found both in Old Irish and in a Bergamesque language from the Italian Alps. All this attests to an extraordinary acceptance of Anatolius' principles in the Latin West during the early Middle Ages, no doubt due to their elegance and usefulness.

As it turns out, Christian authors maintained ambiguous relations with the Jewish calendar. In fact, in most cases the debates and agreements were held not against the *true* Jewish calendar—either the contemporary one or the one practiced by Jews at the time of the Crucifixion—but rather with *imagined* Jewish calendars. The complexity of the Jewish-Christian debate thus becomes manifest in calendrical writings. Here, Christian literati mediate the alien Jewish customs to their fellow Christians, while making use of, sometimes even manipulating, the inherent tension between the present-day Jews, the biblical Hebrews, and the Jews from the time of Christ. Philip Nothaft discusses several occasions in which medieval Christians made recourse to the Jewish Calendar, whether a real or imagined one. Paul of Burgos, a 14th century rabbi converted to a Bishop, constitutes the pinnacle of this game of identities, as he applies a christological twist to the Talmudic calendrical sophistry, producing a hybrid of Jewish and Christian dates.

Axioms Revisited

The discussion of lunar calendars is based on a set of agreements that are hardly contested. One of these truisms is that lunar months in the great majority of cultures (except possibly ancient Egypt) begin with the sighting of the first crescent at sunset. That is, day 1 of the lunar month begins with first lunar visibility (in the evening at the western horizon just after sunset) and continues through the following morning until the subsequent sunset. Leo Depuydt, however, contests this notion in his paper. He considers the question of 'beginning of the day – from morning or from the preceding evening', to be irrelevant in daily life since most people were active during daytime only. He then notes the absence of clear statements about the structure of lunar day 1 in any ancient culture, and proceeds to evaluate the evidence for various options for the beginning of the month in the different lunar calendars practiced in Ptolemaic Egypt.

Yigal Bloch uses a wide variety of administrative cuneiform texts to examine the nature of Assyrian calendars of the second millennium. He claims that a purely lunar, non-intercalated calendar was in use in Assyria in the 13th – 12th centuries BCE. Based on his earlier studies on the Assyrian eponym list, he attempts to reconstruct the chronology of this period, while also asking some pertinent questions about the relationship of theoretical time with real life experience—particularly the hard realities of the annual agricultural cycle—in a world where time is regulated (at least in official documents) by a purely lunar calendar.

From Time Indicators to Calendars: Early Lunar Reckonings

Stanislaw Iwaniszewski bases his article on a distinction drawn by M. P. Nilsson in his epoch-making study of primitive time. Nilsson distinguished societies which use markers such as lunar phases as mere indicators for the passage of time from those who use the markers as units in a comprehensive time system. While all other articles address either fully literate or fully illiterate societies, Iwaniszewski's paper examines the transition between these two stages as they find expression in the calendar. His paper draws an exhaustive survey of the uses of lunar phases as time markers throughout the Americas, as divided into the categories mentioned above. The division correlates somewhat with geographical markers, as Mesoamerican and South American societies retained more comprehensive frameworks than those in use in North America. The article supersedes the standard discussions of calendars in the Americas, as it analyzes the evidence by means of the special point of view of the *lunar* calendar in particular. It then continues to draw various ways in which lunar months were aligned with the seasons or with six-month-long half-years.

A curious 'time indicator' in an otherwise fully-calendrical culture is brought forth in the article by Wayne Horowitz on the day of the Sun god in Mesopotamia. While in ancient Mesopotamia time was predominantly marked by the moon, as Horowitz demonstrates from a variety of examples, some time-markers persisted without any clear connection with the lunar calendar. Such is the celebration of a sacred say for Shamash, the Sun God, on the twentieth of each month. Horowitz surveys the religious significance of this day along the use of the signs ^{d}20 to denote the sun god. He suggests some possible directions of deducing astronomical significance from this number, both within the cuneiform tradition and outside it.

Two articles address pre-calendrical societies—or at least societies that lacked the graphical means to express a systematic calendar. Sabine Beckman, in the only contribution in this volume that is explicitly oriented to art and iconography, suggests a 'calendrical' or at least 'seasonal' reading of the iconography of the 'Blue Bird Fresco' from Minoan Knossos. While the Minoan kingdom certainly did employ a calendar of some sort, being an active member in the burgeoning global interaction of the Middle Bronze Age, no datum of astronomy or calendars has reached us in a clear enough way to be interpreted as such. Instead, Beckman offers to read this oeuvre of landscape art as a decoded reference to the Minoan 'calendar', i.e. a depiction of the annual march of the seasons. Saffron, iris, lily, pomegranate and many other plants are analyzed, before she departs from the images of the 'Blue Bird Fresco', buttressing her conclusions with a variety of later classical literary sources. Art history merges with folklore to draw the image of a calendar lost from the grasp of textual scholarship.

James Walton presents the only article in the present collection that uses methods of Archeoastronomy. Basing himself on the more easily analyzable calendar of modern Hopi tribes in New Mexico, Walton reconstructs lunar observations carried out in the area of Chaco and Mesa Verde in what is now New Mexico and Colorado around the 8th – 12th centuries of our era. Tracing the appearance of the moon along appointed times of the year, he shows how the people that constructed these massive adobe housing and administration edifices calibrated the appearance of the moon in conjunction with prominent landmarks, and deduced from them regular calendrical units: months, years, seasons, and even intercalated months.

Conclusion

The themes outlined here are only a sample of the research directions arising from calendrical and astronomical material. This material paves the way for further study by philologists, political historians, theologians, historians of science, and others. But above all the calendar is a primary expression of cultural history, epitomizing in a myriad of ways a plethora of human cultures, cognitive faculties, invented traditions, expressions of identity and nationality, epistemological patterns, and ideology—whether explicit or implicit—which underlie cultural identity. The articles collected here give the reader but a small first taste of these intellectual riches.

Acknowledgements

In addition to the individuals and institutions named in the forward, the editors would like to thank The Bible Lands Museum and its director Amanda Weiss for hosting the Living the Lunar Calendar conference, The Caeno Foundation and Henry Zemel for their contributions to the original meeting, and Rebecca Barclay for her beautiful cover design. Moreover, we would like to thank the speakers at the conference and contributors to this volume for engaging with us in a dialogue about issues of mutual interest and concern over the millennia, and throughout a myriad of cultures, that lived and still live by the rhythms of our shared Moon.

Jonathan Ben-Dov
Wayne Horowitz
John M. Steele

Sunday in Mesopotamia[1]

Wayne Horowitz

When my colleagues and I began to share ideas about the nature of The Living the Lunar Calendar conference, we conceived of a multi-cultural/cross civilization type meeting, where scholars from different fields would come together to discuss a topic of mutual interest—the calendar and the Moon.[2] Thus, it is a small irony that my own paper below deals primarily with the Sun. As such, lets begin then with a review of the Sun and Moon in the world view of the Ancient Near East.

1. The Sun and The Moon in The Ancient Near East, Time Keeping, and Creation

In the Ancient Near East, the Moon-god was the most important deity of the three astronomical siblings. The Moon-god Nanna-Sin was the big brother, and the Sun-god (Utu-Shamash) and Venus (Inanna-Ištar), his little brother and little sister. This reflects, in my opinion, the paramount role of the Moon in time keeping where the Moon has a role in determining not only the month by means of its phases, but also the day and the year. The day, by marking each day of the month by its specific shape, location, and time of rising or setting, for example, the crescent new moon on the western horizon immediately after sunset marking the first of the month; the full moon on the eastern horizon across from the setting Sun on the western horizon marking the middle of the month; and the old Moon and no Moon marking the end of the Month. Thus, any given date of the year, for example the 20th day of the first month Nisan, could be either counted from the new moon of the month, or established by observation of the moon's phases in the sky.

As for the year, it too was primarily measured in Ancient Mesopotamia by means of the Moon. A cycle of 12 new moons marking the passage of regular years, with the occasional leap year of 13 new moons. So, numerous examples of actual observations of the Moon in the context of calendar reckoning in Ancient Mesopotamia, for example in the Neo-Assyrian astronomical reports published by H. Hunger in SAA 8, but which also posit a place for the Sun and the stars in time keeping. For example, for the stars see SAA 8 98, where observations of the stars prompt a call to intercalate the calendar:

> Let them intercalate a month; all the stars of the sky have fallen behind. Adar (Month XII) must not pass unfavourably; let them intercalate.

Likewise, for the Sun, there are numerous examples of astronomical reports relating to the length of the day and night around the time of the spring equinox, and so the new year in

Nisan, for example SAA 8 140–142:

> On the 6th/15th/broken date of Nisan the day and night were in balance: 6 double-hours of daylight, 6 double-hours of night.

Beyond this day to day practical approach, Mesopotamian creation narratives offer a more theoretical, or comprehensive picture of the role of the Moon, Sun, and stars in determining time. Examples include an Akkadian creation vignette that is placed at the end of the lunar section of *Enuma Anu Enlil* as a sort of epilogue to the lunar eclipse omens. Here the Moon-god Sin and Sun-god Shamash take part in the creation of day and night, month and year:[3]

> When Anu, Enlil, and Ea, the great gods,
> heaven and earth built, fixed the astronomical signs;
> established the stellar-positions, [se]t fast the stellar-locations;
> the gods of the night they . [. .]., divided the paths;
> the stars, the likenesses [of them they drew, the constellations;]
> night (and) day, as equa[ls? they measure] d, month and year they created;
> for Sin and Shamash, . . [, the decisions of heave]n and earth they (Anu, Enlil, and Ea) determined.

Parallels where the gods in effect create both astronomy and time include a passage in the bilingual Exaltation of Ištar, known from late-period tablets from Uruk, where the Sun and Moon-gods are again mentioned in the context of the creation of day and night, and the organization of the stars;[4] the small literary fragment K. 7067;[5] and finally, the opening lines of *Enuma Elish* Tablet V, in what might be called the '*Astronomical Book of Enuma Elish*', for which my teacher W.G. Lambert has recently provided a full new English translation in Lambert (2008). Here in *Enuma Elish* V, it is the Babylonian Marduk who assigns the stars, Moon-god Nanna-Sin, and Sun-god Shamash to their jobs in the sky; first ordering the stars to mark the passage of the year (*Enuma Elish* V: 1–8), then the Moon-god Nanna-Sin to mark the days of the month by his phases (beginning with *Enuma Elish* V 11), and finally Shamash to do his duties in a passage that begins somewhere around line 23, in what unfortunately remains the last long broken passage of the epic. Yet, despite the breaks, what is expected of the Sun-god seems clear. The Sun-god must concern himself with justice (lines 23–26):

> 23. I have [....] . the sign, follow its track,
> 24. Draw near. . (. . . . .) give judgment.
> 25. . [. . .] . Shamash, constrain [murder] and violence,
> 26. . [. . . . .] . me.

With the Sun's calendrical astronomical duties relating to day and night expressed much later, in lines 41–42, and 45–46:

> 41. At the new year [. . .
> 42. The year . . . . [. . .

45. After he (Marduk) had [. . .
46. The watches of night and day [. . .

In this context, the reference to the 29th day back in line 36 may perhaps refer to the 29th of Adar, Mesopotamian New Year's Eve, at the end of the 12th and last month of the year:[6]

35. At the end [. . .
36. Let there [be] the 29th day [. . ."

Thus, here in *Enuma Elish* V, as is generally the case in Ancient Mesopotamian civilization, the Sun-god comes in third as a timekeeper, after the Moon and the stars, but does seem to have some responsibility for the passage of day and night, and also the year, perhaps by means of the determining when to intercalate by noting the length of day and night at the spring solstice as in the reports from SAA 8 noted above.[7]

But what exactly is the Sun's role as timekeeper in Ancient Mesopotamia. For day and night, the answer seems obvious. When the Sun-god is present in the sky, it is daytime. When he is absent, it is night. Further, one can mark the parts of the day by observing the Sun's progression across the sky from east to west, and even keep a closer record of solar time by means of a gnomon or sundial.

For the month and year, things are not so simple and not so solar. In this regard, often the Sun's role seems to be defined in terms of the Moon. For example, Mul-Apin Tablet II offers a section correlating the north-south position of the Sun with the seasons of the year; here with north and south defined in terms of the three traditional paths of the stars in the sky—the northern Path of Enlil, central Path of Anu, and southern Path of Ea:

From the 1st of Adar to the 30th of Iyar, the Sun travels in the Path of Anu,
breeze and warm weat[her]
From the 1st of Sivan to the 30th of Ab, the Sun travels in the Path of Enlil,
harvest and heat.
[Fr]om the 1st of Elul to the 30th of Arahsamnu, the Sun travels in the
Path of Anu, breeze and warm weather.
[From the 1st] of Kislev to the 30th of Shevat, the Sun travels in the
Path of Ea, cold weather.[8]

This passage observes quite correctly that the Sun is located in the north (the Path of Enlil) in summer, south (Path of Ea) in winter, and in between (the Path of Anu) in spring and fall. Yet, this observation regarding the Sun cannot be detached from the Moon and stars. The Paths of Anu, Enlil, and Ea are most properly stellar paths,[9] while the dates given for the change of the seasons are ideal lunar dates; the first of the month, the day of the new moon for Adar, Sivan, Elul, and Kislev (Months XII, III, VI, and IX).

Further, the seasons of Mul-Apin are different than our own. Ours start with the equinoxes and solstices, for example the first day of winter on December 21st and the first day of Summer on June 21st, while the solstices and equinoxes in the Mul-Apin system fall midway through the Mul-Apin seasons. For example, the spring equinox on the 15th Nisan in Mul-Apin (Mul-Apin II i 19–21, ii 21) falls midway into the period that the Sun

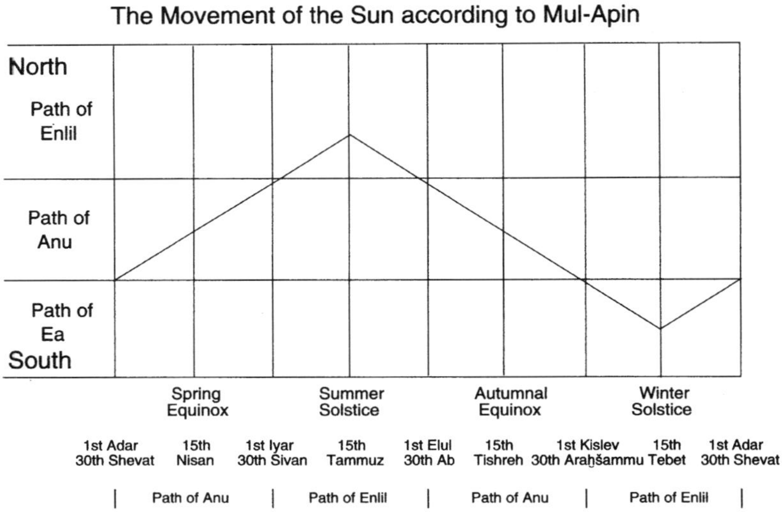

is in the Path of Anu in late Winter/early Spring. Thus, even the solstices and equinoxes in Mul-Apin, what we would imagine to be purely solar phenomena, are in fact solar-lunar, or even solar-lunar-stellar phenomena, here marked by the position of the Sun, with reference to new or full moons, and the stellar paths.

There are, however, a set of examples in cuneiform where the Sun may be liberated from the Moon, where the Sun-god has his own day which is not the day of the new moon or full moon, and so is not a key day in the monthly lunar cycle. This day, is the 20th of the month, with the 20th of the first month of the year, Nisan the 20th, being the Sun's day par excellence, what we might call Sunday in Mesopotamia.

2. The Sun and the 20th of the Month

Our examination of the Sun, Nisan the 20th, and the 20th of each month, follows in the footsteps of a number of number of previous studies which have noted this special relationship. These include Benno Landsberger way back in 1915 in his *Kultische Kalendar*,[10] W.G. Lambert in 1960 in *Babylonian Wisdom Literature*,[11] Rivka Harris in her 1975 book *Ancient Sippar*,[12] M.E. Cohen in his 1993 *Cultic Calendars of The Ancient Near East,*[13] A. Livingstone's study in the same year,[14] and finally Stefan Maul in the 1999 Festschrift of Johannes Renger in a discussion of rituals for the Sun-god in his holy city of Sippar.[15]

Our first piece of evidence will be a little piece of cuneiform gammatria, the obvious observation to those trained in cuneiform studies, that the Sun god's special number (20) matches his day—this according to a tradition whereby the major gods of Ancient Mesopotamia each had a special number, for example 60 for the King of Heaven Anu, 50 for the King of Earth Enlil, 40 for Ea, 30 for the Moon-god (= the number of days of the month), 15 for Ištar, and so on.[16] Hence, when writing a god's name, one could write the divine determinative DINGIR, and then the number. For example ^{d}30 for the Moon-god and ^{d}20 for the Sun-god.

Now the evidence for the 20th of the month, more specifically the 20th of Nisan, and the Sun-god. We begin with the great Akkadian hymn to the Sun-god, *The Shamash Hymn*, itself consisting of 200 (i.e. 20 × 10 lines), that was edited by Lambert in *Babylonian Wisdom Literature*, where the Sun's day on the 20th is a day of joy, when the Sun partakes of beers and ale, and then delivers petitioners from harm:

> On the 20th day (UD.20.KAM) you exult with mirth and joy,
> You eat, you drink their pure ale, the bartender's beer from the market.
> They pour out the bartender's beer for you and you accept.
>
> You deliver safely people surrounded by mighty waves.
> In return you receive their pure, clear libations
>
> You drink their mild beer and ale,
> The you fulfil the desires they conceive.[17]

Two more passages in Lambert's *Babylonian Wisdom Literature* confirm this date as the Sun's day. First a passage included by Lambert in his group 'Popular Sayings', which alludes to the prayer to the Sun-god, 'Shamash, the 20th day is your bright day".[18]

> The fowler cast his net persistently prayed to Shamash: "Shamash, the 20th day (UD.20.KAM) is your bright day."

Our final example gives the text of this prayer, actually an incantation:[19]

> Incantation: "Shamash, the 20th day (UD.20.KAM) is your bright day"
> The 20th day is bright, Ebabbar (your temple) is brig[ht]
> Just as on the 20th day your eyes[ight] is bright
> May their eyesight [too be bright?. . . . .]
> An Enu[ru] incantation

What is happening in this incantation, apparently to cure eye problems, is a little unclear, but I think we can agree with Lambert who writes:[20]

> A comparison of the three passages suggests that the twentieth day was the occasion of asking favours of the Sun-god when he could least refuse," apparently because of his beer induced mirth in The Shamash Hymn.

Yet what is happening in these passages is for us now of less concern than the observation that the three passages all make clear that it was popular belief that the 20th of each month was the Sun's day. This common knowledge also finds reflection throughout the width and breath of cuneiform tradition. For example the late-Babylonian ritual tablet BM 50503 from the Sun's city Sippar from the first millennium (Neo or Late-Babylonian), edited by Stefan Maul in the Festschrift of Johannes Renger, which speaks of rituals of the Sun-god on the 8th and 15th of the month, which can be taken as the time of, or just after, quarter and full moon, and then later in the month on the 20th, a date which has no obvious lunar resonance.[21] Likewise, in the *Lipšur* litanies it is the Sun-god who is invoked on the 20th of the month to release, absolve, free (*pašāru*) a supplicant from sin, divine anger, a curse, an oath etc:[22]

> [U]D.20.KAM *lip-šur šá* ^{d}*Šamaš* (UTU)
> May the 20th day absolve, that of Shamash

3. The 20th of Nisan and Astronomy

A further connection between the Sun-god, his festivals, and the 20th of the month is suggested in Mesopotamian astronomical texts where the 20th of Nisan marks the beginnings of annual astronomical cycles. For example, in another passage from Mul-Apin, this time Mul-Apin I iv 10–14, where the beginning of a series of annual observations of *ziqpu*-stars (stars which culminate overhead an observer of the sky),[23] is set on the 20th of Nisan:

> If you are to observe the *zipqu*, you stand
> in the morning before sunrise, West to your right,
> East to your left, your face directed towards South;
> on the 20th of Nisan the *kumāru* of the Panther stands in the middle of the sky
> opposite your breast, and the Crook rises.

So too the end of another *ziqpu*-star list on the Neo-Babylonian astronomical fragment BM 38269+77242, which I myself edited in Horowitz (1994):

20. [A tota]l$^{?}$ of 12 leagues (360°) of the circle of the *zi*[*qpu*-(stars)]
21. amidst the stars of the Path of [Enlil]

---

22. From (the constellation) ŠU.PA to . [. .]
23. which the observer of the sky [sees] at [night]
24. and the risings and settings of the s[tars in their midst]

---

25. Each day, one degree, the star[s from the morning]
26. into the evening go [in]

---

27. Each day, one degree, the star[s]
28. from the evening into the [morning go out]
29. In the month of Nisan, on the 20th . [. . .
30. . [. . .

(end of fragment)

In both these *ziqpu*-star texts, it would appear that the annual cycle of the *ziqpu*-stars begins on the 20th of Nisan, with the second example describing the sequence of *ziqpu*-stars as a circle (12 leagues in Babylonian geometry = 12 × 30° = 360°); this stellar circle being realized at a rate of 1° per day of change in stellar position. This, without doubt, refers to the Ancient Mesopotamian ideal 360 day year of 12 months × 30 days, with this year somehow culminating, or at the very least somehow connected with the 20th of Nisan.

Again our line 29:

> 29. In the month of Nisan, on the 20th . [. . .

In any case, both here and in Mul-Apin, I would argue, the 20th of Nisan marks the start of an annual stellar sequence, thus making the 20th of Nisan, given its connections with the Sun-god, a sort of solar-stellar New Year's Day.

This is also the case, for the Sun-god Shamash at least, in two Old Babylonian period oracular inquiries of a certain Ur-Utu, a man whose name means literally 'The Dog of the Sun', i.e. 'The Servant of the Sun-god'. This Ur-Utu comes from Tel-ed-Der (Ancient Sippar Amnanum), a suburb of the Sun-god's main Mesopotamian city, Sippar, what one might call Sun-city Mesopotamia. In Ur-Utu's inquiries, one again finds an ideal annual cycle of 360 days that here begins and ends on the 20th of Nisan:[24]

> O God, my lord Ninsianna, accept this offering, stand by me when this offering is made, place there an oracle of well-being and life for Ur-Utu, your servant! Concerning Ur-Utu, your servant, who is now standing by this offering, from the 20th of Nisan until the 20th of Nisan of the coming year, six times sixty days, six times sixty nights, . . .

This formula, which reminds me in part of the Jewish Kol Nidre prayer where one speaks of a year from Yom Kippor *zeh* (this Yom Kippor) *ad yom kippor haba* (until the Yom Kippor to come), would seem to presume that the 20th of Nisan from one year to the next marks an annual solar cycle of 360 days. Given the above, I would propose that the 20th of Nisan, the 20th of the first month of year, is in effect the first Sunday of the year—'Sunday in Mesopotamia'.

4. 'Sunday' and 'The Blessing of the Sun'

The Mesopotamian celebrations of the 20th of Nisan came to my mind when I was a witness and participant in the traditional Jewish ceremony '*Birkat Hachamah*' (The Blessing of the Sun), just outside my synagogue in the Judean Desert facing east looking at the Sun rising over the Dead Sea on the 8th of April 2009 (= the 14th of Nisan 5769 in the Jewish calendar). This is one of the few dates in the Jewish calendar which is determined by the Sun, and not the Moon, with the Sun being blessed as it completes its 28 year cycle, dating back in time and place to the creation of the Sun which Jewish tradition holds to be sunrise on the 4th day (Wednesday) of the first week of the first month, Nisan, in line with Genesis 1: 14–19:[25]

> 14 And God said: 'Let there be lights in the firmament of the heaven to divide the day
> from the night; and let them be for signs, and for seasons, and for days and years;
> 15 and let them be for lights in the firmament of the heaven to give light upon the earth.'
> And it was so.
> 16 And God made the two great lights: the greater light to rule the day, and the lesser light
> to rule the night; and the stars.

Thus, without going into this in greater detail, one might assume that the date of the

Blessing of the Sun would be pegged to Nisan the 4th, in the lunar calendar, but a look at the dates of the Blessing of the Sun in recent and coming times demonstrates that this is not so:

* Wednesday, 7 April 1897 (5 Nisan 5657)
* Wednesday, 8 April 1925 (14 Nisan 5685 – Erev Pesach/Passover Eve)
* Wednesday, 8 April 1953 (23 Nisan 5713)
* Wednesday, 8 April 1981 (4 Nisan 5741)
* Wednesday, 8 April 2009 (14 Nisan 5769 – Erev Pesach/Passover Eve)
* Wednesday, 8 April 2037 (23 Nisan 5797)
* Wednesday, 8 April 2065 (2 Nisan 5825)
* Wednesday, 8 April 2093 (12 Nisan 5853)
* Wednesday, 9 April 2121 (21 Nisan 5881)

Clearly, the system is pegged to the solar year, not the lunar year, or as I put it earlier in the context of the 20th of Nisan, the Sun here in Jewish calendrical tradition too is liberated from the Moon, perhaps even echoing what might have been a presumed solar year or solar cycle of some sort that the Jewish community of Babylonia might have learned from their neighbours.

Yet, in all honesty, I must admit that I am still working on all this and cannot for now demonstrate a connection between the aforementioned Babylonian and Jewish practices, even though the connections between late-Babylonian astronomy and calendrical practices, and those of early Judaism are now becoming ever more clear.[26] Thus, I leave as a working hypothesis that the materials which I presented in the first parts of my presentation point to an often overlooked solar aspect in the Babylonian calendrical tradition, that later found much fuller expression in Jewish tradition, most likely at the site famous for its solar calendar in Jewish tradition, Qumran.

Notes

1. The following includes some material that was omitted at The Living the Lunar Calendar conference due to lack of time. Standard Assyriological abbreviations are as in The Chicago Assyrian Dictionary (CAD) and/or The Pennsylvania Sumerian Dictionary (PSD).
2. I would like to take this opportunity to thank all those who attended the conference, and in particular the organizers of the conference: Amanda Weiss of The Bible Lands Museum and Henry Zemel of the Caeno Foundation, as well as my partners on the academic organizing committee Yonatan Ben-Dov of Haifa University and John Steele of Brown University.
3. K 5981 (+) 11867:1–7 // VAT 9805+ 14'–17' (collated). Horowitz (1998), pp. 146–147, Rochberg-Halton (1988), pp. 270–271.
4. Horowitz (1998), pp. 144–145 with previous bibliography.
5. Horowitz (2010).
6. The 29th of Adar (or of intercalary Adar in leap years) is the last day that must occur in a Mesopotamian lunar year. If the month was a hollow month, then the 29th was the last day of the year. Whether the following day was 30th or New Year's Day (Nisan 1) depended on when the new moon became visible.
7. Note that *Enuma Elish* itself places the stars first, instead of the Moon, as might have been expected. This, however can be explained in two ways. First, because the stars here are regulated by Marduk himself, in his astronomical guise as his star ^{mul}*Nēberu* ('The Crossing'), and second because *Enuma Elish* V seems to

consider time-units from the longest to shortest—the year first, then the month, then the day, and finally the parts of the day, namely the watches of the night and day in line 46.

8. Mul-Apin II Gap A: 1–7.
9. Horowitz (1998), pp. 252–258.
10. Landsberger (1915), pp. 137–138.
11. See below.
12. Harris (1975), pp. 199–202.
13. Cohen (1993), pp. 274–275.
14. Livingstone (1993), p. 110. The author also thanks Prof. Livingstone for access to some unpublished materials regarding the place of the Sun-god on the 20th day in the hemerologies.
15. Maul (1999); see also Zawadzki (2005).
16. K. 170+ (Livingstone (1986), pp. 30–31). See the commentary on p. 48 for the two comrades (*tappû*) of the Sun-god, the fire-gods Gibil and Nuska in rev. 5 (Livingstone (1986), pp. 32–33), whose number is 10 each. Hence the 2 gods × 10 = 20 = Shamash.
17. Shamash Hymn: 156–162; Lambert (1960), pp. 136–137 with commentary on p. 323.
18. Lambert (1960), p. 221.
19. Lambert (1960), p. 341.
20. Lambert (1960), p. 341
21. Maul (1999), pp. 303–305 discusses the Sun-god's ceremonies on the 20th of the month and suggests a connection with a '*Schema der Sieben-Tage-Woche*'.
22. Wiseman (1969), p. 178. Cf. George (1992), p. 152 § 12 9' in The Nippur Compendium: UD.20.KAM ^{d}UTU.
23. For the *ziqpu*-stars and *ziqpu*-star texts see Hunger-Pingree (1999), pp. 84–90 and Al-Rawi-Horowitz (2001).
24. Adapted from the translation in Foster (1993), p. 153.
25. The number of years in the 28 year cycle is determined by the fact that the solar year lasts 365 1/4 days = 52 weeks (364 days) + 1 day + 1/4. Thus, for the Sun to return to a specific Wednesday at sunrise requires 7 cycles of 4 years: the 7 for the days of the week to get back to Wednesday, the 4 to get back to the right time of day given the difference of 1/4 day per year (7 × 4 = 28).
26. See, e.g., Ben-Dov (2008).

References

Al-Rawi, F. and Horowitz, W., 2001, "Tablets from The Sippar Library IX. A Ziqpu-Star Planisphere", *Iraq* 63, 171–181.

Ben-Dov, J. 2008, *Head of All Years: Astronomy and Calendars at Qumran in their Ancient Context* (Leiden: Brill).

Cohen, M. E., 1993, *The Cultic Calendars of The Ancient Near East* (Bethesda: CDL Press).

George, A., 1992, *Babylonian Topographical Texts* (Leuven: Uitgeverij Peeters).

Harris, R., 1975, *Ancient Sippar, A Demographic Study of an Old Babylonian City, (1984–1595 B.C.)* (Istanbul: Nederlands Historisch-Archaeologisch Instituut Te Istanbul).

Horowitz, W., 1994, "Two New Ziqpu-Star Texts and Stellar Circles", *Journal of Cuneiform Studies* 46, 89–98.

——, 1998, *Mesopotamian Cosmic Geography* (Winona Lake: Eisenbrauns).

——, 2010, "They Multiplied the SAG by the UŠ in The Sky", in J. Stackert, B. N. Porter and D. P. Wright (eds.), *Gazing on the Deep: Ancient Near Eastern and Other Studies in Honor of Tzvi Abusch* (Bethesda: CDL Press), 51–61.

Hunger, H., 1992, *Astrological Reports to Assyrian Kings*, SAA 8 (Helsinki: Helsinki University Press).

——, and Pingree, D., 1989, *MUL.APIN. An Astronomical Compendium in Cuneiform* (Vienna: Berger & Söhne).

——, and Pingree, D., 1999, *Astral Sciences in Mesopotamia* (Leiden: Brill).

Lambert, W. G., 1960, *Babylonian Wisdom Literature* (Oxford: The Clarendon Press).

——, 2008, "Mesopotamian Creation Stories", in M. Geller and M. Schnipper (eds.), *Imagining Creation* (Leiden: Brill), 14–59.

Landsberger, B.,1915, *Der Kultische Kalender der Babylonien und Assyrer* (Leipzig: J. C. Hinrichs'sche Buchhandlung).

Livingstone, A., 1986, *Mystical and Mythological Explanatory Works of Assyrian and Babylonian Scholars* (Oxford: Clarendon Press).

——, 1993, "The Case of the Hemerologies: Official Cult, Learned Formulation and Popular Practice", in E. Matsushima (ed.), *Official Cult and Popular Religion in the Ancient Near East* (Heidelberg: Universitätsverlag C. Winter), 97–113.

Maul, S., 1999, "Gottesdienst im Sonnenheligtum zu Sippar", in B. Böck, E. Cancik-Kirschbaum, and T. Reichter (eds.), *Munuscula Mesopotamica, Festschrift für Johannes Renger* (Münster: Ugarit Verlag), 285–316.

Rochberg-Halton, F., 1988, *Aspects of Babylonian Celestial Divination: The Lunar Eclipse Tablets of Enūma Anu Enlil* (Vienna: Berger & Söhne).

Wiseman, D., 1969, "A Lipšur Litany from Nimrud", *Iraq* 31, 175–183.

Zawadzki, S., 2005, "Šamaš Visit to Babylon", *N.A.B.U.* 2005: 8–9, no. 9.

Middle Assyrian Lunar Calendar and Chronology

Yigal Bloch[1]

I. Introduction

Reconstructing the chronology of an ancient civilization depends on reconstructing its calendar. In a series of earlier studies, the present author has dealt with two aspects of the chronology of Assyria in the 13th–12th centuries B.C.E.: the order of the yearly eponyms during the reigns of Shalmaneser I and Tukultī-Ninurta I,[2] and the problems posed by the text of the Assyrian King List, solving which allows one to reconstruct a continuous chronology of the kings of Assyria in the second half of the second millennium B.C.E. (the Middle Assyrian period).[3] In this article, we will discuss the structure of the Middle Assyrian calendar.

Our discussion will proceed in five stages. First, we will present the fundamentals of our topic: the distinction between the original Assyrian calendar and the southern Mesopotamian (Babylonian) calendar, which was adopted in Assyria during the reign of Tiglath-pileser I; the basic characteristics of those calendars (both of which were based on lunar months); and the different proposals raised by scholars with regard to the question whether intercalation (i.e., addition of a thirteenth month to the year once in a while in order to keep the calendar in pace with the solar year cycle) was practiced in the original Assyrian calendar.

At the second and the third stages of our discussion, we will consider two specific proposals for the mechanism of intercalation that might have been employed in the original Assyrian calendar. We will argue that these two proposals cover all the reasonable possibilities of intercalation that could have been practiced in the Middle Assyrian period, but none of them was actually employed in that calendar (at least in the 13th–12th centuries B.C.E.), judging by the available evidence.[4] Thus, we will conclude that during the 13th–12th centuries B.C.E., the Assyrian calendar must have been purely lunar, without intercalation (similar to the modern Islamic calendar).

At the fourth stage of our discussion, we will utilize the above conclusion, along with some more information, to determine the precise absolute date—i.e., a Julian date expressed in the Common Era frame of reference—for the beginning of the first regnal year of Tiglath-pileser I (1114/3 B.C.E.). That, in turn, will enable us to convert any Middle Assyrian calendar date into an absolute date with almost full precision (allowing for a margin of error of a day or two at most). A table listing the absolute dates for the first day of each Assyrian calendar year in the 13th–12th centuries B.C.E. will be provided at the end of the present article.

At the fifth stage of our discussion, we will raise the question when the lunar calendar without intercalation was first adopted in Assyria, and we will point to a prospective direction of study which may supply at least a part of the answer to this question. However, we will not be able to pursue that direction in our present study, and will therefore leave the question open. We will conclude by considering the question of how the inhabitants of ancient Assyria in the 13th–12th centuries B.C.E. could have determined the timing of agricultural works (which would be naturally dependent on the solar year cycle) in a purely lunar calendar.

II. Assyrian and Babylonian Calendars – Fundamentals

The reign of Tiglath-pileser I was a turning point in the history of the calendar in the ancient Assyrian kingdom. During his reign, Assyria adopted a calendar, which had already existed by that time for several centuries in southern Mesopotamia—the so-called Standard Mesopotamian calendar.[5] Since during the second half of the second millennium B.C.E., the most important polity using this calendar was Babylonia (under the rule of the Kassite, and then the Second Isin, dynasties), it is justified to call it the Babylonian calendar.[6]

The Babylonian calendar had twelve lunar months: Nisannu, Ayyāru, Simānu, Duʾūzu, Abu, Ulūlu, Tašrītu, Araḫšamnu, Kissilīmu, Ṭebētu/Kanūnu, Šabāṭu and Addaru. The names of these months were normally rendered in writing by Sumerian logograms, originating from the Nippur calendar of the third millennium B.C.E.[7] Each month began with the sighting of the new crescent of the moon and lasted normally 29 or 30 days, dependent on the length of the particular synodic month—the period between two subsequent identical phases of the moon (the average length of the synodic month is ca. 29.53 days). Since the beginning of the new month depended on the sighting of the new lunar crescent, rather than on the moment of the astronomical New Moon conjunction, occasional occurrences of months consisting of 28 or 31 days would be possible, although such months must have been very rare.[8] Twelve lunar months amount on the average to ca. 354.36 days, which is 10.89 days less than the tropical solar year (ca. 365.25 days). Consequently, in order to keep the calendar in pace with the solar year, the years in the Babylonian calendar were occasionally intercalated. That was done by adding a thirteenth month to the year—either the second Ulūlu (recorded as ^{ITU}KIN.^{d}INANNA.2.KAM) or the second Addaru (recorded as "the additional Addaru", ^{ITU}DIRI.ŠE.KIN.KUD, or occasionally simply as "the additional month", ^{ITU}DIRI).[9] In other words, the Babylonian calendar was not purely lunar but luni-solar. Since in the long run, intercalation was intended to keep the lunar calendar dates in pace with the solar year, one can assume that on the average, the thirteenth month would be added once in 29.53 / 10.89 = ca. 2.7 years.

As mentioned above, the Babylonian calendar was adopted in Assyria in the reign of Tiglath-pileser I.[10] Since the average length of the Babylonian calendar year was kept in pace with the average length of the solar (Julian) year through the practice of intercalation, and since precise continuous chronology of the kings of Assyria in the 11th–8th centuries B.C.E. can be established based on the evidence of the Assyrian King List, the lists of Assyrian yearly eponyms and some other sources, one can determine the absolute dates of the reign of Tiglath-pileser I, with near certainty, as 1114–1076 B.C.E.

Before the reign of Tiglath-pileser I, the calendar practiced in Assyria consisted of twelve months with wholly different names (see table 1 below). Those months were lunar, as is clear from the fact that they lasted either 29 or 30 days.[11] The question is whether the Middle Assyrian calendar, practiced before the adoption of the Babylonian calendar by Tiglath-pileser I, employed any mechanism for intercalation.

Among ca. two thousand dated Assyrian documents from the second half of the second millennium B.C.E., published until today,[12] none mentions an intercalary Assyrian month. Hence, it appears that intercalation by the addition of a specifically designated thirteenth month to a year, as known from the Babylonian calendar, was not practiced in Assyria in the relevant period. This situation was clear already by the 1920s, and scholars working on reconstruction of the Middle Assyrian calendar had to propose other mechanisms of intercalation, if they thought that intercalation was practiced in Assyria in the second half of the second millennium B.C.E.

The first scholar to propose a specific mechanism of intercalation for the Middle Assyrian calendar was Ernst Weidner. According to his proposal, first made in the 1920s and re-iterated a decade later, the twelve months of the Middle Assyrian calendar followed one after the other in an unbroken cycle without any month being ever added to that cycle for the purpose of intercalation; however, the starting point of the calendar year could move from one month to another. Thus, if a regular year began, e.g., with the month Ṣippu, it would contain twelve months exactly and end with the month Ḫibur, the next year starting again on the first day of Ṣippu. If, however, a year starting with the month Ṣippu was to be intercalated, then the month Ṣippu following after the twelfth month of the year, Ḫibur, would be included in the same year as the thirteenth month, and the next calendar year would begin not on the first day of Ṣippu but on the first day of the following month, Qarrātu.[13] In other words, in Weidner's reconstruction, the sequence of Middle Assyrian months would move throughout the solar year, but the beginning point of the Middle Assyrian calendar year would always be limited to a specific season of the solar year cycle. Before the mid-1990s, scholars generally adopted Weidner's proposal.[14]

A different mechanism of intercalation for the Middle Assyrian calendar was proposed by Johannes Koch in 1989. According to Koch's proposal, intercalation in the Middle Assyrian calendar was intended to keep each calendar month, though not the beginning point of the year, in a more or less fixed position within the solar year cycle. Hence, when during a given year, the decision about intercalation was made, the intercalary month would be added at the beginning of the following year, bearing the same name as the last month of the current year; the new year, containing twelve months just as the preceding one, would end with a month that occupied, in the terms of the twelve-months cycle, one position earlier than the last month of the preceding year. E.g., if a given calendar year began on the first day of the month Ṣippu, ended twelve months later with the last day of the month Ḫibur, and the decision about intercalation was made during that year, then the month following Ḫibur would be not Ṣippu but yet another Ḫibur, reckoned as the first month of the next calendar year. That next year would now end not on the last day of Ḫibur (located twelve months after the end of the intercalary month) but one month earlier, on the last day of the month Abu-šarrāni; the subsequent year would then begin on the first day of Ḫibur, and so on.[15] Starting from the late 1990s, several scholars expressed their support for Koch's proposal.[16]

The intercalation mechanisms proposed by Weidner and Koch can be summarized in table 1.

Weidner	Koch
1. Ṣippu	1. Ṣippu
2. Qarrātu	2. Qarrātu
3. Kalmartu	3. Kalmartu
4. Sîn	4. Sîn
5. Kuzallu	5. Kuzallu
6. Allānātu	6. Allānātu
7. Bēlat-ekalli	7. Bēlat-ekalli
8. Ša-sarrāte	8. Ša-sarrāte
9. Ša-kēnāte	9. Ša-kēnāte
10. Muḫur-ilāni	10. Muḫur-ilāni
11. Abu-šarrāni	11. Abu-šarrāni
12. Ḫibur	12. Ḫibur
13. Ṣippu	
1. Qarrātu	*1. Ḫibur*
2. Kalmartu	2. Ṣippu
3. Sîn	3. Qarrātu
4. Kuzallu	4. Kalmartu
5. Allānātu	5. Sîn
6. Bēlat-ekalli	6. Kuzallu
7. Ša-sarrāte	7. Allānātu
8. Ša-kēnāte	8. Bēlat-ekalli
9. Muḫur-ilāni	9. Ša-sarrāte
10. Abu-šarrāni	10. Ša-kēnāte
11. Ḫibur	11. Muḫur-ilāni
12. Ṣippu	12. Abu-šarrāni
1. Qarrātu	1. Ḫibur
...	...

TABLE 1. Mechanisms of intercalation proposed for the Middle Assyrian calendar by Ernst Weidner and Johannes Koch (the intercalary month appears in italics, the horizontal line marks the beginning of a new year).

In contradistinction to the abovementioned proposals, several scholars maintained that intercalation was not practiced at all in the Middle Assyrian calendar.[17] Yet, those scholars did not substantiate their proposal to any satisfactory degree, and their view came under considerable criticism during the last decade.[18] In this study, we will offer sufficient evidence to substantiate the view, which holds that the Middle Assyrian calendar was purely lunar and did not employ any mechanism for intercalation.

III. Why Middle Assyrian months were not confined to a single season of the solar year cycle

Once we accept the idea that there were no specifically designated intercalary months in the Middle Assyrian calendar (given that the available evidence offers no indication of their existence), there are only two possibilities for intercalation in that calendar. Either the beginning point of the year would be fixed to a specific season of the solar year, at the expense of allowing the months to move backward through the solar year cycle (with each given day of a given calendar month occurring ca. 10.89 days earlier, in the terms of the solar year cycle, than it occurred in the preceding year), or the months would be fixed to specific seasons of the solar year, at the expense of allowing the beginning point of the year to move backward through the solar year cycle. Hence, the mechanisms of intercalation proposed by Weidner and Koch are the only possible mechanisms of intercalation that could exist in the Middle Assyrian calendar. Demonstrating that neither of these mechanisms was actually employed in the Middle Assyrian calendar is tantamount to proof that the Middle Assyrian calendar was purely lunar, with a year consisting always of twelve lunar months (ca. 354.36 days on the average). We will begin our demonstration by analyzing the intercalation mechanism proposed by Johannes Koch.

1. The unjustified foundations of Koch's proposed mechanism of intercalation

Koch's proposal was explicitly grounded in the idea that each month of the Middle Assyrian calendar was associated with the heliacal rising (just before the sunrise) of a given group of stars—which, of course, would necessitate that each Middle Assyrian month be confined to a more or less fixed position within the solar year cycle.[19] However, there is absolutely no evidence that Middle Assyrian months were associated with the heliacal rising of specific groups of stars. Such association is known for the Babylonian months; even though the earliest text clearly expressing it—*KAV* 218, known as Astrolabe B—was found in the excavations of the Assyrian capital city of Aššur, that text operates exclusively in the terms of Babylonian months and reflects a cosmological view that also finds expression in the Babylonian creation epic, Enūma eliš.[20] Astrolabe B is, in all likelihood, a copy from a Babylonian original, executed by the scribe Marduk-balāssu-ēriš son of Ninurta-uballissu, a member of a scribal family stemming from Babylonia that was active in Aššur during the second quarter of the 12th century B.C.E.[21] There is no evidence that would allow to postulate the association of months with the heliacal rising of specific groups of stars for the original Assyrian calendar.

2. The movement of Assyrian months through the solar year cycle in the reign of Tiglath-pileser I.

In fact, Koch's proposed mechanism of intercalation contradicts one of the most important elements of evidence for reconstructing the character of the Middle Assyrian calendar. This evidence consists of a group of texts from the reign of Tiglath-pileser I, which include date formulae identifying Assyrian months with specific months of the Babylonian calendar. The correspondences between the Assyrian and the Babylonian months change over the different eponym years dating the relevant documents. The texts containing double-date

VAT 16400 (*MARV* I 73)[23] 5) *i+na* ^{ITU}*ḫi-bur* ^{ITU}NE UD.20.KÁM 6) *li-me* ^{m}IZKIM-IBILA-É.ŠÁR.RA 7) MAN ^{KUR}Aš-šur	In the month Ḫibur (which is) the month Abu, day 20, the eponym year of Tiglath-pileser (I), the king of Assyria
VAT 20159 (*MARV* IX 42) 1') [^{ITU}]*ḫi-bur* ^{ITU}NE [UD.x.KÁM] 2') [*l*]*i-mu* ^{m}IZKIM-[IBILA-É.ŠÁR.RA] 3') [LUGA]L? KUR[*Aš-šur*]	[The month] Ḫibur (which is) the month Abu, [day x], the [ep]onym year of Tiglath-[pileser (I), the kin]g of [Assyria]
VAT 130984 (*MARV* I 62) 9) ⸢^{ITU}⸣*šá-ke-na!-a-tu* ⸢^{ITU}⸣BÁR UD.6.KÁM 10) [*l*]*i-mu* ^{m}*Ḫi-ia-ša-iu-ú*	The month Ša-kēnāte (which is) the month Nisannu, day 6, the [ep]onym year of Ḫiyašāyu
VAT 15468 (*MARV* V 42) 17) ⸢^{ITU}*a-bu*-MAN⸣.ME[Š-*ni*] ⸢^{ITU}SIG$_4$⸣ UD.24.KÁM 18) *li-mu* ⸢^{m}*Ḫi-ia-ša-iu-ú*⸣	The month Abu-šarr[āni] (which is) the month Simānu, day 24, the eponym year of Ḫiyašāyu
VAT 17921 (*MARV* V 43) 10) ^{ITU}*kal-mar-tu* ^{ITU}⸢BÁR⸣ 11) ⸢UD.18⸣.KÁM ⸢*li-mu*⸣ 12) ^{md}⸢*A-šur*-MU⸣-KAM	The month Kalmartu (which is) the month Nisannu, day 18, the eponym year of Aššur-šuma-ēriš
RIMA 2, A.0.87.4 94) ^{ITU}*ḫi-bur ša tar-ṣi* ^{ITU}GAN UD.18.KÁM *l*[*i-mu* ^{m}*Tà*]*k-lak-a-na-*^{d}*A-šur*	The month Ḫibur, which is during the month Kissilīmu, the ep[onym year of Ta]klāk-ana-Aššur

TABLE 2. A selection of date formulae of documents and transactions, employing parallel Assyrian and Babylonian month names, in the reign of Tiglath-pileser I.

formulae of the abovementioned kind have been extensively surveyed by Helmut Freydank,[22] and a sample of those formulae appears in table 2.

As can be seen from the table, in the eponym year of Tiglath-pileser I (which was probably his first regnal year, 1114/3 B.C.E.),[24] the Assyrian month Ḫibur corresponded to the Babylonian month Abu (the 5th month of the Babylonian calendar). In the eponym year of Ḫiyašāyu, the Assyrian months Ša-kēnāte and Abu-šarrāni corresponded to the Babylonian months Nisannu and Simānu, respectively; and given the order of the months in the two calendars, this means that Assyrian Ḫibur corresponded in that year to Babylonian Duʾūzu (the 4th month of the Babylonian calendar). In the eponym year of Aššur-šuma-ēriš, the Assyrian month Kalmartu corresponded to the Babylonian month Nisannu, which means that Assyrian Ḫibur corresponded in that year to Babylonian Ṭebētu/Kanūnu (the 10th month of the Babylonian calendar). Finally, in the eponym year of Taklāk-ana-Aššur, Assyrian Ḫibur corresponded to Babylonian Kissilīmu (the 9th month of the Babylonian calendar).

In other words, double-date formulae in Assyrian documents from the reign of Tiglath-pileser I indicate that the Assyrian calendar months moved all the way through the Babylonian year. This suggests that the Assyrian months also moved all the way through the solar year cycle, which would contradict Koch's proposal. In order to counter this evidence, Koch assumed that the recording of Babylonian months in Assyrian documents reflected an artificial procedure of the Assyrian scribes, which did not take into account the actual intercalation practice in Babylonia and treated the twelve months of the Babylonian calendar as forming an unbroken cycle without intercalary months.[25] Since Koch assumed that intercalary months were added upon necessity to the Assyrian calendar, his proposal means that every 2.7 years on the average, the Assyrian months would move one month forward in the terms of correspondence to the Babylonian months.

Koch even went so far as to claim that the double-date formulae in some documents from the reign of Tiglath-pileser I constitute evidence in favor of his proposal. He cited the document VAT 16400 (*MARV* I 73), which indicates that in the eponym year of Tiglath-pileser I the Assyrian month Ḫibur corresponded to the Babylonian month Abu, and compared it to another document, VAT 16394 (*MARV* II 2), which indicates that in the same eponym year, the Assyrian month Muḫur-ilāni corresponded to the Babylonian month Du'ūzu,[26] so that the Assyrian month Ḫibur would correspond to the Babylonian month Ulūlu. Koch explained this contradiction by arguing that for the purposes of intercalation, the Middle Assyrian "normal year" ("*Normaljahr*") was counted from spring equinox to spring equinox, but the eponym year began at a different point of time. Thus, according to Koch, the eponym year of Tiglath-pileser I began in the month Ḫibur (corresponding to Babylonian Abu) of a "normal year" that began in the month Ša-sarrāte (corresponding to Babylonian Nisannu); however, during that "normal year" a decision about intercalation was made, and the month Bēlat-ekalli (corresponding to Babylonian Addaru) was followed by another month of the same name, which now corresponded to Babylonian Nisannu and began a new "normal year," so that the month Muḫur-ilāni, at the end of the eponym year of Tiglath-pileser I, corresponded not to Babylonian Simānu but to Babylonian Du'ūzu.[27]

However, Koch's proposal cannot be sustained. First, there is no evidence whatsoever that the Assyrians used in their calendar any concept of a year different from that of the eponym year (whatever the beginning and end points of the latter).[28] Second, Koch's analysis of the sources lead him to admit that a given eponym year could include two adjacent months by the same name (in our case, the eponym year of Tiglath-pileser I would include two months by the name Bēlat-ekalli, the first parallel to Babylonian Addaru, the second parallel to Babylonian Nisannu, and each belonging to a different "normal year"). Now, if every two or three years, there was an eponym year that included two adjacent months by the same name, it is hardly conceivable that administrative recording could function without making an explicit distinction between those two months (as was done in the Babylonian calendar by means of designating the intercalary month with the logogram DIRI or recording it as ITU.MN.2.KAM, where MN is a month name).[29] Yet, no distinctions of this kind are ever attested in Middle Assyrian documents.

In fact, it appears that the correspondence between Assyrian Muḫur-ilāni and Babylonian Du'ūzu in *MARV* II 2 is nothing more than a scribal error, reflecting an insufficient familiarity of the Assyrian scribe who produced this document with the Babylonian calendar (in a period when the Babylonian calendar was in the process of being introduced into Assyrian administrative recording).[30] This interpretation of *MARV* II 2, proposed by Freydank,[31] finds a clear parallel in documents dated to the eponym year of Ḫiyašāyu. As mentioned in table 2 above, the document *MARV* V 42 specifies the Assyrian month Abu-šarrāni as corresponding to the Babylonian month Simānu in the eponym year of Ḫiyašāyu. On the other hand, the document VAT 20118 (*MARV* IX 16) specifies the Assyrian month Abu-šarrāni as corresponding to the Babylonian month Du'ūzu in the eponym year of Ḫiyašāyu.[32] Koch's proposed mechanism of intercalation can account for two Assyrian months by the same name belonging to a single eponym year if those months occur around the spring equinox,[33] but not in summer (as happens with the months corresponding to Babylonian Simānu and Du'ūzu). Thus, the specification of Babylonian months corresponding to the months of the Assyrian calendar in both *MARV* II 2 and *MARV* IX 16 is most reasonably explained as a scribal error.[34]

As for Koch's more general proposal concerning the procedure of recording the parallel Assyrian and Babylonian dates in Assyrian documents from the reign of Tiglath-pileser I, it is refuted by the following considerations. As noted above, Koch's proposal implies that every 2.7 years on the average, the Assyrian months moved one month forward in the terms of correspondence to the Babylonian months. Since in the eponym year of Tiglath-pileser I, Assyrian Ḫibur corresponded to Babylonian Abu, and in the eponym year of Aššur-šuma-ēriš, Assyrian Ḫibur must have corresponded to Babylonian Ṭebētu/Kanūnu, it follows that during the period, which elapsed between these two eponym years, the Assyrian months moved five months forward in the terms of correspondence to the Babylonian months (or seven months backward, dependent on the direction in which one counts). Furthermore, since in the eponym year of Taklāk-ana-Aššur, Assyrian Ḫibur corresponded to Babylonian Kissilīmu, it follows that during the period, which elapsed between the eponym years of Aššur-šuma-ēriš and Taklāk-ana-Aššur, the Assyrian months moved eleven months forward (or one month backward) in the terms of correspondence to the Babylonian months.

Now, the inscription RIMA 2, A.0.87.4, written in the name of Tiglath-pileser I, speaks of a military campaign carried out by Tiglath-pileser I against the Babylonian king Marduk-nādin-aḫḫē in the eponym years of Aššur-šuma-ēriš and Ninu'āyu; the inscription itself is dated to the eponym year of Taklāk-ana-Aššur.[35] This means that both the eponym years of Aššur-šuma-ēriš and Taklāk-ana-Aššur belonged to the reign of Tiglath-pileser I. But if we assume, with Koch, that the Assyrian months moved one month forward, in the terms of correspondence to the Babylonian months as recorded in Assyrian documents during the reign of Tiglath-pileser I, each 2.7 years on the average, then it follows that from the eponym year of Tiglath-pileser I to the eponym year of Taklāk-ana-Aššur there elapsed $5 \times 2.7 + 11 \times 2.7 =$ ca. 43 years. Which is, of course, impossible, since the eponym year of Tiglath-pileser I belongs clearly to his reign (and is probably to be identified with his first regnal year),[36] and the whole reign of Tiglath-pileser I lasted, according to the Assyrian King List, 39 years.[37]

On the other hand, if we assume that the Babylonian months recorded in Assyrian documents from the reign of Tiglath-pileser I were real months of the Babylonian calendar, with the Babylonian practice of intercalation taken into account, and the Assyrian months recorded in the same documents followed one another in an unbroken cycle, without insertion of any intercalary months,[38] then the Assyrian months must have moved one month backward, in the terms of correspondence to the Babylonian months, each 2.7 years on the average. This means that from the eponym year of Tiglath-pileser I to the eponym year of Taklāk-ana-Aššur there elapsed 7 × 2.7 + 1 × 2.7 = ca. 21 years. This result fits very well the evidence of the inscription RIMA 2, A.0.87.4.

In other words, the intercalation mechanism proposed by Koch cannot have been employed in the Assyrian calendar in the reign of Tiglath-pileser I, and the Babylonian dates recorded in the documents from his reign must have been real, rather than schematic, Babylonian dates. This conclusion is also supported by the liver omen text Assur 4530, which is dated by a Babylonian date only: "Month Tašrītu, day 11, the eponym year of Tiglath-pile[ser] (I)" (^{ITU}DU$_6$ UD.11.KÁM *li-mu* ^{mGIŠ}TUKUL-*ti*-IBILA-É.[ŠÁR.RA], Assur 4530, rev. 49).[39] It stands to reason that if a scribe working in Aššur could date a text that he produced by a Babylonian date only, it was a real and not a schematic Babylonian date.[40]

3. The movement of Assyrian months through the solar year cycle from the late 13th to the late 12th century B.C.E.

Moreover, it can be demonstrated now that the practice of Assyrian calendar months moving all the way through the solar year cycle existed for at least a whole century prior to the reign of Tiglath-pileser I. The crucial evidence to this effect comes from the letter DeZ 3320,[41] discovered at Tell Šēḫ Ḫamad (ancient Dūr-Katlimmu) on the Lower Ḫabūr.[42] This letter is only partly preserved, so that the names of the sender and the addressee are broken away, but it appears that the sender was Sîn-mudammeq, a high Assyrian official active in the area of the Upper Ḫabūr and the Upper Balīḫ, and the addressee was Aššur-iddin, the Assyrian Grand Vizier and the governor of the Assyrian territories in northeastern Syria.[43] The letter is dated to day 27 of the Assyrian month Allānātu, the eponym year of Ina-Aššur-šumī-aṣbat[44] (which, as the present author has argued in an earlier study, was the 18th regnal year of Tukultī-Ninurta I, belonging thus to the last quarter of the 13th century B.C.E.).[45]

For the purposes of calendrical reconstruction, the following passage from the letter DeZ 3320 is important:

> 8') TÚG.GAD.MEŠ *ša ša-at-te an-ni-te* ⸢*e-ṣi*⸣*-du-ni* 9') 10 GÚ.UN *i-ba-áš-ši iš-tu* TÚG.GAD.MEŠ *la-áš-šu-ni* 10') *i-ga-ra-te-ma ù tar-ba-ṣa ša* TÚG *li-qa-al* 11') *lu né-pu-uš tu-ur* UD.MEŠ *i-ka-ṣu-ú* 12') *a-na ma-sa-e la-a i-lak pa-ni ṭé-mi ša* EN-*ia* 13') *a-da-gal a-na ka-ṣa-ri* EN *li-iš-pu-*⸢*ra*⸣ 14') *li-ik-ṣu-ru li-im-ḫi-ṣu* 15') *a-di* UD.MEŠ *ṭa-bu-ni li-im-si-ú*

> The flax harvested this year amounts to 10 talents (ca. 300 kg). When no flax remains (in storage), should (someone) take care about the walls and the enclosure (intended for storage) of flax, or should we do (it)? The days will become cold again; then it (the flax) will not be fit for washing. I am waiting for the decision of my lord. May my lord write me concerning the tying! Let them tie and crush (the flax); let them wash (it) as long as the days are fine![46]

The key verbs in this passage, denoting the activities, which had to be performed with relation to the harvested flax, are *masā'u* (Assyrian form of *mesû* "to wash"), *kaṣāru* ("to tie, bind together") and *maḫāṣu* ("to beat").

There are several descriptions of the growing, harvesting and processing of flax, written by authors who lived in or visited the Near East from the ancient times down to the

Month(s) of the solar year cycle	Stage in the processing of flax
October-November	Sowing
March	Harvesting (pulling the stalks out of the ground)
March-April	Preliminary combing
April-June	Initial retting (in order to clean, shrink and bleach the stalks, for ca. two weeks)
(subsequently)	The stalks are dried in the open air and turned over several times
July	Crushing of the stalks with hammers
August-September	The stalks are retted in water again, for about a month
October	The stalks are crushed again with hammers and separated into fibers; the fibers are tied in bundles

TABLE 3. Stages in the processing of flax in pre-modern Near East, according to Aḥmad bin ʾAlī al-Maqrīzī, as summarized by Moshe Gil.

Late Middle Ages. Probably the most detailed of those is the account of Aḥmad bin ʿAlī al-Maqrīzī, an Egyptian author of the early 15th century C.E., which was recently summarized by Moshe Gil.[47] Gil's summary can be expressed in table 3.

Since the letter DeZ 3320 speaks of the need to wash the flax before the days get cold, it appears that the washing referred to is the second retting of flax described by al-Maqrīzī, which would take place in August-September (the average summer temperatures in north-eastern Syria are quite similar to those in Egypt; the colder and wetter winter in north-eastern Syria would be irrelevant for activities that would take place in the summer). This interpretation is further supported by the reference to the tying of flax, which, according to al-Maqrīzī, followed after the second retting.[48]

Now, we have seen that in the eponym year of Tiglath-pileser I (which was probably his first regnal year) the Assyrian month Ḫibur corresponded to the Babylonian month Abu—i.e., day 1 of the month Ḫibur corresponded to day 1 of the month Abu (see above, table 2). Day 1 of Abu, in turn, must have occurred four months after day 1 of the month Nisannu—the beginning point of the Babylonian year.

Assyrian records from 747–625 B.C.E.—a period when the Babylonian calendar had been long established in Assyria but the regular 19-year cycle of intercalation, known from the second half of the first millennium B.C.E., was not yet introduced—indicate that Nisannu 1 fell from 35 days before the spring equinox to 17 days after the spring equinox.[49] However, the calendrical theory accepted in Assyria during the 8th–7th centuries B.C.E. demanded the spring equinox to take place on Nisannu 15, whereas the original Babylonian calendrical theory, known from the first half of the second millennium B.C.E. onwards, demanded the spring equinox to take place on Addaru 15 (of course, these are ideal dates for the spring equinox, and in actual practice, both in Babylonia and in the first-millennium B.C.E. Assyria, the dates of the spring equinox varied due to the shifting positions of the lunar months in relation to the solar year cycle).[50] In the light of this, one is to allow for the placement of Nisannu 1 up to a month later, in relation to the spring equinox, during the second millennium B.C.E., compared to the placement of the same date reflected in the Assyrian documents of the 8th–7th centuries B.C.E. At the bottom line, it appears justified to assume that during the second millennium B.C.E., day 1 of Nisannu could occur anytime from ca. 35 days before the spring equinox to ca. 45 days after the spring equinox.[51]

During the 12th century B.C.E., the spring equinox would occur on March 31 or April 1 (Julian dates).[52] This means that day 1 of Nisannu fell in the first regnal year of Tiglath-pileser I from late February to mid-May. Day 1 of Abu must have fallen in the same year four months later—i.e., from late June to mid-September—and that was also day 1 of Ḫibur. Given the order of the months in the Assyrian calendar, day 27 of Allānātu must have occurred in the first regnal year of Tiglath-pileser I a little over five months before day 1 of Ḫibur—i.e., from late January to mid-April. This is 4–8 months earlier than the placement of Allānātu 27 as indicated by the letter DeZ 3320 (around the time proper for the second retting of harvested flax, in August-September). Consequently, it is impossible to maintain, as does Koch, that each of the Middle Assyrian months always belonged to the same season of the solar year cycle.

On the other hand, if we assume that the Assyrian months moved all the way through the solar year (10.89 days backward for each cycle of twelve Assyrian lunar months) during the whole period spanning the 13th–12th centuries B.C.E., then we can calculate the magnitude of this movement from the 18th regnal year of Tukultī-Ninurta I to the first regnal year of Tiglath-pileser I. Given the results of our recent reconstruction of the continuous chronology of the kings of Assyria based on the Assyrian King List,[53] we can calculate that from the beginning of the 18th regnal year of Tukultī-Ninurta I to the beginning of the first regnal year of Tiglath-pileser I, there elapsed 114 years (20 out of the 37 years of reign of Tukultī-Ninurta I, 4 regnal years of Aššur-nādin-apli, 6 regnal years of Aššur-nērārī III, 5 regnal years of Ellil-kudurrī-uṣur, 13 regnal years of Ninurta-apil-Ekur, 46 regnal years of Aššur-dān I, 1 regnal year of Ninurta-tukulti-Aššur, 1 regnal year of Mutakkil-Nusku, and 18 regnal years of Aššur-rēša-iši I).

The years recorded in the Assyrian King List must have been Assyrian calendar years. If the Assyrian calendar year consisted always of twelve lunar months, then, during 114 years, the discrepancy between the solar year and the Assyrian calendar year would have accrued to $114 \times 10.89 = 1241.46$ days. Since the number of 1241.46 days includes the duration of three full solar year cycles ($365.25 \times 3 = 1095.75$ days), we can express this discrepancy, in the terms of the placement of a given Assyrian calendar date within the solar year cycle, as $1241.46 - 1095.75 = 145.71$ days. In other words, in the 18th regnal year of Tukultī-Ninurta I, day 27 of the month Allānātu would occur ca. five months later, in the terms of the solar year cycle, than it occurred in the first regnal year of Tiglath-pileser I. Consequently, Allānātu 27 would occur in the 18th regnal year of Tukultī-Ninurta I between late June and mid-September. The later part of this period fits the season of the solar year cycle, in which day 27 of Allānātu must have occurred in the 18th regnal year of Tukultī-Ninurta I according to the letter DeZ 3320.

If we consider, for the sake of the argument, the mechanism of intercalation for the

Middle Assyrian calendar proposed by Weidner,[54] we will have to reckon with the existence of a 13-month year every 2.7 years, on the average, in the Middle Assyrian calendar, which would however not disturb the backward movement of the Assyrian calendar months through the solar year cycle. This means that every period of ca. 32 Assyrian calendar years would include 33 complete cycles of twelve lunar months. The 114 years that elapsed from the beginning of the 18th regnal year of Tukultī-Ninurta I to the beginning of the first regnal year of Tiglath-pileser I comprised three complete periods of 32 years; consequently, following Weidner's proposal, this period would include 114 + 3 = 117 complete cycles of twelve lunar months. The displacement of Allānātu 27 in the 18th regnal year of Tukultī-Ninurta I, compared to the first regnal year of Tiglath-pileser I, would then have to be increased by 10.89 × 3 = 32.67 days, i.e., by slightly more than a month. In this case, Allānātu 27 would occur in the 18th regnal year of Tukultī-Ninurta I between late July and mid-October. This period fits almost entirely the season of the solar year cycle, in which day 27 of Allānātu must have occurred in the 18th regnal year of Tukultī-Ninurta I according to the letter DeZ 3320.

Thus, the evidence of the letter DeZ 3320 fits both the possibility that the Middle Assyrian calendar years were purely lunar and the possibility that the mechanism of intercalation proposed by Weidner was employed in the Middle Assyrian calendar. What this evidence does not fit is the possibility that the intercalation mechanism proposed by Koch was employed in the Middle Assyrian calendar. In the light of this conclusion, and in the light of the evidence furnished by the documents from the reign of Tiglath-pileser I considered above, it is clear that the mechanism of intercalation proposed by Koch could not have been employed in the Middle Assyrian calendar in the 13th-12th centuries B.C.E.[55]

IV. Why the beginning point of the Middle Assyrian year was not confined to a single season of the solar year cycle

Now we can turn to the intercalation mechanism for the Middle Assyrian calendar proposed by Weidner. As mentioned above, this mechanism allows the movement of Assyrian months through the solar year but demands that the beginning point of the Assyrian calendar year be confined to a single season of the solar year cycle. Below, we will consider several Middle Assyrian documents, which provide evidence for establishing the beginning point of several specific years, and we will demonstrate that the evidence of these documents refutes Weidner's proposal.

1. *MARV* II 19

The first document to be considered is VAT 19193 (*MARV* II 19). This document is a table recording, month by month, for the duration of two years, the numbers of hides of different kinds of sacrificed sheep and goats. The subscript to the table records that the accounting of the hides was performed by the chief knacker Amurru-šuma-uṣur.[56] In the table itself, the record for each year begins with the month Ṣippu and ends with the month Ḫibur;[57] thereafter follows the summary record for the relevant year (followed by a double dividing line). The first year recorded in the table is the eponym year of Usāt-Marduk; the second is the eponym year of Ellil-ašarēd.[58]

From *MARV* II 19 it is clear that the eponym years of Usāt-Marduk and Ellil-ašarēd began with day 1 of the month Ṣippu and ended with the last day of the month Ḫibur. This was recognized already by Weidner,[59] who, regrettably, categorized this document with several other Middle Assyrian documents cited in an earlier study by Hans Ehelolf and Benno Landsberger as alleged evidence that a calendar year in the Middle Assyrian period could begin with any of several different months.[60] In fact, the documents cited by Ehelolf and Landsberger indicate merely that different calendar months in the Middle Assyrian period could correspond to the Babylonian month Nisannu; the assumption that the Assyrian calendar year began in a month corresponding to Nisannu cannot be justified without explicit evidence to that effect, which neither Ehelolf and Landsberger nor Weidner did provide.[61] The contribution of *MARV* II 19 is valuable precisely because it provides evidence of two Middle Assyrian calendar years beginning with the month Ṣippu regardless of the correspondences between the Assyrian and the Babylonian months.

Can we determine, in what season of the solar year cycle the first day of the month Ṣippu occurred in the eponym year of Ellil-ašarēd? This eponym belongs to the reign of Shalmaneser I, and a previous study by the present author concerning the order of the eponyms belonging to his reign has reached the conclusion that the eponym year of Ellil-ašarēd I was the 28th regnal year of Shalmaneser I.[62] Given our recent reconstruction of the continuous chronology of the kings of Assyria in the 13th–12th centuries B.C.E., based on the Assyrian King List,[63] we obtain that from the beginning of the 28th regnal year of Shalmaneser I to the beginning of the first regnal year of Tiglath-pileser I there elapsed 134 years (3 out of the 30 years of reign of Shalmaneser I, 37 regnal years of Tukultī-Ninurta I, 4 regnal years of Aššur-nādin-apli, 6 regnal years of Aššur-nērārī III, 5 regnal years of Ellil-kudurrī-uṣur, 13 regnal years of Ninurta-apil-Ekur, 46 regnal years of Aššur-dān I, 1 regnal year of Ninurta-tukulti-Aššur, 1 regnal year of Mutakkil-Nusku, and 18 regnal years of Aššur-rēša-iši I).

In section III of the present article, we have demonstrated that the Assyrian calendar months must have moved all the way through the solar year (10.89 days backward for each cycle of twelve Assyrian months). We have utilized this datum to calculate the displacement of a specific Assyrian calendar date over a given number of Assyrian years, both with regard to the possibility that the Assyrian calendar year consisted always of twelve lunar months and with regard to the possibility that the Assyrian calendar year was intercalated every 2.7 years on the average by the addition of a thirteenth month, without disturbing the backward movement of the Assyrian calendar months through the solar year, in accordance with the mechanism of intercalation proposed by Weidner. Since the present section of our article deals with Weidner's proposed mechanism of intercalation, all our calculations in this section will be based on the assumption that Weidner's proposal is correct, with the aim to demonstrate that even on this assumption, it is impossible to assign the beginning points of all Assyrian calendar years in the 13th–12th centuries B.C.E. to a single season of the solar year cycle. The sole aim of our calculations in the present section of the article is this kind of 'proof by contradiction'; therefore, those calculations will be admittedly invalid for determining the actual absolute dates of the beginning points of Middle Assyrian calendar years, once Weidner's proposed mechanism of intercalation is refuted.

According to Weidner's proposed mechanism of intercalation, every period of ca. 32 Assyrian calendar years would have included 33 complete cycles of twelve lunar months. The period of 134 Assyrian calendar years that elapsed from the beginning of the 28th regnal year of Shalmaneser I to the beginning of the first regnal year of Tiglath-pileser I comprised four complete periods of 32 years; following Weidner's proposal, this period must have included 134 + 4 = 138 complete cycles of twelve lunar months. Hence, the displacement of day 1 of the month Ṣippu in the 28th regnal year of Shalmaneser I, compared to the first regnal year of Tiglath-pileser I, must have amounted to 138 × 10.89 = 1502.82 days. Since the number of 1502.82 days includes the duration of four full solar year cycles (365.25 × 3 = 1461 days), we can express the discrepancy, in the terms of the placement of a given Assyrian calendar date within the solar year cycle, as 1502.82 – 1461 = 41.82 days. In other words, in the 28th regnal year of Shalmaneser I, day 1 of Ṣippu would occur ca. 42 days later, in the terms of the solar year cycle, than it occurred in the first regnal year of Tiglath-pileser I.

In section III of the present article, we have established that day 1 of the month Ḫibur in the first regnal year of Tiglath-pileser I must have occurred between late June and mid-September. Day 1 of the month Ṣippu must have occurred in the same year one month later (or eleven months earlier)—i.e., between late July and mid-October. Consequently, day 1 of the month Ṣippu must have occurred in the 28th regnal year of Shalmaneser I between early September and late November.

This result is sufficient to disprove Weidner's original contention that the Middle Assyrian calendar year began regularly around the spring equinox.[64] However, the evidence of *MARV* II 19 still makes it possible to maintain that the Middle Assyrian calendar year began regularly around the autumn equinox[65] (in the 13th–12th centuries B.C.E., the autumn equinox would occur on October 3-5).[66] We need more evidence concerning the beginning points of calendar years in the Middle Assyrian period to determine whether this possibility can be substantiated.

2. *MARV* II 17+

One element of the relevant evidence is provided by the document VAT 18007+ (*MARV* II 17+). This is a summary document recording delivery of rations to various groups of people serving the Assyrian state. For some of those groups, the periods of time, during which they consumed their rations, are specified. The periods of interest for calendrical matters in *MARV* II 17 are cited below (with the mention of the line numbers in the text):

61) . . . *iš-tu* ^{ITU}*mu-*[*ḫ*]*ur-*DINGIR.MEŠ UD.21[+5.KÁM]
62) [*a-di* ^{IT}]^{U}*ḫi-bur* UD.20.KÁM 1 ⸢ITU⸣ 24 UD.MEŠ 1 SÌLA.TA.ÀM *e-*⸢*ták*⸣*-*[*lu*]
. . . from the month Mu[ḫ]ur-ilāni, day 2[6, until the mon]th Ḫibur, day 20, for 1 month and 24 days, one *qû* (ca. 1 liter, of barley) daily th[ey] consu[med].[67]

67) [. . . *iš-tu* ^{ITU}]*a-bu-*LUGAL.MEŠ UD.29.KÁM *a-di* ^{ITU}*ḫi-bur* UD.20.KÁM 22 UD.ME[Š]
[. . . from the month] Abu-šarrāni, day 29, to the month Ḫibur, day 20, for 22 days.

78) . . . *iš-tu* ^{ITU}*mu-ḫur-*DINGIR.MEŠ UD.27[+1].K[ÁM *li-me* ^{m}*A-bi-*DINGIR]
79) [*a-di* ^{ITU}*ṣ*]*i-ip-pi* UD.]⸢9⸣.KÁM *li-me* ^{md}SILIM-*ma-nu-*⸢MU-PAP⸣ 2 ITU 11 UD.M[EŠ. . .]
. . . from the month Muḫur-ilāni, day 2[8, the eponym year of Abī-ilī, until the month Ṣ]ippu, [day] 9, the eponym year of Salmānu-šuma-uṣur, for 2 months and 11 day[s. . .].[68]

84) [. . . *iš-tu* ITU]⸢*ṣi-ip*⸣*-pi* UD.9.KÁM *a-*⸢*di*⸣ ^{ITU}*qa*[*r-r*]*a-a-te*
85) [UD.x.KÁM *li-me* ^{md}]SILIM-*m*[*a-nu*-MU-PAP]. . .
[. . . from the month] Ṣippu, day 9, until the month Qa[rr]ātu, [day x. . . of the eponym year of] Salm[ānu-šuma-uṣur]. . .[69]

89) *iš-tu* ^{ITU}*mu-ḫur*-DINGIR.MEŠ UD.27.KÁM *li-me* ^{m}*A-bi*-DINGIR *a-*⸢*di*⸣ ^{ITU}*k*[*al-mar-te*] ⸢UD⸣.20!.KÁM]
90) *li-me* ^{md}SILIM-⸢*ma*⸣*-nu*-MU-PAP 4 ITU 20 UD.MEŠ 1 SÌLA.TA.ÀM ⸢*e*⸣-[*ták-lu*]
From the month Muḫur-ilāni, day 27, the eponym year of Abī-ilī, until the month K[almartu], day 20, the eponym year of Salmānu-šuma-uṣur, for 4 months (and) 20 days, one *qû* (of barley) daily th[ey] c[onsumed].[70]

93) *iš-*⸢*tu*⸣ ^{ITU}*mu-*⸢*ḫur*⸣-DI[NGIR.MEŠ] UD.25[+x].KÁM ⸢*li*⸣*-me* ^{m}*A-bi*-DINGIR *a-di* ^{ITU}*qar-r*[*a-te* UD.x.KÁM]
94) 3 ⸢ITU⸣ 23 U[D.MEŠ] 1 SÌLA.⸢TA.ÀM⸣ ⸢*e*⸣-[*tá*]*k-l*[*u*]
From the month Muḫur-i[lāni], day 25[+x], the eponym year of Abī-ilī, until the month Qarr[ātu, day x], for 3 months (and) 23 d[ays], one *qû* (of barley) daily th[ey] c[on]sum[ed].[71]

What can we learn from these records? From lines 89–90 of *MARV* II 17 it is clear that the eponym year of Salmānu-šuma-uṣur must have begun between the months Abu-šarrāni (the next month after Muḫur-ilāni, which still belonged to the eponym year of Abī-ilī) and Kalmartu. Lines 93–94 further reduce the available time-span for the beginning of the eponym year of Salmānu-šuma-uṣur, indicating that the new eponym year must have begun already by the month Qarrātu, which preceded Kalmartu in the Middle Assyrian cycle of months. (Although the eponym year of Salmānu-šuma-uṣur appears not to have been mentioned explicitly in lines 93–94, there would be no reason to mention the month Muḫur-ilāni as belonging to the eponym year of Abī-ilī unless the end point of the time-span recorded in these lines belonged already to the eponym year of Salmānu-šuma-uṣur.) Lines 61–62, which record the time-span from the month Muḫur-ilāni to the month Ḫibur, without mentioning the change of the eponym during this period, indicate that all those months belonged to one and the same calendar year, which in the context of *MARV* II 17 can only be identified as the eponym year of Abī-ilī. This is further supported by line 67, which records the time-span from the month Abu-šarrāni to the month Ḫibur as belonging to one and the same calendar year (i.e., without a change of the eponym during the relevant period). Lines 84–85 record the time-span from the month Ṣippu to the month Qarrātu as belonging already to the eponym year of Salmānu-šuma-uṣur. This leaves no other possibility but to conclude that the eponym year of Salmānu-šuma-uṣur began with the month Ṣippu. Finally, lines 78–79 confirm that the eponym year of Salmānu-šuma-uṣur began indeed with the month Ṣippu.

An earlier study by the present author concerning the order of the eponyms in the reign of Tukultī-Ninurta I has reached the conclusion that the eponym year of Salmānu-šuma-uṣur (following immediately after the eponym year of Abī-ilī son of Katiri) was the 22nd regnal year of Tukultī-Ninurta I.[72] Since Shalmaneser I is recorded in the Assyrian King List to have reigned 30 years, from the beginning of his 28th regnal year (the eponym year of Ellil-ašarēd) to the beginning of the 22nd regnal year of Tukultī-Ninurta I there must have elapsed 24 years. This means that in the 22nd regnal year of Tukultī-Ninurta I, day

1 of the month Ṣippu must have occurred 24 × 10.89 = 261.36 days earlier, in the terms of the solar year cycle, than it occurred in the 28th regnal year of Shalmaneser I. Since the solar year is a cyclical phenomenon, the last result is tantamount to the statement that in the 22nd regnal year of Tukultī-Ninurta I, day 1 of the month Ṣippu must have occurred 365.25 – 261.36 = 103.89 days later, in the terms of the solar year cycle, than it occurred in the 28th regnal year of Shalmaneser I.

Now, if one divides the solar year into four seasons based on the astronomical points of reference (the spring and the autumn equinoxes, the summer and the winter solstices), the average length of each of those seasons is 365.25 / 4 = ca. 91.3 days. The temporal distance, in the terms of the solar year cycle, between the positions of the beginning points (day 1 of Ṣippu) of the eponym years of Ellil-ašarēd and Salmānu-šuma-uṣur is ca. 12 days greater than the duration of an average season of the solar year.[73] In other words, the beginning points of these two eponym years, belonging to a relatively short period in the history of Assyria (which was, insofar as we can judge, free of large-scale political cataclysms that could interfere with the regular procedure of intercalation), were separated by a period longer than a single astronomical season of the solar year cycle. This constitutes clear evidence against Weidner's proposed mechanism of intercalation for the Middle Assyrian calendar.

3. *MARV* V 8

But the matters do not end here. The document VAT 19908 (*MARV* V 8) records quantities of oil (Ì, *šamnu*) and sesame (ŠE.Ì.GIŠ.MEŠ, *šamaššammū*) associated with different persons. The final line of the document dates it to day 15 of the month Bēlat-ekalli, the eponym year of Marduk-aḫa-ēriš.[74] Some of the preceding lines mention that the produce recorded in the document was "received instead of their fruit syrup" (*ki-mu* LÀL-*šu-nu ma-aḫ-ru-*⌜*ú*⌝, l. 66) or "instead of the [fruit syr]up (?) of the Assyr[ians]" (*ki-m*[*u* LÀ]L? *ša*! *Aš-šu-ra-*[*ie-e*], l. 64).[75] The excavation number of the document (Assur 18767 b) indicates that it belongs to the archive of the administration of the regular offerings in the temple of the god Aššur in the city of Aššur.[76] Sesame (used for production of oil) and fruit syrup were ordinarily delivered for the regular offerings in the temple of Aššur by different provinces and cities of Assyria.[77] Hence, it stands to reason that *MARV* V 8 records an arrangement, whereby several residents of the city of Aššur delivered sesame or oil in lieu of the supply of fruit syrup incumbent on the city.

The dates of some deliveries are recorded in the document. Thus, for the baker Būniya (m⌜*Bu*⌝*-ni-ia a-láḫ-ḫe-nu*, l. 19),[78] delivery of some quantity of sesame is recorded generally for the month Ša-kēnāte ([x ŠE].⌜Ì⌝.GIŠ.MEŠ ⌜ITU*šá*⌝*-ke-na-a-te*, l. 17), then another quantity of sesame is recorded specifically for day 19 of the same month (x [x ŠE].Ì.GIŠ.MEŠ ITUKI.MIN UD.19.KÁM, l. 18), and another quantity, perhaps of oil, is recorded for day 28 of the month Ḫibur (⌜x Ì⌝? ITU*ḫi-bur* UD.28.KÁM, l. 19).[79] Further below, there is a record of the delivery of some quantity of oil by Aššur-rēmanni, a baker, for the month Bēlat-ekalli ([x] ⌜Ì m⌝*Áš-šur-re-*⌜*ma*⌝-[*ni a-láḫ*]-*ḫi-i-nu* ITUdNIN-É.⌜GAL⌝-[*li*], ll. 28-29). Since in the Middle Assyrian cycle of months, the month Bēlat-ekalli is located two months before Ša-kēnāte and five months before Ḫibur, the record for Aššur-rēmanni indicates that individual deliveries, in this part of the document, are not arranged in the chronological order.

Another, very poorly preserved record, mentions day 19 of some month (whose name is broken) in the eponym year of Erīb-Aššur: [. . . ^{ITU}x x (x x)] UD.19.KÁM ⸢*li-mu*⸣ ^{m}S[U-*Aš*]-*šur* (l. 44). Then comes what appears to be the summary record of the total quantity of produce (certainly sesame, and perhaps also oil, which may have been mentioned in the broken beginning of the line) for the same eponym year: [. . . x+]6 ANŠE ŠE.Ì.GIŠ.MEŠ ⸢*ša*⸣ *li-me* ^{m}SU-*Aš-šur*, "[. . .x+]6 homers of sesame of the eponym-year of Erīb-Aššur" (l. 45).

Following this record, there is a double horizontal dividing line, after which more deliveries are recorded, some of them with dates: day 2 of the month Ṣippu, the eponym year of Marduk-aḫa-ēriš (^{ITU}*ṣi-ip-*⸢*pu* UD.2⸣.KÁM *li-mu* ^{md}⸢KU.A-ŠEŠ⸣-KAM, l. 48); day 16 of the month Qarrātu (^{ITU}*qar-ra-*⸢*tu*⸣ UD.16.KÁM, l. 49); day 10 of the same month (^{ITU}KI.MIN[80] UD.10.KÁM, l. 50); day 6 of the month Kalmartu, the eponym year of Marduk-aḫa-ēriš (^{ITU}*kal-mar-tu* UD.6.KÁM *li-mu* ^{md}KU.A-⸢ŠEŠ⸣-KAM, l. 52); the same month, without specification of the day (^{ITU}KI.MIN, l. 54; [^{IT}]^{U}⸢*kal-mar*⸣-*tu*, l. 57); day 4 of the month Sîn, the eponym year of Marduk-aḫa-ēriš ([^{ITU}]$^{⸢d⸣}$XXX UD.4.KÁM *li-mu* ^{md}⸢KU⸣.A-ŠEŠ-KAM, l. 59); and the month Bēlat-ekalli, with the specification of the day broken ($^{ITUfd!}$NIN-É.GAL-*lì* U[D . . .], l. 62). Although some of these records do not mention the eponym, it is clear that all of them refer to the eponym year of Marduk-aḫa-ēriš. This is supported also by the summary note at the end of the document, which specifies the total time-span covered by the deliveries: "[From the month Allan]ātu (?), the eponym year of Erīb-Aššur, to the month Bē[lat-ekalli, the eponym year of Mardu]k-aḫa-ēriš" ([*iš-tu* ^{ITU}*al?-la*ʾ*-n*]*a-a-*⸢*te*⸣ *li-me* ^{m}SU-*Aš-šur a-di* ^{ITUd}NI[N-É.GAL-*lì li-me* ^{md}AMAR.UT]U-⸢ŠEŠ⸣-KAM, ll. 65-66).[81]

Unlike the deliveries of the eponym year of Erīb-Aššur, those of the eponym year of Marduk-aḫa-ēriš are arranged in *MARV* V 8 almost completely in the chronological order (the only exception is the mention of the delivery on day 16 of the month Qarrātu before the delivery on day 10 of the same month). It appears that for the eponym year of Erīb-Aššur, the deliveries were recorded according to the administrative status (or some other order) of the persons who made them,[82] whereas the order of the recording for the eponym year of Marduk-aḫa-ēriš was, in principle, chronological. This is evidently the reason why, in the part of *MARV* V 8 corresponding to the eponym year of Marduk-aḫa-ēriš, deliveries by Būniya the baker (^{m}*B*[*u*]-⸢*ni-ia*⸣ *a-láḫ-ḫe-nu*, l. 46) and Ziʾānu, the priest of the goddess Nunnaʾītu ([^{m}]⸢*Zi-*ʾ*a-a*⸣*-nu* SANGA *šá* $^{⸢d⸣}$*Nun-na-i-te*, l. 47), are first recorded together, accompanied by a single date (day 2 of Ṣippu, l. 49), but then appear as separate entries with different dates: the entries for Ziʾānu appear on ll. 49–53, dated to the months Qarrātu and Kalmartu, while the entries for Būniya appear on ll. 56–58 and 60–62, dated to the months Kalmartu and Bēlat-ekalli. Thus, it is reasonable to conclude from the document *MARV* V 8 that the transition from the eponym year of Erīb-Aššur to that of Marduk-aḫa-ēriš occurred with the beginning of the month Ṣippu.[83]

As demonstrated by the present author in an earlier study, the eponym years of Erīb-Aššur and Marduk-aḫa-ēriš are the antepenultimate and the penultimate years, respectively, in a group of eponyms belonging to the period from the death of Tukultī-Ninurta I to the death of Ninurta-apil-Ekur, and from the chronological viewpoint, that group is the latest among the several groups of eponyms belonging to the relevant period (all the groups together comprise twenty-eight eponyms, corresponding to the twenty-eight regnal years of the kings from Aššur-nādin-apli, the direct successor of Tukultī-Ninurta I, to Ninurta-apil-Ekur).[84] This means that the eponym years of Erīb-Aššur and Marduk-aḫa-ēriš are to be identified with the antepenultimate and the penultimate years, respectively, of the reign of Ninurta-apil-Ekur—i.e., his 11th and 12th regnal years.

From the beginning of the 22nd regnal year of Tukultī-Ninurta I (the eponym year of Salmānu-šuma-uṣur, which, as shown above, also commenced with the month Ṣippu) to the beginning of the 12th regnal year of Ninurta-apil-Ekur there elapsed 42 years (16 out of the 37 years of reign of Tukultī-Ninurta I, 4 regnal years of Aššur-nādin-apli, 6 regnal years of Aššur-nērārī III, 5 regnal years of Ellil-kudurrī-uṣur, and the first 11 regnal years of Ninurta-apil-Ekur).

According to Weidner's proposed mechanism of intercalation, every period of ca. 32 Assyrian calendar years would have included 33 complete cycles of twelve lunar months. The period of 42 Assyrian calendar years that elapsed from the beginning of the 22nd regnal year of Tukultī-Ninurta I to the beginning of the 12th regnal year of Ninurta-apil-Ekur comprised one complete period of 32 years; following Weidner's proposal, this period must have included $42 + 1 = 43$ complete cycles of twelve lunar months. Hence, day 1 of the month Ṣippu in the 12th regnal year of Ninurta-apil-Ekur must have occurred $43 \times 10.89 = 468.27$ days earlier, in the terms of the solar year cycle, than it occurred in the 22nd regnal year of Tukultī-Ninurta I.

Since 365.25 days constitute a full solar year cycle, the last result is tantamount to the statement that in the 12th regnal year of Ninurta-apil-Ekur, day 1 of the month Ṣippu must have occurred $468.27 - 365.25 = 103.02$ days earlier, in the terms of the solar year cycle, than it occurred in the 22nd regnal year of Tukultī-Ninurta I. This displacement is ca. 12 days longer than the duration of an average season of the solar year cycle (ca. 91.3 days), and it constitutes additional evidence against Weidner's proposed mechanism of intercalation for the Middle Assyrian calendar.

4. VAT 9410

Finally, it has been long known that the document VAT 9410 (excavation number Assur 6096 b) testifies to a Middle Assyrian calendar year beginning with the month Ṣippu. This document, published over seventy years ago by Weidner himself,[85] records the consumption of sheep in the court of Ninurta-tukulti-Aššur, and includes the summary record:

39) . . . ⸢PAP-*ma* 3⸢2 UDU⸣.MEŠ
40) *iš-tu* ^{ITU}*ḫi-bur* UD.11.KÁ[M] *li-me*
41) ^{m}*Aš-šur*-KAR-*a-ni* DUMU *Pa-ʾu-z*[*i*]
42) a-di ^{ITU}*ṣi-pi* UD.28.K[ÁM]
43) *li-me* ^{md}XXX-*še-ia*
44) DUMU ÌR-DINGIR.MEŠ

. . . Total: 32 sheep from the month Ḫibur, day 11, the eponym year of Aššur-šēzibanni son of Paʾuz[u], to the month Ṣippu, day 28, the eponym year of Sîn-šēya son of Urad-ilāni, (were consumed).

As argued by Weidner, the number of sheep mentioned in this record, compared to numbers of sheep listed elsewhere in the same document for the individual months Ḫibur and Ṣippu, indicates that the period covered by the summary record could not have lasted more than ca. two months.[86] In other words, the document VAT 9410 indicates that the eponym year of Sîn-šēya began with the month Ṣippu.

The document VAT 9410 belongs to the archive Assur 6096, which dates to the reign of Ninurta-tukulti-Aššur. An earlier study by the present author included analysis of the chronological framework of the reign of Ninurta-tukulti-Aššur, based on the recent proposal of Dr. Heather D. Baker of the University of Vienna, according to which the record *ṭuppišu* for the reign of Ninurta-tukulti-Aššur in the Assyrian king List indicates that he reigned for one official year (the same is true for his brother and successor, Mutakkil-Nusku).[87] The study in question concluded that the reign of Ninurta-tukulti-Aššur must have comprised a few months from the death of his father, Aššur-dān I, to the beginning of the new calendar year, and eleven months after the beginning of that year, which was evidently counted as the first, and the only, regnal year of Ninurta-tukulti-Aššur.[88] Thus, the eponym year of Sîn-šēya was the next year after the year of the death of Aššur-dān I, and it was the first regnal year of Ninurta-tukulti-Aššur.

Now, from the beginning of the 12th regnal year of Ninurta-apil-Ekur to the beginning of the year following the death of Aššur-dān I there elapsed 48 years (two last regnal years of Ninurta-apil-Ekur and 46 regnal years of Aššur-dān I). According to Weidner's proposed mechanism of intercalation, every period of ca. 32 Assyrian calendar years would have included 33 complete cycles of twelve lunar months. The period of 48 Assyrian calendar years that elapsed from the beginning of the 12th regnal year of Ninurta-apil-Ekur to the beginning of the year following the death of Aššur-dān I comprised one full period of 32 years; following Weidner's proposal, this period must have included 48 + 1 = 49 complete cycles of twelve lunar months.

Consequently, in the first regnal year of Ninurta-tukulti-Aššur, day 1 of the month Ṣippu must have occurred 49 × 10.89 = 533.61 days earlier, in the terms of the solar year cycle, than it occurred in the 12th regnal year of Ninurta-apil-Ekur. Since 365.25 days constitute a full solar year cycle, the last result is tantamount to the statement that in the first regnal year of Ninurta-tukulti-Aššur, day 1 of the month Ṣippu must have occurred 533.61 – 365.25 = 168.36 days, or ca. 5½ months, earlier, in the terms of the solar year cycle, than it occurred in the 12th regnal year of Ninurta-apil-Ekur.

This displacement of ca. 5½ months in the position of the beginning point of the Assyrian calendar year within half a century testifies clearly that no mechanism was employed to confine the beginning point of the Middle Assyrian calendar year to a specific season of the solar year cycle. To be sure, in the reign of Aššur-dān I, Assyria had probably suffered from some political crisis, judging by the fact that during four years, following the eponym year of Da''ānī-Ninurta, the orderly nomination of the yearly eponyms was interrupted.[89] However, a political crisis of such a limited duration could hardly produce a five-month displacement of the beginning point of the calendar year, in the terms of the solar year cycle, if an intercalation mechanism designed to confine that beginning point to a specific season of the solar year were at work in the first place.

Based on the evidence presented above, we must conclude that Weidner's proposed mechanism of intercalation could not have been employed in the Middle Assyrian calendar in the 13th-12th centuries B.C.E. In the light of the foregoing discussion, this is tantamount to the conclusion that the Middle Assyrian calendar in the 13th-12th centuries B.C.E. was purely lunar, with the year beginning always on the first day of the month Ṣippu and consisting of twelve lunar months.

V. The precise dating of the beginning points of Middle Assyrian calendar years

Given the above conclusion, can we find a mechanism for converting Middle Assyrian calendar dates into precise absolute dates, expressed in the terms of the Julian calendar in the Common Era frame of reference? In order to do that, we have to consider once more the letter DeZ 3320 from Tell Šēḫ Ḥamad. This letter is dated to day 27 of the Assyrian month Allānātu, the eponym year of Ina-Aššur-šumī-aṣbat—i.e., the 18th regnal year of Tukultī-Ninurta I.[90]

In the light of our conclusion that Middle Assyrian calendar years were purely lunar, we can now uphold the calculation presented in section III of the present article, which demonstrated that given the placement of day 1 of Ḫibur in the first regnal year of Tiglath-pileser I between late June and mid-September, day 27 of Allānātu in the 18th regnal year of Tukultī-Ninurta I must have occurred also between late June and mid-September.[91] As observed above, the letter DeZ 3320 speaks of a retting of flax, which was probably to take place in August-September. Thus, in principle, we could place the Assyrian date of the letter DeZ 3320 anywhere from the beginning of August to mid-September.

However, placing the Assyrian date of the letter DeZ 3320 in the first half of September is problematic. Such a placement would require us to assign day 1 of the month Ḫibur in the first regnal year of Tiglath-pileser I to the latest slot of the solar year cycle, to which it can reasonably belong—i.e., to the first half of September. Yet, such a dating of day 1 of Ḫibur in the first regnal year of Tiglath-pileser I appears to be impossible, since the correspondence between the Assyrian month Ḫibur and the Babylonian month Abu had probably pertained not only in the first regnal year of Tiglath-pileser I, but also in the preceding year, 1115/4 B.C.E.

The evidence for this correspondence of months comes from the document VAT 15581 and from its poorly preserved envelope fragments (published as *MARV* VI 86). The document and the envelope record delivery of produce, due for the regular offerings in the temple of Aššur, by officials responsible for the province of Arbail (modern Arbīl, ca. 80 km east of Mosul). This delivery is recorded to have taken place "in the month Kuzallu (which is) the month Ṭebētu/Kanūnu, day 25, the eponym year of Šamaš-apla-ēriš son of Aššur-šēzibanni" (*i+na* ^{ITU}*ku-zal-li* ^{ITU}AB UD.25.KÁM *li-me* ^{md}UTU-IBILA-KAM DUMU ^{d}*Aš-šur*-KAR-*ni*, *MARV* VI 86, envelope fragment 1, 6'–7'). After the date of the delivery and the impressions of the seals of the officials who carried it out, the envelope features the record: "[. . . the epo]nym year of Tiglath-pil[eser] (I)" ([. . . *li*]-*me* $^{\ulcorner\mathrm{m}\urcorner}$IZKIM-⌜IBILA⌝-[É.ŠÁR.RA], *MARV* VI 86, envelope fragment 2, 5'). In contradistinction to the envelope, the document records the date of the delivery in the eponym year of Šamaš-apla-ēriš,[92] but does not mention the eponym year of Tiglath-pileser I.

Apparently, the eponym year of Tiglath-pileser I forms part of the date of the envelope, and the fact that it does not appear in the document itself suggests that after the original document was drawn up on Kuzallu 25 of the eponym year of Šamaš-apla-ēriš, it was for some reason audited and placed in a new envelope in the eponym year of Tiglath-pileser I; the new envelope recorded both the date of the original transaction and the date on which the envelope itself was inscribed. Consequently, the eponym year of Tiglath-pileser I appears ro have postdated the eponym year of Šamaš-apla-ēriš.

Given the order of the months in the Assyrian and the Babylonian calendars, the correspondence between the months Kuzallu and Ṭebētu/Kanūnu indicates that in the same year, the Assyrian month Ḫibur must have been parallel to the Babylonian month Abu. This correspondence is spelled out explicitly in the date formula recorded in another document, VAT 20171 (*MARV* VII 50): ^{ITU}*ḫi-bur* ^{ITU}⌜NE UD.21.KÁM *li-me*⌝ ^{md}UTU-IBILA-KAM DUMU ^{d}*Aš-šur*-K[A]R-*a-ni*, "The month Ḫibur (which is) the month Abu, day 21, the eponym year of Šamaš-apla-ēriš son of Aššur-šē[zi]banni" (*MARV* VII 50, 4'-5').

The same correspondence between the Assyrian and the Babylonian months must have pertained also in the eponym year of Ištu-Aššur-ašāmšu, which was subsequent to the eponym year of Tiglath-pileser I.[93] A sequence of years with the same correspondences between the Assyrian and the Babylonian months is a sequence of years, during which no intercalary months were added to Babylonian calendar years (with the possible exception of an intercalary month added at the end of the last Babylonian year in the sequence). Since such a sequence is not likely to comprise more than three years, the eponym year of Šamaš-apla-ēriš is most likely to be dated immediately before the eponym year of Tiglath-pileser I, which was probably the first regnal year of that king.[94] In other words, the eponym year of Šamaš-apla-ēriš was, in all likelihood, the year of the death of Aššur-rēša-iši I and the accession year of Tiglath-pileser I,[95] i.e., 1115/4 B.C.E.

The correspondence between Assyrian Ḫibur and Babylonian Abu in the eponym year of Šamaš-apla-ēriš indicates that day 1 of Ḫibur in that year must have fallen between late June and mid-September. Since we have determined above that the Middle Assyrian calendar year began always with the month Ṣippu and consisted of twelve lunar months, it must have ended always with the month Ḫibur. Given that the eponym year of Šamaš-apla-ēriš is to be dated to 1115/4 B.C.E., one has to place day 1 of Ḫibur of that year between late June and mid-September of 1114 B.C.E. (−1113 in the astronomical notation, which, in order to ease calculations for periods that include the beginning of the Common Era, accounts for the year 1 B.C.E. as year 0).

Astronomical data for 1114 B.C.E. indicate the last day of the first visibility of a new moon in Mesopotamia before mid-September in that year as August 25, with the crescent becoming visible at 7:26 PM local time (ca. 40 minutes after the sunset).[96] Although the calendar based on lunar months demanded the day to begin in the evening, we will, for the sake of simplicity, adopt the convention, which expresses the absolute dating of Assyrian calendar dates in the terms of Julian days lasting from midnight to midnight.[97] Thus, we will consider the Assyrian month, which followed the first sighting of the lunar crescent in the evening of August 25, 1114 B.C.E., to have begun on August 26 of the same year.

In the eponym year of Tiglath-pileser I, which was probably also his first regnal year (1114/3 B.C.E.), day 1 of Ḫibur must have occurred ca. 10.89 days earlier, in the terms of the solar year cycle, than it occurred in the eponym year of Šamaš-apla-ēriš. The only suitable date for Ḫibur 1 in the first regnal year of Tiglath-pileser I is August 15, 1113 B.C.E. (−1112 in the astronomical notation, after the new lunar crescent had become visible on August 14 of the same year at 7:54 PM, ca. an hour after the sunset). Placing day 1 of Ḫibur in the first regnal year of Tiglath-pileser I in mid-August requires us to assign day 27 of Allānātu in the 18th regnal year of Tukultī-Ninurta I—the date of the letter DeZ 3320—also to mid-August. This is fully in line with our conclusion presented in section III of the present article, according to which the date of the letter DeZ 3320 must belong to the period appropriate for the second retting of harvested flax, which would take place in August-September.

Having dated day 1 of Ḫibur of the first regnal year of Tiglath-pileser I to August 15,

1113 B.C.E., we must necessarily date the first day of that Assyrian year—day 1 of the month Ṣippu—eleven lunar months earlier. This yields the date of September 25, 1114 B.C.E., for Ṣippu 1 of the first regnal year of Tiglath-pileser I (after the new lunar crescent had become visible on September 24, 1114 B.C.E., or -1113 in the astronomical notation, at 7:22 PM, ca. 70 minutes after the sunset).

Based on the Julian date of September 25, 1114 B.C.E., for the first day of the first regnal year of Tiglath-pileser I, we can now calculate the absolute date of the first day of any Assyrian calendar year in the 13th–12th centuries B.C.E. In the appendix at the end of the present article, we will specify the absolute dates of day 1 of Ṣippu for each Assyrian calendar year from 1301 to 1093 B.C.E.[98] Of course, it must be borne in mind that the actual visibility of the new lunar crescent would be subject to factors of unpredictable nature, such as the weather conditions in Assyria or the eyesight capabilities of the persons charged with the observation; hence, our calculated absolute dates may differ by a day or two from what would be the actual absolute date of the beginning of a given Assyrian year. Still, the degree of precision, attained in converting Assyrian calendar dates into absolute dates in table 4 below, would be unthinkable without a definite reconstruction of the structure of the Middle Assyrian calendar.

VI. The Middle Assyrian lunar calendar and the earlier calendar(s) of Assyria

While we have shown above that in the 13th–12th centuries B.C.E., the calendar practiced in Assyria was purely lunar, it was apparently not the original calendar of Assyria. For the period spanning the 17th–14th centuries B.C.E., very few Assyrian documents are known, and they do not allow any clear conclusion concerning the structure of the calendar used in Assyria in that period. On the other hand, for the 20th–18th centuries B.C.E. (the Old Assyrian period), documentary evidence is abundant. Most of it comes from the mound of Kültepe in Anatolia, which in the early second millennium B.C.E. was the location of the city of Kaneš—a city that included a trading colony (*kārum*) established by merchants from Aššur.[99] Assyrian documents were discovered in the archaeological levels Ib and II of *kārum* Kaneš (the lower city at Kültepe), and the month names mentioned in these documents are almost the same as those attested in Middle Assyrian sources, although, understandably, the forms of the month names used in the Old Assyrian documents are generally more archaic from the grammatical viewpoint. The only substantial difference between the Old Assyrian and the Middle Assyrian month cycles is the fact that instead of "the month of the moon-god" ($^{\text{ITU}\,\text{d}}$*Sîn*), known from Middle Assyrian sources, the level II documents from *kārum* Kaneš utilize the name "the month of figs" (^{ITU}*teʾinātim*).[100]

Already in 1976, Mogens Troll Larsen argued that the documents of *kārum* Kaneš level II (which are older than those of level Ib) reflect a calendar, in which the months were lunar but the beginning point of the year was fixed around the winter solstice. Larsen's argument was based on the fact that the date formulae of some level II documents record eponyms who received their office "(out) of the hand" (*ša qāte*) of other eponyms. Apparently, succession in the office of the eponym was involved, and Larsen suggested that the Assyrian merchants residing at Kaneš identified a year by noting its eponym as the successor of the eponym of the preceding year, because the identity of the eponym for the current year was not known to them due to cessation of communications between the city of Aššur and Anatolia. Since the communications would be most likely severed in winter, which in Anatolia was rather cold and snowy, and since the mention of the *ša qāte* formula is almost exclusively confined to dates located in a restricted group of months—the five months from Bēlat-ekallim to Ab-šarrāni[101]—Larsen concluded that the Old Assyrian calendar year began in early winter, and that Bēlat-ekallim was the first month of the year. In this way, the identity of the eponym for the new year would be often unknown in Kültepe several months after the year had begun.[102]

Larsen's proposal was further developed by Klaas Veenhof.[103] Veenhof was able to identify three mentions of a previously unknown month, *zibibirum/zibibarum*, in documents from *kārum* Kaneš level II. Two mentions of the month *zibibirum* appear in documents dated to specific eponym years. In one year—the eponym year of Puzur-Niraḫ—this month probably followed the month Allānātum, judging by the fact that one and the same week spanned in that year parts of the months Allānātum and *zibibirum*.[104] For another year—the year after the eponym Idī-Aššur son of Kubīdi—a date in the month Allānātum is attested, beside a date in the month *zibibirum*, and the *ša qāte* formula in the document dated to the month Allānātum mentions explicitly an eponym "who will take" (*ša ilaqqe'u*) the office "(out) of the hand" (*ša qāte*) of Idī-Aššur. In other words, succession in the office of the eponym was expected but had not yet taken place, and the name of the successor eponym was unknown.[105]

Based on the these data, Veenhof suggested that *zibibirum* was the intercalary month, following after Allānātum, and that the unusual dating formula in the month Allānātum of the year designated as *ša qāte* Idī-Aššur son of Kubīdi "might reflect the uncertainty at the end of the year of a scribe confronted not only with the expected change of year-eponym but also with an upcoming intercalary XIIIth month".[106] In a later study, Veenhof observed that *ša qāte* datings in the month Allānātum (without an indication that succession in the office of the eponym had not yet taken place) are known for several more eponym years of kārum Kaneš level II,[107] and suggested that in all such cases, the reference is to an intercalary Allānātum, appended after the regular month of the same name. According to this proposal, the term *zibibirum* was a variant designation for the intercalary Allānātum.[108]

Veenhof did not address the question whether the intercalary Allānātum belonged to the same year as the preceding month by the same name, or was already reckoned as the first month of the new year. However, the very fact that two adjacent months were known by the same name, without almost any attempt to use a special designation for the intercalary month (except for the three lone mentions of the month *zibibirum*), suggests that the distinction between the months was made according to the eponym year, to which they belonged—i.e., that the year ended with the regular Allānātum, and the intercalary Allānātum, if inserted into the cycle of months, became the first month of the new year.[109]

In this regard, the mechanism of intercalation in the Old Assyrian calendar, reconstructed by Veenhof, is somewhat similar to the mechanism of intercalation proposed for the Middle Assyrian calendar by Koch.[110] However, the similarity is only partial, since, as shown in sections II–III of the present article, the mechanism of intercalation proposed by Koch allows the year—whether the eponym year or the so-called 'normal year' (for which no real evidence exists)—to begin with different months, the beginning point of the year shifting one month backward each 2.7 years on the average. In contradistinction, as pointed out by Veenhof, the evidence from *kārum* Kaneš level II suggests that the Assyrian calendar year in this period began regularly with the month Bēlat-ekallim, around the beginning of the winter. In order for the beginning point of the year to remain in the vicinity of the winter solstice, a year that began with an intercalary Allānātum must have ended with the regular Allānātum and consisted of 13 months rather than 12. In other words, with regard to the fundamental question of the relation between the calendar year and the solar year, the intercalation mechanism of *kārum* Kaneš level II reconstructed by Veenhof must have been more similar to the mechanism employed in the Babylonian calendar than to the mechanism proposed by Koch.

Whether the intercalation mechanism reconstructed by Veenhof for the Old Assyrian calendar was employed during the period of *kārum* Kaneš level Ib, is unclear. Larsen observed that documents belonging to this period utilize the *ša qāte* formula with more month names than documents from level II, and no limitation of the *ša qāte* formula to a specific part of the Assyrian calendar year is evident. Hence, Larsen argued that during the period of *kārum* Kaneš level Ib, the Assyrian calendar was purely lunar, without intercalation.[111] Veenhof, on his part, voiced skepticism concerning this argument, noting the paucity of dates with the *ša qāte* formula from the period of *kārum* Kaneš level Ib and the possibility that political problems of some kind interrupted the orderly communication between Aššur and Anatolia in other seasons beside the winter. That possibility would lead to a situation, in which the Assyrian merchants at Kaneš would remain ignorant of the name of the eponym for a new year all the way down to the year's end (so that the *ša qāte* formula would be utilized even in the concluding months of a year).[112]

Currently, it may be possible to establish, which of the two opinions—Larsen's or Veenhof's—concerning the calendar of *kārum* Kaneš level Ib was closer to the truth. This can be done by collecting the eponyms from the relevant period, for which dates with the *ša qāte* formula and specific month names are known,[113] and locating those eponyms in the chronologically ordered list of eponyms, which has been recently discovered at Kültepe and which covers, for the most part, the period of *kārum* Kaneš level Ib (this list is known as KEL G).[114] By doing so, one will be able to observe the progression of the months, mentioned in the *ša qāte* formulae, through the Assyrian calendar year with time, and to compare it with the expectable progression of months belonging to the winter season in a purely lunar calendar (in which every month moves backward, on the average, 10.89 days per year in relation to the solar year cycle).

This task, however, cannot be accomplished in the framework of the present article, and hence the question of the origin of the purely lunar calendar employed in Assyria in the 13th–12th centuries B.C.E. must be currently left open. What should be stressed is the fact that the transition from the Old Assyrian calendar of *kārum* Kaneš level II to the Middle Assyrian calendar of the 13th–12th centuries B.C.E. involved not one but two elements of change: the transition from a luni-solar calendar to a purely lunar one (i.e., the abandonment of intercalation), and the shift from a year beginning in the month Bēlat-ekallim to a year beginning in the month Ṣippu (Old Assyrian Ṣipʾim). When each of these changes was introduced, and what the temporal relation between them was, are questions that demand further study.

Finally, it should be pointed out that the transition from a luni-solar to a purely lunar calendar, such as postulated here for Assyria sometime between the 18th and the 13th centuries B.C.E. is, in a sense, counter-intuitive, because for a society heavily dependent on agriculture—as all the societies in the Ancient Near East were—a calendar correlated with the solar year would make it much easier to figure out the timing of agricultural works. However, the authorities determining the character of the calendar were obviously not the peasants but the higher echelons of the administrative and cultic hierarchy, who may have had their own reasons for different kinds of calendrical changes.[115]

Moreover, it is well possible to figure out the timing of agricultural works in a purely lunar calendar, provided that one regularly observes some stellar phenomena, either in relation to the cycle of the day (e.g., a series of heliacal or acronychal risings of certain stars) or in relation to the progression of the moon through the sky (e.g., observing the day, on which the moon approaches a certain group of stars, during a sequence of lunar months). A system of observation combining the two abovementioned methods, for different parts of the year, is used, for example, by the peasants of the Munnebih tribe in northwestern Yemen, who utilize for administrative purposes the purely lunar Islamic calendar.[116]

There is nothing to preclude that the Assyrian peasants of the late second millennium B.C.E. would do something similar, if the administrative calendar practiced by their rulers demanded that. Since the evidence amassed in the present study supports clearly the conclusion that the Middle Assyrian kingdom used, at least in the 13th–12th centuries B.C.E., a purely lunar calendar without intercalation, neither the use of intercalation in Assyria in the 20th–18th centuries B.C.E. nor the lesser convenience of a purely lunar calendar for an agricultural society can undermine this conclusion. For a present-day scholar, at least, the use of a purely lunar calendar in the Middle Assyrian period is a blessing, since it allows a simpler conversion of Middle Assyrian calendar dates into absolute dates (see the Appendix).

Appendix. Julian dates for the first day (Ṣippu 1) of Assyrian calendar years, 1301–1093 B.C.E.[117]

Regnal year of a king	Eponym (if known)[118]	Julian date of Ṣippu 1 (B.C.E.)[119]
Adad-nērārī I, year 1		1301, June 25
Adad-nērārī I, year 2		1300, June 15
Adad-nērārī I, year 3		1299, June 4
Adad-nērārī I, year 4		1298, May 24
Adad-nērārī I, year 5		1297, May 12
Adad-nērārī I, year 6		1296, May 1
Adad-nērārī I, year 7		1295, April 21
Adad-nērārī I, year 8		1294, April 10
Adad-nērārī I, year 9		1293, March 30
Adad-nērārī I, year 10		1292, March 20
Adad-nērārī I, year 11		1291, March 9
Adad-nērārī I, year 12		1290, February 26
Adad-nērārī I, year 13		1289, February 17
Adad-nērārī I, year 14		1288, February 3
Adad-nērārī I, year 15		1287, January 24
Adad-nērārī I, year 16		1286, January 14
Adad-nērārī I, year 17		1285, January 3
Adad-nērārī I, year 18		1285, December 23
Adad-nērārī I, year 19		1284, December 12
Adad-nērārī I, year 20		1283, December 1
Adad-nērārī I, year 21		1282, November 20

Adad-nērārī I, year 22		1281, November 8
Adad-nērārī I, year 23		1280, October 28
Adad-nērārī I, year 24		1279, October 19
Adad-nērārī I, year 25		1278, October 8
Adad-nērārī I, year 26		1277, September 26
Adad-nērārī I, year 27		1276, September 15
Adad-nērārī I, year 28		1275, September 6
Adad-nērārī I, year 29		1274, August 25
Adad-nērārī I, year 30		1273, August 13
Adad-nērārī I, year 31		1272, August 3
Adad-nērārī I, year 32		1271, July 23
Shalmaneser I, year 1	Shalmaneser I[120]	1270, July 12
Shalmaneser I, year 2	Mušabši'u-Sibitti (?)	1269, June 30
Shalmaneser I, year 3	Šerrīya	1268, June 20
Shalmaneser I, year 4	Aššur-kāšid	1267, June 9
Shalmaneser I, year 5	Aššur-mušabši son of Iddin-Mēr	1266, May 30
Shalmaneser I, year 6	Aššur-mušabši son of Anu-mušallim	1265, May 19
Shalmaneser I, year 7	Qibi-Aššur son of Šamaš-aḫa-iddina	1264, May 8
Shalmaneser I, year 8	Aššur-nādin-šumāte	1263, April 27
Shalmaneser I, year 9	Abī-ilī son of Aššur-šumu-lēšer	1262, April 16
Shalmaneser I, year 10	Aššur-ālik-pāni	1261, April 5
Shalmaneser I, year 11	Mušallim-Aššur	1260, March 25
Shalmaneser I, year 12	Ilī-qarrād (?)	1259, March 15
Shalmaneser I, year 13	Qibi-Aššur son of Ṣilli-Marduk	1258, March 4
Shalmaneser I, year 14	Ina-pî-Aššur-lišlim (?)	1257, February 22
Shalmaneser I, year 15	Adad-šamšī son of Adad-šumu-lēšer	1256, February 10
Shalmaneser I, year 16	Kidin-Sîn son of Adad-tēya	1255, January 30
Shalmaneser I, year 17	Bēr-šumu-lēšir	1254, January 19
Shalmaneser I, year 18	Aššur-dammeq son of Abī-ilī	1253, January 9
Shalmaneser I, year 19	Ištar-ēriš son of Salmānu-qarrād[121]	1253, December 28
Shalmaneser I, year 20	Bēr-bēl-līte	1252, December 18
Shalmaneser I, year 21	Lullāyu	1251, December 8
Shalmaneser I, year 22	Aššur-kettī-īde	1250, November 27
Shalmaneser I, year 23	Ekaltāyu	1249, November 15
Shalmaneser I, year 24	Aššur-da'issunu	1248, November 4
Shalmaneser I, year 25	Rīš-Adad (?)	1247, October 24
Shalmaneser I, year 26	Nabû-bēla-uṣur	1246, October 14

Shalmaneser I, year 27	Usāt-Marduk	1245, October 3
Shalmaneser I, year 28	Ellil-ašarēd	1244, September 22
Shalmaneser I, year 29	Ittabši-dēn-Aššur	1243, September 11
Shalmaneser I, year 30	Ubru	1242, August 31
Tukultī-Ninurta I, year 1	Tukultī-Ninurta I[122]	1241, August 19
Tukultī-Ninurta I, year 2	Qibi-Aššur son of Ibašši-ilī	1240, August 8
Tukultī-Ninurta I, year 3	Mušallim-Adad son of Salmānu-qarrād	1239, July 29
Tukultī-Ninurta I, year 4	Adad-bēl-gabbe son of the king	1238, July 19
Tukultī-Ninurta I, year 5	Šunu-qardū	1237, July 7
Tukultī-Ninurta I, year 6	Libūr-zānin-Aššur	1236, June 27
Tukultī-Ninurta I, year 7	Aššur-nādin-apli son of the king	1235, June 16
Tukultī-Ninurta I, year 8	Urad-ilāni (?)	1234, June 5
Tukultī-Ninurta I, year 9	Adad-umaʾʾi	1233, May 24
Tukultī-Ninurta I, year 10	Abattu son of Adad-šamšī	1232, May 14
Tukultī-Ninurta I, year 11	Abattu son of Adad-šumu-lēšir	1231, May 3
Tukultī-Ninurta I, year 12	Aššur-daʾān	1230, April 23
Tukultī-Ninurta I, year 13	Etel-pî-Aššur son of Kurbānu	1229, April 11
Tukultī-Ninurta I, year 14	Uṣur-namkūr-šarri	1228, March 31
Tukultī-Ninurta I, year 15	Aššur-bēl-ilāni	1227, March 20
Tukultī-Ninurta I, year 16	Aššur-zēra-iddina	1226, March 10
Tukultī-Ninurta I, year 17	Bēr-nādin-apli (?), Aššur-mušabši (?; son of Adad-bān-kala?)[123]	1225, February 27
Tukultī-Ninurta I, year 18	Ina-Aššur-šumī-aṣbat son of Aššur-nādin-šume	1224, February 16
Tukultī-Ninurta I, year 19	Ninuʾāyu (?)	1223, February 6
Tukultī-Ninurta I, year 20	Adad-šamšī (son of Mariannu?)	1222, January 26
Tukultī-Ninurta I, year 21	Abī-ilī son of Katiri[124]	1221, January 15
Tukultī-Ninurta I, year 22	Salmānu-šuma-uṣur	1220, January 3
Tukultī-Ninurta I, year 23	Ellil-nādin-apli (?)	1220, December 24
Tukultī-Ninurta I, year 24	(?)[125]	1219, December 13
Tukultī-Ninurta I, year 25	Kaštiliašu (?)	1218, December 3
Tukultī-Ninurta I, year 26	Bēr-išmanni (?)	1217, November 21
Tukultī-Ninurta I, year 27	Ilī-padâ son of Aššur-iddin (?)[126]	1216, November 11
Tukultī-Ninurta I, year 28		1215, October 31
Tukultī-Ninurta I, year 29		1214, October 20

Tukultī-Ninurta I, year 30		1213, October 8
Tukultī-Ninurta I, year 31		1212, September 28
Tukultī-Ninurta I, year 32		1211, September 17
Tukultī-Ninurta I, year 33		1210, September 7
Tukultī-Ninurta I, year 34		1209, August 27
Tukultī-Ninurta I, year 35		1208, August 16
Tukultī-Ninurta I, year 36		1207, August 5
Tukultī-Ninurta I, year 37		1206, July 25
Aššur-nādin-apli, year 1	Aššur-nādin-apli	1205, July 13
Aššur-nādin-apli, year 2		1204, July 2
Aššur-nādin-apli, year 3		1203, June 22
Aššur-nādin-apli, year 4		1202, June 12
Aššur-nērārī III, year 1	Aššur-nērārī III	1201, May 31
Aššur-nērārī III, year 2		1200, May 20
Aššur-nērārī III, year 3		1199, May 9
Aššur-nērārī III, year 4		1198, April 28
Aššur-nērārī III, year 5		1197, April 17
Aššur-nērārī III, year 6		1196, April 7
Ellil-kudurrī-uṣur, year 1	Ellil-kudurrī-uṣur	1195, March 27
Ellil-kudurrī-uṣur, year 2		1194, March 17
Ellil-kudurrī-uṣur, year 3		1193, March 5
Ellil-kudurrī-uṣur, year 4		1192, February 22
Ellil-kudurrī-uṣur, year 5		1191, February 11
Ninurta-apil-Ekur, year 1	Ninurta-apil-Ekur	1190, February 1
Ninurta-apil-Ekur, year 2		1189, January 21
Ninurta-apil-Ekur, year 3		1188, January 10
Ninurta-apil-Ekur, year 4		1188, December 30
Ninurta-apil-Ekur, year 5		1187, December 19
Ninurta-apil-Ekur, year 6		1186, December 8
Ninurta-apil-Ekur, year 7		1185, November 26
Ninurta-apil-Ekur, year 8		1184, November 16
Ninurta-apil-Ekur, year 9		1183, November 6
Ninurta-apil-Ekur, year 10		1182, October 26
Ninurta-apil-Ekur, year 11	Erīb-Aššur[127]	1181, October 15
Ninurta-apil-Ekur, year 12	Marduk-aḫa-eriš	1180, October 4
Ninurta-apil-Ekur, year 13	Pišqīya	1179, September 23
Aššur-dān I, year 1	Aššur-dān I	1178, September 12
Aššur-dān I, year 2		1177, September 1
Aššur-dān I, year 3		1176, August 21
Aššur-dān I, year 4		1175, August 11
Aššur-dān I, year 5		1174, July 31
Aššur-dān I, year 6		1173, July 20
Aššur-dān I, year 7		1172, July 9
Aššur-dān I, year 8		1171, June 28
Aššur-dān I, year 9		1170, June 17

Aššur-dān I, year 10		1169, June 6
Aššur-dān I, year 11		1168, May 26
Aššur-dān I, year 12		1167, May 16
Aššur-dān I, year 13		1166, May 5
Aššur-dān I, year 14		1165, April 23
Aššur-dān I, year 15		1164, April 12
Aššur-dān I, year 16		1163, April 2
Aššur-dān I, year 17		1162, March 22
Aššur-dān I, year 18		1161, March 11
Aššur-dān I, year 19		1160, March 1
Aššur-dān I, year 20		1159, February 18
Aššur-dān I, year 21		1158, February 7
Aššur-dān I, year 22		1157, January 27
Aššur-dān I, year 23		1156, January 15
Aššur-dān I, year 24		1155, January 5
Aššur-dān I, year 25		1155, December 26
Aššur-dān I, year 26		1154, December 15
Aššur-dān I, year 27		1153, December 4
Aššur-dān I, year 28		1152, November 23
Aššur-dān I, year 29		1151, November 12
Aššur-dān I, year 30		1150, November 1
Aššur-dān I, year 31		1149, October 20
Aššur-dān I, year 32		1148, October 10
Aššur-dān I, year 33		1147, September 29
Aššur-dān I, year 34		1146, September 19
Aššur-dān I, year 35		1145, September 7
Aššur-dān I, year 36		1144, August 27
Aššur-dān I, year 37		1143, August 16
Aššur-dān I, year 38		1142, August 5
Aššur-dān I, year 39		1141, July 25
Aššur-dān I, year 40		1140, July 15
Aššur-dān I, year 41		1139, July 4
Aššur-dān I, year 42		1138, June 23
Aššur-dān I, year 43		1137, June 11
Aššur-dān I, year 44		1136, June 1
Aššur-dān I, year 45		1135, May 21
Aššur-dān I, year 46	Pišqīya son of Kaššu, Aššur-šēzibanni son of Pa'uzu[128]	1134, May 11
Ninurta-tukulti-Aššur, year 1	Sîn-šēya son of Urad-ilāni	1133, April 30
Mutakkil-Nusku, year 1		1132, April 19
Aššur-rēša-iši I, year 1	Aššur-rēša-iši I	1131, April 8
Aššur-rēša-iši I, year 2		1130, March 28
Aššur-rēša-iši I, year 3		1129, March 17
Aššur-rēša-iši I, year 4		1128, March 6

Aššur-rēša-iši I, year 5		1127, February 24
Aššur-rēša-iši I, year 6		1126, February 13
Aššur-rēša-iši I, year 7		1125, February 3
Aššur-rēša-iši I, year 8		1124, January 22
Aššur-rēša-iši I, year 9		1123, January 11
Aššur-rēša-iši I, year 10		1123, December 31
Aššur-rēša-iši I, year 11		1122, December 21
Aššur-rēša-iši I, year 12		1121, December 10
Aššur-rēša-iši I, year 13		1120, November 29
Aššur-rēša-iši I, year 14		1119, November 19
Aššur-rēša-iši I, year 15		1118, November 8
Aššur-rēša-iši I, year 16		1117, October 27
Aššur-rēša-iši I, year 17		1116, October 16
Aššur-rēša-iši I, year 18	Šamaš-apla-ēriš son of Aššur-šēzibanni	1115, October 5
Tiglath-pileser I, year 1	Tiglath-pileser I	1114, September 25
Tiglath-pileser I, year 2	Ištu-Aššur-ašāmšu	1113, September 13
Tiglath-pileser I, year 3		1112, September 3
Tiglath-pileser I, year 4		1111, August 23
Tiglath-pileser I, year 5		1110, August 12
Tiglath-pileser I, year 6		1109, July 31
Tiglath-pileser I, year 7		1108, July 20
Tiglath-pileser I, year 8		1107, July 10
Tiglath-pileser I, year 9		1106, June 29
Tiglath-pileser I, year 10		1105, June 18
Tiglath-pileser I, year 11		1104, June 8
Tiglath-pileser I, year 12		1103, May 28
Tiglath-pileser I, year 13		1102, May 17
Tiglath-pileser I, year 14		1101, May 5
Tiglath-pileser I, year 15		1100, April 24
Tiglath-pileser I, year 16		1099, April 14
Tiglath-pileser I, year 17		1098, April 4
Tiglath-pileser I, year 18		1097, March 23
Tiglath-pileser I, year 19		1096, March 13
Tiglath-pileser I, year 20		1095, March 2
Tiglath-pileser I, year 21		1094, February 19
Tiglath-pileser I, year 22		1093, February 8

Notes

1. The argument underlying this article was partly presented by the author in a lecture at the conference Living the Lunar Calendar in the Bible Lands Museum, Jerusalem, on January 31, 2010, and in a lecture at the 56th Rencontre Assyriologique Internationale in Barcelona, on July 27, 2010. The article uses the following abbreviations: *AHw* = von Soden, W. F., 1956-81, *Akkadisches Handwörterbuch* (Wiesbaden: Harrassowitz); *CAD* = Oppenheim, A. L. et al. (eds.), 1956- , *The Assyrian Dictionary of the Oriental Institute of the University of Chicago* (Chicago: The Oriental Institute); *GAG* = von Soden, W. F., 1995, *Grundriss der akkadischen Grammatik*, 3rd edn., with cooperation of W. R. Mayer, Analecta Orientalia 33 (Rome:

Pontificium Institutum Biblicum); *KAV* = Schroeder, O., 1920, *Keilschrifttexte aus Assur verschiedenen Inhalts*, Wissenschaftliche Veröffentlichungen der Deutschen Orient-Gesellschaft 35 (Leipzig: Hinrichs); *MARV* I–II = Freydank, H., 1976–82, *Mittelassyrische Rechtsurkunden und Verwaltungstexte*, I–II, Vorderasiatische Schriftdenkmäler der Staatlichen Museen zu Berlin 19, 21 (Berlin: Akademie-Verlag); *MARV* III = Freydank, H., 1994, *Mittelassyrische Rechtsurkunden und Verwaltungstexte*, III, Wissenschaftliche Veröffentlichungen der Deutschen Orient-Gesellschaft 92 (Berlin: Gebr. Mann); *MARV* IV–VII = Freydank, H., 2001–2006, *Mittelassyrische Rechtsurkunden und Verwaltungstexte*, IV–VII, Wissenschaftliche Veröffentlichungen der Deutschen Orient-Gesellschaft 99, 106, 109, 111 (Saarbrücken: Saarbrücker Druckerei und Verlag); *MARV* VIII–IX = Freydank, H., 2007–2010, *Mittelassyrische Rechtsurkunden und Verwaltungstexte*, VIII–IX, Wissenschaftliche Veröffentlichungen der Deutschen Orient-Gesellschaft 119, 125 (Wiesbaden: Harrassowitz); RIMA 2 = Grayson, A. K., 1991, *Assyrian Rulers of the Early First Millennium*, I (1114–859 B.C.E.), The Royal Inscriptions of Mesopotamia, Assyrian Periods 2 (Toronto: University of Toronto Press).

2. Bloch (2008); Bloch (2010a). Some corrections to those reconstructions are required now (see below, nn. 121, 123, 125), but the essential arguments presented in the abovementioned studies remain valid. In addition, the present author has reconstructed the order of the eponyms during the brief reign of Ninurta-tukultī-Aššur, about seventy years after the death of Tukultī-Ninurta I, and figured out several groups of eponyms belonging to the 28 years that elapsed from the death of Tukultī-Ninurta I to the death of Ninurta-apil-Ekur, as well as the chronological relations between those groups (Bloch 2010c). The specific order of the eponyms within most groups of eponyms belonging to the period from the death of Tukultī-Ninurta I to the death of Ninurta-apil-Ekur cannot yet be established. However, for some groups such order has been reconstructed, and it will prove important for our present study.
3. Bloch (2010c).
4. Since out of the whole Middle Assyrian period, evidence shedding light on the structure of the Assyrian calendar is currently available only for the 13th–12th centuries B.C.E. (and the early 11th century B.C.E.), we will use the term "Middle Assyrian calendar" with reference to the calendar practiced in Assyria during this period. When calendars practiced in Assyria before the 13th century B.C.E. are discussed, that will be specified explicitly.
5. For the emergence of the Standard Mesopotamian calendar between ca. 1750 and ca. 1500 B.C.E., see Cohen (1993), pp. 297–305.
6. Just as the adoption of the Standard Mesopotamian calendar by the Jews in the 6th century B.C.E. is commonly perceived as adoption of the Babylonian calendar.
7. See Cohen (1993), pp. 297–298.
8. According to the calculations of Huber et al. (1982), p. 7, one in every 1000 lunar months observable in Babylonia would consist of 31 days; yet, it is not clear whether the administrative calendrical practice allowed for a 31-days month, or rendered any day following the 30th day of a given month as the first day of the next month, regardless of the actual observation of the lunar crescent. A month lasting 28 days is known from a record dating to the 7th century B.C.E.—see Parker and Dubberstein (1946), p. 4.
9. For the practice of intercalation and the recording of intercalary months in the Babylonian calendar in the second half of the second millennium B.C.E., see Brinkman (1976), pp. 400–401.
10. Cohen (1993), pp. 300–301. The full-scale adoption of the Babylonian calendar occurred probably only late in the reign of Tiglath-pileser I, since the earliest administrative document from his reign bearing only a Babylonian date is dated to the eponym year of Ipparšidu, which was the 32nd regnal year of Tiglath-pileser I (out of 39 years of his reign)—see Borger (1957–58).
11. Freydank (1991), p. 81. The evidence for 29-day months is limited, since in administrative recording, Assyrian months were normally reckoned as consisting of 30 days, probably regardless of the actual state of the calendar (but see below, nn. 67–68). The same principle was at work in Mesopotamian administrative practices from the late third millennium B.C.E. onwards—see Englund (1988), pp. 124–125.
12. For a list of ca. 900 documents from the Middle Assyrian period, almost all of them bearing dates and published before the late 1970s, see Saporetti (1979), pp. 175–184. Since then, more than a thousand new documents from that period were published, mostly in volumes II–IX of the *MARV* series, and a

substantial part of those documents include date formulae.

13. Weidner (1928–29); Weidner (1935–36), pp. 28–29.
14. See, e.g., Hunger (1976–80), p. 299; Cohen (1993), pp. 239–241.
15. Koch (1989), pp. 132–141.
16. Huber (1999–2000), p. 287; Veenhof (2000), pp. 141–142.
17. Deller (1987), p. 62; Freydank (1991), p. 81; Gasche et al. (1998), p. 50.
18. See, e.g., Huber (1999–2000); Reade (2000); Sassmanshausen (2006), p. 165.
19. Koch (1989), p. 135.
20. For the text of Astrolabe B and a discussion of its origin, content and significance in the development of Mesopotamian astronomy and cosmology, see Horowitz (forthcoming). The present author is grateful to Prof. Wayne Horowitz for allowing consultation of this work.
21. For the Babylonian origin and the date of the activity of the sons of Ninurta-uballissu, see Bloch (2010b); for the eponym year of Ikkāru, to which *KAV* 218 is dated, see also Bloch (2010c), pp. 28–29, n. 18.
22. Freydank (1991), pp. 82–86. The following table is based on Freydank's publication, updated in accordance with the hand-copies of the relevant documents, which have been published meanwhile in the *MARV* series.
23. The siglum VAT indicates registry numbers of documents kept in the Vorderasiatisches Museum zu Berlin.
24. In the Middle Assyrian period (in fact, until the second half of the 10th century B.C.E.), Assyrian kings normally carried out the office of the eponym in their first full regnal year—see Bloch (2010c), pp. 24–25, n. 9, and the earlier literature cited there.
25. Koch (1989), pp. 140–141.
26. [^{ITU}*mu-ḫ*]*ur*-DINGIR.MEŠ *ša tar-ṣi* ^{ITU}ŠU ⸢UD⸣.[x.KÁM *li-mu*] ⸢^{m}IZKIM⸣-IBILA-É.⸢ŠÁR⸣.R[A], "[The month Muḫ]ur-ilāni, which is during the month Du'ūzu, day [x, the eponym year of] Tiglath-pileser (I)" (*MARV* II 2, 1'–2').
27. Koch (1989), pp. 139–140.
28. In fact, the mechanism of intercalation proposed by Koch does not even allow his "normal year" to begin always around the spring equinox (or any other fixed point of the solar year), because the beginning point of the "normal year" still moves backward in relation to the solar year cycle, with the movement of the lunar months, and it is only the name of the month beginning the "normal year" that changes with intercalation. For the "normal year" to begin at a more or less fixed point of the solar year cycle, the possibility of a "normal year" consisting of 13 months should be provided for, and such possibility is absent from Koch's proposed mechanism of intercalation.
29. See Brinkman (1976), pp. 400–401.
30. Theoretically, one could argue that the correspondence between Assyrian Ḫibur and Babylonian Abu in *MARV* I 73 is a scribal error. However, the fact that this correspondence is also supported by the document *MARV* IX 42, which was unknown to Koch, suggests that the correspondence between Assyrian Ḫibur and Babylonian Abu in the eponym year of Tiglath-pileser I is valid.
31. Freydank (1991), p. 84.
32. ^{ITU}*a-bu*-MAN.MEŠ-*ni* ^{ITU}ŠU UD.⸢26.KÁM *li-mu*⸣ ^{m}*Ḫi-ia-ša-iu-*⸢*ú*⸣$^{?}$, "The month Abu-šarrāni (which is) the month Du'ūzu, day 26, the eponym year of Ḫiyašāyu" (*MARV* IX 16, 7–8).
33. Or around the autumn equinox, if one assumes that the "normal year" serving as the base for intercalation in the Middle Assyrian calendar lasted from autumn to autumn, as proposed by Veenhof (2000), p. 142.
34. The conclusion that among the documents from the eponym year of Ḫiyašāyu, the scribal error appears in *MARV* IX 16, rather than in *MARV* V 42, is based on the fact that the correspondence between the Assyrian and the Babylonian months expressed in *MARV* V 42 is confirmed by the correspondence between the Assyrian and the Babylonian months expressed in *MARV* I 62—see Freydank (1991), p. 85, and cf. above, n. 30.
35. RIMA 2, A.0.87.4, 44–51, 94.
36. See above, n. 24.
37. Grayson (1980–83), p. 112.

38. Note that this assumption disagrees with the intercalation mechanism for the Middle Assyrian calendar proposed by Koch, but agrees with the intercalation mechanism proposed by Weidner (that mechanism will be refuted in section IV of the present article).
39. The text known by the excavation number Assur 4530 was published by Tschinkowitz (1968–69); for the reading of the date formula, see Saporetti (1979), p. 159.
40. The scribe who produced the text Assur 4530 was Šamaš-zēra-iddina son of Šamaš-šumu-lēšir. This scribe stemmed from a family, which had been probably active in Aššur for at least four generations—see Saporetti (1978). Yet, insofar as the use of Babylonian dates is concerned, Šamaš-zēra-iddina son of Šamaš-šumu-lēšir appears to have been several years ahead of the practice introduced into Assyrian administrative recording during (or shortly before) the reign of Tiglath-pileser I. The earliest Assyrian text, bearing a double-date formula that utilizes both an Assyrian and a Babylonian month—the liver omen text VAT 13798—was copied by Šamaš-zēra-iddina son of Šamaš-šumu-lēšir ca. five years before the first regnal year of Tiglath-pileser I; see Freydank (1991), p. 86.
41. The siglum DeZ indicates registry numbers of documents kept in the museum of Dēr ez-Zōr, Syria.
42. This letter was puiblished by Cancik-Kirschbaum (1996), no. 6.
43. Cancik-Kirschbaum (1996), p. 117.
44. ^{ITU}*al-la-na-tu* UD.27.KÁM *li-mu* ^{m}*I+na-*^{d}*A*$^{!}$*-šur-*MU*-aṣ-bat* (Cancik-Kirschbaum 1996, no. 6, 2''–3'').
45. Bloch (2010a), pp. 19–25, 31–32. In fact, our argument in that study was partly based on establishing the chronological relationship between the eponym years of Aššur-zēra-iddina and Ina-Aššur-šumī-aṣbat on the assumption that the Middle Assyrian calendar was purely lunar; thus, using the conclusions of that argument to establish the character of the Middle Assyrian calendar might lead to circular reasoning. However, in another study the present author has offered a different line of evidence for identifying the eponym year of Ina-Aššur-šumī-aṣbat with the 18th regnal year of Tukultī-Ninurta I, based on the document VAT 18100 (*MARV* IV 34) and on the chronological data for Babylonia following the capture of Kaštiliaš IV by Tukultī-Ninurta I. That line of evidence, which does not depend on the question of intercalation in the Middle Assyrian calendar, indicates that three years had probably elapsed between the eponym year of Ina-Aššur-šumī-aṣbat and the eponym year of Abī-ilī son of Katiri, which was identified by the present author with the 21st regnal year of Tukultī-Ninurta I (Bloch [2010c], pp. 70–73, and see below, n. 72). It should be noted that the date formula of *MARV* IV 34 (ll. 24'–25') reads ^{ITUd}NIN-É.GAL*-lì* UD.16.KÁM *l*[*i-mu* ^{m}]⸢*A-bi*⸣-DINGIR, "The month Bēlat-ekalli, day 16, the ep[onym year of] Abī-ilī"; in Bloch (2010c), p. 70, n. 37, the determinative preceding the name of the eponym was mistakenly restored as ^{d} rather than ^{m}.
46. Cancik-Kirschbaum (1996), no. 6, 8'–15'. Our translation differs from the German translation of Cancik-Kirschbaum in interpreting the verbal form *limḫiṣū* in l. 15' as meaning "may they crush" rather than "may they weave" (for the meanings "to beat, crush" and "to weave" of the Akkadian verb *maḫāṣū*, see *AHw*, pp. 580a-581b; *CAD* M/1, pp. 71b–78b). The present author is indebted for his translation to a personal remark of Prof. Shalom M. Paul of the Hebrew University of Jerusalem. This translation will be substantiated in n. 48 below.
47. Gil (2004), pp. 82–83.
48. For this reason, we translate the verbal form *limḫiṣū* in DeZ 3320, 15', as "let them crush" rather than "let them weave": the crushing of stalks of flax into fibers was closely associated with the tying of the fibers in bundles after the second retting, whereas weaving would demand preliminary spinning of the fibers into thread (see Vogelsang-Eastwood [1992], p. 13), and spinning is not mentioned in the letter DeZ 3320. Our interpretation implies that the sequence of actions expressed by the verbal forms *likṣurū limḫiṣū* is put in the letter in the reverse order, compared to the order in which those actions were actually carried out during the procession of flax (crushing of the stalks into fibers would be done before tying the fibers in bundles). However, the sender of the letter could have placed a greater emphasis on tying the fibers in bundles, because he requested from the addressee instructions related specifically to the tying.
49. Huber et al. (1982), pp. 8–9.
50. Horowitz (1996), pp. 42–44. The earliest evidence for the Babylonian calendrical theory, demanding the spring equinox to take place on Addaru 15, is the text BM 17175+17284, dating to the first half of the second millennium B.C.E.—see Hunger and Pingree (1989), pp. 163-164, pl. XIIIa.

51. As argued, on somewhat different grounds, by Huber et al. (1982), pp. 9–10.
52. All dates, employing names of the months that are used in both the Julian and the Gregorian calendars, are specified in the present article in the terms of the Julian calendar. The dates of astronomical phenomena, cited in the present article, have been obtained through Alcyone Astronomical Software (http://www.alcyone.de), for the locality of Babylon. The distance from Aššur to Babylon is negligible for the purposes of calculating the basic variables pertaining to the astronomical phenomena discussed in the present article.
53. Bloch (2010c).
54. We should account for the possibility that this mechanism was indeed employed in the Middle Assyrian calendar, until that possibilty is refuted (which will be done in section IV of the present article).
55. The evidence furnished by the correspondences between the Assyrian and the Babylonian months in the first regnal year of Tiglath-pileser I, on the one hand, and by the letter DeZ 3320, on the other hand, also supports the conclusion of our earlier study (Bloch [2010c], pp. 64–78), according to which Aššur-dān I reigned 46 years, as recorded in the Khorasabad and the SDAS manuscripts of the Assyrian King List, rather than 36 years (as assumed by many scholars following Boese and Wilhelm [1979]). If Aššur-dān I reigned 36 years, then the displacement of Allānātu 27 in the 18th regnal year of Tukultī-Ninurta I, compared to the first regnal year of Tiglath-pileser I, must be reduced by 10.89 × 10 = 108.9 days, i.e., by ca. 3.5 months, both with regard to the possibility that the Assyrian calendar year consisted always of twelve lunar months and with regard to the possibility that a thirteenth month was added to the Assyrian calendar year once in every 2.7 years, on the average, in accordance with Weidner's proposed mechanism of intercalation. On the first possibility, day 27 of Allānātu would occur in the 18th regnal year of Tukultī-Ninurta I between early March and late May; on the second possibility, day 27 of Allānātu would occur in the 18th regnal year of Tukultī-Ninurta I between early April and late June. Neither of these results fits the season of the solar year cycle, in which day 27 of Allānātu must have occurred in the 18th regnal year of Tukultī-Ninurta I according to the letter DeZ 3320 (August-September). Consequently, the proposal that Aššur-dān I reigned 36 years must be rejected.
56. [10')] ⸢PAP⸣ 5 LIM 9 ME 49 <KUŠ UDU.NÍTA.MEŠ> 1 ME 28 <KUŠ UDU.NÍTA.MEŠ *pár-ga-ni-ú-tu*> 35 <KUŠ UDU.NIM> 89 <KUŠ UDU.MÁŠ> KUŠ UDU.MEŠ [11')] *ša* 2 MU.MEŠ *ša pi-i le-a-né ša* SISKUR.MEŠ [12')] *ša ša ku-ru-ul-ti-e* ⸢*ša*⸣ *im-ta-ḫi-ru-ni* [13')] NÍG.ŠID.MEŠ *ṣa-ab-tu-tu ša* ^{md}MAR.TU-⸢MU⸣-PAP GAL *sa-pi-e*, "Total: 5949 <hides of sheep>, 128 <hides of meadow-fed sheep>, 35 <hides of spring-lambs>, 89 <hides of male goats>. (These are) hides of small livestock for two years, (recorded) according to the writing-boards of the sacrifices belonging to the animal-fatteners, which have been received. Accounts made by Amurru-šuma-uṣur, the chief knacker" (*MARV* II 19, rev. 10'–13'). The total numbers for the different categories of sheep and goats appear in the columns of the table, whose captions (obv., 1) refer to the relevant categories; those captions have been added to the above citation in pointed brackets, for the sake of clarity. For *immeru* (UDU) *pargāniʾu* "meadow-fed sheep" and *urīṣu* (UDU.MÁŠ) "billy-goat," see Jakob (2003), p. 368; *AHw*, pp. 833a–b, 1430b–1431a; *CAD* P, p. 184a, *CAD* U/W, pp. 227b–230b.
57. ^{ITU}*ṣi-pu* (*MARV* II 19, obv. 2, 16); ^{ITU}*ḫi-bur* (*MARV* II 19, obv. 14, rev. 8').
58. *li-mu* ^{m}⸢*Ú-sa-at*⸣-^{d}AM[AR.UTU] (*MARV* II 19, obv. 15); *li-mu* ^{m}⸢^{d}ÉN.LÍL-SAG⸣ (*MARV* II 19, rev. 9').
59. Weidner (1928–29), who referred to this document by its excavation number, Assur 13058 kl.
60. Ehelolf and Landsberger (1920).
61. The same, and erroneous, interpretation of the evidence cited by Ehelolf and Landsberger as indicating that the Middle Assyrian calendar year could begin with any of several different months was adopted by Koch (1989), pp. 132–138.
62. Bloch (2008), p. 147.
63. Bloch (2010c).
64. Weidner (1935–36), p. 29.
65. As proposed by Veenhof (2000), p. 142.
66. See above, n. 52.
67. The restoration "day 2[6]" is based on the assumption that the total time-span, 1 month and 24 days, was

in accordance with the record of its limit points, and on the common practice of Middle Assyrian administrative texts to include both limit points of a recorded time-span in the specified duration of that span (for these recording practices, see Freydank [1991], p. 81). If the record were based on reckoning the duration of both the month Muḫur-ilāni and the month Abu-šarrāni (immediately preceding Ḫibur) as 30 days each, the beginning point of the time-span recorded in *MARV* II 17+, 61–62 should have been day 27 of Muḫur-ilāni. Although the number of the day in l. 61 is damaged, the hand-copy of the tablet shows it clearly as consisting of two angular wedges (*Winkelhaken*), which record the number 20, and of at least one vertical wedge following them. The height of that wedge indicates that the vertical wedges following the two Winkelhaken were arranged in two rather than three rows. This, however, does not fit the recording of the number 7 in the document *MARV* II 17+, which consists of seven vertical wedges arranged in three rows, with only one wedge in the bottom row, in the leftmost part thereof (compare, e.g., the recording of day 27 of the month Muḫur-ilāni in l. 89). Consequently, it is most likely that not more than six vertical wedges appeared after the two *Winkelhaken* at the end of the preserved part of l. 61, i.e., that the time-span recorded in ll. 61–62 began on day 26 of the month Muḫur-ilāni. In this case, the only way for the limit points of this time-span to fit its recorded duration (1 month and 24 days) is on the assumption that in the record of the duration of the time-span, "one month" stood for 30 days, whereas in reality, either the month Muḫur-ilāni or the month Abu-šarrāni was 29 days long. The record in l. 67 indicates that the month Abu-šarrāni (in the eponym year of Abī-ilī) was 30 days long: only on this condition could the time-span from day 29 of Abu-šarrāni to day 20 of Ḫibur, with both limit points included, number 22 days as recorded. Hence, one must conclude that the month Muḫur-ilāni in the eponym year of Abī-ilī was 29 days long, and the records for this month are to be added to other evidence of 29-day months in the Middle Assyrian period (for which see above, n. 11).

68. Reading based on the join of *MARV* II 17 and the fragment VAT 18007f (*MARV* IV 171); see Freydank, *MARV* IV, p. 14. From the text preserved on the tablet (as shown on the hand-copy) it appears that the recorded time-span began on day 28 of the month Muḫur-ilāni, because the record for the number of the day (at the end of l. 78) consists of two *Winkelhaken* followed by at least seven vertical wedges; the arrangement of the vertical wedges in two rows indicates that the number of the day could not be 27 or 29 (see the preceding note). The number of the day in the end point of the recorded time-span (l. 79), although slightly damaged, can only be 9, judging by the fact that this number consists of three rows of vertical wedges, with three wedges in the bottom row. The limit points of the recorded time-span (day 28 of Muḫur-ilāni and day 9 of Ṣippu) can match its total duration, 2 months and 11 days, only if it is assumed that in the record of the duration of the time-span, "one month" stood for 30 days, whereas in reality, one of the months included in that time-span was 29 days long. This fits the conclusion reached in the preceding note, according to which the month Muḫur-ilāni in the eponym year of Abī-ilī consisted of 29 days only.

69. Reading based on the join of *MARV* II 17 and the fragment 18109c (*MARV* IV 167); see Freydank, *MARV* IV, p. 14. Enough of the name of the eponym is preserved to identify it as Salmānu-šuma-uṣur (cf. ll. 89–90 below).

70. Reading based on the join of *MARV* II 17 and *MARV* IV 167. In the end point of the recorded time-span, the restoration of the month name as K[almartu] is secured by the mention of the starting point (day 27 of the month Muḫur-ilāni) and of the total length of the time-span (4 months and 20 days). The total length of the time-span also indicates that the number of the day in the month Kalmartu is to be read as 20 rather than 2. That number is written on the right edge of the tablet, and this may have caused the wedges forming it to appear vertical rather than (properly) angular.

71. The absence of the eponym name in the specification of the end point of the recorded time-span is surprising, but the broken space at the end of l. 93, as indicated on the hand-copy of the tablet, is insufficient for restoring the name of the eponym.

72. Bloch (2010a), pp. 27–32. The placement of the eponym year of Abī-ilī son of Katiri no earlier than the 21st regnal year of Tukultī-Ninurta I is based on the reconstruction placing 20 eponyms from the reign of that king before the eponym year of Abī-ilī. The placement of the eponym year of Abī-ilī son of Katiri no later than the 21st regnal year of Tukultī-Ninurta I is based on ration lists from Tell Šēḫ Ḥamad (Dūr-

Katlimmu), which indicate that Mannu-bal-Ištar daughter of Piradi, born no later than the eponym year of Qibi-Aššur son of Ibašši-ilī (the 2nd regnal year of Tukultī-Ninurta I), was still registered in the age-category of *talmittu*, "apprentice," in the eponym year of Salmānu-šuma-uṣur. Analysis of the age-categories, to which members of the working class were assigned in Middle Assyrian documents, by the present author has led to the conclusion that a girl would pass from the category of *tārītu*, "adolescent," to that of *talmittu* at the age of 14 years, and from the category of *talmittu* to that of *ša šipri*, "adult worker," at the age of 20 years or, prior to that age, upon marriage (Bloch [2008], pp. 159–165). If the eponym year of Abī-ilī was the 21st regnal year of Tukultī-Ninurta I, and the eponym year of Salmānu-šuma-uṣur was his 22nd regnal year, and if Mannu-bal-Ištar was born in the eponym year of Qibi-Aššur son of Ibašši-ilī, then by the eponym year of Salmānu-šuma-uṣur she would be exactly 20 years old—the upper boundary of the age-category *talmittu* for unmarried working-class women. However, as admitted by the present author, the specific age for the passage from one age-category to another may have varied by a few months, or even a year, in each direction (Bloch [2008], pp. 161, 165). In this regard, another set of data, not accounted in our previous studies, must be pointed out. According to the ration list DeZ 3082+ from Tell Šēḫ Ḥamad, Rabât-Nisaba, a daughter of the peasant Aššur-ubla, was registered as *ša irti*, "suckling," in the eponym year of Aššur-nādin-apli, which was the 7th regnal year of Tukultī-Ninurta I, and according to the ration list DeZ 3272, she was registered as *tārītu* in the eponym year of Salmānu-šuma-uṣur (Röllig [2004], pp. 39, 42). Assuming that the Rabât-Nisaba was born in the eponym year of Aššur-nādin-apli and that the eponym year of Salmānu-šuma-uṣur was precisely the 22nd regnal year of Tukultī-Ninurta I, Rabât-Nisaba would be registered as *tārītu* at the age of 15 years. This possibility still fits the general definitions of Middle Assyrian age-categories, which emerge from the analysis presented in Bloch (2008), pp. 159–165, although it stretches the margin of variation for the passage from the category of *tārītu* to that of *talmittu* to its apparent maximum (a whole year after the age of 14). It should be observed that according to the data for southern Mesopotamia in the 22nd–21st centuries B.C.E., children passed into the category of full-fledged adult workers between the age of 13 and 15 years (Waetzoldt [1987], p. 133). Such model cannot fit the data of Middle Assyrian ration lists from Tell Šēḫ Ḥamad, which indicate that the passage from the category of apprentice (*talmīdu*/*talmittu*) to that of adult worker (*ša šipri*) took place between 14–15 and 20 years of age. However, the placement of any more eponyms, beyond those indicated in Bloch (2010a), in the reign of Tukultī-Ninurta I before the eponym year of Abī-ilī son of Katiri (with the exception of the correction suggested below, n. 123), would result in allowing a pre-apprentice status (*tārītu*) for Rabât-Nisaba daughter of Aššur-ubla past her 15th birthday, and that appears to be precluded.

73. To be sure, the duration of ca. 91.3 days for a season of the solar year is an average figure. The actual astronomical data for Mesopotamia in the 13th–12th centuries B.C.E. (see above, n. 52) indicate that the seasons of summer and winter (from the summer solstice to the autumn equinox and from the winter solstice to the spring equinox) consisted indeed of ca. 91.3 days, with the margin of deviation not greater than a few hours. On the other hand, the season of autumn (from the autumn equinox to the winter solstice) consisted in the same period of ca. 88.3 days, and the season of spring (from the spring equinox to the summer solstice) consisted of ca. 94.3 days, with margins of deviation of the same magnitude. However, the displacement of ca. 103.89 days between the beginning points of the 28th regnal year of Shalmaneser I and the 22nd regnal year of Tukultī-Ninurta I is still greater (by ca. 9.6 days) than the duration of the longest season of the solar year cycle.
74. [$^{\text{ITU d}}$NIN-]⸢É⸣.GAL-*li* UD.15.KÁM *li-mu* $^{\text{m⸢d⸣}}$AMAR.UTU-⸢ŠEŠ.KAM⸣ (*MARV* V 8, 67).
75. For the translation of Akkadian *dišpu*, rendered by the logogram LÀL, as "fruit syrup" rather than "honey," see Freydank (2007), p. 70, n. 1, and the earlier literature cited there.
76. For a description of the archive Assur 18767, see Pedersén (1985), pp. 43, 49.
77. Freydank (1997); Jakob (2003), pp. 177–178.
78. For the interpretation of the term *alaḫḫennu* in Middle Assyrian sources as "baker," see Jakob (2003), pp. 391–394. However, while professionals recorded in Middle Assyrian sources as *alaḫḫennū* would be normally occupied with baking, they could be sometimes entrusted with responsibility for other tasks (e.g., fattening animals).
79. Following this line, there is another record, which sheds some light on the procedure recorded in the docu-

ment: 5 SÌLA Ì *ki-mu* Ì.ŠUR *ša* ⸢*ki-mu-šu*⸣ *tab-ku-ni* ŠE.Ì.GIŠ.MEŠ *la+a*! *id-din ṭup-*⸢*pu-šu*⸣ *ṣa-ab-ta-*⸢*at*⸣, "5 *qû* of oil instead of (what) the oil-presser, his representative, spilled; he did not give sesame; his tablet is executed (in an officially valid manner)" (*MARV* V 8, 20–21; for the collocation *ṭuppa ṣabātu* "to execute a tablet (in such a way as to render it officially valid)," see Postgate [1986], pp. 18-21). The 3 m. sg. pronominal suffix *-šu* refers here, in all likelihood, to Būniya, who is mentioned at the end of l. 19 of the document. The mention of an oil-presser (Ì.ŠUR, *ṣāḫitu*) acting as a representative of Būniya (*ša kīmūšu*—cf. *AHw*, p. 477a, s.v. *kīma*, 7a; *CAD* K, p. 369a, s.v. *kīma*, a 4') indicates that beside supplying sesame for the temple of Aššur, Būniya was required, or found it preferable on his own initiative, to provide the temple with the end-product, for which sesame was intended—i.e., with oil. He evidently hired a professional oil-presser to produce oil from the sesame, but the latter spilled the oil (for the use of the stative form *tabkuni*, in the subjunctive, with an active meaning, cf. *GAG* §77e), and Būniya was held responsible for restoring the lost amount of produce to the temple. For some reason, he preferred to compensate the temple with oil rather than with sesame (for the indicative use of *lā* in the main clause [*šamaššammē lā iddin*], characteristic of Middle and Neo-Assyrian usage, cf. *GAG* §122a). However, the rest of the produce, which Būniya had to supply to the temple, beside the compensation for the oil spilled by his representative, he evidently delivered in the form of sesame. The only other person mentioned in *MARV* V 8 to deliver sesame (three homers thereof) is [m]*Si-ia-ú-tu* Ì.⸢ŠUR⸣, "Siyaūtu the oil-presser" (l. 25), which indicates that a person's profession did not determine the form, in which the produce was supplied. It is not clear why most persons mentioned in the document supplied the end-product (oil) rather than the raw material (sesame), but it stands to reason that at least some of them also hired professional oil-pressers to produce the oil.

80. Written over erasure.
81. The month name is restored as [Allan]ātu based on the fact that in the Middle Assyrian cycle of months, this month immediately precedes the earliest month, for which a record of delivery appears in the part of *MARV* V 8 detailing the deliveries made in the eponym year of Erīb-Aššur—viz., the month Bēlat-ekalli (ll. 28–29). For the comparable instances of the month-name Allānātu appearing in the genitive case ([ITU]*allānāte*), when demanded by the syntax, in Middle Assyrian sources, see, e.g., Röllig (2008), no. 22, l. 34, and no. 28, l. 15. Our restoration assumes that a record of a delivery in the month Allānātu appeared somewhere in the damaged part of *MARV* V 8, which covers the eponym year of Erīb-Aššur (ll. 33–44). Since the mention of a delivery in the month Allānātu would come after the mention of the delivery in the month Bēlat-ekalli, it would be also out of chronological order with the preceding records, as is characteristic for the records of the deliveries made in the eponym year of Erīb-Aššur.
82. However, it is unclear why in the part corresponding to the eponym year of Erīb-Aššur there are two records of deliveries by Mannu-bal-Aššur, the priest of Ištar-of-the-stars ([m]*Ma-nu-bal-Aš-šur* SANGA *šá* [d]*Iš*$_8$*-tár* MUL, *MARV* V 8, 23, 26). Since these records are not accompanied by dates, nothing can be said about their temporal arrangement.
83. The conclusion that the eponym years of Erīb-Aššur and Marduk-aḫa-ēriš were two successive calendar years is supported by the document VAT 15472 (*MARV* III 30; Freydank [1992], no. 29), which speaks of missing produce for the regular offerings in the temple of Aššur "for two years, for the eponym year of Erīb-Aššur and the eponym year of Marduk-aḫa-ēriš" (*ša* 2 MU.MEŠ *ša li-me* [m]SU-[d]*A-šur ù li-me* [md]AMAR.UTU-ŠEŠ-KAM, *MARV* III 30, 3–5).
84. Bloch (2010c), pp. 30–39, esp. p. 39, n. 39. The eponym years of Erīb-Aššur and Marduk-aḫa-ēriš belong to group 7, out of the eight groups of eponyms listed there.
85. Weidner (1935–36), pp. 32–33. The text is cited here according to Weidner's hand-copy of the tablet.
86. Weidner (1935–36), p. 28.
87. Dr. Baker's proposal was presented at the 56ème Rencontre Assyriologique Internationale in Barcelona, July 28, 2010, and is based on interpreting the term *ṭuppi(šu)*, as a reference to a period of time, in the light of its use in Neo-Babylonian and Neo-Assyrian administrative and economic documents, where the evidence for the meaning "one year" is compelling. A pre-publication version of Dr. Baker's study is now available online: http://iowp.univie.ac.at/?q=node/192.
88. Bloch (2010c), pp. 55–64. Mesopotamian chronological practice, as established from the 14th century B.C.E. onwards, counted the official regnal years of a king from the first New Year's Day after his enthrone-

ment. The period of time from the actual enthronement of a king to the next New Year's Day was known as the accession year (*rēš šarrūti, šurrû šarrūti*)—see Brinkman (1976), p. 403; Tadmor (1958), pp. 27–28.

89. For the interruption of the orderly nomination of the yearly eponyms following the eponym year of Da''ānī-Ninurta, see Freydank (1991), pp. 101–102, 130. For the placement of the eponym year of Da''ānī-Ninurta in the reign of Aššur-dān I, see Freydank (2007), pp. 74–75.
90. See above, nn. 44–45.
91. Another calculation presented in section III of the present article, which demonstrated that if the mechanism of intercalation proposed by Weidner were employed in the Middle Assyrian calendar, Allānātu 27 would occur in the 18th regnal year of Tukultī-Ninurta I between late July and mid-October, can now be discarded, after we have refuted Weidner's proposed mechanism of intercalation.
92. ⸢*i+na* ^{ITU}⸣*ku-zal-li* ⸢^{ITU}⸣AB [U]D.⸢25⸣.KÁM *li-me* ^{md}UTU-⸢IBILA-KAM⸣, "In the month Kuzallu (which is) the month Ṭebētu/Kanūnu, [d]ay 25, the eponym year of Šamaš-apla-ēriš" (*MARV* VI 86, document, 12–13).
93. For the correspondences between the Assyrian and the Babylonian months in the eponym year of Ištu-Aššur-ašāmšu, see Freydank (1991), p. 84. For the placement of the eponym year of Ištu-Aššur-ašāmšu immediately after the eponym year of Tiglath-pileser I, see Freydank, *MARV* VII, p. 10 (on *MARV* VII 42).
94. See above, n. 24.
95. See above, n. 88.
96. For the source of the astronomical data utilized in the present article, see above, n. 52.
97. The same convention was adopted by Parker and Dubberstein (1946), p. 24, and Parpola (1983), p. 382, n. 672.
98. For the reasons, due to which we take specifically these years as the limit points of table 4, see below, n. 117.
99. For the Assyrian trading colony at Kaneš, see Veenhof (2008), esp. pp. 41–55, 131–146.
100. For the month names of the Old Assyrian calendar, see Cohen (1993), p. 239.
101. In this section of the present article, month names are specified in their Old Assyrian form.
102. Larsen (1976), p. 53, n. 18, and p. 193.
103. Veenhof (1995–96), pp. 13–15; Veenhof (2000).
104. Some documents from *kārum* Kaneš level II employ, beside the calendar months and the eponym years, an additional dating device named *ḫamuštum* and identified, like a year, by names of eponymous persons. It is clear that the *ḫamuštum* period comprised a few days; we accept the view of Veenhof (1995–96) that it was a seven-day week.
105. Veenhof (1995–96), pp. 13–14.
106. Veenhof (1995–96), p. 14.
107. Those dates, correlated with the numbers of the eponym years according to the order of the Kültepe Eponym List (KEL)—a list known from several manuscripts and presenting the order of yearly eponyms during the period of *kārum* Kaneš level II—are specified in Veenhof (2000), p. 145. For the publication of the KEL, see Veenhof (2003).
108. Veenhof (2000), p. 146.
109. Conversely, the fact that in the eponym year of Puzur-Niraḫ one and the same week spanned part of the month Allānātum and part of the month *zibibirum* suggests that in this case the intercalary month belonged to the same year as the preceding, regular Allānātum. However, it would be precarious to reach a conclusion concerning the placement of the beginning point of the new year, in relation to the intercalary month, based on this occurrence alone. In the date formula involving the month *zibibirum*, the scribe may have simply forgotten to specify the year as *ša qāte* of the previous eponym. Similar cases of scribal negligence to note the change of the eponym in the first month of a new year are known from documents discovered at Mari and Tell Leilan and dating roughly to the 18th century B.C.E.—see Charpin and Ziegler (2003), p. 161.
110. As recognized by Veenhof (2000), p. 146.
111. Larsen (1976), p. 53, n. 18. Larsen's proposal was accepted by Cohen (1993), pp. 238–239.

112. Veenhof (2000), p. 147.

113. Such eponyms can be gleaned from the lists of the eponyms attested in documents from the period of *kārum* Kaneš level Ib, which have been assembled by Veenhof (2003), pp. 61–66, and Günbattı (2008), pp. 129–131.

114. The list KEL G, published by Cahit Günbattı (2008), overlaps only to a small degree with other manuscripts of KEL, which cover the period of *kārum* Kaneš level II (see above, n. 107). The first 20 eponyms of KEL G are the last 20 eponyms known from the earlier manuscripts of KEL.

115. Thus, reservation is due, e.g., with regard to the following remark of Julian Reade: "Since seasons and harvests outside the tropics depend on the sun, most cultures have developed and kept a solar year, to which the months have been adjusted. . . A comparison with Islamic practice in the Middle East is irrelevant, since the Qurʾanic ban on intercalation was an innovation, related to the occasion when Medina was attacked from Mecca during what was thought to be a month of sacred truce; the attackers must have postponed the sacred month" (Reade [2001], p. 2). What the Islamic example shows is that considerations originating from a specific historical situation could override the general utility of a luni-solar calendar for an agricultural society. The fact that we do not know, which historical circumstances caused the transition to a purely lunar calendar in Assyria, does not mean that no such circumstances existed.

116. Gingrich (1989), pp. 356–358.

117. For the source of the astronomical data, on which this table is based, see above, n. 52. We begin the table with the year 1301 B.C.E. because it is a convenient starting point: a year proximate to the beginning of the 13th century B.C.E. and the first regnal year of an Assyrian king (Adad-nērārī I, whom the Assyrian King List reports to have reigned 32 years). We finish the table with the year 1093 B.C.E. because this is probably the latest year, for which an explicit correspondence between Assyrian and Babylonian months appears in an Assyrian document: in the inscription RIMA 2, A.0.87.4, dated to the eponym year of Taklāk-ana-Aššur (for the conclusion that this year is to be dated ca. 21 years after the eponym year of Tiglath-pileser I—i.e., ca. the 22nd regnal year of that king—see section III of the present article). The dates specified in this table are one year lower, for the period starting with the reign of Aššur-rēša-iši I, than the dates provided by Gasche et al. (1998), p. 63, because the latter group of scholars schematically dated the first regnal year of Aššur-rēša-iši I to 1132 B.C.E., instead of the correct date 1131 B.C.E. On the other hand, for the period preceding the reign of Aššur-rēša-iši I, our dates are one year higher than those provided by Gasche et al., because we account for the *ṭuppišu* reigns of Ninurta-tukulti-Aššur and Mutakkil-Nusku as one year each (see above, n. 87). The dates for the reigns of Shalmaneser I and Tukultī-Ninurta I provided in Bloch (2008) and Bloch (2010a), which were based on the chronological chart of Gasche et al. (1998), p. 63, must be adjusted accordingly.

118. It is to be assumed, unless the evidence suggests otherwise, that each Assyrian king of the 13th–12th centuries B.C.E. carried out the office of the eponym in his first regnal year—see Bloch (2010c), pp. 24–25, n. 9, and the earlier literature cited there. Nevertheless, we cite eponym years of kings only for those kings, for whom they are actually attested. For such attestations in the 13th-12th centuries B.C.E., see Llop (2008), p. 22; for the eponym year of the king Aššur-nādin-apli, see Bloch (2010c), pp. 31–32, n. 21.

119. From the midnight subsequent to the sighting of the new lunar crescent until the next midnight.

120. For the order of the eponyms in the reign of Shalmaneser I, see Bloch (2008); but cf. the following note.

121. In an earlier study, the present author accepted the reconstructions of Freydank (2005), pp. 49–50; Röllig (2004), p. 43; and Röllig (2008), p. 4, who dated the eponym Bēr-bēl-līte (not attested in documents from Tell Šēḫ Ḥamad) before the eponym Ištar-ēriš son of Salmānu(Šulmānu)-qarrād—see Bloch (2008), p. 147. However, the document VAT 18900 (*MARV* III 4), which records an official assessment of harvest from a certain group of fields in the province of Nēmad-Ištar, mentions "old barley from the eponym year of Ištar-ēriš [. . .], which has been dug out of the granary and given out in the eponym year of Lullāyu" (ŠE SUMUN *ša li-me* ^{md}*Iš*$_8$*-tár*-KAM x$^?$ *iš-*⸢*tu* É⸣ *ḫa-ši-ma-te ḫa-aṭ-ṭa i+na li-me* ^{m}*Lu-la-ie-e ta-din*, rev. 3'–4') alongside with "the yield of the harvest of the eponym year of Bēr-bēl-līte, which Aššur-apla-iddina son of Appayūtu, the (royal) commissioner, and Adad-mu/šu[ma . . .] assessed in the eponym year of Lullāyu" (*te-*⸢*li*⸣*-it e-bu-ri* ⸢*ša li*⸣*-me* ^{md}*Be-er-*EN*-li-i-*⸢*te*⸣ *ša i+na* ⸢*li*⸣*-me* ^{m}*Lu-la-ie-e* ^{md}*A-šur-*⸢IBILA-SUM*-na*⸣ DUMU *Ap-pa-iu-ú-te qe-pu ù* ^{d}IM-M[U$^?$- . . .] *ip-šu-ru-ú-ni*, rev. 6'–9'); transliteration and translation of the text

follow Freydank (1994), pp. 21-22. These records indicate clearly that the eponym year of Ištar-ēriš (which produced barley already considered "old" in the eponym year of Lullāyu) must have predated that of Bēr-bēl-līte (a recent harvest of which was officially assessed in the eponym year of Lullāyu). Thus, our reconstruction of the order of the eponyms in the reign of Shalmaneser I must be corrected accordingly.

122. For the order of the eponyms in the reign of Tukultī-Ninurta I (up to his 26th regnal year), see Bloch (2010a); but cf. below, nn. 123, 125.

123. The placement of an eponym named Aššur-mušabši (son of Adad-bān-kala?) in the reign of Tukultī-Ninurta I, and more specifically, within the period when Aššur-iddin son of Qibi-Aššur carried out the office of the Grand Vizier and the ruler of Assyrian possessions in northeastern Syria (Ḫangalbat), has been proposed by the present author (Bloch [2010a], pp. 4–5, 25; Bloch [2010c], pp. 73–74, n. 47), based on the document DeZ 3847/2 from Tell Šēḫ Ḥamad, in which the name of the eponym was identified as Aššur-mušabši by Jakob (2003), p. 56. However, the study of Llop (forthcoming) suggests that the identification of the eponym in DeZ 3847/2 as Aššur-mušabši is uncertain. The same study also provides convincing arguments for dating the eponym year of Bēr-nādin-apli before the eponym year of Abī-ilī son of Katiri. On the other hand, Llop's proposal to place the eponym Bēr-nādin-apli after the eponym Ninuʾāyu has been rightly called into question by Jakob (forthcoming). A final verdict on the existence of an eponym named Aššur-mušabši in the reign of Tukultī-Ninurta I requires publication or collation of the document DeZ 3847/2. However, even before this condition is met, it appears that one of the years shortly preceding the eponym year of Ina-Aššur-šumī-aṣbat was known, at least for the most part thereof, as the eponym year of Bēr-nādin-apli (it cannot be ruled out that the eponym was changed during the year, and part of the same year may have been known under the eponym Aššur-mušabši, if his name is indeed recorded in DeZ 3847/2). The eponym Bēr-nādin-apli cannot be placed before the eponym Aššur-zēra-iddina (the 16th regnal year of Tukultī-Ninurta I), because of the evidence indicating a comprehensive sequence of eponyms from Tukultī-Ninurta I to Aššur-zēra-iddina (for that evidence, see Röllig (2004), pp. 43–49, with corrections by Freydank (2005), pp. 45–50, and supplemented by Bloch (2010a), pp. 4–5, 14–15, n. 48). On the other hand, the eponym Bēr-nādin-apli is also not likely to be placed between the eponyms Ina-Aššur-šumī-aṣbat and Abī-ilī (in addition to the eponyms mentioned in that position in Bloch (2010a), pp. 31–32), because of the considerations presented above, n. 45. Consequently, the most likely position for the eponym Bēr-nādin-apli seems now to be between the eponym year of Aššur-zēra-iddina and that of Ina-Aššur-šumī-aṣbat. The present author is grateful to Dr. Jaume Llop and Dr. Stefan Jakob for providing him with pre-publication drafts of their forthcoming studies.

124. For the dating of the eponym Abī-ilī son of Katiri to the 21st regnal year of Tukultī-Ninurta I, see above, n. 72.

125. In Bloch (2010a), p. 32, the eponym Bēr-nādin-apli was tentatively placed between the eponyms Ellil-nādin-apli and Kaštiliašu, the latter preceding the eponym year of Bēr-išmanni. In the light of the considerations presented above, n. 123, such placement of the eponym Bēr-nādin-apli appears now unlikely. In any event, the sequence of eponyms from the 23rd regnal year of Tukultī-Ninurta I onwards is very much hypothetical.

126. For tentative dating of this eponym to the 27th regnal year of Tukultī-Ninurta I, see Bloch (2010c), p. 76, n. 53.

127. In the reign of Ninurta-apil-Ekur it is possible to date precisely, beside the eponym year of the king himself, only the last three eponyms belonging to the chronologically latest group out of the eight groups of eponyms, which are to be placed in the period from the death of Tukultī-Ninurta I to the death of Ninurta-apil-Ekur—see Bloch (2010c), p. 39, n. 39.

128. For these two eponyms (both of whom carried out the office within one year) and for the following eponym, see Bloch (2010c), pp. 61–64.

References

AHw = von Soden, W., 1956-81, *Akkadisches Handwörterbuch* (Wiesbaden: Harrassowitz).

Bloch, Y., 2008, "The Order of Eponyms in the Reign of Shalmaneser I", *Ugarit-Forschungen* 40, 143–178.

——, 2010a, "The Order of Eponyms in the Reign of Tukultī-Ninurta I", *Orientalia*, N.S. 79, 1–35.

——, 2010b, "The Period of Activity of the Scribal Family of the Sons of Ninurta-Uballissu in Aššur", *Nouvelles Assyriologiques Brèves et Utilitaires* 2010, no. 36.

——, 2010c, "Solving the problems of the Assyrian King List: Toward a Precise Reconstruction of Middle Assyrian Chronology: Part I and Part II", *Journal of Ancient Civilizations* 25, 21–87.

Boese, J. and Wilhelm, G., 1979, "Aššur-dān I, Ninurta-apil-Ekur und die mittelassyrische Chronologie", *Wiener Zeitschrift für die Kunde des Morgenlandes* 71, 19–38.

Borger, R., 1957–58, "Der Eponym Ipparšidu", *Archiv für Orientforschung* 18, 299.

Brinkman, J. A., 1968, *A Political History of Post-Kassite Babylonia, 1158-722 B.C.,* Analecta Orientalia 43 (Rome: Pontificium Institutum Biblicum).

——, 1976, *Materials and Studies for Kassite History, I: A Catalogue of Cuneiform Sources Pertaining to Specific Monarchs of the Kassite Dynasty* (Chicago: The Oriental Institute).

CAD = Oppenheim, A. L. et al. (eds.), 1956–2010, *The Assyrian Dictionary of the Oriental Institute of the University of Chicago* (Chicago: The Oriental Institute).

Cancik-Kirschbaum, E., 1996, *Die mittelassyrischen Briefe aus Tall Šēḫ Ḥamad*, Berichte der Ausgrabung Tall Šēḫ Ḥamad/Dūr-Katlimmu 4 (Berlin: Dietrich Reimer).

Charpin, D. and Ziegler, N., 2003, *Mari et le Proche-Orient à l'époque amorrite: essai d'histoire politique,* Florilegium Marianum 5 (Antony: Société pour l'Étude du Proche-Orient Ancien).

Cohen, M. E., 1993, *The Cultic Calendars of the Ancient Near East* (Bethesda, Md.: CDL Press).

Deller, K., 1987, Review of MARV II, *Archiv für Orientforschung* 34, 58–66.

Ehelolf, H. and Landsberger, B., 1920, "Der altassyrische Kalender", *Zeitschrift der Deutschen Morgenlandischen Gesellschaft* 74, 216–219.

Englund, R. K., 1988, "Administrative Timekeeping in Ancient Mesopotamia", *Journal of the Economic and Social History of the Orient* 31, 121–185.

Freydank, H., 1991, *Beiträge zur mittelassyrischen Chronologie und Geschichte*, Studien zur Geschichte und Kultur des Alten Orients 21 (Berlin: Akademie-Verlag).

——, 1992, "Das Archiv Assur 18764", *Altorientalische Forschungen* 19, 276–321.

——, 1994, "Drei Tafeln aus der Verwaltung des mittelassyrischen Kronlandes", *Altorientalische Forschungen* 21, 13–30.

——, 1997, "Mittelassyrische Opferlisten aus Assur", in H. Waetzoldt and H. Hauptmann (eds.), *Assyrien im Wandel der Zeiten: XXXIXe Rencontre Assyriologique Internationale, Heidelberg 6.–10. Juli 1992*, Heidelberger Studien zum Alten Orient 6 (Heidelberg: Heidelberger Orientverlag), 47–52.

——, 2005, "Zu den Eponymenfolgen des 13. Jahrhunderts v. Chr. in Dūr-Katlimmu," *Altorientalische Forschungen* 32, 45–56.

——, 2007, "'Honig'-Lieferungen für den Gott Aššur", *Altorientalische Forschungen* 34, 70–77

GAG = von Soden, W. F., 1995, *Grundriss der akkadischen Grammatik*, 3rd edn., with cooperation of W. R. Mayer, Analecta Orientalia 33 (Rome: Pontificium Institutum Biblicum).

Gasche, H. et al., 1998, *Dating the Fall of Babylon: A Reappraisal of Second-Millennium Chronology (A Joint Ghent-Chicago-Harvard Project)* (Ghent: University of Ghent, Chicago: The Oriental Institute).

Gil, M., 2004, "The Flax Trade in the Mediterranean in the Eleventh Century A.D. as Seen in Merchants' Letters from the Cairo Geniza", *Journal of Near Eastern Studies* 63, 81–96.

Gingrich, A., 1989, "Kalender, Regenzeit und Stieropfer in Nordwest-Jemen", in B. Scholz (ed.), *Der orientalische Mensch und seine Beziehungen zur Umwelt: Beiträge zum 2. Grazer morgenländischen Symposion (2.–5. März 1989)* (Graz: Grazkult), 353–370.

Günbattı, C., 2008, "An Eponym List (KEL G) from Kültepe", *Altorientalische Forschungen* 35, 103–132.

Horowitz, W., 1996, "The 360 and 364 Day Year in Ancient Mesopotamia", *Journal of the Ancient Near Eastern Society* 24, 35–44.

——, forthcoming, *"Three Stars Each": The Astrolabes and Related Texts.*

Huber, P. J., 1999–2000, Review of Gasche et al. (1998), *Archiv für Orientforschung* 46–47, 287–290.

——, et al., 1982, *Astronomical Dating of Babylon I and Ur III,* Monographic Journals on the Near East, Occasional Papers 1/4 (Malibu: Undena).

Hunger, H., 1976–80, "Kalender", in E. Ebeling et al. (eds.), *Reallexikon der Assyriologie und vorderasiatischen Archäologie* 5 (Berlin: De Gruyter), 297–303.

——, and Pingree, D., 1989, *MUL.APIN: An Astronomical Compendium in Cuneiform,* Archiv für Orientforschung Beiheft 24 (Vienna: Ferdinand Berger).

Jakob, S., 2003, *Mittelassyrische Verwaltung und Sozialstruktur: Untersuchungen,* Cuneiform Monographs 29 (Leiden: Brill, Boston: Styx).

——, forthcoming, "Sag mir quando, sag mir wann. . ." *Proceedings of the 56ème Rencontre Assyriologique Internationale.*

KAV = Schroeder, O., 1920, *Keilschrifttexte aus Assur verschiedenen Inhalts,* Wissenschaftliche Veröffentlichungen der Deutschen Orient-Gesellschaft 35 (Leipzig: Hinrichs).

Koch, J., 1989, *Neue Untersuchungen zur Topographie des babylonischen Fixsternhimmels* (Wiesbaden: Harrassowitz).

Larsen, M. T., 1976, *The Old Assyrian City-State and Its Colonies* (Copenhagen: Akademisk Forlag).

Llop, J., 2008, "MARV 6,2 und die Eponymenfolge des 12. Jahrhunderts", *Zeitschrift für Assyriologie* 98, 20–25.

——, forthcoming, "The Eponym Bēr-nādin-apli and the Documents Referring to the Expeditions to the City of Tille in the Reign of Tukultī-Ninurta I", *Proceedings of the 56ème Rencontre Assyriologique Internationale.*

MARV I = Freydank, H., 1976, *Mittelassyrische Rechtsurkunden und Verwaltungstexte,* Vorderasiatische Schriftdenkmäler der Staatlichen Museen zu Berlin 19 (Berlin: Akademie-Verlag).

MARV II = Freydank, H., 1982, *Mittelassyrische Rechtsurkunden und Verwaltungstexte,* II, Vorderasiatische Schriftdenkmäler der Staatlichen Museen zu Berlin 21 (Berlin: Akademie-Verlag).

MARV III = Freydank, H., 1994, *Mittelassyrische Rechtsurkunden und Verwaltungstexte,* III, Wissenschaftliche Veröffentlichungen der Deutschen Orient-Gesellschaft 92 (Berlin: Gebr. Mann).

MARV V = Freydank, H. and Feller, B., 2004, *Mittelassyrische Rechtsurkunden und Verwaltungstexte,* V, Wissenschaftliche Veröffentlichungen der Deutschen Orient-Gesellschaft 106 (Saarbrücken: Saarbrücker Druckerei und Verlag).

MARV VI = Freydank, H. and Feller, B., 2005, *Mittelassyrische Rechtsurkunden und Verwaltungstexte,* VI, Wissenschaftliche Veröffentlichungen der Deutschen Orient-Gesellschaft 109 (Saarbrücken: Saarbrücker Druckerei und Verlag).

MARV VII = Freydank, H. and Feller, B., 2006, *Mittelassyrische Rechtsurkunden und Verwaltungstexte,* VII, Wissenschaftliche Veröffentlichungen der Deutschen Orient-Gesellschaft 111 (Saarbrücken: Saarbrücker Druckerei und Verlag).

MARV VIII = Freydank, H. and Feller, B., 2007, *Mittelassyrische Rechtsurkunden und Verwaltungstexte,* VIII, Wissenschaftliche Veröffentlichungen der Deutschen Orient-Gesellschaft 119 (Wiesbaden: Harrassowitz).

MARV IX = Freydank, H. and Feller, B., 2010, *Mittelassyrische Rechtsurkunden und Verwaltungstexte,* IX, Wissenschaftliche Veröffentlichungen der Deutschen Orient-Gesellschaft 125 (Wiesbaden: Harrassowitz).

Parker, R. A. and Dubberstein, W. H., 1946, *Babylonian Chronology, 626 B.C. – A.D. 45.* (Chicago: University of Chicago Press).

Parpola, S., 1983, *Letters from Assyrian Scholars to the Kings Esarhaddon and Assurbanipal, II: Commentary and Appendices,* Alter Orient und Altes Testament 5/2 (Neukirchen-Vluyn: Butzon & Berker Kevelaer).

Pedersén, O., 1985, *Archives and Libraries in the City of Assur: A Survey of the Material from the German Excavations,* I, Studia Semitica Upsaliensia 6 (Uppsala: Almqvist and Wiksell).

Postgate, J. N., 1986, "Middle Assyrian Tablets: The Instruments of Bureaucracy", *Altorientalische Forschungen* 13, 10-39.

Reade, J., 2000, "Absolute Dates and Assyrian Calendars", *Akkadica* 119–120, 151–153.

——, 2001, "Assyrian King-Lists, the Royal Tombs of Ur, and Indus Origins", *Journal of Near Eastern Studies* 60, 1–29.

RIMA 2 = Grayson, A. K., 1991, *Assyrian Rulers of the Early First Millennium, I (1114–859 B.C.E.)*. The Royal Inscriptions of Mesopotamia, Assyrian Periods 2 (Toronto: University of Toronto Press).

Röllig, W., 2004, "Eponymen in den mittelassyrischen Dokumenten aus Tall Šēḫ Ḥamad/Dūr-Katlimmu", *Zeitschrift für Assyriologie* 94, 18–51.

——, 2008, *Land- und Viehwirtschaft am unteren Ḫābūr in mittelassyrischer Zeit*, Berichte der Ausgrabung Tell Šēḫ Ḥamad/Dūr-Katlimmu 9 (Wiesbaden: Harrassowitz).

Saporetti, C., 1978, "Ein mittelassyrischer *bāriʾu*", *Assur* 1/8, 175–177.

——, 1979, *Gli eponimi medio-assiri*, Bibliotheca Mesopotamica 9 (Malibu: Undena).

Sassmannshausen, L., 2006, "Zur mesopotamischen Chronologie des 2. Jahrtausends", *Baghdader Mitteilungen* 37, 157–177.

Tadmor, H., 1958, "The Campaigns of Sargon II of Assur: A Chronological-Historical Study", *Journal of Cuneiform Studies* 12, 22–40, 77–100.

Tschinkowitz, H., 1968–69, "Ein Opfershautext aus dem Eponymenjahr Tiglatpilesers I", *Archiv fü Orientforschung* 22, 59–62.

Veenhof, K. R., 1995–96, "The Old Assyrian *ḫamuštum* Period: A Seven-Day Week," *Jaarberichrt van het vooraziatisch-egyptisch genootschap Ex Oriente Lux* 34, 5–26.

——, 2000, "Old Assyrian Chronology", *Akkadica* 119–120, 137–150.

——, 2003, *The Old Assyrian List of Year Eponyms from Karum Kanish and Its Chronological Implications* (Ankara: Turkish Historical Society).

——, 2008, "The Old Assyrian Period", in K. R. Veenhof and J. Eidem, *Mesopotamia: The Old Assyrian Period*, Orbis Biblicus et Orientalis 160/5 (Fribourg: Academic Press, Göttingen: Vandenhoeck & Ruprecht), 13–264.

Vogelsang-Eastwood, G., 1992, *The Production of Linen in Pharaonic Egypt* (Leiden: Stichting Textile Research Center).

Waetzoldt, H. 1987. "Compensation of Craft Workers and Officials in the Ur III Period", in M. A. Powell (ed.), *Labor in the Ancient Near East*, American Oriental Series 68 (New Haven: American Oriental Society), 117–141.

Weidner, E. F., 1928–29, "Der altassyrische Kalender", *Archiv für Orientforschung* 5, 184–185

——, 1935–36, "Aus den Tagen eines assyrischen Schattenkönigs", *Archiv für Orientforschung* 10, 1–52.

Beyond the Moon: Minoan 'Calendar'-Symbolism in the 'Blue Bird Fresco'[1]

Sabine Beckmann

> To every thing there is a season, and a time to every purpose under the heaven: A time to be born, and a time to die; a time to plant, and a time to pluck up that which is planted ...
>
> Ecclesiastes 3:1,2

The possible knowledge of astronomy in the Aegean Bronze Age as well as the existence of Minoan calendars has been treated by various scholars[2] and with different approaches over time, but as there are no preserved astronomical texts or objects clearly recognizable as astronomical tools, the discussion must remain on a speculative basis until future findings can change these premises.

Whereas those approaches are based on astronomical theories and possibilities, this paper follows a different line, based on the fact that Minoan art and especially the well known 'Blue Bird Fresco' (figure 1) from Knossos seem to have had not just a decorative, but a combined practical calendrical and spiritual/medical/magical[3] function not using astronomy, but agricultural phases as expressed in the meanings implied by certain plant images.[4]

Since the beginnings of what we might call 'Old World civilization' in the Eastern Mediterranean some 4000 years ago, flowers and trees have played a special role in our cultural and spiritual realms, whatever religion they might have belonged to over time. From the unfolding of life at birth to the fear of doom by eternal withering at death humans have seen plants for eons as symbols of flourishing hope in life and the return to a new growth after death, as powerful emblems of the promise of fertility, be it for man, beast or field. As many of the revered plants had also healing or conserving properties their practical value justified the special position given to them—a fact that is nearly forgotten nowadays in a time of synthetical remedies and preservatives, even though people still cherish flowers and trees as they always have. This may be the reason why Minoan landscape art seems to have had a special appeal ever since the various sites they came from were excavated from the beginning of the 20th century. Many a scholar has tried to describe this appeal as no more than the Minoan artists' special ability to depict nature in a playful, dynamic way. Still the following pages might show that most admirers of Minoan art may have been feeling more of the underlying symbolism that in many a small way has survived until today.

The Minoan wall painting panel shown in the Iraklion museum and known as the 'Blue Bird Fresco' represents only part of the many fragments excavated by Evans in 1923 in the

FIGURE 1. The Blue Bird Fresco, House of Frescoes, Knossos (Iraklion Archaeological Museum).

'House of Frescoes':[5] A rocky and floral general setting containing blue birds, blue monkeys and various plants, as well as a panel with heraldically positioned *agrimia* (wild Cretan goats). Mark Cameron's restoration for the latter and the fresco with birds and monkeys[6] includes more of the excavated fragments (see figure 2) and thus plants not represented in the exhibit (e.g. lilies).

N. Marinatos[7] and J. Schäfer[8] rightly noted that the 'Blue Bird Fresco' was not just a decorative wall-painting as Evans believed who stated 'religious themes are avoided',[9] but to the contrary showed a 'religious landscape'.[10] This idea is confirmed by the obvious ritual objects (a Linear A-inscribed libation-table, ritual ladle and vessels decorated with double axes, cf. figure 11) found near the room the fresco fragments were found in. A recent analysis of the architecture and setting of the Knossian house containing the fresco (the 'House of Frescoes') as reflecting a possible religious ceremonial function is developed by Shaw,[11] coming to a similar conclusion.

Whereas scholars suspected the heraldically positioned *agrimia* of Cameron's reconstruction to have a 'religious connection',[12] the exhibited part of the fresco has mostly been seen as decorative (probably in comparison to modern wall-paper). Where S. Immerwahr suspected the implication of a Goddess of Nature,[13] N. Marinatos saw 'a symbolic landscape suggestive of the renewal of nature',[14] because of the plants known to have a 'sacred use' elsewhere,[15] and while A. Chapin recently did not deny the religious significance of

FIGURE 2. Mark Cameron's reconstruction of the Blue Bird Fresco (Cameron 1968, fig. 13, detail).

Aegean landscape painting in general, she saw it also as part of the Minoan 'elite's claim to power and high status'.[16] N. Marinatos elaborated in calling the fresco a 'conceptual rather than realistic' image because of the different terrains shown next to each other in a way she took for impossible in reality, thus reaching her interpretation of a painting showing 'a compression of nature' into an image she calls 'a symbolic depiction of ideal spring'.[17]

While I would agree concerning the 'compressed' image (possibly showing a kind of planted garden),[18] I will show in the following that it is not just spring, but the whole cycle of the year the Minoan painter has expressed in this symbolical yet still realistic fresco.[19] While the symbolism of growth in general can be seen in ancient Cretan art from the beginning of Middle Minoan times,[20] the 'Blue Bird Fresco' can also be interpreted in a more specific sense.

By changing the usual perspective here in declaring what has been so often seen as decorative background a main subject of the painting (i.e. ignoring birds and monkeys for this approach), and with the help of relevant texts from later antiquity the possible 'calendrical' function of the Bronze Age plant images shall be shown here. Even though several centuries and the so-called Greek Dark Age (the ca. 400 years after the collapse of Minoan civilization where archaeological knowledge of religious symbolism is scarce) lie between the latest Minoan art and the first Greek literary sources, already the earliest comparisons of Minoan and Mycenaean (and later) Greek cultures and religious symbolism showed many similarities.[21] Literary sources may provide evidence of traces of a continuity of Minoan symbolism that survived in later Greek mythology, but there is also the possibility that both go back to the same or similar common early Aegean sources.

The Plants of Time: A Perennial 'Epochologio'

The practical ('epochial' rather than 'calendrical', to be precise, as no astronomical factors are involved) and the symbolical connotations of the plants together generate in the Blue Bird Fresco a special kind of 'epochologio': nature's cycle complete in all its elements—possibly for use in ritual and agricultural reality.

The Aegean year has seasons totally different from what northern Europe knows as such, a rather obvious fact rarely taken into account in archaeology. There are only three seasons: Winter, Spring, Summer. In the Eastern Mediterranean summers are hot and dry—all sensitive plants wither and dry up, and (without water being available for irrigation) death by thirst is their fate from the end of May. Only after the first heavy rains, when the soil has become well saturated (usually in November, when days are clearly colder and darker) does the green season, the 'winter', begin. Only then is the land ready for the sowing of winter seeds (cereals and leguminous plants), and saffron crocus is among the first plants to appear from the dead soil of summer.

Saffron – out of the dead

> Therewith the son of Cronos clasped his wife in his arms, and beneath them the divine earth made fresh-sprung grass to grow, and dewy lotus, and crocus, and hyacinth, thick and soft, that upbare them from the ground.
> Homer, Iliad, 14, 345–348[22]

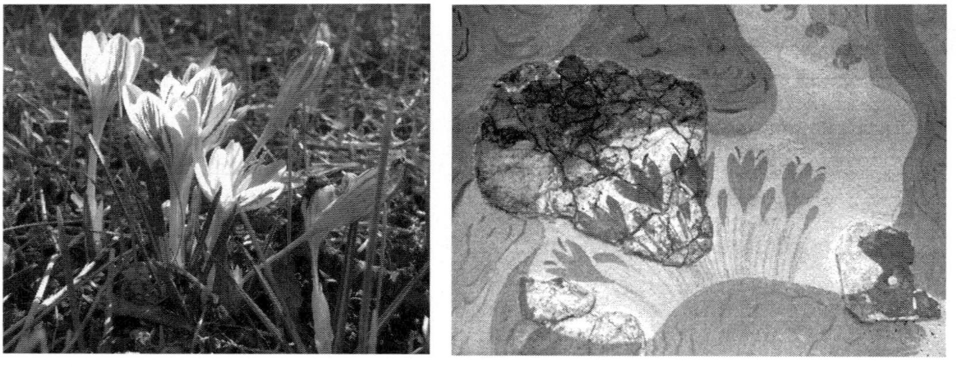

Figure 3. Crocus in nature and on fresco (House of Frescoes, Knossos; Iraklion Archaeological Museum).

The typical surroundings of wild saffron in Crete are the rocky landscapes also known from the so-called Saffron-Gatherer frescoes from Bronze Age Knossos and Thera (figure 3).[23] Saffron crocus flowers only after the first deep-reaching rains. Like called with a magic wand, and long before the leaves, the tiny flowers break from the outwardly still barren soil and thus—even more so in uncultivated areas—show all the world that the dry season of death is over. While their appearance used to be seen as the joyful event of something like the first spring-flower in more northern countries, in Crete even nowadays it always coincides with farmers tilling their fields and sowing.

As ancient sacred plant, saffron images can be seen in continued use since the times of the first Minoan palaces around 1800 BC,[24] painted on pottery and frescoes found in ritual spaces.

Cretan saffron in nature rarely grows in clusters as shown on Minoan frescoes, but

FIGURE 4. Linear A/B 33 variations (based upon Platon & Brice 1975).

mostly in single blooms, or small groups, a fact that suggests the depicted scenes might show planted 'wilderness-gardens'. The dried flowers are reminiscent of the colours croci are painted with in Minoan iconography. It was also employed as a symbol in Linear A/B writing (sign 33 RA3 = RAI, probably corresponding to the first syllable of its lost Minoan name; see figure 4).

Through later antiquity saffron was used mainly in gynaecology and obstetrics and was often mentioned explicitly in connection with deities. As we do not know the flower's name in Mycenaean Greek, the earliest textual evidence for Greek κρόκος (krokos) comes from the Iliad (see above), and the Homeric Hymn to Demeter also mentions it several times: 'soft crocuses mingled with irises and hyacinths, and rose-blooms and lilies, marvellous to see, and the narcissus which the wide earth caused to grow yellow as crocus'.[25] This is Persephone's report of her abduction by Hades when collecting different flowers—note that these, too, are not flowering at the same time of year—amazingly similar to the collection of the 'Blue Bird Fresco', especially if the here mentioned ροδέας κάλυκας (rodeas kalykas), the rose-bloom, was possibly misread from a ροιάς κάλυκας (roias kalykas) a pomegranate-bloom, to be discussed below for the fresco.

Saffron is mentioned here in a context of a *hieros gamos* (the sacred marriage)—symbolizing fertility—as well as in a context of the underworld-goddess (fertility in another aspect).

Its earlier importance might also be hinted at by the later existence of the Κροκωνίδαι (krokonidai), a cast of priests in the Eleusinian Mysteries. They tied a saffron-dyed ribbon (κρόκης – krokes) around the mysts' right hand and left foot. This seems amazingly similar to depictions on Bronze Age frescoes from Thera in Xeste 3, where saffron-coloured ribbons can be seen. The saffron flower next to the lady's blood[26] (she seems to be wounded at her foot) recalls other myths where flowers sprang from blood as a symbol of sacred renewal.

All in all it seems the importance of saffron before all other plants in Minoan iconography can only be understood if one takes into account that after months of draught and heat saffron was the plant to impressively announce that the green, fertile season was coming back, fresh seeds could be sown and the cycle of life would carry on. Thus although we are dealing with a cyclical understanding in general, most probably this phase (winter) would have been seen as the beginning of the agricultural year.

Iris – power and fragility

FIGURE 5. Iris cretica in nature and on fresco (House of Frescoes, Knossos; Iraklion Archaeological Museum).

Spring in Crete is the most passionate time of the year. Although Cretan winters are green, around spring equinox especially powerful growth happens with the first really warm days.

Even humble stretches of phrygana (also typical for saffron) are now conquered by an amazing flower: *Iris Cretica* (or *Iris unguicularis cretensis*—figures 5–6). In grass-like tufts of pale green leaves the low, rainbow-bright flowers often open even before the actual spring begins in March, indicating the the intensive time of tending gardens and fields for agriculture. Iris' blooms are strictly three-fold, in colours between mauve, purple and blue, including some white and yellow, with an intricate design of dark mauve lines on them.[27]

These flowers are well-known in Minoan art, together with their taller and more widely known relative, *Iris Germanica* (appearing e.g. in the Iris-fresco from Amnissos),[28] which is also the plant used for the production of orris-perfume, already attested for early Middle Minoan times in Crete.[29] Its ritual function must be connected to the three-fold graphic clearness of *Iris* (cf. the three-partite Cretan year: summer, winter, spring)—as with other important tripartite features in Minoan art: Shrines in architecture, but also the omnipresent lilies—often hardly distinguishable from iris.

On the other hand there is the obvious identity of the Greek 'ἶρις' (iris) with the plant named by the pre-Greek word '*Hyakinthos*' (or *Iakinthos*). Already in the 19th century scholars maintained that it could be identified with some kind of small *Iris*[30]—at a time well before the excavation of Knossos made *Iris cretica* a well-known feature of Aegean Bronze Age art.

With the identity of the Knossian iris and the mythical *hyakinthos*,[31] an important element of Minoan spiritual iconography can now be seen more clearly.

Myth remembers the story of Apollo and his young lover Hyakinthos who is accidentally killed by the god's diskos. Where his blood touches the ground at his death a flower springs up that henceforth carries the youth's name.

This story is obviously an allusion to the Iris-flowers' fate in spring: as soon as the sun

FIGURE 6. Iris cretica, detail.

(the light-god Apollo's disk) gets too strong, the delicate flowers wither away. Then also Iris germanica blooms and withers with the first heat of summer drying up all soft green life.

The medical use of iris in antiquity was also mainly obstetric. But the external application as well as the fact that modern pharmacology has not found any proof to verify the said effect seem to show that its use in obstetrics goes back more to *Hyakinthos*' ritual or symbolical than its actual pharmacological efficiency, and a certainly notable fact is that it must have been used for funerary rituals as well.[32]

Here, again, we met a flower connected with life and death, with an appropriate epochological function: its drying up announced the nearing end of the fertile season.

Lily – light in light

FIGURE 7. Figure 7: Madonna lily in nature and on Amnissos fresco (after Evans 1921–1935, vol. IV,2, Pl. LXVIIb).

The blinding white of the Madonna lily (*Lilium candidum*—figure 7), together with its sweet perfume and the stunning beauty of its symmetrical flowers has always fascinated humans. Like the iris it shows a symphony of threefold symmetry (for the possible ritual connection in the Minoan Bronze Age as a tripartite entity see above), and over time its name has often been mixed up with the names of similar flowers, too: The French 'Fleur de lys' was not a lily but an iris, for instance, and the 'LEI' of the Greek *leirion* (lily) might be the same syllable the saffron-flower denotes in Minoan Linear A/B writing, where 'R' and 'L' were not differentiated (see above under saffron).

The lily blooms in the beginning of the hot part of summer, when all other plants show signs of withering and fading, another manifestation of power of nature's growing forces that seems to defy climate. It stands for the beginning of summer, bright in beauty and light and deadly in heat, when no other green plant can survive and the time for harvest has come.

The mythical attachment of the lily to Zeus' wife Hera must be an expression of its powers for healing women's diseases, still it has always also been a symbol of immortality which made it one of the most favourite flowers for decorating graves in Christian and Muslim tradition.[33]

In the Bronze Age Aegean, too, it seems to have been seen as a symbol of opposite powers, expressed in the way it was painted in Minoan Thera with a white lily's shape but blood red in colour[34] (blood red was usually the white flowers' background colour, as for instance in the fresco from Amnissos—the dark background in figure 7—where lilies are clearly shown as growing from a man-made base), a powerful combination of spiritual and magical attributes, marking the time of harvest in the Blue Bird Fresco's flower *epochologio*.

Pomegranate – from life to life

> Let us see whether the vine has budded
> And its blossoms have opened,
> And whether the pomegranates have bloomed.
> There I will give you my love.
>
> Song of Solomon 7.13

FIGURE 8. Pomegranate in nature and on fresco (Iraklion Archaeological Museum).

When, around the middle of May, the pomegranate-trees begin to bloom, the dry half of the year is beginning and it is too late for further sowing. Even though all green nature seems to shrivel and die, the pomegranate blooms until far into summer, while the first fruit already develop (figure 8).

The bright orange flowers with five to eight petals open up from a succulent rounded base in the form of the later fruit, sitting between leave-assemblies that show a puzzling tendency to changing sizes and numbers, although often a pattern of threes seems to emerge.

While summer develops its dry heat the tiny, rounding green fruits still will not let go of the withered but nevertheless bright orange rest of their flower, seemingly attempting to be flowering and fruiting at the same time.

When looking again now at the Blue Bird Fresco we might reconsider the so-called 'roses' depicted there. Apart from their orange colour they show six petals (the Minoans certainly knew roses had only five), and the slightly oblong green spheres with orange tops called 'rosebuds', even by a rather doubtful Möbius,[35] are obviously the tiny fruit still carrying their dried flowers. Even the threefold shape of the leaves fits the pomegranate more than the rose.[36]

But the pomegranates' ability to survive and thrive with only a tiny amount of water isn't its only quality to make it a paragon of the power of life and overcoming seemingly certain death. Only to be compared with the vine it even grows fruit undeterred over all the dry summer.

The fruits' shape alone has been seen as a symbol of fertility all over the Eastern Mediterranean. When pomegranates are ripe—in October—the dry brown-red paring will break open, revealing the blood-red, juicy inside. When harvested in a good condition pomegranates keep into the next year due to the hard skin protecting the humid kernel-berries. But the pomegranate-peel does not only keep the fruit fresh: it contains so much tannic-acid that it dries even living skin as if to preserve it for eternity—the reason why it was used for tanning. One might wonder if that would have worked possibly on human skins too, e.g. to prepare mummies.

Different gods are said to have suffered and the pomegranate sprung from their blood. It was holy to the goddesses Aphrodite, Hera and mainly Persephone, in modern Greece it is still being used as fertility symbol. And it is also still one of the ingredients of *kolliba* (the traditional mixture of cooked cereals, sweet and pomegranate-kernels perfumed with mint distributed at obsequies in Greece, see also mint below).

The pomegranate came to Egypt only with the beginning of the New Kingdom but was very successful there, too (as in Mesopotamia), it seems, as in the papyrus Turin it is praised as the erotical tree of all seasons.[37] So we know that in antiquity already it was a symbol for all of life from love to death.

As we have seen in the above, this plant does not symbolize a certain instance in time in the Blue Bird's 'plant epochologio', but rather stands for a longer period, reaches over the dry summer from one phase of fertility (spring) to the next (winter), just as the plant is shown reaching from its high position in between red rocks down into the more humid (blue) ranges of water on the fresco.

Mint – defiying draught and death

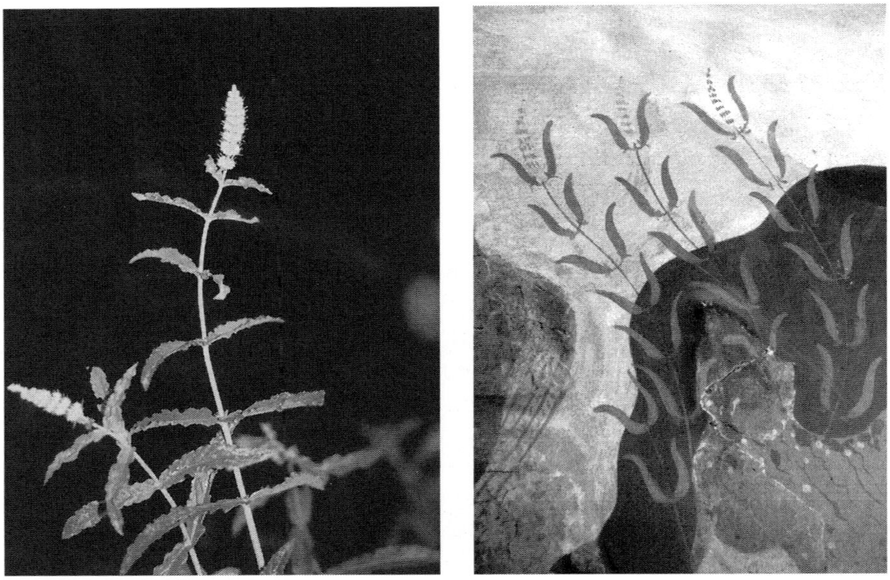

FIGURE 9. Mint in nature and on fresco (Iraklion Archaeological Museum).

The last plant from the Blue Bird Fresco to be mentioned here is the one Evans called 'wild peas or vetches'.[38] They are the same as those rising between two bushes of large iris in the Amnissos fresco called 'with irises and mints'.[39] The opposite paired leaves have side-shots where they spring from the stem, the flowers are leafless light-mauve spikes, thus exactly looking like spear-mint (and other kinds close to *Mentha spicata*—figure 9). Similar wild mints can be found in Crete growing along springs or even in the water.[40]

So obviously in Knossos, too, we have mints, flanking the central blue bird. As the fresco properly shows, mints prefer watery ground to grow from (the blue line probably shows a small creek). There they bloom during the height of summer when all other greens have dried up. To know where mints grow means to know where water is (see the fresco); in Cretan summer only on such a spot a farmer could hope for something to grow at all, and usually regions with enough water for wild mints are used for summer vegetable gardens.

Mint has always been a wide-range remedy: aphrodisiac and treatment against headache, antibiotic and stomachicum. It was also used for different ends in gynecology (e.g. as contraceptive, Dioscorides III, 41), and last not least even for symbolical purification,[41] although nowadays it is usually known mainly for the purification of breath in tooth-pastes and chewing-gums.

Mint has been attested in Linear B tablets (figure 10),[42] and unfortunately its name, too, has been called pre-Indo-European, even though the suffix -nt has long since been declared perfectly Indo-European.[43] The basical meaning of its root 'MI'- must be something like 'growth' or 'sprout', cognate to Greek 'βλάστηση' (vlastisi) but also to Hittite 'mijatar', translated as 'Wachstum', growth.[44] This is especially enlightening if we have a look now at the Linear A/B sign 73 MI (also used as an ideogram for mint), even more in its older

FIGURE 10. Linear A/B 73, MI variations (based upon Platon & Brice 1975).

form. This can easily be recognized as a sprout. To understand why mint has been named like the prototype of 'growth' or 'sprout' there are not only its water-indicating qualities to be thought about, but also its typical way of spreading by shoots. I have argued elsewhere that this word was also part of the religious (in the widest sense) vocabulary inherited in Greek from the Minoan language.[45]

Like the pomegranate, mint defies the dry summers and does not just loose its leaves but vanishes into the earth in winter when the rest of nature is green. Thus later myth has the nymph Mentha trod into the ground by jealous Persephone: Mint sent into the 'underworld'—and possibly seen to be taking Persephone's place as lover of Hades—while the goddess brings fresh growth up into the world.

This important double image of keeping up hope for new growth to come must also have been the reason for the Minoans to give mint the highest position in the rank of sacred plants in the Amnissos fresco: it is shown to grow or rise (flanked by iris) from a tripartite platform[46] as it is usually occupied only by the 'goddess' in Minoan art, symbolizing herself or her powers. It is her emblem, may be even divine in its own right, presiding the room from the goddess' dais as she does in her human form in Thera where her sacred plant saffron is presented to her.[47]

This plant, in terms of the Blue-Bird plant *epochologio*, symbolizes the height of summer, the time where agriculture can bear fruit only near the rare humid spots this plant indicates.

Epochologio and Minoan synopsis

The fresco discussed here shows a seemingly natural assembly of birds, monkeys and plants, deceptively like a realistic part of nature, although, as shown above, the plants grow and bloom not synchronous but one by one during the whole year. The arrangement is depicted as embedded in various background colours that by themselves seem to hint at various seasons present in this pictorial arrangement.[48] The artist's vision thus creates from what looks like a merely decorative mural a highly symbolic image, a synoptic view, of an ideal cyclical age, where flowers of all times of the year bloom simultaneously. The onlooker is transported from the here and now into a virtual 'year-round-garden', showing all nature's bounties and powers together with the very special plants denoting the various phases of this year (as explained above).

Time is presented as a continuous succession of growth periods over the year or rather the years. Such a continuity is possible only if life and death are both accepted as part of a spiritual whole where life springs from death and death creates new life,[49] thus transcending

FIGURE 11. Synoptically decorated jug, from House of Frescoes (from Evans 1921–36, vol. II p. 437).

what might at first gaze resemble a Bronze Age farmers' almanac into a highly spiritual object with *epochological* meanings beyond the simple counting of days, months or even seasons. Thus where Evans suspected simple decoration, a Minoan observer may have felt the spiritual awe of immortality that a Christian feels when contemplating images of the Via Dolorosa and resurrection.

Various kinds of synopsis are well known elsewhere in Minoan iconography, too, when for example double-axes are shown in a union with butterflies (e.g. on a jug from the House of Frescoes, found close to the Blue Bird Fresco, figure 11) thus combining the lethal weapon with the mysterious metamorphosis of the insect. There are also images of plants and animals obviously containing peculiarities of two (or more) different species (as the flowers shaped like white lilies painted in red in Thera).[50] All those are usually called 'hybrids' in archaeology,[51] even though they don't just show the artists' joy in 'random combination',[52] but should rather be seen as various symbols emblematically connected and so mutually enhanced.[53] The mint-fresco from Amnissos is thus metamorphosed from a decorative flower-pot to an image of holy growth promising new life, possibly even symbolizing a goddess, and the Blue Bird Fresco does not show an arbitrary, picturesque part of wilderness, but an inspired and transcendent view of the growth cycle of a whole year or even eternity.

During the Bronze Age, Eastern Mediterranean agriculture was still a hard and numinous undertaking of humans interacting with Earth, certainly also an act of man ritually treating the fields so as to produce food for himself and his family. He would have to be following nature's cycles very closely and take them much more as an expression of a oneness of the divine and nature than can be easily imagined today.

Minoan art reflects the wonder, awe and reverence people living in the Aegean of the 2nd millennium BC felt towards nature and the richness of her gifts, symbolized by typical and important plants of various seasons of the year.

Thus it becomes more understandable, too, why all these plants were in later mythology remembered as symbols of life *and* death: only thus a continuum of growth can be symbolized, only if nature dies she can grow again.[54]

The practical and the symbolical connotations of the plant images together generate in the Blue Bird Fresco a multiple kind of Minoan Synopsis: Crocus for autumn and life returning from death, iris for the power and fragility of spring, lily for the fatal brightness of early summer, mint for the mystery of water in the deadly dry surroundings of an Aegean high summer, pomegranate for the bridge between one phase of life (spring) to the next (autumn)—nature's cycle complete. Life leads to death and death to life, both inevitably connected to constitute a Bronze Age spiritual entity.

Notes

1. An older version of this paper was published as Beckmann (2006a). Since then some aspects of the subject have become clearer—especially with several other publications concerning the subject—and thus it seems appropriate to approach the subject again here. The discussion of the lily motif is new. Unless otherwise noted, photographs are by the author.
2. See, for instance, Beckmann (2006c), Blomberg and Henriksson (2000), Goodison (2001), Gregoriades (2008), Henriksson and Blomberg (1996), MacGillivray (2004).
3. These areas of knowledge were often combined or even exchangeable in ancient times, cf. Scarborough (1991).
4. As in the case of astronomical hypotheses this theory can only express a tentative interpretation that has to speak for itself until further archaeological work might produce findings to enable less speculative approaches. See the author's earlier approach (similar in the conclusions presented here) in Beckmann (2006a).
5. Cf. Evans (1921–36), vol II, p. 446.
6. Cf. Cameron (1968) and Evely (1999).
7. Cf. Marinatos (1984), p. 92.
8. Cf. Schäfer (1992), p. 109, cf. also Schäfer (1977), p. 12.
9. Evans (1921–36), vol II, p. 446.
10. A term from Frankfort (1961), p. 157; cf. Marinatos (1984), p. 92.
11. Chapin and Shaw (2006).
12. Cf. Hood (1997), p. 111.
13. Immerwahr (1990), p. 50.
14. Marinatos (1984), p. 92.
15. Marinatos (1984), p. 89.
16. Chapin (2004), p.61.
17. Marinatos (1984), p. 92.
18. As described e.g. by Shaw (1993).'
19. Cf. also Chapin (2004), p. 58.
20. Cf. Schmidlin (2004): 'Es läßt sich bei einem großen Teil der als vegetabil erkennbaren Motive der sogenannten 'Kamares-Ware' thematisch und strukturell eine Wachstumssymbolik erschließen: thematisch durch eine Akzentuierung der Darstellung auf die jeweils neuen Triebe, strukurell durch Betonung der Kraftlinien und -zentren der Pflanzen durch Verwendung roter Farbe.' Received as personal communication from the author via e-mail.
21. E.g. Nilsson (1927).
22. This, too, is a tableau meaning fertility in general, not spring.
23. Evans (1921), vol. I, Doumas (1996). On these frescoes blue monkeys (Knossos) and women (Thera) are shown collecting saffron in a seemingly natural or wild landscape. In Thera a monkey is shown presenting a basket of saffron flowers to an enthroned, goddess-like figure (cf. also Marinatos (1984)).
24. Stylistically this is Early Kamares, cf. Niemeier (1985), p.61, citing Möbius (1933), p. 10. Niemeier declares the threefold pistils as 'ornamental', although Cretan crocus clearly looks exactly like that. Unfortunately Möbius is still much too often taken as a serious botanical reference (see also other notes below). Cf. for the subject also Faraone and Obbink (1991), passim.

25. Homeric Hymn to Demeter, 425; cf. also ibid. 6 and the similarity with Zeus' and Hera's love-bed in the Ilias cited above, all showing the symbolical landscape of early Aegean fertility.
26. Cf. Doumas (1996), p. 137. The flower is interpreted as fallen by Marinatos (1984), p. 79.
27. Möbius is not reliable here (1933, p. 10): he doesn't seem to have known Cretan iris, an error misleading even Niemeier (1985), p. 63. Although Evans had actually recognized the plants correctly, (Evans 1921–1936, vol II, p. 454) Niemeier writes, following Möbius: 'kommen in der Natur nicht vor'.
28. In a Late Minoan elite building or villa also called 'House of Lilies', close to the beach 7 km east of Iraklion. Frescoes of lilies and other plants decorated walls on the first floor.
29. A workshop for aromatics from MM I A has been excavated at Chamalevri. Vessels from there chemically analyzed prove to have contained oil of orris, cf. Tzedakis and Martlew (1999), p. 48.
30. Cf. Roscher 1886–90, I, col. 2764. Murr (1890), p. 256, n.4 refused the possible identity of the (large) *iris* known to him and *hyakinthos* because Dioskorides described the latter as tufted and with many flowers—exactly looking like *iris cretica*, though. Compare for this hyakinthos-image already Od. VI, 231.
31. Cf. for example Liddell-Scott: hyacinthus or -os, i, *m.*, *the hyacinth*, not, however, our hyacinth, but either the *blue iris* or *fleur-delis*, Iris Germanica.
32. May be its use for purification in antiquity is also a point for this, especially for funerary means: 'the vessels with Iris motifs [from Armenoi, LM III cemetery] in the light of what we now know was the use of oil of iris in the production of aromatics and unguents [...] suggests that these products were used in funerary practice'. Tzedakis and Martlew (1999), p. 55. Until recently graves on (Christian and Muslim) cemeteries in the Eastern Mediterranean have still been planted with iris – and lilies, to be discussed below. Cf. also Dafni et al. (2006).
33. See ibid.
34. For this 'synoptical"'view see also below.
35. Möbius (1933), p. 11.
36. Medieval images of pomegranates also depicted them with this kind of young fruit and threefold leaves.
37. Cf. e.g. Keel (1994), p. 146.
38, Cf. Evans (1921–36), vol. II, p. 454. Möbius called them 'Lupines' and the flowers 'ears', cf. Möbius (1933), p. 23.
39. E.g by Schäfer (1977).
40. For instance in the spring region of the Rouvas creek on the eastern slopes of Mt. Psiloritis above Zaros.
41. For the magical-medical connections typical for antiquity in this context cf. also Daux (1957). He cites Hippocrates, where the subject is 'préscriptions formulées par des "magoi"' (p. 4), in which mint is explicitly forbidden (together with onions and garlic). The interdiction seems to be based on a taboo, probably for the original holiness of the mint (and other plants, the interdiction is dated to 2nd cent. AD) and use for purification referred to, and its use as a funeral plant.
42. From Mycenae and Pylos, together with other aromata.
43. By Kretschmer (1925).
44. mi-ia-tar, in the evocation-ritual KUB XV 34, Vs.II, 22 and 24. Hittite -tar is the nominal element like GR: -ση, ENG: -th, GER: -tum, -(i)ja- meaning ‚to make' (cf. ass-ija-tar).
45. Beckmann (2006b).
46. It does not seem to be necessary to look for prototypes in Egypt as Schäfer (1977), pp. 114, 116 and Shaw (1993) do. The incurved altar in combination with a platform is original enough in Minoan art, cf. also Marinatos (1990).
47. Compare the "goddess" on a tripartite dais in the Xeste 3 frescoes from Thera. Cf. also Warren (1985), p. 201, talking about papyrus on seals flanked by griffins: 'here the plant may be read as a symbol of the goddess since on other seals she occupies this position'.
48. See Cameron's restoration in Evely (1999).
49. As would, in a simple gardener's perspective, be the 'mystery' of a compost heap, where former live plants are turned into fertilizer for new plants in new seasons and one can't exist without the other.
50. Doumas (1996).
51. E.g. Chapin (2004).

52. Warren (1985), p. 192.
53. This seems to change in the Latest Minoan and Mycenaean art, where specific plant characteristics seem to be lost. One might wonder if the artists still knew the mysteries of the older synopsis.
54. Cf. Warren (1985). The mixed flower and plant garlands on a fresco from Knossos he discusses are probably symbolizing a similar kind of cycle.

References

Beckmann, S., 2006a, "Conscious of Time – Minoan 'Calendar'-Symbolism in the 'Blue Bird Fresco'", in *Proceedings of the 9th International Congress of Cretan Studies* (Elounda, Crete: Historical Society), Heraklion, vol. I, 3, 65-82.

——, 2006b, "Root, resin, red and ritual purification – The role of terebinth in Eastern Mediterranean Bronze Age cult", *Proceedings of the IVth International Congress of Ethnobotany* (Istanbul: Yayinlari), 387–396.

——, 2006c, "'The Mystery of Taurus – Knowing the way'. Possibilities for Astronomical Knowledge in the Bronze Age", *Journal of the Washington Academy of Sciences*, 91/1, 55–68.

Blomberg, M. and Henriksson G., 2000, "Further Evidence for the Minoan Origins of the Greek Calendars", in *Proceedings of the 8th International Cretological Congress* (Herakleion) vol. I, 109–128.

Cameron, M., 1968, "Unpublished Paintings from the 'House of the Frescoes' at Knossos", *Annual of the British School at Athens* 63, 1–31.

Chapin, A., 2004, "Power, Privilege, and Landscape in Minoan Art", in A. Chapin (ed.), *CHARIS: Essays in Honor of Sara A. Immerwahr* (Princeton: American School of Classical Studies at Athens), 47–64.

——, and Shaw, M., 2006, "The Frescoes from the House of the Frescoes at Knossos: A Reconsideration of their Architectural Context and a New Reconstruction of the Crocus Panel", *Annual of the British School at Athens* 101, 57–88.

Dafni, A., Lev, E., Beckmann, S., and Eichberger, C., 2006, "Ritual plants of Muslim graveyards in northern Israel", *Journal of Ethnobiology and Ethnomedicine* 2, 38.

Daux, G., 1957, "Interdiction rituelle de la menthe", *Bulletin de correspondance hellénique* 81, 1–5

Dioscorides, 2000, *De Materia Medica: Being an Herbal with many other medicinal materials*, translated by Tess Anne Osbaldeston (Parkhurst: Ibidis Press).

Doumas, C., 1996, *Die Wandmalereien von Thera* (München: Metamorphosis Verlag).

Evans, A. J., 1921–36, *The Palace of Minos at Knossos* (4 volumes) (London: MacMillan & Co).

Evely, D. (ed.), 1999, *Fresco: A Passport into the Past. Minoan Crete through the eyes of Mark Cameron* (Athens: British School at Athens).

Faraone, C. A. and Obbink, D., 1991, *Magika Hiera. Ancient Greek Magic and Religion* (New York and Oxford: Oxford University Press).

Frankfort, H., 1961, *Ancient Egyptian Religion* (New York: Harper).

Goodison, L., 2001, "From Tholos Tomb to Throne Room: Perceptions of the Sun in Minoan Ritual", in R. Laffineur and R. Hägg (eds.) *POTNIA. Deities and Religion in the Aegean Bronze Age. Proceedings of the 8th International Aegean Conference, Göteborg* (Université de Liège – University of Texas at Austin).

Gregoriades, P., 2008, "'Eneoros Minos' and the Minoan Calendrical Abacus", in S.A Paipetis (ed.), *Science and Technology in Homeric Epics* (Berlin: Springer), 319–324.

Henriksson, G. and Blomberg, M., 1996, "Evidence for Minoan Astronomical Observations from the Peak Sanctuaries on Petsophas and Traostalos", *Op Ath* 21, 99-114.

Hood, S., 1997, "The Magico-Religious Background of the Minoan Villa", in R. Hägg (ed.), *The Function of the 'Minoan Villa'* (Stockholm: Aström), 105–116.

Keel, O., 1994, *The Song of Songs: a Continental Commentary* (Minneapolis: Fortress). [English. translation of *Das Hohelied* (Zürich, 1986).]

Kretschmer, P., 1995, "Das nt-Suffix", *Glotta* 14, 84–106.

MacGillivray, J. A., 2004, "The Astronomical Labyrinth at Knossos", in G. Cadogan, E. Hatzaki and A. Vassilakis (eds.), *Knossos: Palace, City, State* (London: British School at Athens), 329–338.

Marinatos, N., 1984, *Art and Religion in Thera: Reconstructing a Bronze Age Society* (Athens: Mathioulakis).

——, 1990, "Minoan-Cycladic Syncretism", in D. A. Hardy, C. G. Doumas, J. A. Sakellarakis and P. M Warren (eds.), *Thera and the Aegean World* III/1, (London: The Thera Foundation), 370–377.

Möbius, M., 1933, "Pflanzenbilder der Minoischen Kunst in botanischer Betrachtung", *Jahrbuch des Kaiserlich Deutschen Archäologischen Instituts* 48, 1–39.

Murr, J., 1890, *Die Pflanzenwelt in der Griechischen Mythologie* (Innsbruck: Wagner).

Niemeier, W.-D., 1985, *Die Palaststilkeramik von Knossos* (Berlin: Gebrüder Mann).

——, 1989, "Zur Ikonographie von Gottheiten und Adoranten in den Kultszenen auf Minoischen und Mykenischen Siegeln", in W. Müller (ed.), *Fragen und Probleme der bronzezeitlichen Ägäischen Glyptik* (Berlin: Gebrüder Mann), 163–186.

Nilsson, M. P., 1927, *Minoan-Mycenaean Religion and its Survival in Greek Religion* (Lund: C. W. K. Gleerup).

Platon, N. and Brice, W. C., 1975, *Enepigraphoi pinakides kai pithoi grammikou systematos A ek Zakrou = Inscribed Tablets and Pithos of Linear A System* (Athens: Athens Archaeological Society).

Roscher, W. H. (ed.), 1886–90, *Ausführliches Lexikon der Griechischen und Römischen Mythologie* (Leipzig: Teubner Verlag).

Scarborough, J., 1991, "The Pharmacology of Sacred Plants, Roots and Herbs", in C. A. Faraone and D. Obbink (eds.), *Magika Hiera: Ancient Greek Magic and Religion* (Oxford and New York: Oxford University Press), 138–174.

Schäfer, J., 1977, "Zur kunstgeschichtlichen Interpretation altägäischer Wandmalerei", *JdI* 92, 1–23.

——, 1992, "Gärten in der bronzezeitlichen ägäischen Kultur? Rituelle Bildsprache und bildliches Konzept Realität", in M. Carroll-Spillecke et al. (eds.), *Der Garten von der Antike bis zum Mittelalter* (Mainz: Von Zabern), 101–140.

Schmidlin, C., 2004, *Ornament und Bedeutung – zur Motivik mittelminoischer Feinkeramik*, PhD thesis, University of Freiburg.

Shaw, M., 1993, "The Aegean Garden", *American Journal of Archaeology* 97/4, 661–685.

Tzedakis, I., and Martlew, H. (eds.), 1999, *Minoans and Mycenaeans, Flavours of their Time* (Athens: Greek Ministry of Culture and National Archaeological Museum).

Warren, P., 1985, "The Fresco of the Garlands from Knossos", in P. Darque and J.-C. Poursat (eds.), *L'iconographie Minoenne* (Paris: Boccard), 187–208.

Early Greek Lunisolar Cycles: The Pythian and Olympic Games

Robert Hannah[1]

We are all familiar today with the celebration of athletics every four years known as the Olympic Games. We are used to the modern version of these games moving around the world, rather than being fixed, like its ancient predecessor, in the one place, Olympia in Greece. We also recognise a Winter Olympics, something the ancient world would have had great difficulty understanding, when the original Olympic Games were synonymous with the hottest time of the year in the Mediterranean.[2] Less familiar to us, however, is the fact that these games, along with others in the ancient Greek world, were based originally not on a four-year cycle, but an eight-year one.

The Pythian Games were celebrated every four years—indeed, it makes its appearance in this format on the recently discovered 'Olympiad dial' of the Antikythera Mechanism, along with other sets of games held also on four-year cycles (the Olympic Games and the Naa at Dodona) or two-year cycles (the Isthmian and Nemean Games).[3] However, writing in the 3rd century CE, Censorinus noted that many Greek cults celebrated their festivals at an interval of eight years, and he expressly mentioned the Pythian Games as an example of just such a festival.[4] We can deduce from other literary evidence that the four-yearly Olympic Games were also run according to an over-arching eight-year cycle, as we shall see.[5]

We have to recall here that the Greek festival year was organised according to the moon. Each year would standardly have twelve lunar months, but festivals were often tied to agricultural or seasonal events, which are governed by the sun. A year run according to the sun lasts 365 1/4 days, whereas a year of twelve lunar months amounts to only 354 days. In order to keep the lunar calendar in sync with the seasons, it is necessary to add to the lunar calendar an extra month every now and then. Just when 'every now and then' should occur is one of the problems of early Greek calendars. Adding it regularly every second year was one approach, but it is ultimately of no use, because the resultant lunar years become longer than the same number of solar years, and one would eventually have to skip a month to make up the difference. At some point it was discovered that it worked better if one added an extra lunar month at irregular intervals of two or three years, but on a regular basis over a set number of years. The eight-year cycle, or *octaeteris*, is one of these methods.[6] Censorinus reports that its invention was sometimes attributed to Kleostratos, whom we may place with some probability towards the end of the sixth century BCE.[7] It is not impossible, as Thomson argued long ago, that the cycle existed in various forms before this time, even as a simple rule-of-thumb, and that what Kleostratos did was to invent a particular form of it.[8] According to this argument, the cycle could have been used from the first celebrations in Olympia from 776 BCE, and in Delphi from 582 BCE.

The *octaeteris* is a precursor to the more accurate 19-year cycle, otherwise known as the Metonic cycle after its Athenian inventor (although the Babylonians knew it too, much earlier).[9] Whether the Metonic cycle superseded the *octaeteris* as the timer for the festivals in Delphi and Olympia, we have no certain way of telling. Elsewhere in Greece, there is reason to suppose that the 19-year cycle may have played a role not only in the organisation of the civil calendar in Athens but perhaps also in the regulation of festivals there from the late fifth century BCE.[10]

But there is a significant difference between the organisation of the Games in Delphi and Olympia even according to an eight-year cycle. Literary and epigraphic sources testify that the Pythian Games were celebrated on the seventh day of the month Boukatios, which was the second month after the summer solstice in the lunar calendar of Delphi.[11] The Olympic Games, on the other hand, were celebrated during 'the days of summer' (Censorinus, *On the Birthday* 21.6), at full moon, and alternately after 49 and 50 months—'now in the month of Apollonios and now in the month of Parthenios, according to the Egyptians in Mesori or Thoth', says an ancient scholion to Pindar's *Olympian Ode* 3.35.[12] The reference to the successive Egyptian months Mesori and Thoth allows us to understand from this scholion that the Greek months of Apollonios and Parthenios were also successive, the one following the other. In addition, we can deduce that as Mesori corresponded to the period 25 July – 23 August, and Thoth to 29 August – 27 September,[13] then the Olympic Games were juggled between late summer and early autumn, at least in the Roman period, when the scholion would have been written.[14] Another scholion to Pindar (on *Olympian Ode* 3.33)[15] gives further clues about the timing of the Olympic Games. Its Greek is now corrupt, so a straightforward translation is impossible, but the gist of the passage is:

- that the first month in Elis fell around the time of the winter solstice;
- that the first Olympic Games were held in the eighth month of the year;
- and that thereafter the Games were celebrated alternately in the season called *opōra* or at the dawn rising of Arcturus.

We shall return to these two scholia and some of their problems,[16] but let us first consider the overall situation regarding the timing of the Pythian and Olympic Games. What is odd about their periodicity is that it can be shown that in order to celebrate the Pythian Games in the same lunar month, Boukatios, every four years, there must also be an alternating interval of 49 and 50 months between successive celebrations, and yet this same alternating interval of 49 and 50 months was used, we are told, to set the Olympic Games, 'now in the month of Apollonios and now in the month of Parthenios'. Somehow, then, the alternation in one site managed to preserve attachment to the same lunar month, whereas in the other site it apparently maintained alternation between one month and the next.

Table 1 illustrates the running of a basic octaeteris. It allows for months of alternately 30 and 29 days, which on average correspond to lunar months of 29.5 days, and which we know the Greeks used. Intercalary months are to be added in this scheme in years 3, 5 and 8, as Geminos reports the cycle,[17] but other patterns are possible. Columns Y1–Y8 are years in the cycle, each comprising months i–xii, alternately of 30 and 29 days, plus an intercalary month of 30 days to be set somewhere in years 3, 5 and 8. In the running of such a cycle the intention would be to maintain as close a congruity as is practicable between the calendar year and the seasonal year over the period of the cycle. The cycle gives 99 lunar

Month	Y1	Y2	Y3	Y4	Y5	Y6	Y7	Y8
i	30	30	30	30	30	30	30	30
ii	29	29	29	29	29	29	29	29
iii	30	30	30	30	30	30	30	30
iv	29	29	29	29	29	29	29	29
v	30	30	30	30	30	30	30	30
vi	29	29	29	29	29	29	29	29
vii	30	30	30	30	30	30	30	30
viii	29	29	29	29	29	29	29	29
ix	30	30	30	30	30	30	30	30
x	29	29	29	29	29	29	29	29
xi	30	30	30	30	30	30	30	30
xii	29	29	29	29	29	29	29	29
Intercalary			30		30			30

TABLE 1. The *octaeteris.*

months, comprising in all (in modern terms) 2,923.52841days. Eight solar years comprise 2,921.93752 days, so congruity between lunar and solar time over the full cycle is close, with a discrepancy of just a day and a half.[18]

There could be occasional years, one would imagine, when the astronomical epoch—the summer solstice, for instance—might end up within the first month of one year rather than in the last month of the previous one, but this anomaly could be immediately corrected in the following year through intercalation. Indeed, the misalignment of solstice and New Year could be taken as a warning for the intercalation as the cycle was gradually developed, but a good deal depends on how these tropical points were defined: equinoxes are notoriously difficult to ascertain by observation, but even the solstices, if judged, for example, by the extreme points for sunrise or sunset on the horizon in summer or winter, or by the shortest or longest shadow cast at noon, can encompass a number of days on either side of the actual astronomical event. It is more likely that a significant misalignment between an expected conjunction of a star and the moon would cause intercalation, as we know was the case in the Near East.[19]

Table 2 shows the same octaeteris but now incorporating the Pythian Games occurring in month ii (standing for Boukatios). We can see that celebrating the Pythian Games always in month ii leads to unequal intervals of alternately 49 and 50 lunar months between celebrations. The festival would be held first in month ii of year 1 of the cycle. Maintaining attachment to month ii, the next celebration would be in year 5, by which time an intercalary month has been added. Then the next Games would occur in year 9, by which time two further intercalary months have been added.

So the use of an alternating system of 49 and 50 months, such as we have deduced

Month	Y1	Y2	Y3	Y4	Y5	Y6	Y7	Y8	Y9
i	30	30	30	30	30	30	30	30	30
ii	**29**	29	29	29	**29**	29	29	29	**29**
iii	30	30	30	30	30	30	30	30	30
iv	29	29	29	29	29	29	29	29	29
v	30	30	30	30	30	30	30	30	30
vi	29	29	29	29	29	29	29	29	29
vii	30	30	30	30	30	30	30	30	30
viii	29	29	29	29	29	29	29	29	29
ix	30	30	30	30	30	30	30	30	30
x	29	29	29	29	29	29	29	29	29
xi	30	30	30	30	30	30	30	30	30
xii	29	29	29	29	29	29	29	29	29
Intercalary			30		30			30	
Sum of months:					49				50

Table 2. The Pythian Games.

for the Pythian Games, ought also to mean that the Olympic Games, which are explicitly described as being run at alternating intervals of 49 and 50 months, were celebrated in the same lunar month each time. Table 3 demonstrates this possibility, with intercalation again in years 3, 5 and 8 by way of example. The principal assumptions underlying the Table are that the start of the year (month i) followed the winter solstice, and that the Games took place in summer, both data given by the literary testimonia. Every four years the Games would repeat in month viii, which is cited as the month for the first celebration of the Games in one scholion.

Yet we are told the Games were not held in the same month, but instead occurred 'now in the month of Apollonios and now in the month of Parthenios'. How can this be so? Even if we assume a different intercalary cycle, such as years 1, 4 and 7, the Games ought to fall in the same month, as Table 4 illustrates. This version of the octaeteris gives gaps of 50 and 49 months between celebrations, which is the reverse of what the scholion reports, but that would seem a small problem. So how can an octaeteris deliver gaps of 49 and 50 months (or the reverse, if necessary) and yet cause the Games to fall in different months in alternate Olympiads? A possible solution is presented in Table 5, where month viii is taken to be Apollonios, and ix Parthenios.[20] By this solution, alternation between months viii and ix (Apollonios and Parthenios respectively) is achieved, along with the intervals of 49 and 50 months between the Games, by imposing a different version of the octaeteris than applied at Delphi. At Olympia the intercalary lunar months might be added in years 1, 4 and 7. Insertion would have to be made after month ix (at least in year 1) in order to maintain the

Month	Y1	Y2	Y3	Y4	Y5	Y6	Y7	Y8	Y9
i	30	30	30	30	30	30	30	30	30
ii	29	29	29	29	29	29	29	29	29
iii	30	30	30	30	30	30	30	30	30
iv	29	29	29	29	29	29	29	29	29
v	30	30	30	30	30	30	30	30	30
vi	29	29	29	29	29	29	29	29	29
vii	30	30	30	30	30	30	30	30	30
viii	**29**	29	29	29	**29**	29	29	**29**	29
ix	30	30	30	30	30	30	30	30	30
x	29	29	29	29	29	29	29	29	29
xi	30	30	30	30	30	30	30	30	30
xii	29	29	29	29	29	29	29	29	29
Intercalary			30		30			30	
Sum of months:					49				50

TABLE 3. The Olympic Games (i).

Month	Y1	Y2	Y3	Y4	Y5	Y6	Y7	Y8	Y9
i	30	30	30	30	30	30	30	30	30
ii	29	29	29	29	29	29	29	29	29
iii	30	30	30	30	30	30	30	30	30
iv	29	29	29	29	29	29	29	29	29
v	30	30	30	30	30	30	30	30	30
vi	29	29	29	29	29	29	29	29	29
vii	30	30	30	30	30	30	30	30	30
viii	**29**	29	29	29	**29**	29	29	29	**29**
ix	30	30	30	30	30	30	30	30	30
x	29	29	29	29	29	29	29	29	29
xi	30	30	30	30	30	30	30	30	30
xii	29	29	29	29	29	29	29	29	29
Intercalary	30			30			30		30
Sum of months:					50				49

TABLE 4. The Olympic Games (ii).

Month	Y1	Y2	Y3	Y4	Y5	Y6	Y7	Y8	Y9
i	30	30	30	30	30	30	30	30	30
ii	29	29	29	29	29	29	29	29	29
iii	30	30	30	30	30	30	30	30	30
iv	29	29	29	29	29	29	29	29	29
v	30	30	30	30	30	30	30	30	30
vi	29	29	29	29	29	29	29	29	29
vii	30	30	30	30	30	30	30	30	30
viii	29	29	29	29	**29**	29	29	29	29
ix	**30**	30	30	30	30	30	30	30	**30**
x	29	29	29	29	29	29	29	29	29
xi	30	30	30	30	30	30	30	30	30
xii	29	29	29	29	29	29	29	29	29
Intercalary	30			30			30		30
Sum of months:					49				50

TABLE 5: The Olympic Games.

balance of 49 and 50 month periods. The main difficulties with this solution are that this version of the octaeteris causes the lunar year to leap ahead of the solar from year 1, a situation which Geminos advised against;[21] and that the cycle begins with Games in month ix, Parthenios, rather than viii, Apollonios, in contradiction to the scholion.

The fact that the literary testimonia cannot really be reconciled without emendation has led to the rejection of the scholia on Pindar, *Olympian Ode* 3.33 and 3.35 as evidence, because they are internally and mutually inconsistent or corrupt. In addition, it has been suggested that the Olympic Games and the Elean year originally began at the same time—therefore, in summer, not winter—and that the festival took place at an astronomically fixed time—e.g. the first full moon after the summer solstice—which would be marked alternately by the months of Apollonios and Parthenios, in a cycle governed by an octaeteris.[22] The periodicity of the Games would then resemble Jewish Passover and Christian Easter, in being timed by a lunar event (full moon) tied to a solar event (the summer solstice for the Olympic Games, the vernal equinox for Passover and Easter). The result would be the same: a festival which would wander from one celebration to the next up and down a limited and determinable span of time following the astronomical event. Samuel found this alternative account attractive, particularly as it solved the problem of how other Greek states could synchronise their individual calendars with the Elean: they did not need to do so, but would instead align with the solstice and moon, to ensure that their athletes arrived at the appropriate time.

By dismissing in particular the scholion to Pindar, *Olympic Ode* 3.33, Samuel also implicitly set aside the astronomical assistance that it offers. Certainly, this information would contradict his preferred timing of the Olympic Games at the first full moon after the summer solstice, since the rising of Arcturus would always occur well after this full moon. Yet on the other hand, Samuel's definition sets up another problem: there would be occasions when the first full moon after the solstice would occur before New Year's Day of the due year for the Games (unless intercalation took place in the previous year to prevent this). I remain inclined to accept the scholia as sources of information, albeit problematic sources. Let us see what they might offer that is of assistance in placing the Olympic Games.

An hypothesis which prioritises the measurement of the solstice and the observation of the moon thereafter may be practicable for those states which began their year after the same astronomical event as Elis (assuming this was the summer solstice), but it leaves the other states seeking the event at some point in the course of their differently organised years. Were there other advance warnings available to the various cities, which could signal the approach of the solstice?

One clue may be provided, ironically, by one of the rejected scholia. The scholiast commenting on Pindar, *Olympian Ode* 3.33, says that after the initial celebration, the Olympic Games were celebrated alternately in the season called *opōra* or at the dawn rising of Arcturus. Here we have seasonal and astronomical signals. LSJ defines *opōra* with disarming simplicity as: 'the part of the year between the rising of Sirius and of Arcturus (i.e. the end of July, all Aug., and part of Sept.)'. The Lexicon does not say so, but the astronomical definition stems from a late source, Galen in the second century CE, in his commentary on the much earlier works of Hippokrates, and specifically the latter's first book of the *Epidemics*.[23] To what extent the definition applied to earlier, Greek periods is hard to say. For the moment, let us run with Galen's definition that *opōra* lies between the rising of Sirius and that of Arcturus. It would suggest that the Olympic Games alternated between a period marked by the rising of Sirius and a period signalled by the rising of Arcturus. Now, over a long period of time in antiquity, Sirius rose in late July, a month after the solstice, while Arcturus rose in mid-September in central Greek latitudes.[24] This is broadly consistent with the information from the other scholion, which placed the Games between July and September, but of course counter to the view that they occurred within a month of the solstice.

The value of these star timings can be appreciated in a world where each Greek state had its own lunisolar calendar, with different month-names and New Year's Days even among ethnically or politically related states. A new month began with the observation of a new moon's crescent setting after sunset, but since this was an observational datum, months could start on different days from one city to another. New Year could be after the summer solstice (e.g. in Athens and Delphi), or the winter solstice (e.g. in Olympia and Boiotia, but also on Delos, an ally of Athens), or at one of the two equinoxes (e.g. the autumn equinox in Phokis and Aitolia).[25] Table 6 shows the correspondence between four state calendars, those of Athens, Delos, Delphi and Aitolia. New Year's Day is underlined in each calendar. The Table illustrates well the potentially confusing nature of the interrelationships, with different beginnings to the year, and with months which bear the same names sometimes coinciding across different states, but at other times not.[26]

Star-rise and star-set could avoid some of the discordance that existed between different state calendars, and which frustrated or bemused historians from Thucydides to Plutarch.[27] The major desideratum of a calendar based on star phenomena would be similarity of physical conditions for observation, particularly the height of the horizon, but we

Athens	Delos	Delphi	Aitolia
Hekatombaion	Hekatombaion	Apellaios	Laphraios
Metageitnion	Metageitnion	Boukatios	Panamos
Boedromion	Bouphonion	Boathoos	Prokyklios
Pyanepsion	Apatourion	Heraios	Athanaios
Maimakterion	Aresion	Daidaphorios	Boukatios
Poseideon	Posideon	Poitropios	Dios
Gamelion	Lenaion	Amalios	Euthaios
Anthesterion	Hieros	Bysios	Homoloios
Elaphebolion	Galaxion	Theoxenios	Hermaios
Mounichion	Artemision	Endyspoitropios	Dionysios
Thargelion	Thargelion	Herakleios	Agyeios
Skirophorion	Panemos	Ilaios	Hippodromios

TABLE 6. The civil calendars of Athens, Delos, Delphi and Aitolia compared.

shall see soon how even grossly disparate horizons might be overcome. The rising of Sirius in late July could therefore have served to warn in advance the coming of the appropriate month of the year somewhere in the equivalent of our mid-August to mid-September. The rising of Arcturus in mid-September would then signal the time by which alternate celebrations of the Games had to take place. While the rise of Sirius provides at least a couple of weeks' notice of the celebrations of the Games in the weeks from mid-August, prospective participants from across the Greek world would ideally have needed more advance warning of both this and more particularly of the alternate celebration at the time of the rising of Arcturus. How could this warning have been given?

In Hesiod's time in the seventh century BCE, the rising of Sirius was given an advance warning by the dawn rising a month earlier of Orion, which itself was preceded a month earlier by the dawn rising of the Pleiades. These signals are preserved down to the fifth century BCE, when they appear in the *parapēgma*, or 'star calendar', of Euktemon, a colleague of Meton. But he adds several intermediate warnings—the evening setting of Capella, two weeks after the rising of the Pleiades; a week later, the evening rising of Aquila, and the evening setting of Arcturus, as well as the dawn rising of the Hyades; three weeks later the shoulder of Orion rises at dawn; and then another three weeks later, all of Orion has risen; two weeks after this, Sirius rises.[28]

Our second star signal is the rising of Arcturus. For Hesiod, this was to be observed when Orion and Sirius were in culmination overhead (*Works and Days* 609–610). There was not much advance warning of this event. But by the end of the fifth century BCE, Euktemon gives us many more signals: one day after the rising of Sirius, the dawn setting of Aquila; then about three weeks later Lyra sets at dawn and Pegasus rises in the evening; three weeks later again Vindemiatrix rises along with Arcturus.

The increase in the number of observations would allow the observant to know well in advance when, for instance, Arcturus should rise and hence when the Olympic Games should start, in the year when it was ordained to begin later in the year. In the alternate years, when the Games were to begin earlier—after the rising of Sirius—again there is plenty of warning in the stars. An interesting parallel case has been made for the timing of visits to Delphi in order to gain an oracle from the god, when this originally occurred only once a year, on the seventh day of Bysios, the eighth month in the Delphian calendar. Salt and Boutsikas have argued that the advance warning of the appropriate time for visitors to start travelling to the shrine from all round the Greek world may have been provided by the observation of the small constellation, Delphinus.[29] Its dawn rising is signalled by Euktemon around the time of the winter solstice, towards the end of our December. Salt and Boutsikas propose that this rising was used as a signal that the next new moon would begin the month sacred to Apollo Delphinus. This works well for Athens, for example, where this month after the winter solstice is Gamelion, in which Apollo Lykaios and Apollo Delphinios were honoured. But at Delphi it is the next month after this again, in which the god is honoured. The reason for this delay is seen by Salt and Boutsikas in the fact that at Delphi the very high local horizon, caused by the surrounding mountains, forces the visibility of the rising of Delphinus to be delayed by about a lunar month in comparison with those places in Greece where the horizon is relatively low, as in Athens. In other words, other parts of Greece would get a month's advance warning of the coming festival month of Apollo, and of the time to consult the oracle, than occurred in Delphi itself—a very handy forewarning in a country where travel to Delphi was not easy.

In a similar fashion, for those many Greek states whose calendars were different from that at Olympia, the risings or settings of stars may have assisted them to know when to set out for the Games. The increased observations, in almanacs like Euktemon's *parapēgma*, would be the type of extra information that would have been useful to such people.

Further temporal clues for prospective participants at the Games could have been provided ecologically. We know, for instance, that the month of Hekatombaion in Athens was associated both with the time of the summer solstice and with the period when the tuna fish breeds, while (for those looking closely!) the months Skirophorion, Hekatombaion and Metageitnion are the period over which the crayfish retains its eggs after conception.[30] Theophrastos records that reeds for playing the pipe in the 'natural' style used to be cut in the month of Boedromion when Arcturus was rising (i.e. in mid-September), but that a change to a more affected style of playing led to the cutting occurring much earlier 'in Skirophorion and Hekatombaion, just a little before the solstice or just after' (i.e. three months earlier, in late June).[31] These instances may seem recondite, and it may appear simply fortuitous that the two astronomical markers observed here—the summer solstice and the rising of Arcturus—are among the ones we are concerned with in regard to the timing of the Games. But however coincidental these references are, they also signify how deeply embedded these and other astronomical signals were in the Greek consciousness of time. Even the olive tree, known in the Games context because its branches provided the victors' wreaths, performed a trick at just the right time: the leaves of the olive, as well as of some other trees, turn their underside over towards the sun at the time of the summer solstice.[32]

George Thomson went so far as to propose a link between the timing of the Olympic Games and olives themselves:

> Apollonios, corresponding to the Attic Metageitnion and the Spartan Karneios, was the month of the fruit harvest, which falls normally towards the end of August. The Olympic prize of victory was

> a crown of wild olive, plucked from the sacred trees that grew in the Altis; and it was said that the Idaean Herakles and his companions used to rest after their exercises on beds of olive leaves. For these reasons it is probable that the primitive nucleus of the Games was a festival consecrated to the fruit harvest.[33]

Unfortunately, he was well astray in his arboriculture. The Games at Olympia certainly required crowns of wild olive for the victors. Pausanias ascribes to Herakles the introduction of the wreath of wild olive (*kotinos*) as the victor's crown at Olympia, noting that it grew in such abundance that Herakles and his brothers could sleep on the green leaves.[34] On the Altis in the sanctuary, there was a wild olive, called, Pausanias tells us, 'the Beautiful Crown olive, and it is customary for the crowns to be given from it to those who win the Olympics'.[35] The olive tree is an evergreen. The main period of growth is from April to September.[36] According to Pliny the Elder in the first century CE, the oil in the olive fruit increases until the rising of Arcturus, but thereafter it is the stone and flesh of the fruit that increase.[37] This change in the development of the olive from oil-secretion in the fruit to growth in the flesh and stone at the time of the rising of Arcturus would suggest that a September harvest would produce green, oil-rich fruit, while later harvests prioritise fleshy fruit. But even the earliest green olives are not ready before well into September and the main olive harvest is November-January, or in a bumper year even into February.

In alternate festivals the Games might coincide with this early harvest in September, but in every other celebration the Games would be long finished by harvest time. It is therefore unlikely that the Games were originally conceived as some kind of harvest festival, if they preceded that harvest in every second celebration. Furthermore, since olive trees fruit on second-year wood, any pruning of that new growth would reduce the next olive crop.[38] So people would certainly not prune potentially fruiting branches in the summer to serve as wreaths.

But is there any connection between the timing of the festival and the production of leaves (as opposed to the fruit) on the tree, which could be harvested for the crowns? Summer is also the time of year when olive trees regularly put out many suckers, often from the base of the trunk and frequently higher up it below any graft, if the tree is a composite of wild trunk and domesticated graft (Figures 1–2).[39] If the olive tree is grown for the production of its fruit, and if these shoots are left to grow, then they will produce inferior quality fruit generally of a 'wild olive' type, and will sap strength from the rest of the tree, particularly the branches of whatever domestic variety was grafted higher up. So what farmers might well do in the summer, as and when they had time, is remove these rogue shoots from around the bases of the trees, so as to encourage growth of desirable vegetation higher up the tree. Furthermore, while acknowledging that farming practices may have differed in this respect between Italy and Greece, Roman agricultural writers imply that deep cultivation of the soil around the olive trees took place at the time of the solstice, so as to prevent the drying out of the roots in the hottest part of the year.[40]

It may be, therefore, that the victors' wreaths were made from the suckers that the trees would shoot out in summer. By removing these, the officials at Olympia performed a useful arboricultural task as well for the future development of the tree.[41] In this scenario, the Games were celebrated over a period between July and September, when some judicious pruning of suckers was likely to take place for the benefit of the olive tree, thus providing shoots of wild olive for the victors' wreaths.

FIGURE 1. Wild olive tree (new shoots at base within box) (photo courtesy A. Pugliese).

FIGURE 2. Wild olive tree, detail from Figure 1 of new shoots at base (photo courtesy A. Pugliese).

So where does this leave us? The summer solstice, if it in fact played a role, is marked by various seasonal signals. We have one scholion (to Pindar, *Olympian Ode* 3.35), now so poorly expressed that it has been dismissed, which nonetheless provides astronomical data which coincide neatly with the calendrical data provided by the other scholion (to Pindar, *Olympian Ode* 3.33). These astronomical data—the observation of the rising of Sirius (implied by the seasonal period *opōra* named in the scholion) and of Arcturus—are not only part of the regular cultural background of the Greek world from the time of Hesiod onwards, but also are core elements of the *parapēgmata*, the star almanacs from the fifth century BCE and later, which usefully provide further star data that could allow people to prepare to go to Olympia well in advance of the opening of the Games. Beyond this dovetailing, we find that ecological data on the cultivation of the olive tree are consistent with the information from both scholia, and help explain how the victors' wreaths could be made from the wild olive on the Altis at Olympia without eventually killing the source. Undoubtedly there are problems with the scholia, especially that to Pindar, *Olympian Ode* 3.35, and emendation seems essential. But by the same token complete dismissal looks to be an over-reaction, and seeking instead to place the Games at the first full moon after the summer solstice appears unjustified. We can also appreciate all the more the close interaction between one mode of time-reckoning—the appearance and disappearance of certain stars—and another—the octaeteris, which sought to integrate the essentially incommensurate cycles of the sun and the moon.

Notes

1. I should like to thank Wayne Horowitz and John Steele, for accepting the conference version of this paper and, more especially, for presenting it in my absence. The topic is difficult enough without leaving colleagues to carry the weight of presenting it. I am also grateful to the anonymous referee for valuable suggestions for improvement.
2. Censorinus, *On the Birthday* 21.6: *ex diebus dumtaxat aestivis, quibus agon Olympicus celebratur* ['. . . only on summer days, when the Olympic Games are celebrated']; Aelian, *Varia Historia* 14.18: Ἀνὴρ Χῖος ὀργιζόμενος τῷ οἰκέτῃ 'ἐγώ σε' ἔφη 'οὐκ ἐς μύλην ἐμβαλῶ, ἀλλ' ἐς Ὀλυμπίαν ἄξω.' πολλῷ γὰρ ᾤετο πικροτέραν ὡς τὸ εἰκὸς εἶναι τιμωρίαν ἐκεῖνος ἐν Ὀλυμπίᾳ θεώμενον ὑπὸ τῆς ἀκτῖνος ὀπτᾶσθαι ἢ ἀλεῖν μύλῃ παραδοθέντα. ["A Chian man, angry with his slave, said, 'I won't cast you to the mill, but I'll take you to the Olympic Games.' For he thought that was probably a much harsher punishment if he was a spectator at Olympia roasting under the sun's rays than being handed over to grind at the mill".]
3. Freeth et al. (2008).
4. Censorinus, *On the Birthday* 18.6: *Ob hoc in Graecia multae religiones hoc intervallo temporis summa caerimonia coluntur; Delphis quoque ludi, qui vocantur Pythia, post annum octavum olim conficiebantur.* ['Therefore in Greece many festivals are celebrated with the greatest veneration at this interval of time; the games at Delphi also, which are called the Pythia, used to be held after the eighth year'.]
5. See Nilsson (1955), pp. 645–647.
6. Cf. Britton (2007), pp. 119–120 for a discussion of the establishment of appropriate intervals between intercalations in various lunisolar cycles.
7. Censorinus, *On the Birthday* 18.5: *Hanc octaeterida vulgo creditum est ab Eudoxo Cnidio institutam, sed alii Cleostratum Tenedium primum ferunt conposuisse.* ['It is generally believed that this octaeteris was instituted by Eudoxus of Cnidus, but others hold that Cleostratus of Tenedus constructed it']. Samuel (1972), pp. 39–40.
8. Thomson (1943), p. 59.
9. The 19-year cycle was in use in Mesopotamia at least from the second half of the sixth century BCE, and systemtically governed intercalations from 485 BCE: Britton (2007). On the Metonic cycle, see Hannah (2005), pp. 27–41, 55–58; Bickerman (1980), p. 29; Samuel (1972), pp. 35–49.
10. Hannah (2009), pp. 31–37; Hannah (2005), pp. 55–70.
11. Farnell (1907), pp. 291, 421 note 256a cites the inscription CIA 2.545, which has the Pythian Games being celebrated in the month of Boukatios at Delphi, and the scholion to the hypothesis to Pindar's *Pythian Odes* (scholion a, Drachmann 1910/1964: 2 line 9), which places the Games 'on the seventh day', but without naming the month. One can readily add to the epigraphic evidence which equates the Games with Boukatios, e.g. *IG* II[2] 1126, *CID* 1:10, 4:1.
12. Drachmann 1903/1964: 115. 35g: γίνεται δὲ ὁ ἀγὼν ποτὲ μὲν διὰ μθ´ μηνῶν, ποτὲ δὲ διὰ ν´, ὅθεν καὶ ποτὲ μὲν τῷ Ἀπολλωνίῳ μηνὶ, ποτὲ δὲ τῷ Παρθενίῳ ἐπιτελεῖται. ['The Games take place now after 49 months, now after 50, and therefore they are celebrated now in the month Apollonios, now in Parthenios'.] See also Drachmann 1903/1964: 115. 35a: διχόμηνις <ὅτι> περὶ τὴν ις´ πανσελήνου οὔσης ἄγεται τὰ Ὀλύμπια, τουτέστι διχομηνίᾳ Παρθενίου ἢ Ἀπολλωνίου μηνὸς, παρ' Αἰγυπτίοις Θὼθ ἢ Μεσωρί. ['At the full moon: because around the 16th day, when it is full moon, the Olympic Games are celebrated, that is at the full moon of the month Parthenios or Apollonios, according to the Egyptians Thoth or Mesori.']
13. Hannah (2005), pp. 89–90; Samuel (1972), p. 177.
14. Samuel (1972), pp. 191, 193–194 n. 3, worried about the equation between lunar months and solar ones, but eventually allowed for the possibility that this might be just an approximation. There was a similar approximating tendency in the various forms of the Easter computus in the medieval period: see Hannah (2005), pp. 37–38. Trümpy (1997), p. 200 accepted Samuel's conclusions regarding the timing of the Olympic Games in summer.
15. Drachmann 1903/1964: 114. 33a: A <ἤδη γὰρ αὐτῷ:> περὶ τοῦ χρόνου καθ' ὃν ἄγεται τὰ Ὀλύμπια καθ' ἑκάστην Ὀλυμπιάδα, καὶ Κώμαρχος ὁ τὰ περὶ Ἠλείων συντάξας φησὶν οὕτως· πρῶτον μὲν οὖν παντὸς περίοδον συνέθηκεν ἐν τῇ ἡμέρᾳ .... ἄρχειν νουμηνίαν μηνὸς ὃς Θωσυθιὰς ἐν Ἤλιδι ὀνομάζεται, περὶ ὃν τροπαὶ ἡλίου

γίνονται χειμεριναί· καὶ π̄α Ὀλύμπια ἄγεται η′ μηνί· ἑνὸς δὲ ὄντος διαφέροντων τῇ ὥρᾳ, τὰ μὲν ἀρχομένης τῆς ὀπώρας, τὰ δὲ ὑπ' αὐτὸν τὸν ἀρκτοῦρον. ὅτι δὲ καὶ ἄγεται ὁ ἀγὼν, καὶ αὐτὸς ὁ Πίνδαρος μαρτυρεῖ.

16. The Elian calendar's first month is called by the scholiast 'Thosuthias'. This name is unique and the text possibly corrupted at this point. Trümpy (1997), pp. 199–200 discusses the issue at length, and promotes the identification of 'Thosuthias' with Boiotian Theilouthios, a summer month. This in turn, however, sets up a contradiction if the Olympic Games were held in summer, but 'Thosuthias' was eight months earlier in winter.
17. Geminos, *Introduction to Astronomy* 8.33: Δι' ἣν αἰτίαν τοὺς ἐμβολίμους μῆνας ἔταξαν ἄγεσθαι ἐν τῷ γῳ ἔτει καὶ ηῳ, δύο μὲν μῆνας μεταξὺ δύο ἐτῶν πιπτόντων, ἕνα δὲ μεταξὺ ἑνὸς ἐνιαυτοῦ ἀγομένου. ['Therefore they arranged that the intercalary months be set in the 3rd, <5th and> and 8th year, two months falling after two years, and one set after one year'.]
18. The difference of just over a day-and-a-half per octaeteris would mount up over time. After just nine octaeterides, i.e. 72 years or a good lifetime, the difference would amount to 14 days, effectively the distance between a new moon and a full one, and this could clearly affect the celebration of a festival attached to a particular phase of the moon. See Hannah (2005), p. 55; Britton (2007), pp. 119–120 comes to the same conclusion, but expresses the error alternatively as 6 days over a 30-year period.
19. Hannah (2005), pp. 30–31; Rochberg-Halton (1992), p. 811.
20. An earlier version of this anaylsis appeared in Hannah (2005), pp. 37–41.
21. Geminos, *Introduction to Astronomy* 8.32: Ἤδη μέντοι γε τοὺς ἐμβολίμους διέταξαν, ὡς ἦν ἐνδεχόμενον μάλιστα δι' ἴσου· οὔτε γὰρ περιμένειν δεῖ, ἕως οὗ μηνιαῖον γένηται παράλλαγμα πρὸς τὸ φαινόμενον, οὔτε προλαμβάνειν παρὰ τὸν ἡλιακὸν δρόμον μῆνα ὅλον. ['Now they arranged the intercalary months so that they were, as much as possible, equally distributed. For it is necessary not to wait until there is a month's variation with respect to the phenomenon, nor to anticipate a whole month with respect to the sun's course'.]
22. Samuel (1972), pp. 192–194, esp. p. 193 note 3.
23. Galen, *In Hippocratis librum primum epidemiarum commentarii iii* (Kühn 17a.17.13–16): καὶ ἀρχή γε τῆς καλουμένης ὀπώρας ἡ ἐπιτολὴ τούτου τοῦ ἀστέρος ἐστίν. καὶ . . . ἄχρι μὲν ἐπιτολῆς τοῦ κυνὸς ἐκτείνουσι τὸ θέρος, ἐντεῦθεν δὲ μέχρις ἀρκτούρου τὴν ὀπώραν. ['And the rising of this star [Sirius] is the beginning of the so-called *opōra*. And . . . they extend summer as far as the rising of the Dog, and then *opōra* through to the rising of Arcturus.']
24. See most readily the tables of star-rise and star-set in Bickerman (1980), pp. 113–114. Computer programs, such as Alcyone Ephemeris (available from http://www.alcyone.de), allow one to make rapid, albeit idealistic, calculations for specific stars in specific years. There is a growing understanding of the ancient criteria for observing first and last risings and settings of stars: see Robinson (2009).
25. Hannah (2005), especially pp. 71–82; Trümpy (1997); Samuel (1972).
26. Cf. also the month-name Apollonios, which we have seen associated in the Elian calendar with the Olympic Games, and which occurs also in Thessaly, where it is, however, a winter month, not a summer one: Trümpy (1997), pp. 200, 216–217.
27. Thucydides 5.20.1–3: αὗται αἱ σπονδαὶ ἐγένοντο τελευτῶντος τοῦ χειμῶνος ἅμα ἦρι, ἐκ Διονυσίων εὐθὺς τῶν ἀστικῶν, αὐτόδεκα ἐτῶν διελθόντων καὶ ἡμερῶν ὀλίγων παρενεγκουσῶν ἢ ὡς τὸ πρῶτον ἡ ἐσβολὴ ἡ ἐς τὴν Ἀττικὴν καὶ ἡ ἀρχὴ τοῦ πολέμου τοῦδε ἐγένετο. σκοπείτω δέ τις κατὰ τοὺς χρόνους καὶ μὴ τῶν ἑκασταχοῦ ἢ ἀρχόντων ἢ ἀπὸ τιμῆς τινὸς ἐς τὰ προγεγενημένα σημαινόντων τὴν ἀπαρίθμησιν τῶν ὀνομάτων πιστεύσας μᾶλλον. οὐ γὰρ ἀκριβές ἐστιν, οἷς καὶ ἀρχομένοις καὶ μεσοῦσι καὶ ὅπως ἔτυχέ τῳ ἐπεγένετό τι. κατὰ θέρη δὲ καὶ χειμῶνας ἀριθμῶν, ὥσπερ γέγραπται, εὑρήσει, ἐξ ἡμισείας ἑκατέρου τοῦ ἐνιαυτοῦ τὴν δύναμιν ἔχοντος, δέκα μὲν θέρη, ἴσους δὲ χειμῶνας τῷ πρώτῳ πολέμῳ τῷδε γεγενημένους. ['This treaty was made as winter was ending, in the spring, immediately after the City Dionysia, just ten years and a few days over having elapsed from when the invasion of Attika and the beginning of this war first took place. This must be considered according to the periods of time, and not by trusting the counting of the names everywhere of those who either from holding office or from some other honour act as markers for past events. For accuracy is not possible, where something may have occurred while they were at the beginning of office, or in the middle, or however it happened to be. But by counting according to summers and winters, as this is written, it will be found that, each having the force of a half of a year, there were ten summers and as many winters in this

first war']. Plutarch, Aristeides 19.8: ταύτην τὴν μάχην ἐμαχέσαντο τῇ τετράδι τοῦ Βοηδρομιῶνος ἱσταμένου κατ' Ἀθηναίους, κατὰ δὲ Βοιωτοὺς τετράδι τοῦ Πανέμου φθίνοντος, ᾗ καὶ νῦν ἔτι τὸ Ἑλληνικὸν ἐν Πλαταιαῖς ἀθροίζεται συνέδριον καὶ θύουσι τῷ ἐλευθερίῳ Διῒ Πλαταιεῖς ὑπὲρ τῆς νίκης. τὴν δὲ τῶν ἡμερῶν ἀνωμαλίαν οὐ θαυμαστέον, ὅπου καὶ νῦν διηκριβωμένων τῶν ἐν ἀστρολογίᾳ μᾶλλον ἄλλην ἄλλοι μηνὸς ἀρχὴν καὶ τελευτὴν ἄγουσιν. ['They fought this battle on the fourth of Boedromion according to the Athenians, but according to the Boiotians on the twenty-seventh of Panemos, when even still now the Hellenic Council meets in Plataiai, and the Plataians sacrifice to Zeus Eleutherios for the victory. The discrepancy between these days should not be wondered at, since even now when people are studying astronomy in greater depth, different people have a different beginning and end of the month'.]

28. Euktemon's data—or at least some of them—are included in the synthesis of *parapēgmata* appended to Geminos's *Introduction to Astronomy*; see Evans and Berggren (2006), pp. 231–240. See also Hannah (2002).
29. Salt and Boutiskas (2005).
30. Aristotle, *Enquiry into Animals* 543b6-13: θέρους δὲ περὶ τὸν Ἑκατομβαιῶνα θυννίς, περὶ τροπὰς θερινάς; Aristotle, *Enquiry into Animals* 549a14-16: Τῶν δὲ μαλακοστράκων οἱ κάραβοι μετὰ τὴν ὀχείαν κύουσι καὶ ἴσχουσι τὰ ᾠὰ περὶ τρεῖς μῆνας, Σκιρροφοριῶνα καὶ Ἑκατομβαιῶνα καὶ Μεταγειτνιῶνα.
31. Theophrastos, *Enquiry into Plants* 4.11.4-5: Τὴν δὲ τομὴν ὡραίαν εἶναι πρὸ Ἀντιγενίδου μέν, ἡνίκ' ηὔλουν ἀπλάστως, ὑπ' Ἀρκτοῦρον Βοηδρομιῶνος μηνός· τὸν γὰρ οὕτω τμηθέντα συχνοῖς μὲν ἔτεσιν ὕστερον γίνεσθαι χρήσιμον καὶ προκαταυλήσεως δεῖσθαι πολλῆς, συμμύειν δὲ τὸ στόμα τῶν γλωττῶν, ὃ πρὸς τὴν διακτηρίαν εἶναι χρήσιμον. ἐπεὶ δὲ εἰς τὴν πλάσιν μετέβησαν, καὶ ἡ τομὴ μετεκινήθη· τέμνουσι γὰρ δὴ νῦν τοῦ Σκιρροφοριῶνος καὶ Ἑκατομβαιῶνος ὥσπερ πρὸ τροπῶν μικρὸν ἢ ὑπὸ τροπάς.
32. A feature that I noted myself recently in Cyprus in the week of the summer solstice; see Varro, *On Agriculture* 1.46: *Horum enim folia cum converterunt se, solstitium dicitur fuisse*. Pliny, *Natural History* 18.266: *vertit oleae ante pedes satae, vertit tiliae ad mille usus petendae, vertit populi albae etiam vitibus nuptae. adhuc parum est. ulmum, inquit, vite dotatam habes: et huius vertam. pabulo folia eius stringis aut deputas: aspice et tenes sidus. alia parte caelum respiciunt quam qua spectavere pridie.*
33. Thomson (1943), p.62. Farnell (1907), p. 291 had earlier wondered about a possible harvest-festival origin for the Pythian Games too, but seemed to have a grain harvest in mind, and saw the festival as ultimately outgrowing this 'low culture' (as he thought it) source.
34. Pausanias 5.7.7: τὸν δὲ Ἡρακλέα παίζοντα—εἶναι γὰρ δὴ αὐτὸν πρεσβύτατον ἡλικίᾳ—συμβαλεῖν τοὺς ἀδελφοὺς ἐς ἅμιλλαν δρόμου καὶ τὸν νικήσαντα ἐξ αὐτῶν κλάδῳ στεφανῶσαι κοτίνου: παρεῖναι δὲ αὐτοῖς πολὺν δή τι οὕτω τὸν κότινον ὡς τὰ χλωρὰ ἔτι τῶν φύλλων ὑπεστρῶσθαι σφᾶς καθεύδοντας. Indeed, we might note in passing that olives, Herakles and Zeus are otherwise connected: Herakles' club was said to be made of the wild olive, though not from a tree in Olympia but from one in the Saronic Gulf region (Pausanias 2.31.10); the statue of Zeus by Pheidias in the temple at Olympia was crowned with a wreath of olive shoots (5.11.1); and the ivory of the statue was preserved against the corrosive atmosphere of the local river by having a pool of olive oil in front of it (5.11.10). There seems to be no way of finding out whether the olive trees at Olympia provided this oil, but Pausanias (2.32.10) says elsewhere of the wild olive (*kotinos*), 'The Troezenians call *rhachos* every kind of barren olive—*kotinos* and *phyllia* and *elaios* [ῥάχους μὲν δὴ καλοῦσι Τροιζήνιοι πᾶν ὅσον ἄκαρπον ἐλαίας, κότινον καὶ φυλλίαν καὶ ἔλαιον]', which would tend to suggest that the wild olive was ultimately fruitless.
35. Pausanias 5.15.3: ἔστι δὲ ἐν τῇ Ἄλτει τοῦ Λεωνιδαίου περᾶν μέλλοντι ἐς ἀριστερὰν Ἀφροδίτης βωμὸς καὶ Ὡρῶν μετ' αὐτόν. κατὰ δὲ τὸν ὀπισθόδομον μάλιστά ἐστιν ἐν δεξιᾷ πεφυκὼς κότινος: καλεῖται δὲ ἐλαία Καλλιστέφανος, καὶ τοῖς νικῶσι τὰ Ὀλύμπια καθέστηκεν ἀπ' αὐτῆς δίδοσθαι τοὺς στεφάνους. τούτου πλησίον τοῦ κοτίνου πεποίηται Νύμφαις βωμός.
36. On olive cultivation in general, see: Foxhall (2007), Isager and Skydsgaard (1992) and Amouretti (1986). I am very grateful to Professor Lin Foxhall for her advice on the cultivation of the olive tree.
37. Pliny, *Natural History* 15.9: *augetur oleum ab arcturi exortu in a. d. XVI kal. Oct., postea nuclei increscunt et caro*; 18.320: *item vindemia facta olivam esse rapiendam et quae ad oleum pertinent quaeque a vergiliarum occasu agi debent.*
38. Foxhall (2007), p.7.

39. I am grateful to my student, Alessandra Pugliese, for the photographs in Figures 1–2, which illustrate a wild olive tree with new shoots on her family property in South Italy.
40. Columella, *On Agriculture* 5.9.12: *nam post solstitium, cum terra aestibus hiat, curandum est, ne per rimas sol ad radices arborum penetret*; cf. 2.2.24: *praesertim in Italia, ubi arbustis atque oleis consitus ager altius resolvi ac subigi desiderat, ut et summae radices vitium olearumque vomeribus rescindantur, quae si maneant, frugibus obsint, et inferiores penitus subacto solo facilius.*
41. One further activity requiring what one supposes would have been a significant number of olive branches was in fact a monthly ritual, in which the Eleans sacrificed on some 32 altars by burning twigs of olive with incense kneaded with honey and libating with wine (Pausanias 5.15.3–10). Pausanias does not specify 'wild olive' (*kotinos*) for this rite, just 'olive' (*elaia*), but the two terms are interchangeable at 5.15.3 in reference to the olive tree which provided the leaves for the victors' crowns.

References

Amouretti, M.-C., 1986, *Le pain et l'huile dans la Grèce antique: de l'araire au moulin* (Paris: Belles Lettres).

Bickerman, E. J., 1980, *Chronology of the Ancient World* (London: Thames and Hudson).

Britton, J. P., 2007, "Calendars, Intercalations and Year-Lengths in Mesopotamian Astronomy", in J. M. Steele (ed.), *Calendars and Years: Astronomy and Time in the Ancient Near East* (Oxford: Oxbow), 115–132.

Drachmann, A. B., 1903/1964, *Scholia Vetera in Pindari Carmina*, vol. 1 (repr. Leipzig: Teubner).

——, 1910/1964, *Scholia Vetera in Pindari Carmina*, vol. 2 (repr. Leipzig: Teubner).

Evans, J., and Berggren, J. L., 2006, *Geminos's Introduction to the Phenomena: A Translation and Study of a Hellenistic Survey of Astronomy* (Princeton, NJ: Princeton University Press).

Farnell, L. R., 1907, *The Cults of the Greek States*, vol. 4 (Oxford: Clarendon Press).

Foxhall, L., 2007, *Olive Cultivation in Ancient Greece: Seeking the Ancient Economy* (Oxford: Oxford University Press).

Freeth, T., Jones, A., Steele, J. M. and Bitsakis, Y., 2008, "Calendars with Olympiad Display and Eclipse Prediction on the Antikythera Mechanism", *Nature* 454, 614–617.

Hannah, R., 2002, "Euctemon's parapegma", in C. J. Tuplin and T. E. Rihll (eds.), *Science and Mathematics in Ancient Greek Culture* (Oxford: Oxford University Press), 112–132.

——, 2005, *Greek and Roman Calendars: Constructions of Time in the Classical World* (London: Duckworth).

——, 2009, *Time in Antiquity* (London: Routledge).

Isager, S., and Skydsgaard, J. E., 1992, *Ancient Greek Agriculture: An Introduction* (London: Routledge).

Nilsson, M., 1955, *Geschichte der griechischen Religion*, vol. 1 (Munich: C.H. Beck'sche Verlagsbuchhandlung).

Robinson, M., 2009, "*Ardua et Astra*: On the Calculation of the Dates of the Rising and Setting of Stars", *Classical Philology* 104, 354–75.

Rochberg-Halton, F., 1992, "Calendars: Ancient Near East", in D. N. Freedman (ed.), *The Anchor Bible Dictionary* (New York: Doubleday), 810–814.

Salt, A., and Boutiskas, E., 2005, "Knowing when to consult the Oracle at Delphi", *Antiquity*, 79, 564–572.

Samuel, A. E., 1972, *Greek and Roman Chronology* (Munich: C.H. Beck'sche Verlagsbuchhandlung).

Thomson, G., 1943, "The Greek Calendar", *Journal of Hellenic Studies*, 63, 52–65.

Trümpy, C., 1997, *Untersuchungen zu den altgriechischen Monatsnamen und Monatsfolgen* (Heidelberg: C. Winter).

What to do on the Thirtieth? A Neo-Platonic Interpretation of Hesiod's *Works and Days* 765–8*

Patrizia Marzillo

Hesiod's *Works and Days* represents the first example in the Greek language of an agricultural calendar. In addition Hesiod offers a list of propitious and unpropitious days based on the days of the lunar month, prescribing which days are good and which bad for various activities. This essay will study how the neo-Platonic philosopher Proclus interpreted Hesiod's prescription not to work on the thirtieth day of the month in terms of neo-Platonic doctrine. An excursus on allegoresis is necessary for that purpose, with special emphasis on the attitude of later interpreters to the work of archaic poets.

1. Allegoresis

Allegoresis, unlike the word allegory,[1] is a modern concept which indicates the praxis of allegorically interpreting what was not conceived of as allegorical originally. The origin of allegoresis is still debated in scholarship. Based on a scholion by Porphyry on *Ilias* 20.67, it is thought to have arisen in the sixth century BCE with the grammarian Theagenes of Rhegium.[2] It was a developed praxis in the fourth century BCE, as the Derveni papyrus confirms.[3]

Opinions are likewise divided as to the reasons for the creation of allegoresis. A possible background is the fact that Homer and Hesiod were harshly criticized and had, therefore, to be defended using new interpretative tools.[4] Tate (1934), in contrast, claimed that allegoresis was not a mere defensive act, but above all an exegetical method and was taken on by philosophers as such.[5]

The Stoics systematically interpreted poetry using physical and etymological allegoresis by having recourse to previous allegorical interpretations which they adapted and revised.[6] They aimed above all to harmonize their philosophy with the ancient poets,[7] but it is possible to find apologetic intentions as well, as in Heraclitus' *Homeric Problems*.[8] By replacing the literal meaning of poets' words with a Stoic one, the Stoics were able to find support for their own doctrine. The goddess Hera, for example, is interpreted by the Stoics as the "air" of their physical system. For this reason, Bernard defines Stoic allegoresis as "substitutive".[9]

Before embarking in a detailed analysis of neo-Platonists' and Proclus' exegesis, it is appropriate to refer to the so-called middle-Platonism which can be summarized by the activity of two authors: Philo of Alexandria and Plutarch.

* The article is dedicated to my friend Marvin Harik. Special thanks go to Jonathan Ben-Dov for his valuable advice and kind patience.

1.1. The middle-Platonists

1.1.1. Philo of Alexandria

An allegorizing technique was known in Jewish Alexandria.[10] Aristobulus (second century BCE), quoted by Eusebius, *Praeparatio Evangelica* 8.10; 13.12, wrote a commentary on the Pentateuch which had an allegorizing character, and the *Letter of Aristeas* 143–50 contains an allegorical explanation of Jewish Dietary laws. Philo (20 BCE/c. 50 CE) further introduced it into Judaism by continuing the Stoic tradition.[11] In Philo, however, it is possible to see a new kind of allegoresis which contains the harbingers of both Christian and neo-Platonic allegoresis. Philo sought to harmonize the Holy Scriptures with Greek philosophy: just as Heraclitus the Stoic saw the source of all wisdom in Homer's poetry, so did Philo see in the Old Testament the source of the entire truth. His allegoresis played an important role in the process leading from Theagenes to Proclus. Many elements in Philo would be further developed by Proclus and the neo-Platonists.[12] Philo speaks of allegory canons or rules.[13] Rather than offering a systematic paradigm of his allegorical activity, Philo only indicates in which circumstances it is appropriate to allegorize.[14] In this sense, he shows conspicuous similarity with Proclus, who would indeed assert that not everything can be allegorized, but only a specific kind of poetry. A second similarity lies in the fact that Philo interpreted agricultural contents allegorically, as did Proclus in his commentary on the *Works and Days*. In the treatise *De agricultura* Philo dealt with *Genesis* 9.20a and especially with the concepts γεωργός and ἀρχή. This treatise, together with *De plantatione* and *De ebrietate*, constitutes a trilogy of dialectic allegoresis.[15] Further, with special relevance for the present essay, Philo (*Quaestiones in Exodum* 2.102) provides an interpretation of gold, silver and iron which, according to Pépin (1958, p. 236), anticipated Hesiod's Myth of Five Ages as interpreted by Proclus in the fragments LXVIII–XCIII: the golden age is the intelligible world, the silver the sky, and the iron the earth.

Whitman (1987, p. 61) believes that "Philo's own exegesis, it is true, often displays the extreme license of Greek allegorization, and at times he even denies the literal meaning of certain scriptural passages". According to Dawson "Philo's allegorical reading transforms Moses' writing (i.e. the Pentateuch) into a rewriting of classical meanings and then paradoxically presents that rewriting as an original writing".[16]

1.1.2. Plutarch of Chaeronea

Scholarship has not yet formed a consensus about Plutarch's attitude to allegoresis. Since it is impossible to deny the presence of allegoresis in *De Iside et Osiride*, many scholars see a contradiction in Plutarch: while he allegorizes in the aforementioned work, in *De audiendis poetis* he embraces Plato's banishment of poetry from the education of the youth.[17] Plutarch furthers Plato's condemnation by writing a treatise on how poetry should be read as propaedeutic to philosophy.[18] His criticism of allegoresis is addressed to the Stoic physical–cosmological allegoresis rather than to allegorical exegesis in general.[19] Bates demonstrated that the allegoresis in Plutarch is the logical consequence of his belief, his demonology and his religiosity. Plutarch believes that there is a truth hidden in myths and religious rites, and considers poets' works as metaphorical, not strictly allegorical; hence he abstains from exaggerations and writes only one allegorizing work, *De Iside et Osiride*, which mainly has an ethical interest.[20] The Stoic allegoresis acquires in Plutarch a new meaning because it is founded on a new procedure. Dawson (1992, p. 60) already noticed that "in the end, Plutarch tends to read the myths as collections of images pointing beyond themselves to a higher mystical reality". Pépin (1958, pp. 182–183) classified Plutarch's exegetical method as *allégorie réaliste*, whereas Bernard (1990, pp. 183–275) considered Plutarch as the first author who allegorized like neo-Platonists.

1.2. The neo-Platonists

Although in Late Antiquity allegoresis was by no means a new phenomenon, neo-Platonic allegoresis introduced some significant innovations in it. Whereas the Stoics and others used to replace the literal meaning with a new one, neo-Platonists maintained both the literal and the allegorical meanings by connecting them on the plane of Platonic philosophy.[21] For this reason I prefer to speak of "complementary allegoresis".[22]

Already Plotinus (*Enneades* 5.1.7; 5.1.4; 5.8.12) interpreted an episode of the Theogony allegorically. Uranus, Cronus and Zeus were explained as the three main hypostases: the One, the Intellect and the Soul. However, a more detailed use of the allegorical exegesis is evident only in his pupil Porphyry (second/third century CE). Scholars debate the dependence of Porphyry's method on the Stoics. Pépin (1958, p. 167) advanced the idea of dependence, whereas Bernard (1990, p. 65) rejected it by referring to a passage from Porphyry's *Against Christians* quoted by Eusebius, *Historia ecclesiastica* 6.19.8. Here Porphyry expresses his opinion about the allegoresis of biblical texts, blaming Origenes and particularly the Stoic origins of Christian allegoresis.[23]

We have only a partial idea of Porphyry's thoughts about allegorical exegesis. In his Περὶ ἀγαλμάτων 24 he allegorically and mystically interprets symbols and attributes of the statues of Greek and Egyptian gods with the scope to come to the knowledge of the divine. The god is perceived by means of a visualization represented by symbols.[25] Such a conception is based on the linguistic theory ὀνόματα-ἀγάλματα, oiginating from the theurgy. The theurgy (lit. "divine-making") is a religious belief which performed various rituals with the intention of uniting with the divine. According to the theurgy, it is possible to trace the names of divinities from an analysis of their statues.[26] Porphyry must have been an admirer of the theurgy in his youth since it is reported that he introduced the Chaldean oracles (a kind of "holy book" for Theurgians) into Plotinus' school.[27] Later, however, he succumbed to Plotinus' religiosity and composed the Letter to Anebus.

Porphyry wrote two works on Homer: *Quaestiones homericae* and *De antro nympharum*.[28] While the *Quaestiones* discuss philological questions, *De antro* focuses on contextual interpretation.[29] Porphyry harmonizes Homer and Plato by interpreting Homer's description of the nymphs' cave (*Odyssey* 13.102-12). It is presented as an ainigma (literally: "riddle") since Homer himself "speaks in riddles" (αἰνίττεται), with αἰνίττομαι in this context meaning "to speak allegorically". Porphyry uses this verb once in a kind of hendiadys with ἀλληγορεῖν (*De antro nympharum* 3). In his explanation of the cave, Porphyry seeks to prove that it is a real place (*De antro nympharum* 4). Allegory and literal meaning are both essential, for him as for the Christians, for the interpretation of a text.[30]

Porphyry has recourse to ethical and physical allegoresis, but above all to the neo-Platonic method. After each element of Homer's description is interpreted according to Platonic doctrine, at the end of chapter 34 Porphyry offers an overall interpretation of the

Odyssey. Here, Odysseus is a symbol for mankind who has to go through all stages of reproduction, the cave being the place where this process ends, as the man returns to himself and his origins.

At the end of *De antro*, Porphyry wants to legitimize his allegorical interpretation by claiming that Homer expressed divine realities in the shape of myths (chapt. 36). It is possible to read this passage as a "manifesto" of Porphyry's allegoresis. Porphyry is convinced that it is impossible to invent a text without an anchor in reality. Following the tradition, Porphyry underlines Homer's wisdom and excellence in every virtue and considers him an authority, even as a philosopher.[31] Despite this outstanding status of the poet, the allegoresis of the nymphs' cave seeks a broader audience.[32]

Other neo-Platonists should be mentioned briefly in this short sketch of the history of allegoresis from Theagenes to Proclus.[33] Iamblichus deepened the analogy between myth and esotericism and attributed to the word "symbol" its full meaning, as can be seen in his *De mysteriis*.[34] This treatise takes the form of the Egyptian archpriest Abammo's answer to Porphyry's *Letter to Anebus*; in the latter, Porphyry expressed his rejection of the theurgy and integrated Greek and Egyptian religious traditions into the neo-Platonic system. Iamblichus, by contrast, defends the theurgy and claims that men have an innate knowledge of the gods that can be rational or beyond the rational; Iamblichus connects the latter with the concept of symbol.[35] Besides σύμβολον, Iamblichus also uses σύνθημα with an identical meaning.[36]

By defending the theurgy, Iamblichus retrieves ancient Eastern wisdom. His treatise Περὶ ἀγαλμάτων is lost, as is his voluminous work (twenty-eight books) on the Chaldean theology,[37] but his *Theologoumena arithmeticae* survived, as well as his commentary on the Pythagorean Nichomachus, where he explains mathematics on the basis of the complex Babylonian and Pythagorean numerology.

The emperor Julian composed a *Speech on the Magna Mater*,[38] and a certain Sallustius proposed a classification of the myths in his treatise *On the Gods and the Universe*. Both works display an interest in allegorical interpretation, which must have been particularly relevant at Syrianus' school, since his students Proclus and Hermeias[39] intensively dealt with it.

Since Syrianus' works are almost completely lost, it is difficult to determine his contribution to allegorical exegesis. His oeuvre is partially reconstructable only through references by his students.[40] He wrote a work *On the Gods in Homer*, a commentary on Homer in seven books (Λύσεις τῶν Ὁμηρικῶν προβλημάτων),[41] four books on Plato's *Republic*, two books on the Orphic theology and ten books *On the Harmony of Orpheus, Pythagoras and Plato with the Oracles*.[42] The latter title mirrors the tendency to harmonize Plato with older "philosopher poets". In his preserved commentary on Aristotle's *Metaphysics* 11.28–36; 43.6–44.17, Syrianus allegorizes Empedocles' doctrine of Love and Strife; 182.26–8 expresses his opinion on philosophy and poetry against Aristotle, *Metaphysics* 1091b: the truth expressed by theologians (that is, Homer, Hesiod, and Orpheus) and philosophers is the same.

2. Proclus as an allegorist

2.1. Entheastic poetry

The sixth essay of Proclus' commentary on Plato's *Republic* focuses on the relationship between Homer and Plato.[43] Attempting to find a solution to Plato's banishment of poetry in *Republic* 10, 595a-608b, Proclus perfected the neo-Platonic allegoresis system by dividing poetry into three categories:[44] divinely inspired, didactic[45] and mimetic.[46] In his opinion, Plato's criticism addressed only mimetic poetry but not the divinely inspired poetry which hides a superior truth.[47] This truth is revealed by allegorical exegesis. Plato's banishment of poetry could thus be conciliated with the neo-Platonic commentaries on Homer and other poets, since Plato nowhere expressly condemns allegorical interpretation. On the contrary, he himself professes to believe in "divine inspiration" (ἐνθουσιασμός).[48]

By "divinely inspired poetry" Proclus refers to three poets: Homer, Orpheus and Hesiod.[49] Their entheastic poetry also contains passages of other kinds of poetry; hence, not everything can be interpreted allegorically (*In Rem publicam* 1.192.22),[50] but only the myths written under divine inspiration.[51]

Proclus' recurrent references to Syrianus in the sixth essay raised the question of his dependence on Syrianus. Sheppard (1980, pp. 39–103) states that Proclus learnt his kind of allegoresis[52] and the need to harmonize Plato with Homer from Syrianus.[53] It is clear, however, that the doctrine of the three kinds of poetry is Proclus' own, since before him only a distinction between inspired and uninspired poetry existed.[54] It seems to me that Proclus' poetical theory results from his readings of Plato: *Phaedrus* 245a for the inspired poetry; *Laws* 1, 630a for the didactic poetry; *Republic* 10, 595a–608b and *Sophist* 235d for the mimetic poetry.

Proclus' theory of interpretation is described in a neglected passage by Marinus, *Life of Proclus* 22:

> Acting according to it (i.e. wisdom), the philosopher penetrated easily the whole Greek and Barbarian theology and the one overshadowed in mythical inventions, and brought it into light for the ones who wanted and were able to understand by interpreting everything in a more inspired way and leading it to harmony; in his analysis of all more ancient works, he accepted critically what was "fruitful" in them, but if he found something "vain", he utterly refused it as blameful; he polemically rejected what was against the right principles by an accurate analysis treating each aspect in his seminars[55] at the same time with competence and wisdom and putting everything as a fundament in his writings.[56]

> Κατὰ ταύτην δὴ ἐνεργῶν ὁ φιλόσοφος πᾶσαν μὲν θεολογίαν Ἑλληνικήν τε καὶ βαρβαρικὴν καὶ τὴν μυθικοῖς πλάσμασιν ἐπισκιαζομένην κατεῖδέ τε ῥᾳδίως καὶ τοῖς ἐθέλουσι καὶ δυναμένοις τε συνέπεσθαι εἰς φῶς ἤγαγεν, ἐξηγούμενός τε πάντα ἐνθουσιαστικώτερον καὶ εἰς συμφωνίαν ἄγων· πᾶσι δὲ τοῖς τῶν παλαιοτέρων συγγράμμασιν ἐπεξιών, ὅσον μὲν ἦν παρ' αὐτοῖς 'γόνιμον', τοῦτο μετ' ἐπικρίσεως εἰσεποιεῖτο, εἰ δέ τι 'ἀνεμιαῖον' ηὕρισκε, τοῦτο πάντη ὡς μῶμον ἀπῳκονομεῖτο· τὰ δέ γε ὑπεναντίως ἔχοντα τοῖς καλῶς τεθεῖσι μετὰ πολλῆς βασάνου ἀγωνιστικῶς διήλεγχε, ἔν τε ταῖς συνουσίαις δυνατῶς ἅμα καὶ σαφῶς ἐπεξεργαζόμενος ἕκαστα καὶ ἐν συγγράμασιν ἅπαντα καταβαλλόμενος.

The metaphor "Proclus brought it into light" can be understood as a reference to Proclus' exegetical method. Marinus, however, defines Proclus' exegesis even more precisely: it is based on "inspiration" and aims to achieve harmony. ἐνθουσιαστικώτερον alludes both to Plato and Proclus' poetical theory; the demanded harmony is the concordance between

Plato and the philosopher (or theologian) poets. The Platonic quotation γόνιμον-ἀνεμιαῖον (*Theaetetus* 151e) can be understood as a reference to Proclus' classification of poetry.

2.2. Linguistic basis for etymological allegoresis

Proclus' linguistic theory can be reconstructed on the basis of his commentary on Plato's *Cratylus*. Attempting to solve the ultimate question whether words are κατὰ φύσιν or κατὰ θέσιν, Proclus claims that words are natural in their meaning (form), and conventional in their sound (matter) because we do not know the analogy between things and phonemes, but can speak properly thanks to the τύχη which is considered by Proclus to be a synonym of "convention". Thus all words are natural because of the analogy between ὀνόματα and πράγματα, and all are at the same time also conventional due to the limits of human knowledge. There is thus no contradiction between nature and convention. Proclus further compares the relation word/thing with the relation gods/gods' statues under the influence of theurgy.[57] The maker of a word has to conform to the nature of things: he should know the nature of things and make the denomination refer clearly to it. There are godly, demonic and human words. The "true" names of the gods are very long and difficult to pronounce; hence Proclus never mentions the true name of a divinity.[58]

Based on his linguistic theory, Proclus builds the etymological component of his allegoresis in order to support and confirm his neo-Platonic, allegorical exegesis. Proclus' interpretation of the term ἔνη καὶ νέα 'old and new (day)', in a passage related to the thirtieth day of the month, is a good example which will be examined closely below.

2.3. The concept of symbol and the theurgy

Proclus' allegorizing method is closely connected to his belief in the traditional gods and the Chaldean Oracles: allegory, symbol and esotericism are interrelated concepts.[59] Proclus very often uses the word "symbol" and other terms originating from mysticism and theurgy. In fact, Proclus and other neo-Platonists preferred the word σύμβολον rather than the word ἀλληγορία.[60] Therefore in the framework of Proclus' allegoresis theory we must analyze his use of the word "symbol".

In Proclus' lost work *On the Mythical Symbols*, which it is partly available through the self-quotations in Proclus' commentaries on *Republic* and *Cratylus* and in his *Platonic Theology*, the term "symbol" means each sign or trace of a higher plane of understanding.[61] The myths of Homer, Hesiod and Orpheus are symbols;[62] Platonic myths are, by contrast, "images". "Symbol" in the Pythagorean system is a higher stage of allegory since it represents ineffable truths which cannot be expressed by means of similes.

Theurgists used symbols to represent invisible and immaterial things, and fell into a trance in order to meet the divinity.[63] Thus, Plotinus initially looked for mystical experiences,[64] and Proclus wrote a work *On the Hieratic Art*.[65] Mysticism was, in fact, the only opportunity for worshippers to know the divinity. Sheppard (1980, p. 153) underlines late neo-Platonists' affinity to the theurgy which can also be seen in their allegoresis. For examples, Proclus explains Achilles' behaviour at the funeral rites for Patroclus (*In Rem publicam* 1.150–3) using the concepts of theurgy.[66]

2.4. *The golden chain*

In his commentary on Plato's *Republic* (1.77.29–79.4), Proclus says that demons are the lowest members of the golden chain. The image of the golden chain comes from the incipit of *Iliad* 8 where Zeus mentions this chain, speaking of his power. The chain was associated with a child's game (like a tug-of-war), but the fact that it is golden suggested different interpretations among allegorists: a cosmological allegory in Orphism; an allegory of the sole sun in Plato, *Theaetetus* 153c; an allegory of the planets in Heraclitus the Stoic; an allegory of Aristotle's unmoved mover; an allegory of the four elements; an allegory of destiny (εἱμαρμένη); a link between gods and men; an anagogic medium of the souls.[67] Proclus, however, used this image to explain his ontological hierarchy. He classified the Homeric gods according to their position (τάξις) in the plane of existence, whereby each divinity exercises his/her power (δύναμις) in a specific field of competence corresponding to his/her own character (*In Timaeum* 1.163.6). The ontological succession of all connected entities builds the chain.[68] Divine chains are a fundamental element in Proclus' theology.[69] Each chain is led by a god. Zeus is the supreme god, but he left a domain of competence to each divinity.[70]

In the commentary on *Alcibiades* I 196, for example, Proclus refers to Hermes' chain as the vehicle of rhetoric inspiration.[71] By means of his theory of the golden chains, Proclus tries to connect philosophical mysticism with Hellenic religion, and the allegoresis of all myths is a necessary outcome of this concern.[72] Proclus, however, also considers the symbolic elements of poetry in their literal meaning.[73] The transition from the literal meaning to the allegorical exegesis of myths through the analogy criterion is due to the myths' incongruence.[74] Proclus rounded off a process which had begun with Porphyry and Plotinus: his allegorical interpretation enables the study of godly inspired poetry in a full Platonic meaning.[75]

Thus, the purpose of Proclus' allegoresis was, on the one hand, to defend Homer, Plato and the pagan cults by resorting to magic, mysticism and eclecticism; while on the other hand he attempted to legitimize his own philosophy through the so-called "theologians".[76] The same intentions can be seen in the exegesis of the *Works and Days* pursued through a systematized use of allegoresis with the characteristics mentioned above: the distinction between didactic and entheastic myths, the linguistic theory of etymological allegoresis, the concept of "symbol" which allows him to interpret entheastic myths in the same way as theurgy and mysteries, and the golden chains, images of Proclus' ontological system.

3. Proclus' commentary on Hesiod's *Works and Days*

3.1. *Philological excursus*

Hesiod's main works were extensively commented on in antiquity.[77] They attracted not only grammarians, but also philosophers: the Stoic Zeno commented on the *Theogony*[78] and Plutarch of Chaeronea on the *Works and Days*.[79] The *Works and Days* was the work most commented on by far. This is easy to understand if we consider the contents of the *Works and Days* and its didactic character, which made it—almost immediately—an essential school book. It is common to divide Hesiod's *Works and Days* into two parts: the *Works*, and the *Days*, with the *Works* divided in turn into two books or sections. In the

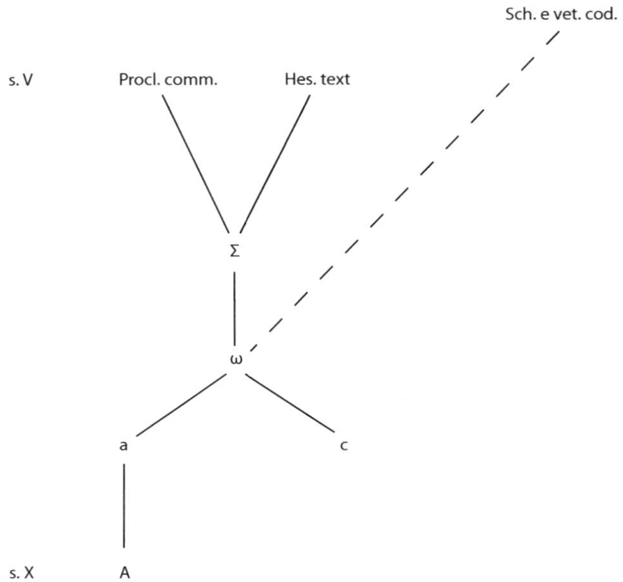

FIGURE 1. The transformation of Proclus' commentary into scholia.

first part, Hesiod recommends work to his brother using general warnings; in the second part (from v. 383 onwards), Hesiod accurately describes the required activities along the seasons of the year. Whereas a small portion is dedicated to seafaring, the larger part is a treatise on agriculture and farming. Hesiod's *Works and Days* represents the most ancient comprehensive rural calendar in the Greek language. Through the interpretation of signs taken from nature and, above all, from the observation of the movements of the stars, Hesiod explained to future generations of Greeks when it was time to sow, to reap, to devote oneself to sailing or to other occupations. Lastly the so-called *Days* suggest which days of the month are convenient or inconvenient for the above mentioned activities according to the influence of the moon.

That the neo-Platonist Proclus was also the author of a ὑπόμνημα[80] on this poem by Hesiod is confirmed by Suda π 2473. Proclus' commentary was the source for the commentaries written later by the Byzantine scholars John Tzetzes (twelfth century), Manuel Moschopulus (thirteenth century) and Maximus Planudes (thirteenth century).

Of all commentaries on Hesiod, no one (including Proclus') has been preserved entirely; they are only available as scholia, i.e., long or short text explanations written in the margin or between the verses of Hesiod's text. Five corpora of scholia exist on the *Works and Days*: those by Proclus (together with *scholia vetera*); Tzetzes; Moschopulus; Maximus Planudes; John Protospatharius (fourteenth century, his scholia only comment on the *Days*).

Figure 1 shows how Proclus' commentary transformed into scholia. Proclus' fifth century commentary on the *Works and Days* was transmitted in its entirety only for a very short time; the next generation of copyists preferred to summarize it and preserve it with Hesiod's text in order to guarantee its survival. Other commentaries written before him had also been condensed and copied in the same manuscripts. To these we give the general name of *scholia vetera*, i.e. ancient scholia. At the end, the summaries of Proclus' and other ancient authors' commentaries fused together. The result of this process is manuscript A of the tenth century that contains Hesiod's *Works and Days* with Proclus' scholia and *scholia vetera*, and which still distinguishes the latter from the former so that it has been mostly possible to identify with certainty genuine scholia by Proclus among the vast amount of scholia.[81]

3.2. Allegorical interpretation

There are many examples of allegorical interpretation of gods and heroes in Proclus' commentary on Plato's *Republic*.[82] They are, however, sporadic explanations which cannot give an exhaustive idea of Proclus' lost works on Homer and Orpheus. For this reason Proclus' commentary on Hesiod gains great relevance.

Hesiod is a godly inspired poet and his myths are for the most part entheastic. In this way Proclus interpreted Prometheus' and Pandora's story (XLIII–LXVII), the myth of Five Ages (LXVIII–XCIII), the Muses (I–III), Eris (XII–XX; CCLXXIII.12–5), Dike (CXII–CXIII), the demonic guardians (CIX–CXI), and Atlas and the Pleiades (CLXII.1–16). He also associated Hesiod's verses with Pythagorean maxims (CCLII), and in general his scholia on the *Days* are influenced by the Pythagorean doctrine. As an example of allegorical interpretation I will illustrate Proclus' explanation of the myth of Prometheus and Pandora.

The story of Prometheus and Pandora had been already allegorized by Plotinus (4.3.14) and Porphyry (*De antro nympharum* 30) to refer to the destiny of the soul. Proclus himself had already claimed in his commentary on Plato's *Republic* (2.53.9–12) that Orpheus and Hesiod referred to the fall of the soul from the intelligible to the becoming world when they spoke of the theft of fire by Prometheus. In the commentary on Hesiod, however, he offers a more detailed explanation both of the individual elements of the myth and of the whole. Here, literal and allegorical interpretations keep pace with each other. Prometheus is a Titan. Titans, in mythology, are usually responsible for "disruption" and, indeed, Prometheus commits a crime by stealing fire from Zeus. But Prometheus is also, as his name indicates, the one who can foresee the consequences of an action, ergo he is, on the allegorical plane, the one who can also confer this ability to men. Fire is, in fact, the intellectual life (XLIV.2–5; XLVI.1–2), an image of practical life and light (XLVI.4–5), and so we return to the literal meaning since fire really "enlightened" and improved human life. Ultimately, fire is the soul itself which comes from above and is moved to another place.[83] This is easy to understand considering that Prometheus has stolen the fire by putting it in a giant fennel stalk (XLV). Apart from the connection between fennel stalk and Dionysius (whose thyrsus is made from a fennel stalk), it is evident in this context that the narthex, the delicate envelope that Prometheus chose to transport fire, is the human body; hence fire can be identified with the human soul. Hesiod rightly spoke of "theft" since Prometheus gave the intellectual life, which belonged to the soul living in the intelligible (= Zeus), to the souls fallen into generation. Not only is falling into a body a punishment for the soul, but Zeus also sent the irrational life (ἄλογος ζωή): Pandora (L, LI, LII).[84] She has human shape, but at the same time she is like the goddesses. This in-between position defines her as a demon (LIV), i.e., as an intermediary being between gods and men. Pandora opens the jar and by so doing sets free the evils and the goods contained in destiny (= the jar). This is, indeed, a demonic task (δύναμις): they assign illness and fate by acting invisibly (LXVII). Proclus interprets all of Pandora's traits, as listed by Hesiod, as characteristics of the irrational life: Pandora is pretty because the irrational life is free from generation, and for this reason contains a certain beauty which is manifest in her body; her concupiscence is, however, ugly (ἀργαλέον) because it is material and baffles the ability to judge (LVIII, LIX). Pandora, and accordingly the irrational life in men, are not able to recognize the truth due to their wrong conceptions (LX).

Irrational life is an evil, but a necessary evil without whom the soul would not live within the body. In order to explain *Works and Days* 58, Proclus admits that irrational life can be a pleasant evil. Thus, on the one hand, the soul suffers less of the pain and the labour coming from the physicality of life; on the other hand, by living within the body, the soul also feels that it does not belong to the sensible world, but to the intelligible.

Prometheus and Epimetheus are presented as two types of thinking: those that occur before and after action respectively, representing the souls that fell upon the human generation and the two possible behaviours for them to adopt.[85]

4. The question of the thirtieth day

On the last day of the month, Hesiod (*Works and Days* 765–8) recommends checking the work done in the previous month and distributing food to the slaves. He writes:

Bear well in mind the days that come from Zeus and point them out according to their portion to the slaves. The thirtieth of the month is the best for watching over the works and distributing the rations: people celebrate it because they distinguish the truth.[86]

Ἤματα δ᾽ ἐκ Διόθεν πεφυλαγμένος εὖ κατὰ μοῖραν
πεφραδέμεν δμώεσσι· τριηκάδα μηνὸς ἀρίστην
ἔργά τ᾽ ἐποπτεύειν ἠδ᾽ ἁρμαλιὴν δατέασθαι,
εὖτ᾽ ἂν ἀληθείην λαοὶ κρίνοντες ἄγωσιν.

Since it is quite logical that the last day of the month is good for checking what the slaves have done and to organize the activities for the coming month, commentators did not have much to duscuss on the first three verses (765–7). West (1978, p. 350) only comments on that the sentence in 766–7 is an "odd conflation of two ideas": on the one hand, "bear well in mind the days coming from Zeus and tell your slaves about them"; on the other, "inspect slaves' work and distribute them their rations on the thirtieth". The montly distribution of food was also common: in ἁρμαλιή we have to see the "incentive" for the work that has to be done in the coming month.[87]

The last verse is more difficult to account for. Since an average lunar month consists of about twenty-nine and a half days, calendar months vary between "hollow" months of twenty-nine days and "full" months of thirty days. West (1978, p. 351) understands the mention of truth in verse 768 as an appeal by the poet to reckon the days correctly.

Non-philosophical scholia on Hesiod interpret:[88]

Sch. 766a Pertusi

<The thirtieth of the month is the best:> because the thirtieth day represents faithfully the truth; for on this day, the moon comes together and is united to the sun. And since it is peculiar of the truth to be univocal unlike the falsehood—for the falsehood is split into many parts—for this reason, the thirtieth day, due to its uniting activity, resembles the univocal truth.

<τριηκάδα μηνὸς ἀρίστην·> διότι ἡ τριακοστὴ ἀπομιμεῖται τὴν ἀλήθειαν· κατὰ γὰρ ταύτην ἡ σελήνη τῷ ἡλίῳ συνέρχεται καὶ συνήνωται. καὶ ἐπεὶ ἴδιον τῆς ἀληθείας τὸ μονοειδὲς ναντίως τοῦ ψεύδους ἐχούσης – τὸ γὰρ ψεῦδος πολυσχιδές – , διὰ ταῦτα ὡς ἑνοποιὸς ἡ τριακοστὴ τῇ μονοειδεῖ ἀληθείᾳ ἀπείκασται.

Sch.768a Pertusi

<When ... the truth:> When men celebrate the days because they distinguish them by truth, i.e. when they understand that there is the real conjunction and that it took place with precision. For everyone has a different opinion since it was common to insert intercalary months. The syntaxis is like this: when people truly examine the works that have been made by their slaves in the entire month.

<εὖτ' ἂν ἀληθείην·> ὅταν μετὰ ἀληθείας <κρίνοντες ἄγωσιν> οἱ ἄνθρωποι τὰς ἡμέρας, ἤτοι ὅταν τὴν ἀληθῆ ἐπίστανται σύνοδον καὶ μετὰ ἀκριβείας γινομένην. ἄλλοι γὰρ ἀλλοίως δοξάζουσιν, ἐπειδὴ καὶ ἐμβολιμαίους μῆνας εἰώθασι τιθέναι. ἡ σύνταξις τοιαύτη, τουτέστι· ὅταν οἱ λαοὶ ἐρευνῶσιν ἀληθῶς τὰ τῶν δούλων ἔργα ἃ τῷ ὅλῳ μηνὶ πεποιήκασιν.

Sch.768b Pertusi

<When ... the truth:> When people celebrate it because they distinguish the real conjunction. He says this: some people say that the conjunction comes to be in twenty-eight days, others in more days; and truly they invented also intercalary months.

<εὖτ' ἂν ἀληθείην·> ὁπόταν τὴν ἀληθῆ σύνοδον οἱ λαοὶ διακρίνοντες ἄγωσιν. τοῦτο δέ φησιν ὅτι ἄλλοι ἐν εἴκοσι ὀκτὼ ἡμέραις, ἕτεροι δ' ἐν πλείοσι τὴν σύνοδον γίνεσθαί φασιν· ἀλλὰ μὴν καὶ ἐμβολιμαίους ἀναπλάττονται μῆνας.

According to the first scholion, the thirtieth day is the best due to its association with the truth mentioned in Hesiod's verses. On the thirtieth the conjunction takes place, which makes the sun and the moon a unity in the same way that the truth is one. Despite its philosophizing tenor, however, this scholion does not explain Hesiod's original intention. The second and third scholia, in contrast, speak of 'real' conjunction, whereby the question returns to the lunar calendar and the intercalary months.

Proclus' explanation is more detailed:[89]

The days coming from Zeus: the instructions given about choosing and avoiding particular days have their origin in observations that men have made, but some instructions have prevailed among some people and others among others; for also in Orpheus a number of differences among the days are spoken about and in the hereditary laws of the Athenians they were exactly determined: some days were good, others bad, yet others neutral. And not only were entire days assumed by people as being convenient for the beginning of certain activities, but also parts of the day: sometimes they praised the earliest hours of the morning, sometimes the latest of the evening; again, people say that the time until noon belonged to the gods, the time after noon to the heroes. Hesiod, then, knew most of the contemporary observations in this field, and he himself took the step of assigning the differences between propitious and unpropitious days to the movements of the sun and the moon and to their mutual aspects, from whose changes the becoming of all that is mortal mostly depends. But for some beings more than for others the revolutions of the sun and the moon are favourable or unfavourable. Indeed this is shown by plants, some of which move in conjunction with the moon, others with the sun. Roses and violets and also the heliotropes turn their leaves to the rising sun, and similarly to the setting sun by inclining towards the west. And the leaves of the olive tell the farmers, by turning over, that the winter or summer solstice has come, having their darker side facing upwards at one time and the lighter side at another. Everybody says, too, that the eyes of the cats and the entrails of mice contract as the moon wanes, and increase as it grows to the full. [....] If <a plant> should be taken up at the full moon it still retains the principle of growth and sprouts again at the proper season, but if taken up when the moon is waning, it is sterile. And in general some things flourish when the moon is waxing, and others when it is waning, since the moisture shed by the increasing light of the moon is beneficial to some things but harmful to others. Hesiod, then, starts from the thirtieth (day on which the real conjunction takes place), that is now the thirtieth day without subtraction of a day, now the twenty-ninth if also the day before it is put aside. That is why he also said "people celebrate it because they distinguish the truth" requiring to take the actual thirtieth. It is reasonable that he has started from this day on which the conjunction takes place. For it is necessary, since this kind of communion between sun and moon is the beginning of both their revolutions, to make it also the beginning of the days we designate according to their mutual aspect. (Hesiod) wants us to survey on this day all works of the month and to hand out food to the workers: the first activity aims to inspect the works of the past month; the second is an incentive for the works that have to be done in the following month.

ἤματα δ᾽ ἐκ Διόθεν· αἱ περὶ τῆς τῶν ἡμερῶν ἐκλογῆς καὶ ἀπεκλογῆς παραινέσεις ἔχουσι μὲν τὰς ἀρχὰς ἐκ τῶν παρατηρήσεων, ἄλλαι δὲ παρ᾽ ἄλλοις ἐκράτησαν· ἐπεὶ καὶ παρ᾽ ᾿Ορφεῖ (Fr. 753 Bernabé) λέγονταί τινες αὐτῶν διακρίσεις καὶ ἐν τοῖς ᾿Αθηναίων πατρίοις διωρίσθησαν, αἱ μὲν ἀγαθαί τινες, αἱ δὲ φαῦλαι, μέσαι δέ τινες εἶναι. καὶ οὐχ ὅλας ἡμέρας μόνον ὑπέλαβόν τινες εὐκαιρίαν ἔχειν πρὸς καταρχάς τινων πράξεων, ἀλλὰ καὶ μόρια τῆς ἡμέρας, ὁτὲ μὲν τὰ ἑωθινὰ ἐπαινοῦντες, ὁτὲ δὲ τὰ περὶ δείλην ὀψίαν, ὅπου δὲ καὶ τοῖς μὲν θεοῖς οἰκεῖα τὰ πρὸς μεσημβρίαν εἰρήκασιν, ἥρωσι δὲ τὰ μετὰ μεσημβρίαν. ὁ γοῦν ῾Ησίοδος τὰς πολλὰς ἐν τούτοις εἰδὼς τῶν κατ᾽ αὐτὸν παρατηρήσεις, εἰς τὰς ἡλίου κινήσεις καὶ σελήνης καὶ τὰς πρὸς ἀλλήλους σχέσεις αὐτὸς ἀποβλέψας ἀνάγει τὰς τῶν ἐπιτηδείων καὶ ἀνεπιτηδείων διαφοράς, ἀφ᾽ ὧν μάλιστα γίνεται πάντα μὲν τὰ θνητὰ κινουμένων. ἄλλα δὲ μᾶλλον ἄλλων πρὸς τάσδε τὰς περιόδους αὐτῶν οἰκείως ἢ ἀλλοτρίως ἔχει τῶν γινομένων. δηλοῖ δὲ καὶ τῶν φυτῶν τὰ μὲν σελήνῃ συγκινούμενα, τὰ δὲ ἡλίῳ. τὰ μὲν γὰρ ῥόδα καὶ ἴα καὶ μετὰ τούτων τὰ ἡλιοτρόπια πρὸς ἥλιον ἀνίσχοντα τρέπει τὰ φύλλα καὶ πρὸς καταδυόμενον ὡσαύτως εἰς ἑσπέραν ῥέποντα. τὰ δὲ τῶν ἐλαιῶν φύλλα διδάσκει καὶ τοὺς γεωργικοὺς γεγονέναι τροπὰς ἢ χειμερινὰς ἢ θερινὰς διὰ τῆς ἑαυτῶν περιστροφῆς, ὁτὲ μὲν ἄνω τὸ μελάντερον ἰσχόντων, ὁτὲ δὲ τὸ λευκόν. τὰ δὲ τῶν αἰλούρων ὄμματά φασι καὶ τὰ σπλάγχνα τῶν μυῶν πάντες φθίνειν μὲν σελήνης ληγούσης, αὔξεσθαι δὲ ἀκμαζούσης. [....] εἰ μὲν περὶ πανσέληνον ἐξαιρεθείη, τὴν γόνιμον ἀρχὴν ἔτι φυλάττει καὶ αὖθις βλαστάνει κατὰ τὴν προσήκουσαν ὥραν, εἰ δὲ φθινούσης, ἄγονον γίνεται. καὶ ἁπλῶς τὰ μὲν πληρουμένης εὐθηνεῖται, τὰ δὲ ληγούσης αὐτῆς, τοῖς μὲν ὠφελίμου τῆς ὑγρότητος οὔσης ἣν διαχεῖ τὸ σεληναῖον φῶς αὐξανόμενον, τοῖς δὲ βλαβερᾶς. ἄρχεται οὖν ὁ ῾Ησίοδος ἐκ τῆς τριακάδος (καθ᾽ ἣν ἡ ἀληθής ἐστι σύνοδος) ὁτὲ μὲν οὖσαν τριακάδα ἄνευ ἐξαιρέσεως, ὁτὲ δὲ εἰκοστὴν ἐνάτην, ὅτε καὶ ὑπεξαιρεῖται ἡ πρὸ αὐτῆς. διὸ καὶ αὐτὸς ἀξιῶν τὴν οὖσαν τριακάδα λαμβάνειν εἶπεν εὖτ᾽ ἂν ἀληθείην λαοὶ κρίνοντες ἄγωσιν. εἰκότως δὲ ἀπὸ ταύτης ἤρξατο, καθ᾽ ἣν ἡ σύνοδος. δεῖ γὰρ τὴν κοινωνίαν <τ>αύτην, ὡς μίαν ἀρχὴν οὖσαν ἀμφοτέρων, ἀρχὴν ποιεῖσθαι καὶ τῶν ἡμερῶν, ἃς ἐκ τῆς πρὸς ἀλλήλους ποιοῦσι σχέσεως. βούλεται δὲ ἐν ταύτῃ τά τε ἔργα τοῦ μηνὸς ἐφορᾶν πάντα καὶ τὴν τροφὴν ἀπομερίζειν τοῖς ἐργάταις· τὸ μὲν εἰς τὴν τοῦ παρελθόντος συντεῖνον ἐπίσκεψιν τῶν πονηθέντων, τὸ δὲ εἰς τὴν τοῦ μέλλοντος συντελοῦν προτροπὴν τῶν πραχθησομένων.

The fragment can be divided into four parts. Proclus first gives a brief account of the instructions given for choosing and avoiding particular days (or even parts of days) of the month in various cultures. He then attributes to the divinely inspired poet Hesiod broad knowledge on this subject, stating that Hesiod's special merit was to have understood that the quality of days depends on the phases of the moon. This is the opportunity for an excursus (part 3) about the beings in the realm of generation which benefit or are damaged by the revolutions of sun and moon. Whereas this section reveals the influence of Plutarch and Iamblichus,[91] the last part is Proclus' own interpretation. Here he finally comes to the thirtieth day, asking which day is the final day of the month: the thirtieth or the twenty-ninth? Based on Hesiod's words, Proclus makes it clear that Hesiod speaks of the "genuine" thirtieth, i.e., of the month with thirty days. This leads us to the core of my article. In fact, the "genuine" thirtieth gains relevance in a neo-Platonic perspective since only thirty is a special number and not twenty-nine. In a Platonic-Pythagorean point of view, thirty is considered a "perfect" number because it is constructed of three and ten.[92] Whereas three developed as a "perfect" number only after Plato,[93] ten was always a number of perfection since it contains in itself all mathematic ratios and all harmonic consonances. Three and ten were very important numbers for Proclus: three represented the triadic structure of the universe, the basic unity of Proclus' metaphysical system; ten is, according to the Pythagorean tradition which Proclus acknowledged, the "perfect" number, in which all proportions are contained. Only on the thirtieth and not on the twenty-ninth could Proclus have founded his neo-Platonic interpretation of Hesiod.

Proclus, at any rate, does not only explain how Hesiod's allusion to the "truth" is to be understood. Rather, he focuses on the question why Hesiod began from this day.

The thirtieth day is the last day of the month. Indeed, it could be logically expected to begin the month with the first day and leave the thirtieth to the end of the account. Proclus, by contrast, reminds us that the thirtieth day is also the beginning of the sun and moon's revolutions, and that already the ancients used to call the final day of the month "the old and new" (ἔνη καὶ νέα).[94] It was thus both the "beginning" and the "end" already in a non-Platonic perspective. Beginning and end coincide in a kind of *Ringkomposition*. Proclus implicitly ascribes to Hesiod a narrative technique that presupposes knowledge of the doctrine according to which the "end" is nothing else but a "come-back" to the beginning. The same can be observed in a fragment at the end of the agricultural prescriptions (*In Hesiodum* CCXXIII.2–7). Hesiod had begun his section on agriculture by mentioning the Pleiades ("When the Atlas-born Pleiades rise, start the harvest—the plowing, when they set", vv. 383–4) and at the end of the same section, he "returns" to this constellation and to plowing ("When the Pleiades and Hyades and the strength of Orion set, that is the time to be mindful of plowing in good season. May the whole year be well-fitting",[95] vv. 614–7). Proclus comments on

Hesiod connected the end of the agricultural works with the beginning, plowing with plowing; for he began from here, from the Pleiades' set and from the plowing and left off with the same topics. For this reason, he also added—after going through everything—that the pleiôn, i.e. the whole year, could have in this way a well-fitting end of the works. The whole year is called pleiôn because it has fulfilled (sumplêrôsas) the seasons, into which Hesiod divided the agriculture.

τῇ ἀρχῇ τὸ τέλος συνῆψε τῶν γεωργικῶν ἔργων, τῷ ἀρότῳ τὸν ἄροτον· καὶ γὰρ ἤρξατο ἐντεῦθεν, ἀπὸ τῆς δύσεως τῶν Πλειάδων καὶ ἀρότου, καὶ κατέληξεν εἰς αὐτά. διὸ καὶ ἐπήγαγε διὰ πάντων διεξελθὼν ὅτι ὁ πλειών, ὅς ἐστιν ὁ ἐνιαυτός, τῶν ἐπὶ γῆς ἔργων οὕτως ἂν ἔχοι τέλος ἐναρμόνιον. πλειὼν δὲ ὁ ἐνιαυτὸς ὡς τὰς ὥρας συμπληρώσας καθ᾽ ἃς τὴν γεωργίαν διεῖλε.

Proclus thus suggests that Hesiod was aware of his "circular" composition technique and deliberately used it.

Circularity is an important concept for Proclus. His basic triad, on which his whole metaphysical system is founded, consists of Remaining, Procession, Reversion (μονή, πρόοδος, ἐπιστροφή), which together constitute a circle. "Cyclic" is, indeed, according to Proclus, the basic sequence of proceeding from what is static and returning to it. From this cyclic process all phenomenal movements emerge, including the rotation of the celestial bodies.[96] The thirtieth means "reversion" since it consists of three. In Fr. CCLXVIII 5–6, Proclus identifies the numbers one, two and three with Remaining, Procession and Reversion: [...] μονὰς δυὰς τριάς (εἰσίν <δ'> οὗτοι μονὴ πρόοδος ἐπιστροφή).

Remaining, Procession and Reversion do not simply represent the main triad alongside other triads, but rather they are *in primis* the reason and constitution of every other triad.[97] They take place in every other triad, in every stage of ontology whose constitutive principle is the κοινωνία. This is precisely the term Proclus uses to describe the conjunction of sun and moon. It is a moment where the two celestial bodies are lined up and both are at the end and at the beginning of their revolutions. Proclus defines it as "communion" (κοινωνία). By saying that Hesiod begins from the end of the month (the thirtieth day), Proclus attributes Hesiod with his own credo, which enables him to interpret him allegorically in neo-Platonic terms.

According to Proclus' exegetical method, the allegorical and literal planes are interlaced and ever present. Accordingly, after suggesting a neo-Platonic reading of Hesiod he returns to the literal meaning, citing the most down-to-earth interpretation: it is proper to check on the thirtieth the works of the recently passed month and to plan for the coming month.

In his explanation of *Works and Days* 769–71, Proclus develops his ideas with further examples (*In Hesiodum* CCLXI.1–10):

These are the days that come from Zeus: (Hesiod) has started from the thirtieth day for the reason we mentioned and considers it reasonable not to work on this day, but to inspect the works. Besides, some people say that even ants rest on this day; manifestly this fact is also, for many, the evidence that it is the thirtieth of the month and that the conjunction between sun and moon is taking place. The Egyptians say that the sow is unholy because it enjoys copulation when the moon is hidden by the sun. And may it not be that this animal, since it belongs to the earth and enjoys procreation, is connected particularly with that conjunctive phase of the moon-goddess, who—they say—is related to the sun as female to male.[98]

αἵδε γὰρ ἡμέραι εἰσὶ Διός· τὴν μὲν τριακάδα πεποίηται ἀρχήν, δι' ἣν εἴπομεν (Fr. CCLX.29-33) αἰτίαν, οὐκ ἐργάζεσθαι ἀξιώσας ἐν αὐτῇ, ἀλλ' ἐποπτεύειν τὰ ἔργα. φασὶ γοῦν τινες τότε καὶ τοὺς μύρμηκας ἡσυχίαν ἄγειν· ἐπιδήλως καὶ τοῦτο πολλοῖς γίνεται τεκμήριον τοῦ εἶναι τριακάδα καὶ τὴν σελήνην σύνοδον ποιεῖσθαι πρὸς ἥλιον. τὴν δὲ σῦν ἀνίερον Αἰγύπτιοί φασιν, ὅτι μίξεσι χαίρει κρυπτομένης ὑπὸ τοῦ ἡλίου τῆς σελήνης, καὶ μήποτε καὶ τοῦτο τὸ ζῷον, ὡς χθόνιον καὶ γεννήσεσι χαῖρον, οἰκεῖόν ἐστι πρὸς ταύτην εἰκότως μάλιστα τῆς θεοῦ τὴν συνοδικὴν φάσιν, ἣν πρὸς ἥλιον λόγον ἔχειν ὡς θῆλεος πρὸς ἄρρενά φασι.

Living beings are influenced by sun and moon and their mutual aspects, as we have seen. On the thirtieth, at conjunction, sun and moon "rest" together in a common, static and unchanging state before their revolutions start again. It is therefore also time for men to rest by inspecting the works done in the past month and by organizing things for the next month.

For the accounts on animals, Proclus' source was again Plutarch, although the doctrine that the "hylic world" was dominated by the moon has a Chaldean origin too.[100] Speaking of the Chaldean doctrine, it must be asked at this point whether Hesiod's prescription was considered by Proclus also as a religious precept. Elsewhere, for example in ancient Israel, the days of the interlunium required an interruption of the daily work and an intensification of ritual activity.[101] Rest on this day has such religious meaning for Proclus too, since he celebrated this day of the month with solemnity, as his biographer, Marinus of Neapolis, *Life of Proclus* 19, reports. In praising Proclus' religiosity and his respect of all possible Greek and foreign rites, Marinus writes:

Every last day of the month he fasted, without even having eaten the night before. And he likewise celebrated the New Moon with magnificence and sanctity.	πᾶσαν γὰρ ἔνην καὶ νέαν τοῦ μηνὸς μηδὲ προδειπνήσας ἠσίτει. ὥσπερ δὴ τὰς νουμηνίας λαμπρῶς ἐπετέλει καὶ ἱεροπρεπῶς.

Hesiod's warning to rest on the thirtieth by checking the work done by the slaves in the previous month and by distributing food to them for the coming month is thus interpreted in a full neo-Platonic sense both in a metaphysical and in a religious perspective.

5. Conclusions

Allegoresis was a common praxis in classical antiquity. Grammarians first made use of it in order to defend Homer's verses; then philosophers borrowed this device and transformed it into a method capable of supporting and legitimizing their own doctrine.

Before embarking on the exegesis of poems, neo-Platonists had to face Plato's banishment of poetry from the ideal state. They based their argumentations on Plato's doctrine of the "divine inspiration" (ἐνθουσιασμός) and on *Republic* 2, 378d, in which Plato seems to assume a neutral position about poets' allegorical interpretation. Proclus, in particular, elaborated a theory of poetics by distinguishing between divinely inspired, didactic and mimetic poetry. In his opinion, Plato's banishment concerned only mimetic poetry which is an imitation of the imitation (since it imitates our world which is, in turn, an imitation of the world of Forms); in contrast, divinely inspired poetry must be studied and interpreted because it hides theological truths. Divinely inspired poets—Homer, Hesiod, and Orpheus—were also called, from Porphyry onwards, "theologians".

Neo-Platonists thus tried to explain ancient poets' works in terms of neo-Platonic dialectics. The literal plane was, however, not replaced, but integrated in an overall interpretation. The influence of the Chaldean theology and of theurgy on the allegoresis is evident, above all in Iamblichus' *De mysteriis* and in Proclus. Despite its "oracular" tone, neo-Platonists' allegoresis aimed to reach a broader audience. It also resorted to the etymological analysis.

Most of Proclus' works on ancient poets have been lost. Preserved are a number of references derived from his commentary on Plato's *Republic* and from his other writings. For this reason, his commentary on Hesiod, despite its fragmentary state, acquires great relevance.

Among Hesiod's works, the *Works and Days* enjoyed the most success throughout the centuries. It presented a way of life, imparted notions of farming and sailing, and offered a concrete and detailed calendar for the work seasons. Because of its encyclopaedic contents, the *Works and Days* was continuously commented on in antiquity up to the Byzantine era. Among the many commentaries written by grammarians and philosophers, which are now only available as scholia, the commentary composed by the neo-Platonist Proclus in the fifth century CE stands out. By having recourse to the so-called allegoresis, Proclus interpreted Hesiod in terms of neo-Platonic doctrine.

Proclus commented on the entire *Works and Days* allegorically. Specifically, we have seen how he understood Hesiod's prescription to rest on the thirtieth day of the month. In the frame of a "full" lunar month he describes the influence of the moon on living beings and its relationship to the sun. In particular, Hesiod's astronomical allusion to conjunction allows the exegete to transfer Hesiod's verses to a metaphysical plane by using a Pythagorean approach regarding the particular value of the number thirty.

Thirty is a "perfect" number because it contains the three and the ten. Perfection is to be regarded as completeness, as limitation and as calmness, as absence of movement. According to Proclus, the number three represents the ἐπιστροφή, the reversion to the remaining after the procession. In other words, three is the "end" which is somehow identical with a new "beginning", a new μονή. That is why Hesiod, in Proclus' eyes, lists the days of the month as *beginning* from the *end* (the thirtieth).

On this day, the conjunction between sun and moon takes place. It is the moment in which they stop their revolutions before commencing the movement again. Since the sun and the moon affect the development of living beings, it is clear that the pause in their revolutions produces rest both for the realm of nature and for mankind. The thirtieth is thus most suitable in neo-Platonic eyes for the activities assigned by Hesiod to the final day of the month: inspecting the works of the previous month and organizing activities for the following month.

By means of his allegoresis Proclus not only adds metaphysical meaning to the most logical explanation of Hesiod's warning, but he also charges the passage with a religious nuance. Indeed, for Proclus the thirtieth was a fast day, and since Nature rests under the influence of the sun and moon, it is reasonable also for men to rest and to respect the moon's conjunction.

Notes

1. Allegory is the English transcription of the Greek ἀλληγορία which comes from the verb ἀλληγορεύω (= ἄλλα ἀγορεύειν), see Whitman (1987), pp. 263–268. This rhetorical figure was probably introduced into the Greek literature by Gorgias of Leontini (cf. Suda s.v. Γοργίας). Originally, it was called ὑπόνοια (see Plato, *Republic* 2, 378d). Plutarch reports how the latter was progressively replaced by the former (*De audiendis poetis* 19, E10–F1). According to Pépin (1987), pp. 15–20 and Buffière (1973^2), pp. 45–48, both concepts could coexist because ἀλληγορία refers to an activity of the language, ὑπόνοια to an activity of the thought.
2. The scholion has been interpreted in different ways, cf. Detienne (1962), pp. 65–67, Tate (1927), p. 215, Tate (1934), p. 108, Wehrli (1929), p. 81, Schibli (1990), p. 27; 56 n. 12; 99 n. 54.
3. Burkert (1968), p. 101 thinks that the allegoresis in the papyrus is due to an eclectic author, who tried to combine Orphic theogony with pre-Socratic physics. Laks (1997), pp. 134–140, by contrast, believes that allegorical interpretation has to be seen in relation to the Orphic religion, especially in relation to its separation between adherents and non-adherents. Finally, Most (1997), p. 128 explains that the allegoresis

aims to support the eschatological theology expressed in the preceding columns.

4. Xenophanes was one of the first authors who cast doubt on the Homeric/Hesiodic theology, especially as for the anthropomorphism, cf. Buffière (1973^2), pp. 13–22, and Pépin (1958), p. 74; 93–94. Although Blönnigen (1992), pp. 20–21 thinks of an apologetic purpose as well, he does not exclude the possibility of another function for the allegoresis.
5. So does Heinemann (1949), p. 17–18 who notwithstanding criticized Tate's position (p. 37).
6. See Thompson (1974), p. 91 and Bates (1906), p. 21.
7. In Cicero, *de natura deorum* 1.41, it is reported that Chrysippus, in the second book of his work on the gods, intended to harmonize his philosophy with the myths of Orpheus, Musaeus, Hesiod and Homer: *Vult Orphei, Musaei, Hesiodi Homerique fabellas accomodare ad ea, quae ipse primo libro de dis immortalibus dixerat, ut etiam veterrimi poetae, qui haec ne suspicati quidem sint, Stoici fuisse videantur*. According to Tate (1929/30), pp. 3–4 the Stoics believed that the most ancient poets were the wisest, and for this reason utilized them in order to support and explain their own doctrine.
8. Cf. Bates (1906), pp. 21–22; Dawson (1992), pp. 38–52; Pépin (1958), pp. 159–167; Most (1989), pp. 2018–2023; Long (1992), pp. 64–66; Ramelli (2003), p. 40.
9. See Bernard (1990), p. 15 et passim.
10. See Dawson (1992), pp. 74–82; Blönnigen (1992), pp. 57–69.
11. Cf. Dörrie (1974), pp. 133–134; Lamberton (1986), pp. 44–54.
12. For the similarities between Philo and Porphyry's *De antro nympharum* see Runia (1990), pp. 115–117.
13. *De somniis* 1.73.2 and 1.102.3.
14. See Christiansen (1969), pp. 4–5.
15. See Christiansen (1969), pp. 99–133.
16. Dawson (1992), p. 73. Thompson (1974), pp. 35–37 and Pépin (1958), pp. 231–242 attribute to Philo an apologetic purpose comprising an attempt to "save" both the Jewish and Greek traditions. Christiansen (1969), p. 28 in contrast, claims that Philo's exegesis could not have existed before the period of middle-Platonism. She underlines that the allegorical method is based on the dihairetic procedure of the Platonic dialectics whose gnoseologic system Philo adapted to his purposes.
17. See Wehrli (1929), pp. 19–20; Heinemann (1949), p. 6.
18. *De audiendis poetis* 15F–16A, cf. Bates (1906), p. 28; Griffiths (1967), p. 81, and Marzillo (forthcoming^3).
19. *De audiendis poetis* 19E–F; 31D–E; cf. Pépin (1958), pp. 179–182, and Babut (1969), p. 376.
20. Bates (1906), p. 32: "Plutarch's allegorical interpretation exhibited the same intellectual habits and mental bias as his other literary activities, an absence of historical perspective and an inclination to moral consideration upon all occasions".
21. In this sense, neo-Platonic allegoresis is very similar to the Christian exegetical method, see De Lubac I (1959), pp. 425–439.
22. Marzillo (2010), p. XIV.
23. *Epistula ad Anebum* 36–38 in Eusebius, *Praeparatio evangelica* 3.4.2; cf. Pépin (1958) pp. 462–466 and (1987), pp. 54ff.
24. In Eusebius, *Praeparatio evangelica* 3.7.1–4; 3.9.1-5; 3.11–13.2; this work was written according to Buffière (1973^2), pp. 535–540 after Porphyry met Plotinus.
25. For an analysis of the symbol in this work see Crome (1970), pp. 124–142.
26. See Hirschle (1979), pp. 42–44.
27. Porphyry also composed a work on the philosophy of the Oracles which is preserved only fragmentarily and is edited by Wolff (1856). As for the influence of the Chaldean theology on Porphyry, see e.g. Des Places (1971), pp. 18–24 and (1984), pp. 2308–2311.
28. Fragments of the lost work Styx are transmitted by John of Stobi; cf. Lamberton (1986), pp. 113–115.
29. Cf. Lamberton (1986), pp. 108–133; Buffière (1973^2), pp. 419–459; Simonini (1986), pp. 30–31. Most scholars ascribe the *De antro* to the period between 250 and the date of Porphyry's death (about 300). A clarification on the date will be particularly interesting in order to reconstruct the development of allegorical interpretation in Plotinus' school.

30. Simonini (1986), p. 94; see also Pépin (1965) pp. 235–240.
31. See e.g. Pépin (1965) p. 235. Porphyry referred to Homer, Hesiod and Orpheus by the expression "the theologians"; Proclus will do it as well.
32. Cf. Pépin (1965) pp. 261ff. The popular character of neo-Platonic allegoresis is evident also in Proclus' commentary on Hesiod, see Marzillo (2010), pp. XIX–XX; LII–LIII.
33. Proclus' contemporaries and later neo-Platonists will not be treated here.
34. For the attribution of this work to Iamblichus, see Hopfner (1922), pp. V–XII; Des Places (1989), pp. 7–8.
35. Crome (1970), pp. 44; 68; 70–77.
36. See Hirschle (1979), p. 48.
37. Damascius, *De principiis* 1.86; 1.154 refers to this work by Iamblichus. The Chaldean Oracles play an important role in neo-Platonic philosophy. They were composed at the end of the second century BCE by a certain Julian in hexameters. The theosophy professed in them united Platonic and Oriental elements in a demonology based on the dualism between the intelligible (the so-called "paternal intellect") and the material. The followers of this doctrine practiced different theurgical rites by using magical objects (e.g. the ἴυγξ) in order to join the divine.
38. Attis and Cybele must have been a common topic in neo-Platonic allegoresis. In Eusebius, *Praeparatio evangelica* 3.11.12-6, Porphyry interprets this myth allegorically, and Marinus reports (*Life of Proclus* 33) that Proclus wrote a book on the mother of the gods in which he explained all pertinent rites and Attis' myth allegorically.
39. Cf. Hermias, *In Platonis Phaedrum scholia* 30.10–31.2 and Bernard (1990), pp. 51–58.
40. See Sheppard (1980), pp. 39–103.
41. Quoted by Proclus, *In Rem publicam* 1.95.30–1.
42. Suda σ 1662.
43. Cf. Sheppard (1980), pp. 27–38. The essay is divided into two books which respectively treat Homer (1.69–124) and Plato (154–205).
44. Cf. also Gallavotti (1933), pp. 44–54; Friedl (1936), pp. 54–59; Koster (1970), pp. 111–114; Coulter (1976), pp. 107ff.; Kuisma (1996), pp. 122–134; Pichler (2006), pp. 73–79.
45. Beierwaltes (1985), p. 304 considers this definition as a little ambivalent since all kinds of poetry are didactic; hence, he proposes to call it "epistemic poetry" because it addresses the rational part of the soul.
46. Mimetic poetry is, in turn, divided into eicastic (imitation of the things as they are) and phantastic poetry (imitation of the things as they appear to the most).
47. See Bernard (1990), pp. 35–50; Sheppard (1980), pp. 162–202.
48. See *Apologia* 22b–c; cf. also *Phaedrus* 245a and *Ion* 533d; cf. Giuliano (2005), pp. 137–218.
49. Cf. Kuisma (1996), p. 143 and n. 26.
50. Cf. Bernard (1990), p. 43. Allegoresis' theoretical justification is in Proclus, *In Rem publicam* 1.71.21–86.23.
51. *In Rem publicam* 1.76.24–77.4, cf. Kuisma (1996), pp. 103–106; see also Trouillard (1981), p. 300.
52. Cf. also Friedl (1936), p. 65; Kuisma (1996), p. 70.
53. *In Rem publicam* 1.71.3ff. mentions a lecture by Syrianus about the κοινωνία δογμάτων between Plato and Homer.
54. Cf. Sheppard (1980), pp. 95–103; 168.
55. Συνουσία is the evening seminar in the Academy, see Schissel (1926), pp. 268–272.
56. I have translated Greek passages into English, unless otherwise stated.
57. Cf. Erler (1987), p. 192.
58. Cf. Hirschle (1979), pp. 3–31; Erler pp. 193–197.
59. See Dörrie (1975), pp. 276–281; Kuisma (1996), pp. 86–87; Festugière (1971), pp. 575–596; Rist (1964), pp. 220–225, Sheppard (1980) pp. 145–161.
60. Cf. Proclus, *In Hesiodum* CCLII.
61. See Trouillard (1981), pp. 297–299; for the meaning of symbol see Crome (1970), pp. 159–196.
62. *In Rem publicam* 1.71–96.

63. Cf. also Dodds (1951), pp. 291–299.
64. Porphyry, *Vita Plotini* 23.14–8.
65. For this work see Bidez (1928), pp. 137–151; translation in Brémond (1933), pp. 102–106 and (partially) in Festugière (1950), pp. 133–136; commentary in Corbin (1955), pp. 199–205 and 263–267.
66. Cf. Lewy (1978[3]), p. 184–185; 207; Sheppard (1980), pp. 74–78; Friedl (1936), pp. 61–62; 93; 101; Buffière (1973[2]), pp. 483–485; Kuisma (1996), pp. 92–93.
67. Cf. Lévêque (1959), pp. 13–34.
68. Cf. Friedl (1936), pp. 78–86; see also Coulter (1976), p. 51–52; Bernard (1990), pp. 165–178.
69. Cf. Lévêque (1959), pp. 61–62.
70. Proclus, *In Hesiodum* CIX.10–1.
71. Further examples from the commentary on Hesiod will be illustrated in the next paragraph.
72. *Platonic Theology* 1.4 I 22.11–23.11.
73. Cf. Kuisma (1996), pp. 7–8.
74. See Friedl (1936), pp. 86–88, Kuisma (1996), pp. 107–109, Whitman (1987), p. 96.
75. Cf. Sheppard (1980), p. 9.
76. According to Lamberton (1986), p. 232 "the focus of Proclus' attention lies far beyond the text, whether Homer or Plato is before him. He has convictions regarding the structure of the universe and its hierarchies of meticulously subdivided, mutually dependent entities that find illustration in virtually any authoritative text".
77. See West (1978), pp. 63–71.
78. Von Arnim (1903–1905), Frr. 100, 103, 104, 105, 167, 276 pp. 28–29; 43; 63. Proclus' commentary on Hesiod also offers the evidence of Zeno's exegetical activity on Hesiod: *In Hesiodum* CXXVI reports that the Stoic Zeno inverted the order of *Works and Days* 293–295.
79. For Plutarch's commentary see Marzillo (2010), p. XLIX, and Marzillo (forthcoming[3]).
80. For the different meanings of ὑπόμνημα see Marzillo (forthcoming[2]).
81. According to an alphabetic/numeric system. For a more detailed analysis of Proclus' authorship, see Marzillo (2010), pp. LXV–LXX.
82. For Homer's allegoresis by Proclus, I relegate to Friedl's, Wehrli's, Sheppard's, Kuisma's, Bernard's and Pichler's detailed monographs.
83. See *In Hesiodum* LI. Musäus (2004), p. 147 n. 64 blames Proclus to be inconsequent in his terminology.
84. Cf. Musäus (2004), pp. 147–149.
85. *In Hesiodum* L: σοί τ᾽ αὐτῷ μέγα πῆμα· τὴν τοῦ πυρὸς κλοπὴν μέγα πῆμα λέγει τῷ Προμηθεῖ ἔσεσθαι, λαβὼν τὸ αἴνιγμα ἀπὸ τῶν προνοουμένων ὑπ᾽ αὐτοῦ ψυχῶν. ταύταις γὰρ ὄντως ἡ κάθοδος πῆμα καὶ τοῖς ἀνθρώποις, οἳ γεγόνασιν ἐκ τούτων τῶν πεσουσῶν ψυχῶν.
86. Translation by Most (2006).
87. Proclus, *In Hesiodum* CCIX.8–9 explains why Hesiod calls the food ἁρμαλιὴ by resorting to etymological allegoresis: "because it is agreeable and suitable (εὐάρμοστον) for the ones who are nourished by it".
88. I will not compare Proclus' interpretation with Tzetzes', Moschopulus' and Protospatharius' exegesis since they depend on Proclus.
89. Proclus, *In Hesiodum* CCLX.1–36. Translation into English is adapted from Sandbach (1969) and my own German translation.
90. Proclus refers here to the Athenian calendar, cf. Marzillo (2010), pp. 361–262.
91. Cf. Plutarch, *de Iside et Osiride* 8, 353F; 63, 376E–F, *Quaestiones conviviales* 3.10, 658E–659C; 4.5, 670B, *Aetia phyica* 24, 917F–918A, fr. 102; Iamblichus, *de mysteriis* 5.8.
92. 'Perfect number' is actually in mathematics a number which is the sum of its proper positive divisors excluding itself. For example, 6 is the first perfect number because 1, 2, and 3 are its proper positive divisors, and 1 + 2 + 3 = 6 (other perfect numbers are 28, 496 and 8128). Here, on the contrary, what is meant by a 'perfect number' is one considered as perfect according to Pythagorean criteria.
93. Cf. for example Plutarch, *Aetia Romana* 2, 264A; 102, 288C–D, Iamblichus, *Theologoumena arithmeticae* 4.19, Proclus, *In Hesiodum* CCLXVIII, 1–4, Martianus Capella 2.105; 7.735; 736.
94. Proclus, *In Hesiodum* CLXIX.6–7.

95. Translations by Most (2006).
96. Siorvanes (1996), p. 107.
97. Beierwaltes (1965), p. 118.
98. Translation adapted from Most (2006) and Sandbach (1969) with my own German version.
99. Cf. *de Iside et Osiride* 8, 353F–354A, fr. 103, *Amatorius* 24, 770A, *de facie in orbe lunae* 30, 944E.
100. Cf. Lewy (19783), pp. 144; 300–301.
101. See Van der Toorn (1996), p. 213.

References

Adler, A., 1928–38, *Suidae Lexicon*, 5 vols. (Leipzig: Teubner).

Arnim, H. von, 1903–05, *Stoicorum veterum fragmenta*, 4 vols. (Leipzig: Teubner).

Babut, D., 1969, *Plutarque et le stoïcisme* (Paris: Presses Univ. de France).

Bates Hersman, A., 1906, *Studies in Greek Allegorical Interpretation* (Chicago: University of Chicago).

Beierwaltes, W., 1965, *Proklos. Grundzüge seiner Metaphysik*, Philosophische Abhandlungen 24 (Frankfurt am Main: Klostermann).

——, 1985, *Denken des Einen. Studien zur neuplatonischen Philosophie und ihrer Wirkungsgeschichte* (Frankfurt am Main: Klostermann).

Bernabé Pajares, A., 2004–07, *Poetae Epici Graeci. Testimonia et Fragmenta. Pars II: Orphicorum et Orphicis similium testimonia et fragmenta*, 3 vols. (Munich/Leipzig: Saur/Teubner).

Bernard, W., 1990, *Spätantike Dichtungstheorien. Untersuchungen zu Proklos, Herakleitos und Plutarch*, Beiträge zur Altertumskunde 3 (Stuttgart: Teubner).

Bidez, J., 1928, *Catalogue des manuscrits alchimiques grecs VI, en appendice Proclus, Sur l'art hiératique* (Brussels: Lamertin), 137–151.

Blönningen, C., 1992, *Der griechische Ursprung der jüdisch-hellenistischen Allegorese und ihre Rezeption in der alexandrinischen Patristik*, Europäische Hochschulschriften XV: Klassische Sprachen und Literaturen 59 (Frankfurt am Main et al.: Lang).

Brémond, A., 1933, "Notes et documents sur la religion néo-platonicienne", *Recherches de science religieuse* 23, 102–112.

Buffière, F., 1973², *Les mythes d'Homère et la pensée grecque* (Paris: Belles Lettres).

Burkert, W., 1968, "Orpheus und die Vorsokratiker. Bemerkungen zum Derveni-Papyrus und zur pythagoreischen Zahlenlehre", *Antike und Abendland* 14, 93–114.

Christiansen, I., 1969, *Die Technik der allegorischen Auslegungswissenschaft bei Philon von Alexandrien* (Tübingen: Mohr).

Corbin, H., 1955, "Sympathie et théopathie chez les « fidèles d'amour » en islam", *Eranos-Jahrbuch* 24, 199–301.

Coulter, J.A., 1976, *The literary Microcosm. Theories of Interpretation of the later Neoplatonists*, Columbia Studies in the Classical Tradition II (Leiden: Brill).

Couvreur, P., 1971, *Hermeias von Alexandrien in Platonis Phaedrum Scholia* (Hildesheim: Olms).

Crome, P., 1970, *Symbol und Unzulänglichkeit der Sprache. Jamblichos, Plotin, Porphyrios, Proklos*, Humanistische Bibliothek: Abhandlungen und Texte I, 5 (Munich: Fink).

Dawson, D., 1992, *Allegorical Readers and Cultural Revision in Ancient Alexandria* (Berkeley et al.: University of California Press).

Delatte, A., 1915, *Études sur la littérature pythagoricienne* (Paris: Champion).

De Lubac, H., 1959–1964, *Exégèse médiévale. Les quatre sens de l'écriture*, 3 vols. (Paris: Cerf).

Des Places, E., 1971, *Oracles chaldaïques* (Paris: Belles Lettres).

——, 1984, "Les Oracles chaldaïques", *Aufstieg und Niedergang der römischen Welt* II 17.4, 2299–335.

——, 1989², *Jamblique. Les mystères d'Égypte* (Paris: Belles Lettres).

Detienne, M., 1962, *Homère, Hésiode et Pythagore. Poésie et philosophie dans le pythagorisme ancien*, Collection Latomus LVII (Brussels-Berchem: Latomus).

Diels, H., Kranz, W., 1903, *Die Fragmente der Vorsokratiker* (Berlin: Weidemann).

Di Gregorio, L., 1975, *Scholia vetera in Hesiodi Theogoniam*, Pubblicazioni della Università Cattolica del Sacro Cuore. Scienze filologiche e letteratura 6 (Milan: Vita e Pensiero).

Dillon, J., 1976, "Image, Symbol and Analogy: Three Basic Concepts of Neoplatonical Allegorical Exegesis", in R. Baine Harris (ed.), *The Significance of Neoplatonism, Studies in Neoplatonism: Ancient and Modern* I (Norfolk: International Society for Neoplatonic Studies, Old Dominion Univ.), 247–262.

——, 1987, "Jamblichus of Chalcis (c. 240–325 A.D.)", *Aufstieg und Niedergang der römischen Welt* II 36.2, 862–909.

Dimitrijević, M. R., 1899, *Studia Hesiodea* (Leipzig: Teubner).

Dodds, E. R., 1951, *The Greeks and the Irrational* (Berkeley: University of California Press).

Dörrie, H., 1974, "Zur Methodik antiker Exegese", *Zeitschrift für die neutestamentliche Wissenschaft* 65, 121–138.

——, 1975, "Die Religiosität des Platonismus im 4. und 5. Jahrhundert nach Christus", in H. Dörrie (ed.), *De Jamblique à Proclus*, Entretiens sur l'antiquité classique Tome XXI (Geneva: Fondation Hardt), 257–281.

Erler, M., 1987, "Interpretieren als Gottesdienst. Proklos' Hymnen vor dem Hintergrund seines Kratylos-Kommentars", in G. Boss and G. Seel (eds.), *Proclus et son influence, Actes du colloque de Neuchâtel juin 1985* (Zurich: Ed. du Grand Midi), 179–217.

Esper, M. N., 1979, *Allegorie und Analogie bei Gregor von Nyssa* (Bonn: Habelt).

Esser, H. P., 1967, *Untersuchungen zu Gebet und Gottesverehrung der Neuplatoniker* (Cologne: University of Cologne).

Festugière, A. J., 1950², *La révélation d'Hermès trismégiste I: L'astrologie et les sciences occultes* (Paris: Gabalda).

——, 1971, *Études de philosophie grecque* (Paris: Vrin).

Friedl, A. J., 1936, *Die Homerinterpretation des Neuplatonikers Proklos* (Würzburg: Univ. of Würzburg).

Gallavotti, C., 1933, *L'estetica greca nell'ultimo suo cultore* (Turin: Reale Accademia delle Scienze).

Giuliano, F. M., 2005, *Platone e la poesia. Teoria della composizione e prassi della ricezione*, International Plato Studies 22 (Sankt Augustin: Academia-Verlag).

Gomperz, T., 1912, *Hellenika*. Eine Auswahl philologischer und philosophiegeschichtlicher kleiner Schriften I (Leipzig: Veit), 101–104.

Griffiths, J. G., 1967, "Allegory in Greece and Egypt", *Journal of Egyptian Archaeology* 53, 79–102.

Guthrie, K. S., 1986, *The Life of Proclus, or, Concerning Happiness by Marinus of Samaria* (Grand Rapids: Phanes Press).

Hani, J., 1976, *La religion égyptienne dans la pensée de Plutarque* (Paris: Les Belles Lettres).

Hannah, R., 2005, *Greek and Roman Calendars. Constructions of Time in the Classical World* (London: Duckworth).

——, 2009, *Time in Antiquity* (London/New York: Routledge).

Heinemann, I., 1949, "Die wissenschaftliche Allegoristik der Griechen", *Mnemosyne* 4.2, 5–18.

Hirschle, M., 1979, *Sprachphilosophie und Namenmagie im Neuplatonismus*, Beiträge zur klassischen Philologie 96 (Meisenheim am Glan: Hain).

Hopfner, T., 1922, *Jamblichus. Über die Geheimlehren* (Leipzig: Theosophisches Verlag-Haus).

Hunger, H., 1954, "Allegorische Mythendeutung in der Antike und bei Johannes Tzetzes", *Jahrbuch der österreichischen byzantinischen Gesellschaft* 3, 35–54.

Koster, S., 1970, *Antike Epostheorien* (Wiesbaden: Steiner).

Kuisma, O., 1996, *Proclus' Defence of Homer*, Commentationes Humanarum Litterarum 1996 (Helsinki: Soc. Scientiarum Fennica).

Laks, A., 1997, "Between Religion and Philosophy: The Function of Allegory in the Derveni Papyrus", *Phronesis* 42, 121–142.

Lamberton, R., 1986, *Homer the Theologian. Neoplatonist Allegorical Reading and the Growth of the Epic Tradition* (Berkeley et al.: University of California Press).

Lévêque, P., 1959, *Aurea catena Homeri. Une étude sur l'allégorie grecque* (Paris: Belles Lettres).

Lewy, H., 1978³, *Chaldaean Oracles and Theurgy. Mysticism, Magic and Platonism in the Later Roman Empire* (Paris: Études Augustiniennes).

Long, A. A., 1992, "Stoic Readings of Homer", in R. Lamberton and J. J. Keaney (eds.), *Homer's Ancient Readers. The Hermeneutics of Greek Epic's earliest Exegetes* (Princeton: Princeton Univ. Press), 41–66.

Marzillo, P., 2010, *Der Kommentar des Proklos zu Hesiods 'Werken und Tagen'*, Classica Monacensia 33 (Tübingen: Narr).

——, (forthcoming[1]), "An example of Neoplatonic allegoresis: Proclus on the Prometheus' myth in Hesiod", *Actas del V Coloquio Internacional Mito y Performance de Grecia a la Modernidad, La Plata* (Argentina), June, 16–19th 2009.

——, (forthcoming[2]), "Performing an academic talk. Proclus on Hesiod's 'Works and Days'", in E. Minchin (ed.), *Orality and Literacy in the Greco-Roman World. Composition and Performance* (Leiden: Brill).

——, (forthcoming[3]), "Plutarco esegeta. Il commento a Le Opere e i Giorni di Esiodo", in P. Volpe Cacciatore (ed.), *Gli scritti di Plutarco: tradizione, traduzione, ricezione, commento*. Atti del IX Convegno Internazionale dell' IPS (Naples: D'Auria).

Meritt, D. B., 1974, "The Count of Days at Athens", *American Journal of Philology* 95, 268–279.

——, (1977), "The Hollow Month at Athens", *Mnemosyne* 30,3, 217–242.

Merkelbach, R., West, M. L., 1967, *Fragmenta Hesiodea* (Oxford: Clarendon Press).

Most, G. W., 1989, "Cornutus and Stoic Allegoresis. A Preliminary Report", *Aufstieg und Niedergang der römischen Welt* II 36.3, 2014–2065.

——, 1993, "Die früheste erhaltene griechische Dichterallegorese", *Rheinisches Museum* 136, 209–212.

——, 1997, "The Fire Next Time: Cosmology, Allegoresis, and Salvation in the Derveni Papyrus", *Journal of Hellenic Studies* 117, 117–135.

——, 2006, *Hesiod. Theogony, Works and Days, Testimonia* (London/Cambridge Ma.: Loeb).

Müri, W., 1976, "Symbolon. Wort- und sachgeschichtliche Studien", in E. Vischer (ed.), *Griechische Studien*, Schweizerische Beiträge zur Altertumswissenschaft 14 (Basle: Reinhardt), 1–44.

Musäus, I., 2004, *Der Pandoramythos bei Hesiod und seine Rezeption bis Erasmus von Rotterdam*, Hypomnemata 151 (Göttingen: Vandenhoeck & Ruprecht).

Nestle, W., 1948, "Metrodors Mythendeutung", in W. Nestle (ed.), *Griechische Studien* (Stuttgart: Hannsmann), 164–172.

Nilsson, M. P., 1962, *Die Entstehung und religiöse Bedeutung des griechischen Kalenders*, second revised and supplemented edition, Scripta minora / Regia Societas Humaniorum Litterarum Lundensis 61,1 (Lund: Gleerup).

Pertusi, A., 1955, *Scholia vetera in Hesiodi opera et dies*, Pubblicazioni dell'Università del Sacro Cuore 13 (Milan: Vita e Pensiero).

Pépin, J., 1958, *Mythe et allégorie. Les origines grecques et les contestations judéo-chrétiennes* (Paris: Aubier).

——, 1965, "Porphyre, exégète d'Homère", in H. Dörrie (ed.), *Porphyre*, Entretiens sur l'antiquité classique Tome XII (Geneva: Fondation Hardt), 229–272.

——, 1987, "Terminologie exégétique dans les milieux du paganisme grec et du judaisme hellénistique", in: C. Curti (ed.), *La terminologia esegetica nell'antichità, Atti del Primo Seminario di antichità cristiane, Bari 25 ottobre 1984* (Bari: Edipuglia), 9–24.

Pfeiffer, R., 1970, *Geschichte der klassischen Philologie von den Anfängen bis zum Ende des Hellenismus* (Reinbek bei Hamburg: Rowohlt).

Pichler, R., 2006, *Allegorese und Ethik bei Proklos. Untersuchungen zum Kommentar zu Platons Politeia* (Berlin: Frank & Timme).

Pritchett, W. K., 1976, "The Athenian Count of Days", *California Studies in Classical Antiquity* 9, 181–195.

Ramelli, I., 2003, *Anneo Cornuto. Compendio di teologia greca* (Milano: Bompiani).

Reale, G., 1993[8], *Storia della filosofia antica. II. Platone e Aristotele* (Milan: Vita e Pensiero).

Rist, J. M., 1964, "Mysticism and Transcendence in Later Neoplatonism", *Hermes* 92, 213–225.

Runia, D. T., 1990, *Exegesis and Philosophy. Studies on Philo of Alexandria* (Aldershot: Variorum).

Saffrey, H. D., Westerink, L.G., 1968–97, *Proclus. Théologie platonicienne*, 6 vols. (Paris: Belles Lettres).

Sandbach, F. H., 1967, *Plutarchi Moralia VII* (Leipzig: Teubner).

——, 1969, *Plutarch's Moralia XV. Fragments* (London/Cambridge Ma.: Loeb).

Schibli, H., 1990, *Pherekydes of Syros* (Oxford: Oxford Clarendon Press).

Schissel, O., 1926, "Der Stundenplan des Neuplatonikers Proklos", *Byzantinische Zeitschrift* 26, 265–272.

Sheppard, A., 1980, *Studies on the 5th and the 6th Essays of Proclus' Commentary on the Republic*, Hypomnemata 61 (Göttingen: Vandenhoeck & Ruprecht).

——, 1991, "Phantasia and mental Images. Neoplatonist Interpretations of De anima, 3.3", in H. Blumenthal and H. Robinson (eds.), *Oxford Studies in Ancient Philosophy, supplementary volume: Aristotle and the later Tradition* (Oxford: Oxford Clarendon Press), 165–173.

Simonini, L., 1986, *Porfirio. L'antro delle ninfe* (Milan: Adelphi).

Siorvanes, L., 1996, *Proclus: Neoplatonic Philosophy and Science* (New Haven et al.: Yale University Press).

Tate, J., 1927, "The Beginnings of Greek Allegory", *Classical Review* 41, 214–215.

——, 1929/30, "Plato and Allegorical Interpretation I", *Classical Quarterly* 23, 142–154; II, *Classical Quarterly* 24 (1930), 1–10.

——, 1934, "On the History of Allegorism", *Classical Quarterly* 28, 105–114.

Thompson, C.L., 1974, *Stoic Allegory of Homer. A critical Analysis of Heraclitus' Homeric Allegories* (Ann Arbor: Univ. Yale microfilms).

Trouillard, J., 1981, "Le symbolisme chez Proclos", *Dialogues d'Historie Ancienne* 7, 297–308.

——, 1982, *La Mystagogie de Proclos* (Paris: Belles Lettres).

Van der Toorn, K., 1996, *Family Religion in Babylonia, Syria and Israel* (Leiden: Brill).

Wehrli, F., 1929, *Zur allegorischen Deutung Homers* (Borna-Leipzig: University of Basle, diss.).

Weinstock, S., 1927, "Die platonische Homerkritik und ihre Nachwirkung", *Philologus* 82, 121–153.

West, M. L., 1978, *Hesiod, Works and Days* (Oxford: Clarendon Press).

Whitman, J., 1987, *Allegory. The Dynamics of an Ancient and Medieval Technique* (Oxford: Oxford Clarendon Press).

Wolff, G., 1856, *Porphyrii de philosophia ex oraculis haurienda librorum reliquiae* (Berlin: Springer).

Why Greek Lunar Months Began a Day Later than Egyptian Lunar Months, Both Before First Visibility of the New Crescent

Leo Depuydt

1. Statement of Purpose: First Visibility of the New Crescent Never Served as Marker of the Beginning of the Lunar Month in Egypt, Greece, or Mesopotamia

1.1. A Prevalent Assumption

Hardly any assumption about the ancient world has been as dominant, or as stubborn, from time immemorial as the one that ancient months—in as far as they were lunar, which they mostly were—began with first visibility of the new crescent (short: first crescent visibility). Strictly speaking, beginning the month with first crescent visibility means—or at least ought to mean—that daylight of lunar Day 1 immediately follows the evening in which the new crescent is for the first time sighted. This strict principle is nowhere explicitly enunciated in ancient sources and hardly ever in modern publications. According to the strict understanding of the principle, it should not be possible to begin a lunar month at any time before the new crescent has been sighted.

In conjuring up the ancient world and picturing daily life in, say, Athens, are we to imagine, as is now universally done, that the minds of ancient Greeks like Pericles, Socrates, and Aristotle were anxiously anticipating the first sighting of the thin new crescent as marker of the beginning of a new month and also of the year's first month and therefore of new year?

Quite to the contrary, all kinds of evidence scattered over multivaried sources support the assumption that first visibility *nowhere* in the ancient world, not in Greece nor anywhere else, marked the beginning of the month. And that includes in all probability Babylonian astronomical texts, that bastion of chronological exactitude, in relation to which probably no one would now dare to deny that lunar months always began after the new crescent had been sighted.

Nothing has had a greater influence on the modern interpretation of ancient lunar months, I believe, than the general awareness of the custom undeniably prevalent in Islam since the time of the Prophet in the lands where Egyptian and Babylonian civilizations once flourished to begin months with the sighting of the new crescent. In other words, daylight of lunar Day 1 begins in the morning that follows the evening in which the new crescent is first sighted. There is nothing wrong with beginning months with first crescent visibility. But that is just not how the ancients did it.

I have elsewhere already undertaken provisional efforts to question the role generally attached to first visibility of the new crescent in the modern study of the ancient world[1] and have wanted to treat it at greater length for some time, expressing my hopeful anticipation on no fewer than four occasions.[2]

1.2. Egypt — Greece — Mesopotamia

The proposition that lunar months begin before first crescent visibility is somewhat established when it comes to ancient Egypt. Evidence is adduced below to confirm that Greek lunar months too began on average before first crescent visibility. The Greek evidence in fact comes from Egypt, which was home in the Late period to a large immigrant Greek-speaking population. In addition, it will be argued that Egyptian lunar months began a day or so before Greek lunar months.

As regards Mesopotamia, first crescent visibility rules absolutely. Kugler, the pioneering decipherer of much of Babylonian astronomy, summed it up as follows in his *Von Moses bis Paulus* after more than two decades of research on the subject: "Das erstmalige Wiedererscheinen der feinen Sichel am abendlichen Westhimmel ist das Zeichen für den Anfang des Monats".[3] Then again, in more recent years, it has been noted off and on that the picture that emerges from the sources is not nearly as neat as Kugler's terse statement suggests, for example in relation to the need for actually sighting the crescent to begin a new month.[4]

Babylonian astronomical texts are thought to provide the premier evidence of first crescent visibility's role. But it will be argued below that, in astronomical texts, Babylonian months might just as well begin with the earliest possible measurable interval between sunset and moonset after new moon, called NA in Babylonian. A NA-calendar, if the term is permitted, would be the result of adding strict regularity to a lunar calendar used in society at large. The aim would be to organize empirical data in a rigorous manner that is more suitable to the creation of Babylonian lunar theory. It is true that the interval NA, at least when it is measured and not estimated or predicted, is one facet of a set of events of which first crescent visibility is another facet. But it is not because "big" and "green" are two facets of the phenomenon "big green car" that the mind cannot focus on the facet "green" at the exclusion of the facet "big".

As regards texts other than Babylonian astronomical texts, much evidence will be adduced in support of a newly postulated type of ancient lunar calendar. The new calendar will be called here the *new crescent (in)visibility calendar*, which must have included a great measure of *ad hoc* determinations on whether to make a lunar month 29 days or 30 days long (see section 10.13). It cannot be said that the months of this calendar begin at first crescent visibility. They may or they may not. Instead, the contrast between visibility and invisibility of the new crescent on Day 29 is exploited to determine whether the day that follows Day 29 of the month will be the last of the present lunar month or the first of the next lunar month. The aim is not to begin the month with first crescent visibility or with any other lunar phenomenon for that matter. The aim is to determine how long the present lunar month will last, 29 days or 30 days. In works on calendars, the new crescent is normally associated with visibility. But in the new crescent (in)visibility calendar, both the visibility and the invisibility of the new crescent play equally important roles. Each has its own distinct consequence. Visibility on Day 29 produces a month of 29 days. Invisibility on Day 29 produces a month of 30 days. Importantly, invisibility on Day 29 by no means guarantees visibility on Day 30. If there is invisibility also on Day 30, then daylight of lunar

Day 1 begins before first crescent visibility. Accordingly, first crescent visibility is not the marker. The toggle of (in)visibility on Day 29 is. That being said, there are many indications in the sources that the new crescent was sighted too early or too late. Such irregularity must have had a domino effect affecting ensuing lunar months. Strict observation of the Day 29-rule must therefore often have been impossible.

Along the same lines, it will be argued below (see section 6.7) that, strictly speaking, the Egyptian and Greek lunar calendars are not old crescent invisibility calendars, but rather *old crescent (in)visibility calendars*.

A comprehensive survey of lunar time-reckoning in all nations of the ancient world remains desirable. But producing such a survey would be no small feat. Then again, it is advisable, and even indispensable, to securely establish individual points before moving on to a grand synthesis.

Much of the present paper revolves around two basic calendrical concepts: (1) the beginning of the lunar month; (2) first visibility of the new crescent as marker of the beginning of the lunar month. Before outlining this paper's line of argument in section 4, it will be useful to dwell briefly with these two concepts in sections 2 and 3.

2. The Beginning of the Lunar Month

Ancient lunar months began around new moon. New moon, also called conjunction, is the point in time when the moon is right between the earth and the sun. New moon cannot be observed because the moon is invisible at the time. Its light is drowned out by the sun that is right behind it. The exception is when sun, moon, and earth are on the exact same line. A solar eclipse then takes place. The moon is seen as a black disk passing from right to left across the sun disk, covering it wholly or partly.

It is clear what one sees of the moon shortly before and shortly after new moon. One morning roughly 25 to 50 hours before new moon, the old crescent appears for the last time in the eastern horizon right before sunrise. That is last visibility of the old crescent. Or short: last crescent visibility. Then, for a period lasting mostly either one and a half or two and a half days, the moon remains out of sight. Next, one evening roughly 25 to 50 hours after new moon, and weather permitting, the new crescent can be seen in the western horizon right after sunset. That is first visibility of the new crescent. Or short: first crescent visibility.

In order to fix the beginning of a lunar month, a certain daylight period needs to be picked and assigned the number one or some equivalent designation. But when it comes to establishing by which methods a certain daylight period was made into daylight of lunar Day 1 in various ancient nations, a problem arises: *nowhere in any ancient source is there an explicit report as to what exactly was done to determine which daylight is that of lunar Day 1.*

To my knowledge, no ancient source says more about events relating to the beginning of the lunar month than the Talmud. A number of scenarios of what happens around new moon are discussed in detail. Yet, these scenarios are not quite detailed enough to establish what exactly was done to fix the beginning of the month. One would like to picture—as if watching a video recording—a specific person making specific decisions and uttering specific statements. I am not referring to celebrations or sanctifications of the decision, to which the Talmud makes ample reference, but rather to the precise circumstances of the decision itself as a rational act. The sources do not bring us close enough to that critical event. What really happened therefore remains the subject of speculation.

From the prominent ancient civilizations of Egypt, Greece, and Mesopotamia, we do not even have anything close to detailed descriptions such as those in the Talmud. True, quite a few cuneiform texts refer in one way or another to events that take place around the turn of the month and somehow relate to fixing the beginning of the month. Moreover, there are various references to the beginning of the lunar month in classical authors and rare references in hieroglyphic sources. But all this still leaves us in the dark as to how exactly Day 1 was picked in Egypt, Greece, and Mesopotamia as an act of the free will.

It is generally assumed that observation either of the old crescent or of the new crescent played some role in determining the beginnings of lunar months in the ancient world. The old crescent is observed in the morning and the new crescent in the evening. In the case of the old crescent, the natural thing would seem to be to wait for the morning of first *in*visibility. The previous morning is then the one of last visibility. In principle, last visibility itself can only be established after the fact; one cannot know with certainty beforehand whether the old crescent will or will not be visible the next morning, even if visibility may be so limited one morning that invisibility the next morning seems all but certain. In the case of the new crescent, the natural thing would seem to be to wait for the evening of first visibility. The previous evening is then obviously the one of last *in*visibility. In principle, last *in*visibility itself can only be established after the fact; one cannot know with certainty beforehand whether the new crescent will or will not be visible the next evening.

Much attention has been paid to the problem of the visibility of the crescent. This problem is scientific and astronomical in the strict sense. It involves many factors. All agree that certain decisive factors are forever lost. These factors include local weather conditions and the location of the observers as well as their state of mind, including focus and attention span.

However, visibility is just one facet of fixing the beginning of the lunar month. What was no longer seen one morning or what was again seen one evening needed to be exploited in order to fix daylight of lunar Day 1. Such exploitation involved specific thoughts, words, and acts on the part of flesh-and-blood people. It is with those thoughts, words, and acts, regarding which no explicit evidence survives, that the present paper is concerned.

3. First Visibility of the New Crescent as Marker of the Beginning of the Lunar Month

Strictly speaking, a lunar month can be said to begin with first visibility of the new crescent only if the daylight period that follows the evening in which the new crescent is first seen is daylight of lunar Day 1. In other words, the morning of lunar Day 1 ought to immediately follow the evening of first visibility. Nights were probably not counted in antiquity, except very rarely in certain technical contexts. Nighttime was just a numberless stretch of darkness and of human inactivity separating numbered episodes of light and of human activity that begin with dawn and end with dusk. Activities started before dawn or continued after dusk were annexed to the adjacent daylight period.

This definition of first visibility of the new crescent as marker of the beginning of lunar months may seem slightly cumbrous. But my personal experience in reading works on calendars and chronology is that more explicitness in defining first principles would help promote the study of the subject as a mature discipline in its own right. One advantage of the above definition as I see it is that it avoids the vexed question of when the day began in various ancient nations, in the morning, in the evening, at midnight, and so on. The beginning of the day is called its epoch. Accordingly, a day whose beginning falls in the morning is said to have a morning epoch. This is not the place to address the issue of the epoch of the day at length. But briefly here (see also section 8), my view on the matter is that for most people most of the time, the day began in the morning and ended in the evening and included activities started before dawn and continued after dark. It seems like the most natural view. In earlier papers I have myself, like most every student of chronology before me, eagerly embraced the relevance of the concept of the epoch of the day. But I have now become rather convinced that the epoch may well be a ghost concept, and its study hence the pursuit of a ghost problem. Even the fact that, in certain Babylonian astronomical texts, what happens at night is described before what happens during the day does not necessarily imply an evening epoch (see section 10 below).

Both in ancient sources and in modern writings, one occasionally finds statements to the effect that the new crescent first appears on lunar Day 1. But such statements often critically leave unsaid whether the new crescent is first seen in the evening that immediately follows daylight of lunar Day 1 or in the evening that precedes daylight of lunar Day 1. One possibility that needs to be contemplated is that, in some ancient nation, one aimed more or less for the new crescent to appear just after the sunset ending daylight of lunar Day 1. That sunset is after all closer to daylight of lunar Day 1 than the sunset of the previous evening. However, the result of such a procedure is in effect that daylight of lunar Day 1, and hence the month itself, begins *before* first visibility. Again, strictly speaking, it would seem wise to delimit the concept of beginning lunar months with first visibility of the new crescent to the practice of having daylight of lunar Day 1 *follow* first visibility in the evening, as is the case in the religious Muslim calendar. However, if anyone insists on describing a calendar in which first crescent visibility falls in the evening following daylight of lunar Day 1 also as beginning lunar months with first crescent visibility, then it should at least be made explicitly clear in which evening in relation to daylight of lunar Day 1 the new crescent was first seen.

4. Outline of the Argument

The ultimate aim of this paper and its capstone is to make a contribution to the theme of the conference at which an extract of it was read. This capstone constitutes the paper's final section, section 11. The conference theme is evoked in the question: What does it mean to live the lunar calendar? This paper's line of argument will lead to a slightly different but related question: What does it mean to live three lunar calendars? Indeed, an attempt is made below to show that, during a few decades in the third century BCE, no fewer than three different lunar calendars were operative all at the same time in Egypt: (1) the native Egyptian lunar calendar; (2) the Greco-Macedonian lunar calendar; and (3) the Hebrew calendar. In 332 BCE, Alexander had conquered Egypt. In the last three centuries before the common era, rulers descending from Alexander's Macedonian general Ptolemy ruled Egypt as Pharaohs. Most rulers of this dynasty were called Ptolemy. This phase of Egyptian history is therefore known as the Ptolemaic period.

Naturally, it is not possible to seek an answer to the question as to what it meant to live with three lunar calendars if there were no three lunar calendars. Much of this paper is therefore devoted to the preliminary effort of proving that there were three distinct lunar calendars in the earlier third century BCE in Egypt. The existence of the Hebrew calendar hardly requires proof. The calendar must have been used by the Jewish diaspora in Egypt. But it has not been demonstrated that the native Egyptian lunar calendar and the Greco-Macedonian lunar calendar differed from one another. It goes without saying that the names of the months of the two calendars differed. But did the two calendars operate according to different calendrical mechanisms?

There are six main steps in this paper's line of argument leading up to the capstone section 11. They are as follows. The first main step is a review of pertinent past research pertaining to the Egyptian and Greek lunar calendars (see section 5). There is of course no lack of references in the literature to the effect that ancient lunar months began with first crescent visibility. Gathering such references would be fastidious. The search is instead for references that ancient lunar months did not so begin. It has been known for some decades now that native Egyptian lunar months simply begin too early for the new crescent to have become visible before daylight of lunar Day 1. First Ludwig Borchardt in anecdotal and incipient manner and then more systematically Richard A. Parker (earlier perhaps also Heinrich Brugsch, but only he) have done much to establish this fact. As there is nothing new to be said about the fact in question, the evidence will not be systematically revisited here; still, some of it will receive mention below. The Greek calendar is another matter. To my knowledge, only one scholar has seriously challenged the view that Greek lunar months began with first crescent visibility, namely W. Kendrick Pritchett. The aim of this paper's step one is to analyze Pritchett's evolving views regarding the matter. Pritchett refers to his ideas as a "theory". The design of the present paper is to go beyond theory and adduce proof.

The second main step of this paper is an investigation of the core empirical data (see section 6). The set consists of 32 Julian dates of lunar Day 1 derived from 32 double dates. The focus is on demonstrating that two key numerical relations apply. First, native Egyptian lunar months began on average roughly a day before Greco-Macedonian lunar months. Second, both began on average before first visibility of the new crescent. Daylight of Day 1 of Egyptian lunar months began on average roughly half a day before conjunction or new moon. Daylight of Day 1 of Greek lunar months began on average roughly half a day after conjunction or new moon. Half a day after conjunction is a point in time that is simply too soon for the new crescent to have become visible, let alone half a day before conjunction.

Owing to all the complexities characterizing the course of the moon, there was much fluctuation from month to month in the afore-mentioned distances. However, what matters here is the averages. There is also much inherent irregularity in how lunar calendars based on observation of the moon operate. This irregularity is bound to distort the averages somewhat. However, what matters here is that the chances that certain crucial characteristics of the empirical data adduced below are pure coincidence would appear to be statistically insignificant. Therefore, the possibility that the data do not support what needs to be demonstrated can be rejected with high probability. The irregularity also accounts for the fact that lunar months might on occasion begin after first visibility of the new crescent. However, what matters here is that, when they do, such a beginning was not intentional. It cannot be excluded that the ancients sometimes considered it significant that the new crescent often became visible for the first time in the evening following daylight of lunar Day 1. However, what matters here is that the new crescent played no role in fixing lunar Day 1.

The documentation of the two afore-mentioned numerical relations will need to be supplemented by an explanation of which calendrical mechanisms could have produced them. These calendrical mechanisms will be described in detail and historical evidence that points to their existence will be adduced.

In the third main step, additional evidence pertaining to the native Egyptian lunar calendar is provided (see section 7). Additional evidence pertaining to the Greek lunar calendar follows in step four (see section 8).

Steps two, three and four are devoted to a presentation of the empirical data and a description of the calendrical mechanisms that presumably produced them. The fifth main step presents a control test that forms a bridge between the data and the mechanisms (see section 9). A theoretical test is performed to establish whether, when the mechanisms are applied to any lunar months at any time, they produce modern data that are similar to the ancient data, thus corroborating the validity of the ancient data.

Steps two to five concern the lunar calendars of two of what are, judging by the scope of the surviving sources, the three most prominent ancient civilizations before Rome came onto the scene. The sixth main step concerns the third, ancient Mesopotamian civilization, and especially its cultural capital, Babylon (see section 10). In steps two to five, evidence is adduced to the effect that neither Egyptian nor Greek lunar months began after first crescent visibility. Step six extends this same tenet to the Babylonian lunar calendar. In trying to knock two hallowed tenets off their pedestals, namely that the day in Babylon began in the evening and that the month began with first crescent visibility, step six will naturally be perceived as more radical than the preceding steps.

5. Voice Crying in the Desert: W. Kendrick Pritchett

5.1. Pritchett in 1959

The new crescent has an enduring place in Classicists' intellectual, and perhaps also romantic, imagination of the past. This state of affairs has persisted to the present day. One reads in the most recent survey of what is known about Greek and Roman calendars that, in Athens, the first month of the year began "on the evening of the first sighting of the new moon's crescent following the summer solstice".[5] Yet, in 1959, Pritchett questioned this fundamental assumption when he wondered whether "the Athenians used some observable phase other than the thin crescent of the new moon to determine the length of their lunar months".[6] I am not aware of any other dissenting voices in the history of Classical scholarship besides Pritchett's. Nor have I searched very systematically for reactions to Pritchett's dissension. If one considers only original reasoned arguments that have appeared in print, Pritchett may well have stood pretty much alone in the last half a century with his view that the new crescent was not used to fix lunar Day 1 in Greek calendars.

Pritchett first took up the study of the Greek calendars during or soon after World War II by collaborating with Otto Neugebauer of Brown University on a joint publication entitled *The Calendars of Athens*.[7] In this work, it is shown that three calendars were used in ancient Athens: (1) a lunar calendar whose lunar Day 1 was determined by observation of the moon; (2) a calendar "according to archon" in which the lunar calendar was modified by adding and subtracting days, but whose year still began on the same new year as the lunar calendar; (3) a calendar that originally reflected the division of the year according to the prytanies of the boule.

The study of the Greek calendar and related topics such as the sequence of the archons is of such complexity that it could easily turn into a full-time occupation. As incomplete as my knowledge of the Athenian calendar may be at this time, I do not hesitate to embrace Alexander Jones's assessment that the central thesis of Pritchett's and Neugebauer's book "has on the whole held up well against a steady barrage of rival models".[8] In the half century that followed his joint publication with Neugebauer, Pritchett wrote regularly about the ancient Greek calendar.[9] His principal opponent in regard to all kinds of facets of ancient Greek calendars was Benjamin Meritt, a skilled student of epigraphy in his own right. I am personally quite comfortable with the notion that there was both a winner and a loser in this debate—and that Pritchett won.

In 1959, influenced by what Parker, also of Brown University, had recently written on Egyptian lunar months,[10] Pritchett suggested that lunar Day 1 of Athenian lunar months was fixed in time by means of observation of the *old* crescent in the morning at the beginning of the 29th daylight period of the month. There can be no doubt that Egyptian lunar months began before first visibility of the new crescent. Some of the empirical evidence to that effect is repeated below. It was therefore inferred that observation of the old crescent must somehow have played a role in fixing lunar Day 1, even if there is no explicit statement anywhere in the sources that it did. Yet, to the present day, the notion that ancient Egyptians were diligently waiting for the new crescent to begin a lunar month, just as all other inhabitants of the ancient world presumably were, has persisted in the popular imagination. There is no doubt that, to the religious Muslims in modern Egypt, the new crescent is everything. The arrival of Ramadan's new crescent is a much anticipated event. But it is time that the view became universally established that, to their fellow Egyptians in antiquity, the new crescent meant nothing at all as far as the strict calendrical structure of the lunar month was concerned.

Pritchett was led to propose that observation of the old crescent may also have played a role in the Athenian calendar because of the specific way in which days are named at the end of the lunar month (see below). Accordingly, he suggested that old crescent observation operated as follows at Athens. If the old crescent was no longer visible in the morning at the beginning of the 29th daylight period of the month, the 29th daylight period that immediately followed became the last daylight period of the month, the next daylight period became that of lunar Day 1, and the month had only 29 days. If the old crescent was still visible that same morning, the 29th daylight period became the penultimate daylight period of the month, the next daylight period became the 30th of the month, and the month had 30 days. That way it was known either one or two days in advance when daylight of lunar Day 1 would begin.

It is to be assumed that there were occasional irregularities in a manmade product such as an observational lunar calendar. These irregularities presumably required *ad hoc* adjustments. Therefore, when it comes to the empirical data presented below, it will be above all the averages that count to reveal the nature of the Greek lunar calendar.

Parker's Egyptian lunar calendar, which had earlier also been mostly Ludwig Borchardt's, differed from Pritchett's Athenian lunar calendar of 1959. In both calendars, lunar Day 1 was fixed by means of observation of the old crescent. But in Parker's Egyptian lunar calendar, old crescent observation took place in the morning at the beginning of *the daylight period that follows* the 29th daylight period of the month—not in the morning at the beginning of the 29th daylight period of the month as in Pritchett's Athenian lunar calendar of 1959. Simply put, old crescent observation took place after 28 days in Athens and after 29 days in Egypt.

In Parker's Egyptian lunar calendar, if the old crescent was no longer visible, the daylight period that follows the 29th daylight period of the month became daylight of lunar Day 1 and the month had 29 days. There was then no advance notice of the start of the new lunar month. If the old crescent was still visible, the daylight period that follows the 29th daylight period of the month became daylight of lunar Day 30, the next daylight period became lunar Day 1 of the following month, and the month had 30 days.

It inevitably follows from the mechanisms described above that the morning of first invisibility of the old crescent on average fell at the beginning of daylight of lunar Day 1 in the Egyptian lunar calendar but at the beginning of daylight of the last day of the lunar month, either the 29th or the 30th, in the Athenian lunar calendar. In other words, the Egyptian lunar calendar began on average a day earlier than the Athenian lunar calendar. At the same time, the same procedures stipulate that the morning of old crescent observation is the 28th daylight period of the month in the Athenian lunar calendar but the day after the 29th daylight period of the month in the Egyptian lunar calendar. In other words, the morning of old crescent observation is a day farther removed from lunar Day 1 in the Egyptian lunar calendar than it is from lunar Day 1 in the Athenian calendar.

On the one hand, the Athenian lunar month on average begins a day *later* than the Egyptian lunar month. On the other hand, the morning of old crescent observation in the next month is one day *less removed* from lunar Day 1 in the Athenian lunar calendar than it is in the Egyptian lunar calendar. Consequently, the morning of old crescent observation will on average fall in the same morning in both calendars. Old crescent observation will then again produce the one-day interval between the beginning of the Athenian lunar month and the beginning of the Egyptian lunar month. One lunar month later, old crescent visibility will again fall in the same morning in both calendars, and so on in perpetuity. Accordingly, if the two calendars are used in the same locale, they will be synchronized—with one running on average a day behind the other. Pritchett's statement that "the day of new crescent visibility . . . is the first day of the month, not the second as in Egypt"[11] clearly implies that Athenian lunar months began *a day later* than Egyptian lunar months. It is a relation that will be confirmed by actual empirical evidence adduced below.

The crucial question remains: What led Pritchett to postulate that the mechanism used to determine the beginnings of lunar months in the morning of old lunar observation was not the same in Athens as in Egypt? Pritchett took note of the fact that, in 29-day Athenian lunar months, the name of the day that was omitted was the name of Day 29 in a 30-day lunar month. In other words, the last day of the lunar month had the same name in both 29-day and 30-day lunar months. No one had ever considered the possibility that this peculiar property of Greek calendars might be relevant to how lunar Day 1 is fixed by means of observation of the new moon. As Pritchett phrases it, "[t]his terminology is used at the end of the [Athenian] months, and it has never received discussion in connection with an observation calendar".[12] From the "terminology" in question, Pritchett implies that "[b]y the end of the 28th day the official in charge of the calendar would have to decide whether the 29th day of the month was [the last-but-one day or the last day of the month]".[13]

Strictly speaking, by the internal logic of Pritchett's lunar calendar of 1959, one ought to read "by the middle of the 28th day" instead of "by the end of the 28th day" in the statement just cited. The day lasted from evening to evening in Pritchett's calendar. Old crescent observation naturally falls in the morning and therefore in the middle of a day. In Pritchett's calendar of 1959, the task at hand was to infer from old crescent observation which day was going to begin in the evening that followed, the last-but-one or the last one. But again, there is no need to bring the issue of when the day began into the discussion. One can strictly operate with successively numbered or named daylight periods that include any extensions to before dawn and to after sunset. As the time of almost all human activity, daylight is most of what matters. Pritchett somewhere calls daylight a "business" day.[14] In sum, Pritchett's above cited statement is here reinterpreted as stating that, in the morning of old crescent observation, it needed to be established which day the immediately following daylight period would be.

Pritchett does not fully explicate the logic that causes old crescent observation, along with the concomitant choice of name for the immediately following daylight period, to take place when it does. What is that logic? The first step is to realize that Athenian and other Greek lunar months can end in one of two ways, as follows: (1) (Day 29) name A, (Day 30) name B; (2) (Day 29) name B. The second step is the observation that there are two options in naming Day 29, name A and name B. Options inevitably imply a choice. And a choice cannot be made later than the time when the options take effect. Since the options undeniably take effect in the morning at the beginning of the 29th daylight period, that morning is the latest possible time for making the choice.

Pritchett's Athenian lunar calendar of 1959 might be called an old crescent calendar. But it would appear that the new crescent was also heeded in some sense. That is because Pritchett assumes that the decision on Day 28 was designed so that "the day of new crescent visibility . . . [would be] the first day of the month".[15] By new crescent visibility, Pritchett apparently means that first visibility of the new crescent immediately follows—rather than precedes, as one strictly speaking expects in the case of a first crescent visibility calendar—daylight of lunar Day 1. In that sense, the lunar calendar proposed by Pritchett in 1959 heeded *both* the old crescent and the new crescent.

Pritchett otherwise also assumed in 1959 that the day began in the evening in Athens. In other words, his 1959 calendar had an evening epoch. However, when the day begins plays no role in the line of argument presented in this paper.

In the end, Pritchett does not seem to have been fully convinced of what he wrote in 1959. In the entry "Calendars" in the second edition of the *Oxford Classical Dictionary*,[16] to which he contributed a description of Greek calendars probably submitted in 1964–1966,[17] it is laconically noted that lunar Day 1 was "determined presumably by observation of the first visibility of the new moon after conjunction".[18] This statement clearly contradicts what he wrote in 1959. Indeed, by 1970, Pritchett's partial about-face of 1964–1966 appears to have become a full one.

5.2. *Pritchett in 1970*

In his monograph entitled *The Choiseul Marble*, Pritchett makes a proposal[19] that is "an alternative explanation to that proposed [in Pritchett (1959)]".[20] Whereas he assumed in 1959 that both the old crescent and the new crescent played a role in fixing lunar Day 1, he returns in 1970 to the universally accepted view that, if the Athenian lunar calendar operated by observation of the moon, only the new crescent could have played a role. Whereas he assumed in 1959 that the day began in the evening, he now believes that it began in the morning.[21] But once again, when the day began will be considered irrelevant in what follows; in a sense, Pritchett's switch from assuming an evening beginning and ending to assuming a morning beginning and ending is a step closer to the definition of the natural day as beginning in the morning and ending in the evening in that at least one of the two limits overlaps.[22]

Pritchett's lunar calendar of 1970 was a first visibility calendar in the strictest sense, as defined above. That is, daylight of lunar Day 1 followed first visibility of the new crescent in the evening before. What one does in a first visibility calendar such as the religious Muslim calendar is to watch in the evenings close to new moon until one first spots the new crescent. When the crescent is spotted, the daylight beginning the next morning becomes daylight of lunar Day 1. In Pritchett's Athenian lunar calendar of 1970, one does something that has the exact same effect. The reason that the mechanism is different is that, as was noted above, there are two options for naming Day 29 in the Athenian lunar calendar. The naming of the 29th daylight therefore needs to involve a choice that is made before the 29th daylight itself. In Pritchett's Athenian lunar calendar of 1959, the choice was made at old crescent observation. Since the old crescent plays no role in Pritchett's calendar of 1970, the choice inevitably needs to be made at new crescent observation, which takes place in the evening. In Pritchett's calendar of 1959, the morning of old crescent observation *begins* daylight of Day 29. In his calendar of 1970, the evening of new crescent observation *ends* daylight of Day 29. In old crescent observation, visibility precedes invisibility. In new crescent observation, the opposite is the case. Therefore, in old crescent observation, visibility *delays* the end of the lunar month by a day. In new crescent visibility, visibility *expedites* the end of lunar month by a day. If the new crescent is visible, the immediately preceding daylight period is given the name that the last day of the month has in both 29-day and 30-day Athenian lunar months and the next daylight is that of Day 1. If the new crescent is invisible, the immediately preceding daylight period is given the name that Day 29 has in 30-day lunar months and one waits a day hoping that the new crescent will be visible the next evening. But anyway, what is seen the next morning is not all that relevant anymore because lunar months are never longer than 30 days.

It follows that the interval between the beginning of the Egyptian lunar month and the Athenian lunar month is no longer one day, as it is in case of the 1959 calendar, but rather two to three days.

Pritchett was aware that one disadvantage of his calendar of 1970 is that daylight of Day 29 was not named until the evening that immediately follows it.[23] How could Athenians spend a day without knowing which day it was?

An even more problematic characteristic of his calendar of 1970 is that its months begin one to two days after those of the 1959 calendar. Such a late beginning is simply contradicted by empirical evidence presented below and all else that is known about the Athenian and other Greek lunar calendars, much of it summarized later by Pritchett himself.[24]

Finally, one of the names of the last day of the month in both 29-day and 30-day calendars is ἕνη καὶ νέα "old and new". In his *Life* of Solon, at 25, Plutarch writes that the "old" portion of that day belongs to the expiring month and the "new" portion to the beginning month. But what divides the two portions? In 1970, Pritchett assumed it was "the visibility of the new crescent by the evening light".[25] Later, in 1982, he took it to be the "synodic conjunction", that is, the moment when the moon is right in front of the sun and therefore invisible except at solar eclipses; conjunction precedes first visibility by very roughly a day on average. The two assumptions of 1970 and 1982 directly contradict one another. But in 1982, Pritchett adduced abundant evidence that the ancient Greeks themselves assumed that conjunction separated the two parts of ἕνη καὶ νέα.

5.3. Pritchett in 1982

In 1970, Pritchett had radically reversed his position of 1959 regarding the beginning of the Athenian lunar month, concluding that "the picture was quite clear".[26] But in 1982, he made a second radical reversal by returning more or less to his position of 1959, except for the fact that he now assumed that the day began in the morning, as he already had in 1970, and not in the evening.[27] These two radical reversals make it easy to lose confidence in Pritchett's discussions concerning how ancient Athenians and Greeks in general determined the beginning of the lunar month.[28]

Then again, the treatment of 1982 is by far the longest of the three and presents significant new evidence. Pritchett adduces an impressive lineup of ancient authors who all state as a matter of course that the Greek lunar months lasted from conjunction to conjunction and that the last day of the month is the day of conjunction. There is not a peep in any of these sources about the first visibility of the new crescent. Importantly, Pritchett adduces certain considerations that, in lunar months whose beginning is determined by old crescent observation in the morning that begins the 29th daylight period of the month, conjunction will indeed on average take place on the last day of the lunar month. Empirical evidence of this fact will be adduced below.

The amount of convincing detail in Pritchett (1982) and the coherent picture it offers do much to favour the view that Pritchett (1970) was a transitory aberration. In my mind, Pritchett emerges as someone whose expertise with calendars, which was inspired by Otto Neugebauer and Richard A. Parker of Brown University, led him to look in one direction when everyone else was and had always been looking in the opposite direction. Rowing upstream into a heavy countercurrent, he apparently needed some time to get to what he was looking for. The fact that it did take some time and included an episode of self doubt is best viewed as evidence of how difficult it is, and still may prove to be, to dethrone the new crescent definitively from its lofty perch atop the modern imagination of the ancient world.

Pritchett was born in 1909, retired from the University of California at Berkeley in 1976, and died at age 98 in 2007. In 1982, when he finally arrived at where he wanted to be in regard to the beginning of the Athenian lunar month, he was well into retirement and into his seventies. However, an Internet obituary reporting that he worked 10 hours a day seven days a week for 30 years after retirement suggests that he was still very much in his prime in 1982. In fact, in 2001, when he was in his early nineties, he published a monograph on the subject of Athenian calendars that "purports to be an evaluation of the criticisms which were brought against *Calendars of Athens* published with the collaboration of Otto Neugebauer in 1947". After 54 years, Pritchett's lifelong study of Greek calendars had come full circle.

In the end, Pritchett deserves great merit for relating the specific way lunar days are named at the end of the Athenian lunar month to the way in which the beginning of the month was determined. It was a seminal insight. In light of the fact that the present paper revolves around how the Greek lunar calendar and the Egyptian lunar calendar relate to one another, it is interesting to note that Pritchett came to his understanding of the Athenian calendar with the help of what he had learned first about the Egyptian lunar calendar.

When this investigation was well underway and some notion had been formed about the difference between the Egyptian lunar calendar and the Greek lunar calendar (especially in light of the fact that both the Greek name of the *last* day of the lunar month, ἕνη καὶ νέα "old and new", and the Egyptian name of the *first* day of the lunar month, *psḏntyw*,[29] apparently refer to day of conjunction), I came upon the following statement by Pritchett: "If we advance the [Julian] dates by one [day], the Greek [lunar] calendar becomes a replica of the Egyptian lunar calendar".[30] This is more or less the same as stating that Greek lunar months begin a day later than Egyptian lunar months. It was encouraging to find oneself in good company. Pritchett offers no hard empirical data in support of his statement, even if he gathers many useful supportive indications from various sources. Accordingly, he describes the statement cited above as a "theory".[31] The aim of what follows is to present empirical data in support of this theory.

6. Core Empirical Data: Modern Dates of Lunar Day 1 Derived from Double Dates

6.1. The Irrelevance of Most Ancient Lunar Dates to the Present Hypothesis

Two sets of data are being sought. The first set ought to allow one to establish when Egyptian lunar months begin. The second set ought to allow one to establish when Greek lunar months begin. Once the two sets of data have been secured, the Egyptian beginning and the Greek beginning can be compared with one another.

There is no lack of Egyptian and Greek lunar month and day dates in ancient sources. In proportion to the totality of the sources that have survived, there are bound to be fewer Egyptian than Greek lunar dates because the Egyptian lunar calendar was restricted mainly to the religious sphere. In Egypt, events outside the confines of life in the temple were as a rule dated according to the so-called civil calendar, which is not lunar but consists of 12 months of 30 days with five added days for a total of 365.

Many Egyptian and Greek lunar dates have survived from many centuries of history. But of hardly any of these dates do we know the exact equivalent date in the modern common calendar now used worldwide, the so-called Julian or Julian-Gregorian calendar. For the sake of simplicity, I will call the Julian date the modern date. Needless to say, there were no Julian dates before 45 BCE, when that calendar was instituted. The modern calendar is simply extended back into the past to before 45 BCE in order to conveniently date events according to a single standard.

It is often possible to determine the modern equivalent of an ancient lunar date to within a margin of two to three days. But that is not good enough for the purpose of establishing when exactly lunar months begin. Exact dates are needed. As it happens, on rare occasions, the exact modern equivalent of an ancient lunar date can be established indirectly. Two cases are as follows. First, a lunar date may date an astronomical event whose modern date can be obtained independently from computations. That modern date naturally also coincides with the lunar date. Second, a lunar date may be equated in a document with a date by another ancient calendar whose modern date is known. That modern equivalent also coincides with the lunar date. Two dates equated with one another are called a double date.

6.2. *Double Dates*

The focus of what follows will be on a set of double dates, some including an Egyptian lunar date and others a Greek lunar date. The lunar dates are all from Egypt, including all the Greek dates. After Alexander's conquest of Egypt in 332 BCE, Greek had become an Egyptian language and the Greco-Macedonian calendar, including the Macedonian month names, were imported into Egypt. For the purpose of comparison, it is a fortunate circumstance that both sets were found at about the same longitude. Lunar months begin at different times at different longitudes.

In all double dates, the other date is an Egyptian civil date about whose modern equivalent there is no doubt. This set of double dates will be deemed sufficient to confirm the theory in whose support they are adduced. Still, some related dates and other additional evidence will be adduced as well below (see sections 7 and 8).

6.3. *How to Derive a Modern Date of Lunar Day 1 from Double Dates*

If the modern date of an ancient lunar date is known, then the modern date of daylight of Day 1 of the ancient lunar month in question is known as well. For example, in one Egyptian lunar date that is part of a double date (see below), an Egyptian lunar Day 15 is equated in a document with an Egyptian civil date. Daylight of the civil date is known to fall on 19 Oct 559 BCE. Daylight of lunar Day 15 is therefore also daylight of 19 Oct 559 BCE. If daylight of lunar Day 15 is daylight of 19 October, then daylight of lunar Day 1 falls 14 days earlier and coincides with daylight of 5 October.

Once the modern date of lunar Day 1 is determined, it can be established where lunar Day 1 falls in relation to the astronomical lunar cycle that passes from new moon to waxing moon to full moon to waning moon and back to new moon. The relation between the calendrical lunar month and the astronomical lunar cycle can be expressed by measuring the interval between a fixed point in time in the calendrical lunar months and a fixed point in time in the astronomical cycle. When several such intervals have been measured, an average interval can be calculated. When an average interval has been obtained both for a set of Egyptian lunar dates and a set of Greek lunar dates, the two averages can be compared.

The fixed point in time in the calendrical lunar month chosen here is 6:00 a.m., the average time of sunrise, in the morning that begins daylight of lunar Day 1. Another fixed point in the calendrical day could equally well have been chosen, for example, noon of lunar Day 1. The real time of sunrise has not been chosen because that would involve the prior assumption that the changing time of sunrise somehow played a role in the relation between the astronomical lunar cycle and the calendrical lunar month, whatever that role may be. Logically speaking, one cannot make that assumption *before* making the comparison. That would be an instance of circular reasoning.

The fixed point in time in the astronomical lunar cycle chosen here is new moon or conjunction. Various programs available on the Internet provide the times of conjunction to within minutes. But as it happens, Goldstine's times published in 1973 will still do for the purpose of historical investigations and they will be used here.[32] One could conceivably choose another point in the astronomical lunar cycle for the purpose of the initial comparison, for example the last visibility of the old crescent or the first visibility of the new crescent. The old crescent is last seen in the morning about one to two days before conjunction. Visibility needs to be distinguished from sighting. It is not because the moon could be seen that it actually was. However, modern computations of crescent visibility are not as accurate as those of conjunction. It is not always possible to establish with very high probability in which morning the old crescent could last be seen or in which evening the new crescent could first be seen. Conjunction is therefore preferred here to last crescent visibility as a fixed point of reference. Still, after the initial comparison, the relation of the empirical data to visibility or invisibility of the old or new crescent will be examined as well.

It is distinctly possible that principles used to determine the beginnings of lunar months were not always strictly applied. Lax application of the rules would produce irregularities in the data. But what matters, in the first instance, is the average interval.

In the example mentioned above, daylight of 5 October of 559 BCE is daylight of lunar Day 1. New moon or conjunction occurred at 2:11 a.m. on 6 October 559 BCE. This is Goldstine's time for Babylon,[33] reduced by 47 minutes to reflect the location of Assuan, whose longitude is about that of Edfu where the date was found. The distance between 6:00 a.m. on 5 October and 2:11 a.m. on 6 October is 20 hours and 11 minutes. The design of what follows is to obtain two sets of such intervals, one set pertaining to Egyptian lunar dates and the other to Greek lunar dates, and to produce two averages.

Once the modern date of Day 1 of an ancient lunar month is obtained, its relation to the morning of last crescent visibility and the evening of first crescent visibility can be established. In the example cited above, daylight of 5 October of 559 BCE is daylight of lunar Day 1. The morning of October 4 was most probably the last morning in which the old crescent could still be seen.[34] Consequently, daylight of lunar Day 1 begins in the morning in which the old crescent is for the first time invisible.

6.4. The Known Modern Dates of Lunar Day 1 Derived from Double Dates (28)

One reads in Genesis 18 that the Lord promised to spare Sodom if ten righteous men could be found. But no ten were found. In the case of instances of modern dates of lunar Day 1 derived from double dates consisting of an Egyptian or Greek lunar date and a civil date, I am relieved to note that at least ten do survive. But not a whole lot more. Exactly 28 are known to me. This is not much for two to three thousand years of the combined histories of Egypt and Greece. Perhaps other double dates will surface at some point.[35] The double dates from which the 28 instances of lunar Day 1 have been derived have been well-known for some time and there is on the whole not much controversy regarding their interpretation. What minor problems the dates pose will be discussed below. I believe these 28 to be sufficient in number to yield results that are statistically significant.

As was noted before, in double dates, two dates are equated with one another. One of the two dates is a civil date, whose modern equivalent is known with certainty in the period of Egyptian history to which the double dates belong. The other date is lunar. Naturally, the double dates containing an Egyptian lunar date are written in Egyptian and the double dates containing a Greek date are written in Greek.

In eight of the 28 modern dates of lunar Day 1 derived from double dates, the lunar date is Egyptian. In 20 instances, the lunar date is Greek. Further below, the eight Egyptian dates will be supplemented by 12 instances of a modern date of lunar Day 1 as derived from temple service dates.

	Year	I Civil or Alexandrian Date	II Lunar Date	III Modern Date of Date in Col. I (Daylight)	IV Modern Date of Lunar Day 1 (Daylight)	V Closest Conjunction	VI Distance in Hours from Conjunction to 6:00 a.m. in Morning of Lunar Day 1
1	12 of Amasis	II *šmw* 13	I *šmw* 15	19 Oct 559 BCE	5 Oct	6 Oct, 2:11 a.m.	– 20h 11m
2	28 of Ptolemy VIII	IV *šmw* 18	III *šmw* 23	10 Sep 142 BCE	19 Aug	17 Aug, 11:29 p.m.	+ 30h 31m
3	30 of Ptolemy VIII	II *šmw* 9	II *šmw* 6	2 Jul 140 BCE	27 Jun	27 Jun, 9:10 p.m.	– 15h 10m
4	10 of Ptolemy III	III *šmw* 7	Day 6	23 Aug 237 BCE	18 Aug	18 Aug, 1:27 a.m.	+ 4h 33m
5	10 of Ptolemy IV	III *šmw* 7	Day 6	17 Aug 212 BCE	12 Aug	11 Aug, 5:01 p.m.	+ 12h 59m
6	6 of Cleopatra VII	III *šmw* 13	Day 5	13 Jul 46 BCE	9 Jul	8 Jul, 5:23 a.m.	– 12h 37m
7	1 of Augustus	IV *prt* 21	Day 16	17 Apr 29 BCE	2 Apr	2 Apr, 2:54 p.m.	– 8h 54m
8	21 of Augustus	III *šmw* 10	Day 16	4 Jul 9 BCE	19 Jun	19 Jun, 6:03 a.m.	– 0h 03m

TABLE 1. Double Dates Consisting of an Egyptian Lunar Date and an Egyptian Civil Date. For details and bibliographical references pertaining to the sources from which the double dates have been derived and the rationale for ordering them in the way that they are, see Depuydt (1997a), pp. 161–175. Dates 1, 2, and 3 in column II have a month date and a day date; dates 4 to 8, a day date only. Dates 1 to 7 in column II are civil dates; date 8, an Alexandrian date. The times for conjunction are Goldstine's (1973) for Assuan, except 6, which is for Memphis.

The 28 lunar dates are all from Egypt. That is to be expected in the case of the Egyptian lunar dates. But the Greek lunar dates are also from Egypt. They belong to the Macedonian calendar used by the Greek-speaking population of Egypt in Ptolemaic times. The Greek lunar dates all belong to the third century BCE. In the later third century BCE, the Macedonian calendar lost its lunar character and was eventually entirely assimilated to the civil calendar.

6.5. Two Average Time Intervals between Conjunction and 6:00 a.m. of Lunar Day 1 Yielded by Double Dates, One Egyptian, the Other Greek

The double dates yield 28 specific time intervals that separate conjunction from 6:00 a.m. in the morning that begins daylight of lunar Day 1. From these 28 specific time intervals, two average time intervals can be derived: one from the eight specific time intervals pertaining to the Egyptian lunar calendar; the other from the 20 specific time intervals pertaining to the Greek lunar calendar.

Details pertaining to the double dates that include an Egyptian lunar date are listed in table 1, along with what can be derived from them. The specific intervals that separate conjunction from 6:00 a.m. in the morning of lunar Day 1 are listed in column VI. These specific intervals yield an average interval of *minus 1 hour 06 minutes 30 seconds*. In other words, 6:00 a.m. in the morning that begins daylight of lunar Day 1 falls on average a little over one hour before conjunction. This means that the previous evening fell on average about half a day before conjunction. There is therefore no way that the new crescent could have been seen on average. Further below, the eight specific intervals will be supplemented by 12 additional specific intervals derived from temple service dates in order to obtain an average interval derived from, not just eight specific intervals, but 20 (see section 7). The average derived from 20 intervals will corroborate the average derived from just eight.

The 20 specific intervals pertaining to the Greek lunar calendar are derived from 32 double dates. That means that some intervals are derived from more than one double date. The 32 double dates are listed by Grzybek.[36] Grzybek's list has been adapted for table 2, which lists the details regarding the double dates and what can be derived from them.

Grzybek was aware of more double dates than those listed in table 2. But he assumed that, in the dates he does not list, the lunar date may well have been artificially derived from the Egyptian civil date by being given the same day number or a day number that is 7 or

Table 2 (following page). Double Dates Consisting of a Greek Lunar Date and an Egyptian Civil Date. For details and bibliographical references pertaining to the sources from which the double dates have been derived, see Grzybek (1990), pp. 135–137. Columns I and II follow Grzybek. The dates in IV may differ from Grzybek's, primarily because he assumes that certain letters were written between dawn and sunrise and therefore had a day date that is one lower than letters written after sunrise. I consider a change in day date at sunrise highly improbable (see Depuydt (2002), p. 474, referring to the insightful but largely forgotten study by Bilfinger (1888)). The regnal years of Ptolemy II Philadelphos are Macedonian regnal years. Nos. 14 to 18 belong in time after number 24. This is because Grzybek's numbering of the double dates runs from Dios to Dios, Dios being Month 1 of the Macedonian year, whereas the regnal year in his view began on 27 Dystros (Grzybek (1990), pp. 157–169). The order of the Macedonian months is as follows: Dios, Apellaios, Audnaios, Peritios, Dystros, Xandikos, Artemisios, Daisios, Panemos, Loios, Gorpiaios, and Hyperberetaios. In nos. 2, 27, and 31, the Macedonian dates stand between parentheses, not because they are absent from the text, but because they may be artificial. In each of the three double dates, the Macedonian day date is seven less than the Egyptian day date. It has been argued that seven may sometimes have been mechanically subtracted without heeding the lunar cycle (cf. Grzybek (1990), pp. 151–154). Grzybek likewise assumes the existence of mechanical subtraction of 10 (Grzybek (1990), 152). The conjunctions in column V are Goldstine's (1973) for Babylon adjusted to Alexandria by subtracting 58 minutes.

Notes to the table:

* The year is 252 BCE

** "36" is the year date in the edition of the papyrus, P. Hib I 77. Grzybek ((1990), p. 154 note 93) emends to "30", following F. Uebel. Uebel's emendation is accepted here. As Grzybek suggests, a dark spot or speck erroneously read as "6" may have followed "30" and an inspection of the original manuscript could clear up the matter. He also assumes (Grzybek (1990), p. 154, note 93) that the document was written between dawn and sunrise and that the day date changed from 23 to 24 at sunrise. Yet, later (p. 155), he refers to the date as "an error or false reading of the scribe". In any event, I accept "23" at face value and consider a change of day date at sunrise highly unlikely (see subscript above).

		I	II	III	IV	V	VI
	Year of Ptolemy II Philadelphos	Civil Date	Lunar Date	Modern Date of Date in Col. I (Daylight)	Modern Date of Lunar Day I (Daylight)	Closest Conjunction	Distance in hours from Conjunction to 6:00 a.m. in Morning of Lunar Day 1
1.	22	Epeiph 12	Loios 19	4 Sep 264 BCE	(1) 17 Aug	16 Aug, 6:21 p.m.	+ 11h 39m
2.	29	Pharmouthi 30	(Artemisios 23)	22 Jun 257 BCE	(2) 31 May	1 Jun, 10:57 a.m.	– 28h 57m
3.	29	Thoth 9	Hyperberetaios 8	3 Nov 257 BCE	(3) 27 Oct	26 Oct 9:27 a.m.	+ 20h 33m
4.	29	Thoth 13	Hyberberetaios 12	7 Nov 257 BCE	same as 3	same as 3	same as 3
5.	29	Thoth 21	Hyperberetaios 20	15 Nov 257 BCE	same as 3	same as 3	same as 3
6.	29	Thoth 24	Hyperberetaios 23	18 Nov 257 BCE	same as 3	same as 3	same as 3
7.	29	Choiak 4	Audnaios 4	27 Jan 256 BCE	(4) 24 Jan	23 Jan, 1:57 p.m.	+ 16h 03m
8.	29	Choiak 24	Audnaios 24	16 Feb 256 BCE	same as 7	same as 7	same as 7
9.	29	Tybi 10	Peritios 10	4 Mar 256 BCE	(5) 23 Feb	22 Feb 8:11 a.m.	+ 21h 49m
10.	29	Tybi 11	Peritios 11	5 Mar 256 BCE	same as 9	same as 9	same as 9
11.	30	Pachons 9	Artemisios 10	1 Jul 256 BCE	(6) 22 Jun	20 Jun 9:55 a.m.	+ 44h 05m
12.	30	Mesore 13	Loios 16	3 Oct 256 BCE	(7) 18 Sep	16 Sep 10:44 a.m.	+ 43h 16m
13.	30	Choiak 10	Apellaios 21	2 Feb 255 BCE	(8) 13 Jan	12 Jan 2:49 p.m.	+ 15h 11m
14.	31	Phamenoth 27	Dystros 20	20 May 254 BCE	(9) 1 May	30 Apr 9:49 a.m.	+ 20h 11m
15.	31	Phamenoth 29	Dystros 22	22 May 254 BCE	same as 14	same as 14	same as 14
16.	31	Phamenoth 30	Dystros 23	23 May 254 BCE	same as 14	same as 14	same as 14
17.	31	Phamenoth 30	Dystros 23	23 May 254 BCE	same as 14	same as 14	same as 14
18.	31	Phamenoth 30	Dystros 23	23 May 254 BCE	same as 14	same as 14	same as 14
19.	31	Pharmouthi 4	Xandikos 15	27 May 255 BCE	(10) 13 May	11 May 8:28 a.m.	+ 45h 32m
20.	31	Pachons 18	Daisios 2	10 Jul 255 BCE	(11) 9 Jul	9 Jul 7:39 a.m.	– 1h 39m
21.	31	Pachons 30	Daisios 14	22 Jul 255 BCE	same as 20	same as 20	same as 20
22.	31	Payni 2	Daisios 16	24 Jul 255 BCE	same as 20	same as 20	same as 20
23.	31	Payni 11	Daisios 25	2 Aug 255 BCE	same as 20	same as 20	same as 20
24.	31	Phamenoth 6	Peritios embol. 28	29 Apr 254 BCE	(12) 2 Apr	31 Mar 6:25 p.m.	+ 35h 25m
25.	32	Mesore 1	Panemos 26	21 Sept 254 BCE	(13) 27 Aug	26 Aug 4:31 p.m.	+ 13h 29m
26.	33	Payni 14	Daisios 20	4 Aug 253 BCE	(14)16 Jul	16 Jul 8:50 a.m.	– 2h 50m
27.	34	Hathyr 29	(Dios 22)	21 Jan 251 BCE	(15) 31 Dec*	30 Dec 5:49 a.m.	+ 24h 11m
28.	34	Phamenoth 3	Peritios 28	25 Apr 251 BCE	(16) 29 Mar	28 Mar 10:08 a.m.	+ 19h 52m
29.	35	Epeiph 30	Panemos 28	19 Sep 251 BCE	(17) 23 Aug	22 Aug 5:07 p.m.	+ 12h 53m
30.	36**	Pachons 22	Artemisios 23	14 Jul 256 BCE	(18) 22 Jun	20 Jun 9:55 a.m.	+ 44h 05m
31.	37	Phaophi 16	(Hyperberetaios 9)	8 Dec 249 BCE	(19) 30 Nov	26 Nov 6:44 p.m.	+ 83h 16m
32.	37	Choiak 21	Apellaios 17	11 Feb 248 BCE	(20) 26 Jan	25 Jan 9:07 a.m.	+ 20h 53m

10 removed from the civil day number.[37] The artificial character of the lunar date of such double dates can sometimes apparently be confirmed by the fact that the lunar date does not accord with the lunar cycle. Whatever may be of Grzybek's proposal, the dates he does not list have not been considered here.

The specific intervals that separate conjunction from 6:00 a.m. in the morning of lunar Day 1 are listed in column VI of table 2. They yield an average interval of plus 22 hours 56 minutes 51 seconds. In other words, 6:00 a.m. in the morning of lunar Day 1 falls on average a little under 23 hours after conjunction. Accordingly, the previous evening falls on average roughly about ten hours after conjunction. That is on average too soon after conjunction for sighting of the new crescent to have become possible. The age of the moon, that is, the time from conjunction, needs to be on average very roughly 24 hours.

6.6. A Comparison of the Egyptian and Greek Average Time Intervals

A simple comparison of the two averages reveals that the Egyptian lunar months in question on average begin about *24 hours 4 minutes 21 seconds* (1h 06m 30s before conjunction + about 22h 56m 51s after conjunction) before the beginning of Greek lunar months. The number of the lunar dates is not very high. But the difference between the set of Egyptian dates and the set of Greek dates is unmistakable, it would appear.

In table 3 below, spatial distribution is used to evoke the difference between the Egyptian set and the Greek set. Each instance of the letter E represents an Egyptian lunar date; each instance of G, a Greek lunar date. The position of the letter indicates where 6:00 a.m. of the morning that begins daylight of lunar day 1 is located in relation to conjunction.

from – 30h	to	– 20h:	E	G
from – 20h	to	– 10h:	E E	
from – 10h	to	0h:	E E	G G
		conjunction		
from 0h	to	+ 10h:	E	
from + 10h	to	+ 20h:	E	G G G G G G
from + 20h	to	+ 30h:		GG G G G
from + 30h	to	+ 40h:	E	G
from + 40h	to	+ 50h:		G G G G
more than 80h:				G

TABLE 3. Locations of 6:00 a.m. of Lunar Day 1 in relation to Conjunction (E = Egyptian lunar date; G = Greek lunar date).

Another way of evoking the sharp difference between the Egyptian set and the Greek set is in relation to a certain point of time before or after conjunction. In no Egyptian date does 6:00 a.m. in the morning of lunar Day 1 fall later than 35 hours after conjunction. By contrast, it does so in six, or 30%, of the Greek dates. Furthermore, in only one Egyptian date, or 12½%, does 6:00 a.m. in the morning of lunar Day 1 fall later than 13 hours after conjunction. But it does so in no fewer than 15, or 75%, of the Greek lunar dates.

In selecting the 20 double dates, Grzybek made certain assumptions that can be seen as weakening their validity. But none of these assumptions is so outrageously unreasonable that calculating averages by taking his choices at face value did not seem worthwhile. Still, it remains a fact that the dates are not entirely without problems. There are three types of problems and the problems affect six of the 20 modern dates of lunar Day 1 (for the sequential numbers, see column IV in table 2), with two types of problems affecting one of the six. First, two of these six dates are the result of emendation. The day date is emended in one (no. 14)[38] and the year date in another (no. 18).[39] Second, one of the six dates is what may be called an egregious outlier (no. 19). In it, the lunar month begins on the fourth day after conjunction, untypically very late. Third, in four of the six dates (nos. 2, 9, 15, and 19), the difference between the day number of the lunar date and the day number of the civil date is 7. There are instances of double dates in later Ptolemaic times in which the lunar dates have by all appearance lost their connection with the moon and the difference between the two day dates is also 7. In Grzybek's view, already mentioned above, there is therefore reason to suspect that, in those later double dates, the lunar date was artificially differentiated from the civil date by subtracting or adding 7. As it happens, in one of the dates in which the difference between the two day dates is 7 (no. 2), the lunar month begins uncharacteristically early; in another, uncharacteristically late (no. 19). It seems therefore tempting to discard all four lunar dates involved as evidence because they may be artificial. Then again, lunar dates from the reign of Ptolemy II Philadelphos otherwise generally retain their connection with the lunar cycle. And it cannot be denied that there must have been cases in which the civil day number and the lunar day number were seven apart.

It will be wise to obtain additional average intervals for the Greek lunar dates by disregarding, say, only the outlier, only the emended date, all six of the slightly problematic dates, and other kinds of selections from the problematic dates. It appears that, however one may play the numbers, the shortest average interval, obtained by omitting just the outlier and the date in which the year is emended, is still *plus 18h 25m 20s*. The difference between the Egyptian and the Greek average intervals is then still *plus 19h 31m 50s* (1h 06m 30s + 18h 25m 20s). If one throws out all six of the slightly problematic dates, the average interval in the case of the Greek lunar dates is about *plus 22h 47m 13s*.

There may be yet other ways of playing the numbers involving yet other specific selections of data based on certain assumptions. But I do not see how the distinctive gap of roughly a day between the Egyptian lunar dates and the Greek lunar dates could be eliminated by any thinkable kind of selection.[40]

One more interesting, and perhaps legitimate, way of comparing the Egyptian and Greek lunar dates is as follows. Each set of dates has one outlier. One Greek lunar date (no. 10 in column IV in table 2) begins more than 80 hours after conjunction. Even if the date was bona fide, special circumstances may have applied, such as bad weather. But the important consideration is as follows. Such a date cannot be typical of any rule-determined way of beginning the lunar month. It is the only one of the 20 Greek dates regarding which this assessment is possible. It is therefore justified to omit it. The resulting average interval is about *plus 19h 46m 22s*. One Egyptian lunar date (no. 2 in table 1) is also something of an outlier. If one omits it, the average interval drops to about *minus 5h 37m 34s*. Disregarding the two outliers, the gap between the two average intervals is then about *25h 23m 56s*, about a day.

6.7. The Egyptian and Greek Lunar Calendars as, Strictly Speaking, Old Crescent (In)visibility Calendars

In what precedes, old crescent invisibility has been described as a key ingredient of Egyptian and Greek lunar calendars. And to a great extent, it is. But strictly speaking, there is more to the story. It is clearly not the case that, in either calendar, it is established that the old crescent is invisible. Rather, it is determined *whether it is still visible or not*. Its invisibility leads to a certain course of action and so does its visibility. In other words, both invisibility and visibility play a role and they do so as a kind of a toggle. What is more, if the old crescent is still visible at the time in question, it does not necessarily mean that it will be invisible in the next evening. There is too much irregularity in the Egyptian and Greek lunar months, in spite of the fact that certain principles seem to be followed in determining the beginning of the lunar month.

Along the same lines, there is every reason to believe that a new crescent (in)visibility calendar was used in Mesopotamia (see section 10.13). The nomenclature of the final days of the Greek months otherwise seems to guarantee that the old crescent is watched in the Greek world and not, as in Mesopotamia, the new crescent.[41] Naturally, the extant evidence for Greek lunar calendars covers only part of the Greek world. Likewise, the extant evidence for Mesopotamian lunar calendars covers only part of the Mesopotamian world. The best that can be done is to extrapolate from this evidence. If different mechanisms, or even if the old crescent sometimes played a role in Mesopotamia and the new crescent in the Greek world, we may well never learn anything about that in the absence of evidence.

6.8. Excursus 1: A Statistical Analysis

What are the chances that the distribution of Egyptian lunar dates and Greek lunar dates described above is completely random? One way in which statisticians answer this type of question is by trying to quantify what the chances are that the null hypothesis needs to be rejected. According to the null hypothesis, the said distribution is random. The higher the chances that the null hypothesis needs to be rejected, the smaller the chances that the research hypothesis—according to which the distribution is not random—is wrong. Basic statistical theory includes the z-test and t-test, in which two numbers z and t are determined by using formulas and then looked up in tables to find a corresponding p-value between 0.00 and 1.00 that expresses how probable it is that a distribution is random. Subtracting that number from 1 naturally yields the probability that the distribution is not random. For example, if a distribution has a probability of 0.99 or 99% of being random, the probability that it is not random is 0.01 (1 – 0.99) or 1%.

In the winter and spring of 2010, David Sheffield, an undergraduate physics major at Brown University (class of '11), took a tutorial with me on calendars and chronology of the ancient world. When the subject of the present paper came up, he volunteered to perform the z-test and the t-test and produced probabilities of less than 1% that the above distribution is random, noting that probabilities of less than 5% are generally considered significant and probabilities of less than 1% highly significant. But he also cautioned that the low number of data may affect the degree of probability and that additional analysis of the statistical kind remains desirable.

At the very least, it needs to be said that statistical considerations do not undermine the value of the empirical data derived from double dates in their capacity as evidence. It seems to me that they rather strengthen it.

6.9. Excursus 2: How Do Individual Modern Dates of Lunar Day 1 Relate to First Invisibility of the Old Crescent?

As has been noted above, when it comes to the average time interval between conjunction and 6:00 a.m. of the morning that begins daylight of lunar Day, there is a distinct difference between the Egyptian lunar calendar and the Greek lunar calendar. Greek lunar months appear to begin on average roughly a day later than Egyptian lunar months. In the footsteps of Pritchett, the difference was explained as follows. It was assumed that ancient Egyptians in principle began lunar Day 1 right away in the morning when the old crescent was no longer visible. At the same time, it was assumed that, that very same morning, ancient Athenians in principle decided instead that daylight of lunar Day 1 would begin the next morning, a day later. The one-day delay in the Athenian calendar inevitably follows from the fact that ancient Athenians gave the same name to the last day of both 29-day and 30-day lunar months. There is no trace of a similar procedure in the surviving day names of the native Egyptian lunar calendar. In 29-day months, Athenians did not use the name that Day 29 had in 30-day months. Lunar day 29 thus was named differently in 29-day and 30-day months. Consequently, it was necessary to decide what to name Day 29 at the latest by the morning that began the 29th period of daylight of the month.

The following crucial question has remained unanswered: Does the morning in which the old crescent is for the first time invisible begin daylight of lunar Day 1 in the Egyptian lunar calendar, on the one hand, and daylight of the last day of the month, either the 29th or the 30th, in the Greek lunar calendar, on the other hand? The date of the morning of first invisibility of the old crescent can be computed with much probability. However, invisibility can be marginal. It then seems advisable to propose two dates, one more probable, one less probable, for the morning of first invisibility in order to achieve something like very high probability. In such cases, it is scientifically impossible to date first invisibility to one single date with very high probability. In addition, historically speaking, even if the old crescent could no longer be seen, certain circumstances may have prevented watchers from establishing whether it was or was not visible. Such circumstances, when they applied, will forever remain unknown.

In what follows, I use the dates for last visibility of the old crescent yielded for the location of Cairo (30°00' N and 31°17' E) and for a visibility arc of 11°30' by a program created by Lange and Swerdlow[42] as well as the dates of first invisibility of the old crescent computed by Peter J. Huber for "a location near Memphis (30° E, 30° N)".[43] Naturally, last visibility falls a day before first invisibility. One is therefore readily derived from the other. Lange's and Swerdlow's program produces a single date for the morning of last visibility of the old crescent. On the other hand, Huber gave a second possible date of first invisibility of the old crescent in instances in which invisibility is marginal, that is, in three of the eight instances of Day 1 of an Egyptian lunar month listed in table 1, namely nos. 1, 2, and 4.

Let us first consider the eight dates of lunar Day 1 according to the Egyptian lunar calendar. If the calendrical mechanism described above were flawlessly applied, one would expect the eight modern dates of Day 1 of an Egyptian lunar month listed in table 1 to be the date of the morning of first invisibility of the old crescent. Is this so? Yes, but only in five out of eight instances, or 62.5%, according to the Lange-Swerdlow dates (nos. 1, 3, 4, 7, and 8) and only in four out of eight instances according to Huber's primary date (nos. 1, 3, 7, and 8). Then again, in no. 4, Huber offers the Lange-Swerdlow date as an alternative date of lower probability. In the three other instances, the lunar month twice begins a day later (nos. 5 and 6) and once two days later (no. 2) according to both the Lange-Swerdlow dates and the Huber dates. However, Huber offers a date that is only one day later as an alternative for no. 2. Incidentally, his third alternative date is a day before first invisibility for no. 1.

In sum, computations indicate that the following picture has a certain undeniable probability of being true: the lunar month begins in the morning of first invisibility in five instances and one day later in three instances. In other words, the score is not perfect. The Egyptian lunar dates begin a little later than one might expect from a perfect execution of the calendrical mechanism, three days in eight instances, or 0.375 days on average. There is every reason to assume that it is possible that the calendrical mechanism was not flawlessly applied. Conditions for watching out for invisibility may have been marginal or even entirely wanting. It cannot be excluded that the inability to make a decision whether the day should be 29 days or 30 days long resulted in choosing 30 over 29 as the length of the day because 30 is the ideal length of the month. Such a mind set would result in a calendar running slightly later than one in which the afore-mentioned calendrical mechanism was perfectly applied. At the next turn of the month, first invisibility of the old crescent would then tend to occur earlier than it otherwise would have, perhaps even before the morning that begins the daylight period that follows the 29th daylight period. The natural choice would then obviously be to make the month 29 days long, thus compensating for a month that may have been longer than it otherwise should be. Nothing is easier to assume the distinct possibility of such scenarios. But in the end, we will never know every detail about what happened at the end of a lunar month in terms of deciding the beginning of the next.

It will also never be known whether fixing the beginning of the lunar month might on occasion in any way have been influenced by trying to have the new crescent be visible in the evening that follows daylight of lunar Day 2, called *ȝbd* in Egyptian. There is evidence that the Egyptians at least sometimes assumed that the new crescent would be visible that evening.[44] But then, the name of lunar Day 3, *mspr*, is perhaps derived from *spr* "arrive"; Parker therefore considered it possible that the name refers to the occasional arrival of the new crescent on that day.[45] Someone who carefully observed the change in distance between the moon and the sun, on average a little over 12° per day, and who also had some idea of how far the moon needed to be removed from the sun to be visible might perhaps have been able to estimate in which evening the new crescent was most likely to be visible for the first time and consequently give advice as to when to begin the lunar month to make sure that the new crescent would be visible in the evening following daylight of *ȝbd*. Any such effort would cause the beginning of the lunar month to fall later than a beginning determined by the calendrical mechanism described above. That is because the interval between the morning of first invisibility of the old crescent and the evening of first visibility of the new crescent is one and a half days in about 54.18% of the cases and two and a half days in about 44.30% of the cases.[46] In the latter 44.30% of the cases, the lunar month would need to begin one day later than first invisibility of the old crescent in the morning, that is, one day later than a beginning determined by the calendrical mechanism described above. Visibility in the evening following daylight of *ȝbd* is possible according to a somewhat meager four of Huber's dates (nos. 3, 4, 7, and 8), of which one (no. 3) is the primary date of two possible dates, and also according to four of the Lange–Swerdlow dates. In other words, there is no recognizable influence from visibility of the new crescent in the evening following daylight of *ȝbd*.

It will also never be known whether aiming to have conjunction fall on *psḏntyw* might have played a role. Conjunction is an invisible event, except at solar eclipses. Still, anyone keenly observing the moon approaching the sun by a certain amount every day might have some notion of the day on which the moon would join the sun. But as it happens, if daylight of *psḏntyw* always begins in the morning of first invisibility of the old crescent, 6:00 a.m. of that morning will on average fall about half a day before conjunction (see below). Lunar Day 1 will therefore typically be the day on which the moon joins the sun and also a day on which the moon is invisible both in the morning before sunrise and the evening after sunset.

Let us next consider the 20 dates of lunar Day 1 according to the Greek lunar calendar. The 20 modern dates of Day 1 of a Greek lunar month yielded by the 32 modern double dates listed in table 2 are as follows (years are BCE): (1) 17 Aug 264; (2) 21 May 257; (3) 27 Oct 257; (4) 24 Jan 256; (5) 23 Feb 256; (6) 22 Jun 256; (7) 18 Sep 256; (8) 13 Jan 255; (9) 1 May 254; (10) 13 May 255; (11) 9 Jul 255; (12) 2 Apr 254; (13) 27 Aug 254; (14) 16 Jul 253; (15) 31 Dec 252; (16) 29 Mar 251; (17) 23 Aug 251; (18) 22 Jun 256; (19) 30 Nov 249; and (20) 26 Jan 248. If the calendrical mechanism described above were flawlessly applied, one would expect these 20 dates to be the date of *the morning that follows one day after* the morning of first invisibility of the old crescent. Is this so? Yes, but only in eight out of 20 instances, or 40%, according to the Lange-Swerdlow dates and Huber's primary dates (nos. 1, 3, 4, 5, 8, 13, 15, and 17).

In 11 of the other 12 dates, the Lange-Swerdlow dates and Huber's primary dates are the same. Six of these 12 dates (nos. 6, 7, 9, 16, 18, and 20) fall two days after the morning of first invisibility, or one day too late to accord with the calendrical mechanism described above. One (no. 10) falls three days after the morning of first invisibility, or two days too late to accord with the calendrical mechanism. One (no. 19) falls four days after the morning of first invisibility, or three days too late to accord with the mechanism. Two (nos. 11 and 14) are the dates of the morning of first invisibility itself and therefore fall one day too early to accord. And one (no. 2) falls one day before first invisibility, or two days too early to accord.

In one of the 13 dates (no. 12), the Lange-Swerdlow date and Huber's primary date differ. The Lange-Swerdlow falls two days after the morning of first invisibility, or one day too late to accord with the calendrical mechanism. Huber's primary date falls three days after the morning of first invisibility, or two days too late to accord. However, Huber's secondary date is the same as the Lange-Swerdlow date.

The fact that only eight of the 20 dates accord with the postulated calendrical mechanism at first sight does not seem to bode well for this paper's research hypothesis, even if these eight dates still constitute the most frequent case, edging out the six cases in which the lunar month begins a day later than the calendrical mechanism requires. However, the picture changes significantly as soon as one does two things. I believe that neither of these two things can be interpreted as manipulating the data. The first thing is to take Huber's secondary dates into consideration. The second thing is to consider what has been said above about the problems that affect six of the 20 dates.

First, Huber's secondary dates. There are five. In four of them (nos. 5, 11, 14, and 15), first invisibility happens a day earlier. In one of them (no. 12), it happens a day later. Huber's secondary dates may be less probable than his primary dates. However, both ultimately exhibit a certain probability and both therefore deserve to be considered as distinct possibilities. It has been noted above that, according to Huber's primary date, two dates (nos. 11 and 14) are those of the morning of first invisibility itself and therefore fall one day too early to accord with the calendrical mechanism. As it happens, according to Huber's secondary date, the dates fall a day later and therefore do accord. The number of dates that accord hence rises to 10 out of 20. Moreover, one date that falls two days too late (no. 12) falls only one day too late according to the secondary date. The following optimal picture therefore has a certain probability of being true. Besides the 10 dates that accord (nos. 1, 3, 4, 5, 8, 11, 13, 14, 15, and 17), seven fall one day too late (nos. 6, 7, 9, 12, 16, 18, and 20); one (no. 2), two days too early; one (no. 10), two days too late; and one (no. 19), three days too late.

The second thing is to consider what has been said above about the problems that affect six of the 20 dates (nos. 2, 9, 14, 15, 18, and 19). For example, the date that is two days too early (no. 2) and the date that is three days too late (no. 19) are either much too early or much too late by any measure. It is not clear what caused these deviations. If one eliminates these two from consideration, the following picture emerges: 10 dates accord (nos. 1, 3, 4, 5, 8, 11, 13, 14, 15, and 17); seven fall one day too late (nos. 6, 7, 9, 12, 16, 18, and 20); and one (no. 10), two days too late. If one eliminates all of the six problematic dates from consideration, the following picture emerges. Eight dates, or about 57.1%, accord (nos. 1, 3, 4, 5, 8, 11, 13, and 17); five, or about 35.7% fall one day too late (nos. 6, 7, 12, 16, and 20); and one (no. 10), two days too late. There is again no compulsion to assume that the calendrical mechanism was flawlessly applied. Bad conditions for watching the old crescent might well have produced 30-day lunar months when a 29-day lunar month would have accorded better with a strict application of the calendrical mechanism.

Could attempts to have first visibility of the new crescent fall in the evening following daylight of lunar Day 1 have played a role? According to 12 of Huber's dates (nos. 1, 3, 4, 5, 8, 9, 12, 13, 15, 16, 17, and 20), of which two are secondary dates (nos. 1 and 12), visibility at that time is a possibility. If one eliminates the six dates affected by problems (see above), visibility is possible in 10 dates (nos. 1, 3, 4, 5, 8, 12, 13, 16, 17, and 20) out of fourteen, or 71.4%. In three of the four others (nos. 6, 7, and 10), visibility was possible the day before; in one (no. 11), the day after.

It was occasionally assumed in ancient Greece that conjunction fell on the last day of the lunar month.[47] Could aiming to have conjunction fall on that day by following the regression of the moon in relation to the sun in the days before conjunction have played a role in fixing the beginning of the lunar month? Possibly. But as it happens, if the calendrical mechanism described above was flawlessly applied, 6:00 a.m. of the morning of the last day would on average fall about half a day before conjunction (see below). Conjunction would therefore typically fall on the last day of the month. Even if the fact that conjunction tends to fall on the last day of the month followed unintentionally from the application of said calendrical mechanism, at least some ancients may have viewed that fact as a welcome consequence.

In six out of 20 dates (nos. 6, 7, 10, 12, 18, and 19), sighting the new crescent in the evening before daylight of lunar Day 1 is possible. But the number is not sufficiently high to justify that the Greek calendar operated by first visibility of the new crescent as is now universally assumed.

7. More Modern Dates of Day 1 of Egyptian Lunar Months

So far, eight modern equivalents of Day 1 of an Egyptian lunar month have been subjected to examination in order to establish when in relation to the lunar cycle Egyptian calendrical months began. The sample is admittedly rather small. Additional data seem desirable. Among other data relevant to establishing when Egyptian lunar months began is an interesting set of 32 complete or incomplete temple service dates of Ptolemaic or Roman times recently assembled and studied in detail by Bennett.[48] A temple service date is a day number signifying the position of the day in question in a specific term of service in a temple. The term of service was called *wrš* in Egyptian. For example, a temple service date may identify a certain day as Day 12 of the *wrš*.

Temple service dates are not Egyptian lunar dates in the strict sense and they make no explicit reference to the Egyptian lunar calendar. But since the terms of service in question began on a certain day at the beginning of the lunar month and lasted a lunar month, temple service dates are no doubt lunar in character. A full understanding of the Egyptian lunar calendar therefore requires an analysis and correct understanding of the related temple service dates.

It is not at once clear whether the temple service dates can be adduced as evidence in support of the theory of the relation between Egyptian lunar calendar and the Greek lunar calendar adopted and defended in this paper. The principal problem is that there is some uncertainty as to whether the temple service began on lunar Day 1, called *psḏntyw* in Egyptian, or on lunar Day 2, called *ꜣbd* in Egyptian. Two facts for which there is some evidence are as follows. First, on the one hand, computation indicates that the beginnings of the temple service terms accord well with the day that follows the day of first invisibility of the old crescent.[49] Second, on the other hand, there is fairly good evidence, including the eight lunar dates derived from double dates above, that Egyptian lunar months began a day earlier with first invisibility.[50] These two facts can be reconciled in one of two ways. Either the temple service begins on Day 2. Or the temple service dates are evidence of a second lunar calendar beginning on *ꜣbd* . The second explanation seems unlikely. First, some of the lunar dates derived from double dates are also found in temples and yet their months begin on *psḏntyw*. It therefore seems improbable that two lunar calendars whose months had different beginnings were operative in temples. And second, Luft has produced valid arguments that the temple service began on *ꜣbd* in the Middle Kingdom.[51] Why would the beginning have been different in later times? In sum, the first explanation is to be preferred and in fact will be in what follows.

Bennett analyzes 32 modern dates of lunar Day 1 derived from temple service dates.[52] Of those 32, five are described as "complete" and 20 as "incomplete". Seven more dates are those that Bennett considers sufficiently certain among all the dates that can be derived from evidence found at Dime.[53] Only the five "complete" dates and the seven sufficiently certain dates from Dime are considered here. They are listed in table 4. Conjunction falls on average 11 hours 13 minutes 45 seconds after 6:00 a.m. in the morning at the

		I Modern Date of Lunar Day 2 (= Service Day)	II Modern Date of Lunar Day 1 (Daylight)	III Closest Conjunction (Daylight)	IV Distance in Hours from Conjunction to 6:00 a.m. in Morning of Lunar Day 1	V Morning of First Invisibility of the Old Crescent
1.	Bennett no. 1	24 Dec 56 BCE	23 Dec 56 BCE	23 Dec, 11:17 a.m.	– 5h 17m	23 Dec
2.	Bennett no. 2	22 Jan 55 BCE	21 Jan 55 BCE	22 Jan, 2:16 a.m.	– 20h 16m	21 Jan
3.	Bennett no. 3	29 Aug 48 BCE	28 Aug 48 BCE	28 Aug, 5:30 a.m.	+ 0h 30m	28 Aug
4.	Bennett no. 4	3 Feb 37 BCE	2 Feb 37 BCE	2 Feb, 10:31 a.m.	– 4h 31m	2 Feb
5.	Bennett no. 5	13 Apr 66 CE	12 Ap 66 CE	13 Apr, 9:52 a.m.	– 25h 52m	12 Apr
6.	Bennett no. 26	7 Feb 24 BCE	6 Feb 24 BCE	7 Feb, 1:24 a.m.	– 19h 24m	6 Febr
7.	Bennett no. 27	8 Mar 24 BCE	7 Mar 24 BCE	8 Mar, 4:33 p.m.	– 32h 33m	7 Mar
8.	Bennett no. 28	15 Nov 68 CE	14 Nov 68 CE	13 Nov, 8:20 p.m.	+ 9h 40m	13 Nov
9.	Bennett no. 29	14 Dec 68 CE	13 Dec 68 CE	13 Dec, 11:45 a.m.	– 5h 45m	13 Dec
10.	Bennett no. 30	20 Mar 90 CE	19 Mar 90 CE	20 Mar, 1:59 a.m.	– 19h 59m	19 Mar
11.	Bennett no. 31	19 Apr 90 CE	18 Apr 90 CE	18 Apr, 3:50 p.m.	– 9h 50m	18 Apr
12.	Bennett no. 32	9 Apr 91 CE	8 Apr 91 CE	8 Apr, 7:28 a.m.	– 1h 28m	7 Apr

TABLE 4. More Modern Dates of Lunar Day 1 of Egyptian Lunar Months. For details and bibliographical references pertaining to the sources from which the modern dates of lunar Day 1 have been derived, see Bennett (2008). The times for conjunction are Goldstine's (1973) for Babylon, adjusted for Memphis by subtracting 53 minutes.

beginning of daylight of Day 1. Clearly, the lunar month begins on average at least a day earlier than the beginnings of the lunar month derived from Greek documents listed in table 2. Furthermore, in 10 cases out of 12, daylight of lunar Day 1 begins in the morning of the most probable date for first invisibility of the old crescent. All these empirical data accord well with the research hypothesis of the present paper.

In the modern dates of Day 1 of Egyptian lunar months listed in table 1, conjunction falls on average 1h 06m 30s after 6:00 a.m. in the morning that begins daylight of lunar Day 1. This is 10h 7m 15s later than the average interval derived from the modern dates of Day 1 of Egyptian lunar months in table 3. This gap seems significant. Then again, of the 20 Egyptian dates listed in tables 1 and 4, no. 2 in table 1 is significantly irregular. If just this one date is omitted from consideration, the 10-hour gap between the average interval derived from the dates in table 1 and the average interval derived from the dates in table 4 is roughly cut in half. I see therefore no reason to assume that different calendrical mechanisms produced the dates of table 1, on the one hand, and those of table 4, on the other hand. The average interval derived from all 20 dates in tables 1 and 4 is minus 7h 10m 51s.

In table 5 below, spatial distribution is used to evoke the difference between lunar Day 1 according to the Egyptian lunar calendar and lunar Day 1 according to the Greek lunar calendar, as in table 3, but now including the temple service dates. That means that 20 Egyptian dates are now compared to 20 Greek dates. As in table 3, each instance of the letter E represents an Egyptian lunar date; each instance of G, a Greek lunar date. The Egyptian dates derived from temple service dates are represented by E^s, rather than by just E. The position of the letter indicates where 6:00 a.m. of the morning that begins daylight of lunar day 1 is located in relation to conjunction. David Sheffield notes that, based on the *z*-test, it is less than 0.0001% probable that this distribution is pure coincidence.[54]

from – 40h	to	– 30h:	E^s	
from – 30h	to	– 20h:	E E^s	G
from – 20h	to	– 10h:	E E E^s E^s E^s	
from – 10h	to	0h:	E E E^s E^s E^s E^s E^s	G G
		conjunction		
from 0h	to	+ 10h:	E E E^s E^s	
from + 10h	to	+ 20h:	E	G G G G G G
from + 20h	to	+ 30h:		G G G G G
from + 30h	to	+ 40h:	E	G
from + 40h	to	+ 50h:		G G G G
more than 80h:				G

TABLE 5. Locations of 6:00 a.m. of Lunar Day 1 in relation to Conjunction (E = Egyptian lunar date; E^s = Egyptian lunar temple service date; G = Greek lunar date).

8. More Modern Dates of Day 1 of Greek Lunar Months

There is additional evidence that pertains to how the beginning of ancient Greek lunar months was determined, but not much. This evidence includes a number of modern dates of lunar Day 1 that can with some probability be derived from the extant sources. Most prominent among them are seven month and day dates of astronomical events found in Ptolemy's *Almagest*. The dates are according to a lunar calendar whose month names are Greek. In four of the dates, the month names are Athenian; in three dates, Macedonian. Ptolemy provides the lunar date's equivalent in the Egyptian civil calendar. Since the modern date of the Egyptian civil date is known, so automatically are the modern date of the lunar month and day date and also the modern date of lunar Day 1 of the month in question. It is then possible to establish how the beginning of the month relates to conjunction and to lunar visibility and invisibility. The seven derivations are presented in table 6. What is in italics is actually in the text.

Owing to certain complications, the value of the seven dates as evidence is mixed. For example, not everyone associates the seven instances of lunar Day 1 with the same modern dates, partly in consequence of differences of opinion as to when the day begins in antiquity.[55] In fact, three of the seven dates are in all probability to be disqualified at once as evidence of the Greek lunar calendar. The Macedonian lunar month names in nos. 1, 2, and 3 in table 6 are in all probability Greek in name only. The three dates are said to

	I	II	III	IV	V	VI	VII
	Location in the *Almagest*	"Greek" Lunar Date	Egyptian Date Equated with Lunar Date in II	Modern Date of the Egyptian Date in III	Date of Daylight of Lunar Day 1	Closest Conjunction	Date of First Invisibility of the Old Crescent (FIOC) or First Visibility of the New Crescent (FVNC)
1.	IX 7.10	*5 Apellaios at dawn* (Babylonian!)	*Thoth (night of) the 27th to the 28th*	night of 18 to 19 Nov 245 BCE (dawn)	15 Nov	13 Nov, 1:22 a.m. (Babylon!)	evening of 14 Nov (FVNC)
2.	IX 7.9	*14 Dios at dawn* (Babylonian!)	*Thoth (night of) the 9th to the 10th*	night of 29 to 30 Oct 237 BCE (dawn)	17 Oct	15 Oct, 10:25 p.m. (Babylon!)	evening of 16 Oct (FVNC)
3.	XI 7	*5 Xanthikos in the evening* (Babylonian month!)	*14 Tybi*	1 Mar 229 BCE (evening)	26 Feb	24 Feb, 11:17 a.m. (Babylon!)	evening of 25 Feb (FVNC)
4.	VII 3	*25 Poseideon*	*16 Phaophi*	20 Dec 295 BCE	26 Nov	26 Nov, 8:31 a.m. (Alexandria)	morning of 26 Nov (FIOC) evening of 27 Nov (FVNC)
5.	VII 3	*15 Elaphebolion*	*5 Tybi*	9 Mar 294 BCE	23 Feb	23 Feb, 2:47 a.m. (Alexandria)	morning of 21 Jan (FIOC) evening of 23 Jan (FVNC)
6.	VII 3	*8 Anthesterion*	*29 Athyr*	29 Jan 283 BCE	22 Jan	22 Jan, 11:51 a.m. (Alexandria)	morning of 21 Jan (FIOC) evening of 23 Jan (FVNC)
7.	VII 3	*25 Pyanepsion*	*7 Thoth*	8 Nov 283 BCE	15 Oct	14 Oct, 2:02p.m. (Alexandria)	morning of 14 Oct (FIOC) evening of 16 Oct (FVNC)

TABLE 6. Seven Modern Dates of Lunar Day 1 Derived from Ptolemy's *Almagest*. Text in italics is in the Greek text of the Almagest. In dating certain night-time events, Ptolemy uses so-called "double dates" of the type "Thoth (night of) the 9th to the 10th" (Almagest IX 7.9). There is more than one interpretation of what exactly Ptolemy means by a double date. But, in any event, double dates never leave any doubt about which night Ptolemy is referring to.

be "according to the Chaldaean calendar". The calendar of the dates in nos. 1, 2, and 3 is therefore presumably the Babylonian calendar whose month names have been replaced by Macedonian month names; after Alexander's conquest, the use of Macedonian month names spread to much of his empire. As it happens, the lunar months of dates nos. 1, 2 and 3 begin one or two days later than the Greek lunar months found in double dates discussed above—at least according to the specific assumptions according to which their beginnings were obtained (see below)—and so do lunar months in Babylonian astronomical texts. As to establishing the modern date of daylight of lunar Day 1, a detailed discussion exceeds the scope of the present paper and it is not clear that there is a certain solution. According to one scenario, the day numbers 5, 14, and 5 of the lunar dates 5 Apellaios, 14 Dios, and 5 Xanthikos were found in Babylonian astronomical texts, but they were accompanied by the equivalent Babylonian lunar month names. Babylonian astronomy was thriving in the third and second centuries BCE. The three day numbers were somehow transmitted to Ptolemy in the second century CE. In the translation from Babylonian to Greek, the Babylonian month names were replaced by Macedonian month names. In Babylonian astronomical reports, the description of what happens at night precedes the description of what happened in the daylight period with the same day number. Accordingly, daylight of 5 Apellaios, 14 Dios, and 5 Xanthikos would be daylight of 19 November, 30 October, and 2 March.[56]

But new evidence contradicts this theoretical scenario in one respect. A Babylonian astronomical tablet reveals that the Babylonian name of the Xanthikos at hand was Addaru, that this Addaru began in the evening of 25 February 229 BCE, and that daylight of 1 Addaru was 26 February (Jones (2006), p. 269; I owe this reference to a personal e-mail communication by Chris Bennett of December 14, 2011). Ptolemy's Egyptian civil date leaves no doubt that the Xanthikos observation took place in the evening of 1 March 229 BCE, which is the evening of *6* Addaru in Babylonian astronomical texts. Yet, Ptolemy dates the observation to lunar 5 Xanthikos; he must have found the date in an earlier source. Jones (2006), p. 269 observes that "the observation report should actually correspond to February 29". Bennett (2011), p. 34 suspects a "conversion error". But he also considers it possible that "the conversion is correct", stating,

> If so, the two calendars cannot be identical, although to all appearances they are closely related. For example, the 'Chaldean calendar' might have been a schematic Greek astronomical calendar. Alternately, the 'Chaldean' dates can be simply reconciled mathematically with both the Babylonian calendar and the Egyptian dates for the recorded observations of Mercury and Saturn by supposing that they used a dawn epoch, i.e. that a 'Chaldean' date began at dawn on the corresponding Babylonian date. (p. 35)

In any event, it is now certain that, in the date 5 Xanthikos, the month can be Babylonian but the day number cannot.

I believe that the difference in lunar day date can be accounted for as follows. In Ptolemy's chronological system, the evening following daylight of Day x is always the evening of Day x, and not the evening of Day $x + 1$ as it is in Babylonian astronomical texts. In other words, the day date does not change at sunset as it does in Babylonian astronomy. Anyone converting Babylonian dates of observations into Egyptian civil dates must have known which Babylonian daylight day numbers corresponded to which Egyptian daylight day numbers and must also have been used to systematically decreasing the day number by one for evening events. This procedure could easily have been applied, perhaps automatically so, to lunar Addaru when converting it into Macedonian Xanthikos. In other words, there can have been no doubt at the time of conversion that daylight of 5 Addaru was daylight of Egyptian civil 14 Tybi and that the Babylonian and Egyptian dates of the observation made in the evening immediately following this daylight period were 6 Addaru and 14 Tybi. The lunar day number changed but the Egyptian civil day number did not. It would have been more than tempting to leave the day date unchanged at sunset in the lunar date as it is in the Egyptian date. Furthermore, it is highly improbable that the day date changed at sunset anywhere outside the highly technical context of Babylonian astronomy. Changing 6 Xanthikos into 5 Xanthikos cannot cause confusion because the lunar date does not date the observation. The Egyptian civil date does.

The dates for daylight of Day 1 in nos. 1, 2, and 3 in table 6 have been obtained on the reasonable assumptions that the day numbers of the three lunar dates and the day number of the Egyptian civil dates refer to the same daylight periods and that day numbers did not change at sunset in the lunar dates as they surely did not in the Egyptian civil dates. Consequently, the dates of daylight of 5 Apellaios, 14 Dios, and 5 Xanthikos are 19 November, 30 October, and 1 March, and the dates of daylight of 1 Apellaios, 1 Dios, and 1 Xanthikos are 15 November, 17 October, and 26 February.

As suggested elsewhere in this article, I consider the epoch to be a ghost concept. Even for astronomers like Ptolemy, the day basically began in the morning and ended in the evening and extensions of activity to before sunrise or dawn and to after dusk or sunset were included. Nighttime was in principle a numberless stretch of darkness. Accordingly, dating nighttime events was always a problem. Ptolemy's use of double dates, as in "(in the night) from the 5th to the 6th", was one way of avoiding all ambiguity as to which night was meant. While avoiding ambiguity was the design of double dates, their linguistic expression suggests that the night was itself not numbered. The "5th" and the "6th" in the expression above may well mainly refer to daylight periods.

I do not believe that, in actual calendrical practice, there was ever any focus at any time in antiquity on a single point in time returning once every 24-hour cycle and serving as the end of one 24-hour cycle and the beginning of another 24-hour cycle, a so-called epoch. Ancient epochs owe their existence largely to ancient reports on the beginning of the day by mostly Latin authors, including Censorinus, Gellius, Pliny, and Varro—with Varro firmly at the origin of them all. It seems to have satisfied Varro's sense of orderliness or delight in exotic detail to assign different beginnings of the day to different nations. I believe that, in doing so, he created a pseudo-concept. True epochs came into existence only later with the advent of midnight as the beginning and end point of a 24-hour system. In the confines of astronomy, a noon epoch was also used to fit all the events of a single night conveniently into the same day. But the concept of the epoch should not be projected back into the ancient past, it appears to me. True, in Babylonian astronomy, nighttime precedes daylight with the same number. But does that have to mean anything more than just that?

The four other lunar dates are Greek, in the sense that they were recorded by the Greek-speaking Timocharis in Greek-speaking Alexandria. It is again assumed that the day numbers of the lunar dates and the day numbers of the Egyptian civil dates refer to the same daylight and that day numbers did not change at sunset in the lunar dates as they surely did not in the Egyptian civil dates. Accordingly, two of the four lunar months begin in the morning of first invisibility of the old crescent, which according to what has been said above accords better with the Egyptian lunar calendar than with the Greek lunar calendar. Two of the four lunar months begin in the morning following the morning of first invisibility of the old crescent, which accords better with the Greek lunar calendar. It is not fully clear what exactly to make of this. In any event, the four dates provide additional evidence that daylight of Day 1 of Greek lunar months did not begin in the morning after the evening of first visibility of the new crescent.

Three more possible modern dates of daylight of Day 1 of Greek lunar months are as follows. First, there is good evidence that the modern date of the first daylight period of the first Callippic cycle is 28 June 330 BCE.[57] The morning of first invisibility of the old crescent is 27 June. Daylight of lunar Day 1 therefore begins in the morning that follows the morning of first invisibility of the old crescent, as is common in the Greek lunar dates derived from double dates studied above. Second, reports survive that Meton observed the summer solstice of 432 BCE on 13 Skirophorion.[58] The solstice occurred in the morning of 28 June. If Meton observed the solstice on the correct day, which seems distinctly possible,[59] then daylight of 1 Skirophorion is 16 June. The morning of first invisibility of the old crescent is 16 June. This lunar date could therefore serve as evidence that Greek lunar months begin before first visibility of the new crescent.

The third date concerns the decisive battle at Gaugamela, in which Alexander defeated Darius III, and Greece turned the tables on Persia after two centuries of strife. According to the best possible chronological information, namely Babylonian astronomical texts, the battle was fought in the morning of Day 24 of the Babylonian lunar month Ululu.[60] But according to Plutarch (Life of Camillus 9.15), the battle took place on Day 26 of the Athenian lunar month Boedromion. The battle was fought far from Greece. But then, one of the two armies waging the battle right there and then was Greek and its soldiers would know what day of the month it was according to a Greek calendar. It was a glorious day that they would want to remember long after. If the date is accepted as historical, the Greek lunar month began two days earlier than the Babylonian astronomical lunar month. This is altogether normal. According to computations I owe to Peter J. Huber,[61] such a two-day interval ought to occur on average in about 44.30% of the cases. But a one-day interval is also normal. It is expected to occur in about 54.18% of the cases.

The Babylonian date leaves no doubt that the battle was fought during daylight of 1 October 331 BCE. Daylight of Day 1 of Boedromion therefore falls on 6 September of that year. The morning of first invisibility of the old crescent was 6 September at Mosul. But visibility in the morning of 5 September seems to have been low. In any event, the month could not have begun with first crescent visibility.

It is true that the view that first crescent visibility meant everything to the ancient Greeks still dominates. Still, there is a certain awareness out there, which I have not tried to document at great length, that the little that can be know about when Greek lunar months begin, especially as it is derived from astronomical texts and as it has been evidenced above by four lunar dates in Ptolemy's *Almagest* and by the beginning of the first Callippic cycle, points in a quite different direction. Among earlier testimonies of this same awareness is Epping's in his *Astronomisches aus Babylon*, a milestone in the study of ancient astronomy. Epping writes that, "bei den Griechen", the beginning of the month was determined "nach dem mittleren Neumond".[62] Ideler, the greatest student of chronology of the nineteenth century, observes about Greek lunar months that "ihr Anfang in der Regel dem ersten Tage nach der Conjunction entsprach, wo sich das Mondlicht zuerst zu zeigen pflegt, ob es gleich [modern *obgleich es*] nach der jedesmaligen Lage der Ekliptik auch wol erst am zweiten oder dritten Tage nach der Conjunction sichtbar werden kann, wie Geminus richtig bemerkt".[63]

Modern dates of lunar Day 1 provide direct empirical evidence relating to the beginning of ancient Greek lunar months. These dates leave no doubt that Greek lunar months began before first crescent visibility. But there are two other types of evidence that indirectly support the same notion. These two types cannot be reviewed in detail here.

The first type of evidence consists of ancient testimonies that, in the Greek lunar calendar, the last day of the month, alternatively called ἕνη καὶ νέα and τριακάς, was the day of conjunction, whereas, in the Egyptian lunar calendar, it was the first day, called *psḏntyw*. Much of the evidence can be found gathered elsewhere.[64]

In fact, conjunction seems to have been viewed on occasion as a point in time heralding the beginning of the month. The last day of Greek lunar months and the first day of Egyptian lunar months sit astride that point in time. In that sense at least, the two belong not only to the end of the previous month but also to the beginning of the following month. They are "straddle" days, as it were. That would explain why Hesiod, in his *Works and Days*, at 768, began his calendar with τριακάς, as Pritchett observes, and not with lunar Day 1.[65] That could also explain why, as J. F. Quack proposes,[66] *wrš* designated at the same time the last day of the temple service and *psḏntyw* or lunar Day 1. Both τριακάς and *wrš* were interpreted as straddling the point in time when the moon joins the sun as marker of the beginning of the month and hence had dual status as both an end and a beginning.[67]

The second type of evidence concerns the fact that the last day of 29-day and 30-day lunar months had the same name. For example, a papyrus kept at Cornell University dating to the mid third century BCE contains a record of the amounts of lamp oil to be distributed every day for the two Macedonian lunar months Apellaios and Audnaios.[68] The editors note that, in the second month, "[t]he entry for the 29th is lacking".[69] Day "30" (λ) therefore immediately follows Day "28" (λη). But evidently, as Pritchett first noted, what happened instead is that the second month had 29 days and the first had 30.[70] In a long and at times heated debate that has lasted decades, some have defended the possibility that a day name at the beginning of the last third of the month could be omitted instead. But Pritchett always maintained that no such omission ever took place and I confidently follow him in this regard.

9. A Test as Bridge between the Data and the Calendrical Mechanism

According to the empirical data pertaining to the Greco-Macedonian calendar presented above, 6:00 a.m. in the morning that begins daylight of lunar Day 1 falls on average somewhere roughly between 17 and 25 hours, that is, about a day or a little less, after conjunction. That means that the evening preceding daylight of lunar Day 1 falls on average roughly only half a day or less after conjunction. No one doubts that an evening falling at such a time generally occurs too soon for the new crescent to have been visible in it. Accordingly, visibility of the new crescent cannot have determined the beginning of the Greek lunar months in question. At the same time, no one would doubt that daylight of lunar Day 1 generally falls too late for the morning that begins it to have been the morning of first invisibility of the old crescent.

In the footsteps of Pritchett, it was suggested that a peculiar calendrical mechanism could account for the distance between 6:00 a.m. in the morning that begins daylight of lunar Day 1 and conjunction and therefore also for the specific beginnings of the lunar months in question, namely on average about a day later than Egyptian lunar months. According to this calendrical mechanism, the morning of daylight of lunar Day 1 falls two days or 48 hours after the morning of last crescent visibility.

The question arises: What kind of average interval between 6:00 a.m. of Day 1 and conjunction does the calendrical mechanism in question produce if it is rigorously applied? And how do these beginnings compare to those of the actual Greek lunar months examined above?

It appears that there is no great difficulty in obtaining an ideal average interval. The morning of last crescent visibility can be computed with fairly high probability. In fact, one could even perform a test by using actual modern day observations. In the present case, I have taken the average of the ages of the moon—that is, the distance in time from true conjunction—in the morning of last visibility of the old crescent over a period of 18 years, namely the years 1900 CE–1917 CE, or roughly one Saros cycle, for the location of Cairo, using the program created by Lange and Swerdlow.[71] After one Saros cycle, many properties of the course of the moon come full circle. This circumstance should make the average more representative.

As average moon age I obtained about *minus 35h 04m 07s*, say minus 35 hours. The specific intervals are between sunrise and conjunction. The average interval is therefore between average sunrise, or 6:00 a.m., and conjunction. If the afore-mentioned calendrical mechanism is rigorously applied, there are 48 hours from 6:00 a.m. in the morning of last visibility of the old crescent to 6:00 a.m. in the morning of lunar Day 1. At the same time, there are only about 35 hours from 6:00 a.m. in the morning of last visibility to conjunction. It may be concluded that 6:00 a.m. in the morning of lunar Day 1 falls on average about 13 hours (48–35) after conjunction. That average interval is fairly close to the average interval of 17 to 25 hours derived from actual dates. What matters most is that both average intervals occupy the same position in relation to three points in time in the lunar month. First, both averages fall well after conjunction, between half a day and a day. Second, they fall too early for first visibility of the new crescent to play a role. Third, they fall too late for daylight of lunar Day 1 to begin immediately in the morning of first invisibility of the old crescent.

It should be realized that the 13-hour interval involves an ideal application of the Greek lunar calendar mechanism whereas an interval such as 17 to 25 hours is derived from actual lunar dates. There is every reason to believe that methods of determining the beginnings of lunar months were not always rigorously applied in actuality. Lack of rigor ought not have been too much of an obstacle. After all, lunar months are always either 29 or 30 days long, except in Rome, where they appear to have been either 29 or 31 days long.

If the beginning of the month strayed a little too far from new moon, it should have been easy to restore the synchronism of the month with the lunar cycle by a specific choice of either 29 or 30 as length of the next one or two months. There must also have been obstacles to a precise application of the rule. Weather conditions may have made observation impossible. Or someone might have neglected to observe the moon or made an error of observation.

10. First Visibility of the New Crescent in Mesopotamia: No Marker of the Month's Beginning

10.1. In Egypt and Greece, No; but also in Mesopotamia?

The three most prominent hubs of civilization in the west in the centuries before the rise of Rome are Egypt, Greece, and Mesopotamia, as is obvious from the sheer number of extant sources.

Evidence presented in the preceding sections shows that both Egyptian and Greek lunar months on the whole began just too early for the new crescent to have been visible in the evening immediately preceding daylight of lunar Day 1, with Greek lunar months beginning about a day later than Egyptian lunar months. In other words, first visibility of the new crescent was not used in principle to mark the beginning of the month. The design of the present section is to argue that the same may well have been the case in ancient West Asia. It may be time to question basic assumptions as to how the beginning of the lunar month was determined in ancient West Asia and abandon centuries old conceptions.

A distinction applies between, on the one hand, watching the moon in all its phases including the new crescent and, on the other hand, using the new crescent to determine which daylight period is that of lunar Day 1. There is no lack of evidence that the moon was watched in Mesopotomia. Many reports to this effect have survived.[72] But to my knowledge, nowhere is it stated explicitly in any source either that lunar months began when the new crescent was first sighted or that the daylight period immediately following that evening was daylight of lunar Day 1. Watching the overall behavior moon around new moon, including the new crescent, could have served as a guarantee that the beginning of the lunar month would not stray too far from new moon so that lunar months and lunar cycles from new moon to full moon and back to new moon would mostly overlap with one another. But that is not quite the same as using the first visibility of the new crescent to determine lunar Day 1.

10.2. The Lunar Day 1 Immediately Following Alexander the Great's Death

On June 11, 323 BCE, at 3:00 p.m.–4:00 p.m., or very close to that hour, Alexander the Great died in Babylon.[73] Babylonian astronomical texts leave no doubt that he died on the last day of the month according to the Babylonian calendar, or that daylight of the next day was daylight of lunar Day 1.[74] If the beginning of the new Babylonian month was determined by first visibility of the new crescent, then the new crescent ought to have been for the first time visible just a few hours after Alexander's death. That same evening, the moon rose at 7:52 p.m., 50 minutes after sunset. However, computation reveals that the new crescent could hardly, if at all, be seen that evening.

Lange and Swerdlow give 12 June as date of the evening of first visibility.[75] They use a visibility arc ("the altitude of the lower limb of the moon plus one third of the distance between the centers of moon and sun") of 11.3° as lower limit for possible visibility. In their opinion, "[t]his appears to work as well as any other method—*no method is perfect*—in distinguishing visible from invisible; a few first visibilities have been seen at lower values and a few not seen at higher values".[76] But it may be useful to test the limits by lowering the visibility arc to below Lange's and Swerdlow's lower limit. In fact, when the arc is lowered from 10.66° to 10.65°, their program switches to June 11 as date of the evening of first visibility. The phase of the moon—expressed as the percentage of the moon disk that is illuminated—would be only 0.7% in the evening of June 11. The age of the moon—that is, the distance in time from conjunction to the sunset immediately preceding first visibility—would be plus 15h 13m.[77]

The fact that visibility of the new crescent seems almost impossible in the evening that follows Alexander's death by a couple of hours presses the question: Did or did not the month begin in Babylon with first visibility of the new crescent? A more general version of this question is as follows: Did or did not the month begin anywhere in ancient West Asia with first crescent visibility?

A number of general observations about the evidence are in order. They are presented in the following sections. Together these observations constitute a line of argument favouring the notion that the beginnings of lunar months in Mesopotamia were not narrowly determined by first visibility of the new crescent.

10.3. Two Ways of Establishing the Relation of First Visibility of the New Crescent to Lunar Day 1

There is no lack of ancient West Asian lunar dates. In fact, there is an abundance of them. But only in the case of relatively few of these dates can it be established when exactly first visibility of the new crescent occurred in relation to daylight of Day 1. There are basically two ways in which the location of first crescent visibility in relation to lunar Day 1 can be established: (1) Either a text explicitly states on what day of the month, as signified by a day number, the moon was first seen again (for the sources transmitting such information, see sections 10.11 and 10.13 below). (2) Or the lunar date and its corresponding lunar Day 1 can be dated somehow in the modern Julian calendar (for the sources in which such lunar dates are found, see sections 10.4–10 and 10.12 below). In other words, the ancient lunar date is absolutely dated. It is now possible to compute with reasonably high probability where exactly first crescent visibility fell in relation to any Julian date in antiquity. The same automatically also applies to any ancient lunar date whose Julian equivalent is known.

Among the sources transmitting lunar dates whose Julian equivalent can be known, it seems appropriate to distinguish between astronomical texts (see sections 10.4–10) and non-astronomical texts (see section 10.12). The astronomical texts provide by far the most such lunar dates. It is not clear to which extent the calendrical techniques used in astronomical texts made an impact beyond the narrow confines of the temples in Babylon and Uruk where the astronomers plied their trade. Babylonian astronomy was a highly sophisticated intellectual art practiced by a secretive clan of astronomers. There is evidence that Babylonian astronomers jealously guarded their science.[78] Still, it is difficult to see how, in Babylon and Uruk in the later first millennium BCE, astronomers could have used day dates that differed from those used by the cities' populations at large. Moreover, Babylon was the cultural capital of Mesopotamia. It is therefore reasonable to assume that its calendar was used outside its city walls. There is a small amount of evidence that calendrical information was sent from major cultic centers.[79] Then again, it seems out of the question that the highly sophisticated calendrical techniques of Babylonian astronomy were applied in communities all over West Asia. How many communities in the vast territory of

West Asia were anywhere close to a center of astronomy anyhow? Furthermore, the rules in question could not have played a role before the time when they were invented in the first millennium BCE

What is wrong with the assumption that, in much of West Asia most of the time, the beginning of the lunar month was often determined by certain officials with the sole vague requirement that lunar Day 1 stray not too far from new moon? Keeping it close should have been easy by making specific choices of 29 or 30 as the length of the lunar month in days. As a result, lunar Day 1 would oscillate around new moon. But such oscillation did not have to cause dysfunction. What is wrong with the assumption that, in much of West Asia most of the time, nothing was probably more common than for one of two months associated with the same lunar cycle in two different cities to last 29 days and the other 30 days? And what is wrong with the assumption that, in much of West Asia most of the time, nothing was probably more common than for two months associated with the same lunar cycle in two different cities to begin one, two, or three days apart, probably rarely more?

Conversely, even if the calendrical techniques of Babylonian astronomy were not widespread, they still might reflect practices prevalent over the centuries in Mesopotamian communities at large.

10.4. The Beginning of the Month in Babylonian Astronomical Texts: The Recent Study of the Subject

Just in the last couple of decades, our knowledge of Babylonian astronomy has advanced considerably. One can almost speak of a Renaissance in the field. These advances have also benefited our understanding of the beginning of lunar months. Even so, it has become ever clearer that much still remains to be done when it comes to establishing how the sophisticated Babylonian theories of the third and second centuries BCE were constructed out of raw empirical data in earlier centuries. This stage of Babylonian astronomy is now called the Intermediate stage, to distinguish it both from the more primitive astronomy of the second and early first millennia BCE and the advanced theories of the third and second centuries BCE.

It should be clear right at the outset that none of all this progress is contested in any essential way in what follows. The present focus is on the relation between first visibility of the new crescent and the beginning of the lunar month. The line of argument that is developed below about this relation is seen as somehow complementary to other statements on the same subject.

A brief outline of recent work on the beginning of the lunar month is as follows. The selection of pertinent studies cited below should lead to most everything else that is relevant to the matter.

Owing in great part to efforts by Brack-Bernsen mainly relating to the astronomical facet of the matter and by Hunger mainly relating to the philological facet of the matter, it is now certain that Babylonian astronomers constructed rules to determine the beginning of the lunar month beforehand and also what these rules are.[80] Steele has established that, "in an overwhelming number of cases", different types of Babylonian of astronomical texts "really did agree" on "the beginning and length of Babylonian months".[81] He concludes from this "that during the Seleucid and Parthian periods the day of the beginning of the month was always determined in advance".[82] Stern has devoted a detailed study to the relation between first crescent visibility of the new crescent and the beginning of the new month as evidenced in the astronomical Diaries.[83] The Diaries are day-by-day records of what was observed in the nighttime and daytime skies. They hold much if not most of the ultimate empirical foundation of Babylonian astronomical theories. Other important observations of recent date on the beginning of the lunar month are found in studies by Huber and Steele[84] and by Britton.[85] A recent survey by Steele appeared in *The Oxford Handbook of Cuneiform Culture.*[86]

10.5. The Beginning of the Month and First Visibility of the New Crescent in Babylonian Astronomical Texts

As to the beginning of the month in Babylonian astronomy, one has the impression that the new crescent should have been generally visible in the evening preceding daylight of the corresponding lunar Day 1. That would seem to confirm the notion that the daylight period immediately following the evening of first visibility of the new crescent was daylight of lunar Day 1.

Then again, the score is not perfect. Take the astronomical Diaries, in which what happens in the sky is reported day by day. It has long been known that there are cases in which the lunar month appears to begin a little too early for the new crescent to have been visible already in the evening preceding daylight of lunar Day 1.[87] What is happening here? Sweeping these cases under the rug does not seem advisable.

It is also quite often stated in the Diaries that the new crescent was not seen in the evening immediately preceding daylight of lunar Day 1 or sometimes also that one did not even care to look out for it. Obstacles such as mist and clouds are often cited as the reason. In regard to such cases, it is generally assumed that the possibility of visibility circumstances permitting was predicted.

10.6. The Time Interval Called NA in Babylonian Astronomical Texts

On the whole, sighting the new crescent is not nearly as prominent as one may have expected it to be if it was indeed the defining principle in determining the beginning of the month. Thus, nowhere is it stated explicitly that it had that function. But instead of visibility, there is something else, related to first crescent visibility yet also distinct, that features much more prominently in astronomical texts than first crescent visibility. It is the time interval between sunset and moonset, called NA in Babylonian. NA is as a rule given at the beginning of every month in the Diaries. What is NA?

When the moon reappears soon after new moon after an absence of mostly one and a half to two and a half days, the following is seen. First the sun sets below the western horizon. Then the thin lunar crescent becomes visible above the western horizon about above where the sun has just set. Soon after, the thin crescent sets itself below the western horizon. The interval between the time when the sun cuts the horizon on its downward trajectory and the time when the moon does so in its wake is roughly one to two hours. It is this time interval that is called NA and is measured in UŠ. One UŠ lasts about four minutes.

In recent decades, it has become abundantly clear that the interval NA and other intervals like it served as fundamental and critically important raw empirical data for Babylonian lunar theory. One principal aim of Babylonian astronomy was to describe the course of the moon in all its complexity. Particularly helpful in this regard was the careful description of how the moon moves in relation to the sun. Every day the moon falls behind, as it were, in relation to the sun by on average about 12 degrees. The sun and the moon are like two runners racing across the sky from the eastern horizon to the western horizon with the sun moving faster and catching up with the moon about every 29.5 days at what is called new moon. To describe this movement, the need is for empirical data.

Among phenomena pertaining to the courses of moon and sun that readily present themselves to observation are six time intervals between a horizon crossing of the sun and a horizon crossing of the moon, that is, between the sun cutting eastern or western horizon by rising or setting and the moon cutting either horizon by rising or setting. These six intervals can easily be measured and Babylonian astronomers most eagerly and diligently measured them. They are now called the Lunar Six (the name is due to Abraham Sachs).

Two of these six intervals fall around new moon when sun and moon are close to one another at the same horizon. One is measured when the moon is for the last time seen rising before the sun rises at the end of the month and the other when the moon is for the first time seen setting before the sun sets at the beginning of the month. The latter is called NA, as was noted above. Four of the six intervals fall close to full moon. The first is measured when the moon is seen setting for the last time before the sun rises, the second when the moon is seen setting for the first time after the sun rises, the third when the moon is seen rising for the last time before the sun sets, and the fourth when the moon is seen rising for the first time after the sun sets.

The present concern is exclusively with NA as an indispensable component of Babylonian lunar theory.

10.7. Estimating or Predicting NA

In the Babylonian astronomical Diaries, it is frequently stated that the moon was not seen or that one did not even look in the evening preceding daylight of lunar Day 1. Yet, the length of NA is still as a rule listed. These cases are so numerous and so well-known that it seems superfluous to adduce any here as evidence. It must be assumed that, in such cases, NA was estimated or predicted.

Rules making it possible to determine NA by prediction have indeed come to light (see section 10.4). It is not clear to what extent exactly these rules were applied. According to one of these rules, NA should not be smaller than 10°; according to another, not smaller than 12°.[88] Then again, it is possible to find instances in which NA is smaller than 10° in the Diaries.[89]

Before rules had been designed or when and if existing rules were not applied, no sophisticated theory was necessarily needed to predict NA. Indeed, the following scenario cannot be dismissed out of hand as impossible. In the days preceding the one and a half to two and a half day period that the moon is not visible, someone carefully observing the sky could easily establish roughly how much the moon moves in relation to the sun in one day and roughly estimate where the moon would be in relation to the sun even during the days in which it cannot be seen. The amount by which the moon moves in relation to the sun each day ranges from about 10°46' to about 14°21', for an average of about 12°11'. But the amount does not change abruptly by large increments or diminutions from one day to the next. One could therefore possibly measure the amount accurately in the days before new moon and reliably project the moon's progression forward for a couple of days. Since the crescent is observed in the morning in the days before new moon but in the evening in the days following new moon, an amount corresponding to a lunar movement of a little more than half a day would need to be entered into the estimate beyond a number of full days.

If NA was estimated or predicted, there is the intriguing possibility that it did not matter to Babylonian astronomers if NA became too small for visibility of the new crescent, as long as it remained positive. NA cannot drop so low that it becomes negative, that is, that moonset precedes sunset. In that case, astronomers would not try to identify the evening of first crescent visibility but rather the evening when moonset for the first time follows sunset and no longer precedes it. A negative NA would upset the empirical foundation of Babylonian lunar theory. But it seems as if a positive NA that was too small for visibility would not. It seems moot to speculate what Babylonian astronomers could have thought of a negative NA because negative numbers entered the realm of mathematics only centuries later.

10.8. The Interval Called NA and the First Visibility of the New Crescent

As phenomena, NA and first crescent visibility are related but also distinct. It is critically important to the present line of argument to establish what sets the two apart and what unites them. NA, at least when it is measured and not estimated or predicted, and first crescent visibility are best seen as two distinct facets of a single larger phenomenon. In this case, the single larger phenomenon is the evening of first crescent visibility. One facet is the fact that the light of the moon reaches the eyes of observers, as an event. Another facet is the interval of time between subset and moonset that evening. In the same way, the fact that it rains and exactly how long it rains are two facets of a single event.

As two facets of a single larger phenomenon, NA and first crescent visibility are inextricably connected. However, there is no such necessary connection in the minds of beholders of the two facets. It is altogether possible for the mind to select one facet for attention at the exclusion of the other. For example, one can measure NA without attributing any significance or function to first crescent visibility.

In sum, the fact that NA and first crescent visibility are connected to one another in reality as two facets of a single larger phenomenon does not necessarily mean that there is any connection between the two in the sense that selecting one for attention necessarily implies also paying attention on the other.

10.9. A Universal Assumption regarding First Crescent Visibility in Babylonian Astronomical Texts

It is now universally accepted, or so it would seem, that the lunar month began with first crescent visibility in Babylonian astronomical texts. What is more, it seems generally accepted that Babylonian astronomical texts offer the best proof that the lunar month began with first crescent visibility throughout the towns and cities of Mesopotamia, or at least, that that was the intent, even if certain circumstances caused deviations.

Logically speaking, an alternative assumption concerning astronomical texts cannot be dismissed out of hand. Clearly, the interval called NA mattered a lot to Babylonian astronomers. It is also the first significant empirical datum of the lunar cycle. It therefore naturally belongs at the very beginning of the account of the events of a month. Furthermore, NA is an evening event. Accordingly, if one begins the month with NA, it is necessary to begin the month in the evening. Also, if one begins the month with NA, one can hardly give the first night of the month the same day number as the immediately preceding daylight period, either 29 or 30. It is only natural to begin the month with the number 1. Accordingly, nighttime periods *precede* daylight periods with the same day number. If the month began with NA, then Babylonian astronomers made the existing calendar much more rigorous and were able to impose it on the rest of Babylon and maybe beyond.

It is true that NA and first crescent visibility are two facets of a single larger phenomenon. But as was suggested in section 10.8, it is possible to pay attention to one facet at the exclusion of the other.

Did the lunar month in Babylonian astronomical texts begin with NA or first crescent visibility? Or perhaps even both at the same time? It is difficult to see what could clinch the matter. The instances in which the month begins too early for visibility to be possible while NA remains positive may be viewed as an argument in favour of NA. One thing seems certain. Beginning the month with measured NA, let alone predicted NA, could hardly reflect a practice followed in society at large. It seems much too theoretical and too rigorous. It is therefore not clear what Babylonian astronomical texts tell us about calendrical practices at large. Consequently, the other evidence gains in importance. There are two main types (see 10.3): reports as to on what the day the moon was first seen (see 10.11); lunar dates whose equivalent in the Julian calendar is known (10.12).

But first, a note on the beginning of the day in Babylon is in order.

10.10. The Beginning of the Day in the City of Babylon

It is not clear to what extent, if at all, beginning the day in the evening was observed outside the temple in Babylon where the astronomers were performing their duties. I doubt that a Babylonian asked to join someone for dinner in the evening of lunar Day 5 would show up in the evening that immediately follows daylight of Day 4. Likewise, no one invited to dinner in modern Jerusalem in the evening of "Day 2" (*yom sheni*), that is, Monday, would show up Sunday evening, which follows daylight of "Day 1" (*yom rishon*), just because holidays and the Sabbath begin in the evening. The evening beginning as a Jewish or Muslim custom is limited to religious practice, just as the evening beginning in Babylon may well have been limited to astronomical practice. For most people most of the time, the day must have begun in the morning and ended in the evening and any extensions of activity to before dawn or sunrise and to after dusk or sunset were included. Nights were numberless episodes of inactivity.

10.11. Reports as to on Which Day the Moon Was First Seen

Letters written by Assyrian and Babylonian scholars[90] and astrological reports to Assyrian kings[91] abound with the conditional clauses "if the moon becomes visible on Day 30" and "if the moon becomes visible on Day 1".[92] On the whole, visibility on Day 30 seems to be something bad; visibility on Day 1, something good. The nature of these cuneiform sources is such that there is every reason to believe that they reflect great care in observing the night sky. These sources are among the best we have in terms of diligent observation outside of Babylonian astronomical and certain astrological texts.

It is tempting to interpret the visibility on Day 30 and the visibility on Day 1 mentioned above along the lines of the structure of Babylonian astronomical texts, more specifically the Diaries. I have the impression that this interpretation is widely accepted, even if mostly tacitly or by the absence of an alternative explanation.

According to this interpretation, visibility on Day 30 means that the new crescent becomes visible for the first time at the beginning of nighttime of Day 30, which precedes daytime of Day 30. As in Babylonian astronomical texts, Day 30 is in this case a designation of Day 1 of a lunar month that immediately follows a lunar month of 29 days. Lunar days are either 29 days or 30 days long. But the standard length was considered to be 30 days. Therefore, either a lunar month ended after 29 days and the first day of the following lunar month was called Day 30, with the next day being called Day 2. Or a lunar month ended after 30 days and the first day was called Day 1.

According to this same interpretation, visibility on Day 1 means that the new crescent becomes visible for the first time at the beginning of nighttime of Day 1, which precedes daytime of Day 1.

Considering how often the conditional clauses "if the moon becomes visible on Day 30" and "if the moon becomes visible on Day 1" are attested, one might assume that daylight of lunar Day 1 as a rule immediately followed the evening in which the new crescent was first seen. Upon closer inspection, however, there is much that contradicts this assumption (see below). In the end, I believe it cannot even be positively excluded that first visibility on Day 30 or Day 1 means that the new crescent became visible in the evening immediately *following* daylight of Day 30 or Day 1. In other words, it is not certain that, outside astronomical texts, the day number changed at sunset. Extensions of the workday to before sunrise and to after sunset would then receive the same day number as the daylight period that they are abutting.

It should first be pointed out that the afore-mentioned clauses are conditional clauses. It is not stated that the moon was in fact seen on Day 30 or Day 1. Indeed, one might ask, why attach special significance to the appearance of the new crescent on those days if the whole point was that it became as a rule first visible on those days?

It is not difficult to find evidence in the letters and reports in question that the moon became visible on days other than Day 30 or Day 1. What is more, once the beginning of the lunar month drifted away from first crescent visibility, and lunar months being either 29 days or 30 days long, it may not have been easy to bring daylight of Day 1 back in line with first crescent visibility, if such was the intent. Therefore if one month does not begin with first crescent visibility, others may not have either. A selection of statements to this effect is as follows. A more in-depth study of what is said about the moon around the turn of the month in West Asian documents from the last three millennia BCE remains desirable.

In one astrological report, one finds the following statement: "If the moon at its appearance is visible early, the month will bring worry".[93] There does seem to be a sense that there was an ideal time for the moon to appear and that its appearance could therefore be early in relation to this standard. In an exorcist's report, one finds the conditional clause "if the moon at its appearance becomes visible on Day 28 as if on Day 1". Apparently, there was an awareness as to what the moon was supposed to look like on Day 1. But it is not clear which evening is meant, the one before daylight of Day 1 or the one after daylight of Day 1.

In a letter to an Assyrian king, an exorcist states: "I observed the (crescent of the) moon on Day 30, but it was high, too high to be (the crescent) of Day 30. Its position was like that of Day 2".[94] Clearly, the lunar month did not begin with first crescent visibility. In this case, it began later because the crescent should have been visible the preceding day, circumstances permitting.

In general, it is fairly regularly stated that the moon was not seen when one attempted to sight it. It is superfluous to cite specific instances. It would seem difficult to maintain a rigid criterion such as beginning the lunar month with earliest possible new crescent visibility if absence or presence of the crescent could not be evaluated when it needed to be. What is more, once one month was off kilter, one or more ensuing ones might be too. One solution, as the best one could do under the circumstances, is to keep the beginning of the month fairly close to new moon, which cannot have been all that difficult. Still, it would have been desirable that the new crescent be visible at the beginning of the month, whether in the evening preceding or in the evening following daylight of lunar Day 1.

An alternative modus operandi that may have been used, though to an unknown extent, is presented in section 10.13 below. It involves the creation of a new concept in calendrics. Among postulated ancient calendars, the old crescent is as a rule associated with invisibility and the new crescent with visibility in that the beginning of the lunar month is marked either by the invisibility of the old crescent or the visibility of the new crescent. But what if both visibility and invisibility were associated with the new crescent? I believe that much evidence outside Babylonian astronomy strictly speaking can be construed in favour of a lunar calendar in which the beginning of months was determined by new crescent (in)visibility.

10.12. Lunar Dates Whose Equivalent in the Julian Calendar Is Known in Non-astronomical Sources

The only sources outside Babylonian astronomical texts that may make it possible to establish the Julian dates of the Babylonian lunar Day 1 are, as far as I know, Aramaic double dates from Egypt.

As J. K. Fotheringham was the first and the only ever to observe, the Julian dates of lunar Day 1 fall just *a little too early* for first crescent visibility to have served as marker of the beginning of the lunar month.[95] Stern more generally notes that "perhaps at Elephantine, visibility of the new moon was *not* used as a criterion to determine when the new month began", without specifying whether first crescent visibility came too early or too late.[96]

There are 14 completely preserved double dates.[97] The average distance between conjunction and 6:00 a.m. of Day 1 is about +21h 27m 51s or, without the aberrant no. 3, +30h 30m 28s. The first average is in tune with that obtained from Greek lunar dates; the second falls a few hours later. In any event, even in the latter case, the average distance from conjunction to the evening before daylight of lunar Day 1 is about 18 to 19 hours. This interval is on average too short for first crescent visibility to serve as marker of the month's beginning. Indeed, any interval larger than 18 to 19 hours, say 24 hours, needs to be compensated by one that is shorter, say, 13 to 14 hours, an interval that is clearly too short for the first crescent to have been seen.

It seems difficult to assume that the Western Asian mercenaries on the Egyptian island of Elephantine adopted the manner of beginning lunar months used behind the walls of

Egyptian temples. The Egyptian lunar calendar was strictly religious and almost secretive.

10.13. A New Ancient Lunar Calendar: The New Crescent (In)visibility Calendar

Among the cuneiform sources that are not strictly astronomical, one finds reports of lunar observation pertaining to the calendar. They include a set of astrological reports to Assyrian kings[98] and a set of letters from Assyrian and Babylonian scholars[99] dated to the Neo-Assyrian period (earlier first millennium BCE). In light of how little has been preserved, these sources are as good as it gets when it comes to illustrating how the beginning of lunar months was determined in Mesopotamia outside of astronomical texts. An attempt is made below to reconstruct the principles according to which the beginning of lunar months was determined according to these sources. It remains unclear, naturally, to which extent exactly these rules were used in society at large, be it in the Neo-Assyrian period or in other epochs of Mesopotamian history.

If the beginning of lunar months was strictly determined by means of first crescent visibility, the following would need to be typical in the sources at hand: daylight of lunar Day 1 immediately followed the evening in which the new crescent was first seen. There are two possibilities at this point.

First, if the day began in the evening as it does in Babylonian astronomical texts, the new crescent would typically first be seen on Day 30 at the end of a 29-day lunar month or on Day 1 at the end of a 30-day period. In Babylonian astronomical texts, Day 30 is the designation for the first day of a lunar month that immediately follows a 29-day lunar month.

Second, alternatively, if human activity taking place before dawn or after sunset was dated by the same day number as the daylight period it abuts, the new crescent would typically first be seen either on Day 29 at the end of a 29-day lunar month or on Day 30 at the end of a 30-day lunar month. The next day would in both cases be Day 1 of the next lunar month.

Instead, what one finds in the sources is something entirely different. The fundamental fact is that there is abundant reference to sighting the crescent on Day 29, on Day 30, and on Day 1, *three* different days and not two. This cannot be reconciled with what is found in Babylonian astronomical texts. But what does it mean?

Step One: Sighting on Day 29 as opposed to Sighting on Day 30 and Day 1.—The point of departure of the present line of argument is a striking difference between the sightings on Day 29 and the sightings on Day 30 and Day 1. The sightings on Day 29 are actual sightings as a rule reported in a statement such as "We saw the moon". By contrast, the sightings on Day 30 and Day 1 are hypothetical sightings as a rule described by the conditional clauses "if the moon becomes visible on Day 30"[100] and "if the moon becomes visible on Day 1".[101] Visibility on Day 30 seems to be something bad; on Day 1, something good. Visibility on Day 1 is occasionally predicted, as in "Now, in intercalary Adar, on Day 1 the moon will become visible",[102] or reported, as in "On Day 1, the moon became visible",[103] but this is rare.

Step Two: The Timing of the Moon Watch.—Step two of the present line of argument is that there can be no doubt about the following fact. The evening in which one as a rule actively watched the sky in search of the new crescent is that of Day 29; the new crescent is either sighted or not sighted at that time.[104]

A second, additional, watch may be held on Day 30, as in the following reports: "[We kept] watch for the moon. On Day 29, there were clouds. [We did not see the moon.] On the next day, [. . .] it was two days old";[105] "We watched on Day 29. The clouds were den[se]. We did not see the moon. We watched on Day 30. We saw the moon. It was very high. The (weather) of Day 29 has to do with it. What is it that the king says?";[106] and "We kept wa[tch on Day 29]. There were clouds. We did not see the moon. T[od]ay, on Day 30, there were c[lou]ds again. When they di[spersed], we saw [the mo]on. [It] was [(not) like the] moon of [Day] 29".[107]

In one case, a watch on Day 30 alone is reported;[108] one assumes, in light of what else is known, that another watch may well have taken place on the preceding day.

So when it is reported that "They watched the moon. The clouds were dense. The moon was not seen. Day 30 became long",[109] it may confidently be assumed in light of the specification "Day 30 became long" (that is, the month had 30 days [see below]) that the watch took place on Day 29.

Step Three: The Evening of Day 29, the Time of Sighting, as the Evening Following Daylight of Day 29.—Before trying to reconstruct how the Mesopotamian lunar calendar functioned according to texts other than Babylonian astronomical texts, an important question needs to be settled. It is clear from the sources that the evening of Day 29 was as a rule the evening in which a watch was kept to look out for the new crescent. There is also no doubt that, in Babylonian astronomical texts, the evening of Day 29 immediately follows daylight of Day 28. The question arises: Does that same evening immediately follow daylight of Day 28 or daylight of Day 29 in other texts?

It seems somehow natural that day numbers would not change at the moment of sunset in society at large. Changing the day number at sunset seems only possible in the highly systematized calendar of Babylonian astronomical texts. But the principal consideration in favour of assuming that the evening of Day 29 immediately followed daylight of Day 29 is as follows.

Let us assume that the evening of Day 29 immediately followed daylight of Day 28. The main problem of a lunar calendar is to determine whether the month will have 29 days or 30 days. This is the same as determining whether the day following Day 29 will be the last of the same month or the first of the next month. It seems difficult to make this decision just after daylight of Day 28. The next daylight period will always be that of Day 29. There is otherwise no indication in the day-names in Mesopotamia, as there is in the Greek world, that a decision was made one day ahead of the time, that is, in the evening just after daylight of Day 28 about the daylight period that is next after daylight of Day 29.

If the evening of Day 29 immediately follows daylight of Day 29, then the evenings of Days 30 and 1 immediately follow daylight of Days 30 and 1 respectively. Since cuneiform texts often refer to the possibility of first visibility of the new crescent on Day 1 (see above), it must be assumed that it was considered altogether possible that the lunar month began before the new crescent was first seen. Still, it may be considered a good omen that the new crescent made its first appearance on Day 1. First appearance on Day 1 seems like a good sign rather than a bad sign according to the sources.

Step Four: A Day-29-Rule.—According to the evidence cited above, no time is anywhere nearly as prominent as a time when a moon watch is kept than the evening of Day 29. It is also clear that the new crescent is either sighted or not sighted at that time. The best assumption is that the next daylight period is that of the last day of the month, Day

30, if the new crescent is sighted and Day 1 of the next month when the new crescent is not sighted, even if this principle is not explicitly stated as such in the sources. The focus on the evening of Day 29 suggests the desirability of the Day 29-rule just postulated. In practice, however, the rule was probably often if not very often not applied and the evidence supports this assumption (see Step Six below).

Step Five: The Evening of Day 29 as a Systematic Time of Sighting and First Crescent Visibility on Days 30 and 1.—The evidence cited above suggests that Days 29, 30, and Day 1 are all three days on which sighting of the moon is considered possible if not normal. The Day 29-Rule described above accords with this impression. The sources leave no doubt that the new crescent could be sighted on Day 29. But the sources also make it clear that it could not be sighted at that time. It is a fact that first visibility of the new crescent falls on the first day after the day of first invisibility of the old crescent in about 55.18% of all cases and on the second day in 44.30% of all cases; it falls on the same day in 1.04% of all cases and on the third day in 0.48% of the cases.[110] According to the Day 29-rule (see Step Five), the lunar month has 30 days if the new crescent is invisible in the evening of Day 29. Accordingly, the moon is seen in about half of all cases in the evening immediately following daylight of Day 30 and in about half of all cases in the evening immediately following Day 1. Both are normal. In the latter case, the lunar month begins after first crescent visibility.

The fact that the new crescent is very rarely first visible on the third day after first invisibility of the old crescent, only in 0.48% of all cases, seems to be confirmed by the following statement: "The moon disappeared on Day 27. On Day 28 and Day 29, it stayed inside the sky. It was seen (again) on Day 30. When (else) should it have been seen? It should stay inside the sky less than four days. It never stayed four days".[111] Presumably, visibility on Day 30 is interpreted as "staying inside the sky" for four days and this is considered a problem. The author of the text apparently expected visibility on Day 28 or on Day 29.

Step Six: Fluctuation.—While it seems clear that there is some kind of focus on keeping the watch on Day 29, and while it is reasonably easy to account for this focus by postulating a Day 29-rule, and while this Day-29-rule easily accords with the fact that both Day 30 and Day 1 were considered thinkable as the day of first crescent visibility, and while there was an explicit sense that the old crescent could disappear and the new crescent appear at an "inappropriate time"[112] and by implication also at an appropriate time, one seriously wonders to what extent any kind of rule was neatly applied. For example, the new crescent could become visible "on Day 28 as if on Day 1"[113] and the new crescent could after all become visible at an "inappropriate time".

All in all, it cannot have been all that difficult to keep the beginning of the lunar month close to new moon. To that aim, astronomers would advise kings in their choices of 29-day and 30-day months, for example by planning a couple of 30-day months in succession beforehand if the new crescent seemed to have begun appearing on the whole a little too early. Historically speaking, there is every reason to believe that the rigorous lunar calendar of the Babylonian astronomers was somewhat isolated as a phenomenon.

11. The Sociology and Politics of Living Three Lunar Calendars

A principal design of the conference "Living the Lunar Calendar", at which an extract of the present paper was read, was to probe the sociopolitical ramifications of life with a lunar calendar. The subject acquires a certain additional complexity if one considers what it must have meant to live with more than one lunar calendar. The aim of the present paper so far has been to demonstrate that there were in fact two lunar calendars in simultaneous use in Ptolemaic Egypt in the third century BCE. Before considering sociology or politics, this tenet needed to be demonstrated first. The coexistence of the two lunar calendars was short-lived, however. The Greco-Macedonian calendar lost its lunar connection later in the third century BCE. In the end, it was assimilated completely to the Egyptian civil calendar of 12 months of 30 days plus five added days for a total of 365 days. Even when it was independent, the Greek lunar calendar played only a secondary role, one subordinated to the civil calendar, even if the Greek lunar date was listed first in double dates in deference to the awareness of who was really in charge in Egypt.

But two lunar calendars is not where it stops. A third lunar calendar was used in Egypt by the Jewish diaspora. The Jewish calendar is a descendant of the lunar calendar of Babylonia, to which the Israelites had been exiled in the sixth century BCE.

How was it possible for three lunar calendars to coexist in a single country? If coexistence was the case, three lunar months beginning on three different days may on occasion have been running alongside one another in Egypt. How was this practicable? The validity of the lunar dates derived above from double dates as evidence critically depends on the practicability of lunar calendars existing alongside one another in close vicinity. Indeed, one might ask: Why would a native Egyptian speaker and a Greek speaker both living in a single small Egyptian town use two different calendars? If the close coexistence of two lunar calendars was not practicable, then the numerical data presented and analyzed at length above must be considered one huge statistical aberration.

There are no sources in which ancient Egyptians explicitly tell us how the lunar calendar affected their lives. The practicability of coexisting lunar calendars therefore will need to be demonstrated in other ways. The key observation that can serve as point of departure of such a demonstration is how completely isolated from one another the world of the Egyptian lunar calendar and the world of the Macedonian-Greek lunar calendar were. The Egyptian lunar calendar was entirely religious and used exclusively in the somewhat dark and secretive confines of temple life and hieroglyphic culture. The Greek lunar calendar is mainly represented in business documents written by townspeople. Egyptian priests and Greek businessmen hardly talked with one another. In all probability, only very few native speakers of Greek also spoke Egyptian, and vice versa. The two power blocs perennially competing with one another in Ptolemaic Egypt were the native Egyptian priesthood and the Greek-speaking state bureaucracy with a Pharaoh of Macedonian descent at the top. Both needed to be financed and relied on taxation. The two staked out their respective territories by signing treaties, such as the Memphis Decree of 196 BCE that is inscribed in Egyptian and Greek on the Rosetta Stone. The Rosetta Stone is a point of contact between two communities that largely led separate existences within the borders of a single country.

An additional question is whether it was possible for the Greek lunar month to begin on the same day everywhere in Egypt. Even if there was no time to dispatch messengers to everywhere in Egypt to communicate the beginning of the month, the principles followed to date a Greek document written in the Fayyum must have been basically the same as those followed in dating a document written in Greek-speaking Alexandria and different from those followed behind the thick stone walls of a nearby Egyptian temple enclosure.

Last but not least, there is the Jewish lunar calendar. Like the Egyptian lunar calendar, it must have been 100% religious. Its coexistence with other calendars in modern times easily serves as an analogy of how it must have coexisted with other calendars in ancient times. The Jewish calendar is used today all over the world solely to regulate religious festivals. But for that pillar of Judaism, the celebration and observation of the Sabbath, it is obviously not needed.

Notes

1. Depuydt (2002), pp. 471–477. Some of what follows has already years before the Jerusalem conference "Living the Lunar Calendar" been part of papers read at Annual Meetings of the American Oriental Society held in March 1997 in Miami, Florida, in March 2002 in Houston, Texas, and in March 2004 in San Diego, California and entitled "The Beginning(s) of Day and Month in Egypt and the Ancient World", "The Day in Antiquity", and "First Crescent Visibility's Irrelevance: Additional Evidence, including from Babylonian Astronomical Texts" respectively. These papers mostly concern the moon. But one often travels or moves to the cities where they were read in search of the sun. I thank Peter J. Huber for performing critical calculations, not only in relation to the present paper, but also on various other occasions in relation to investigations on Middle and New Kingdom chronology that have not—and, one hopes, not yet—come to fruition. I also thank Chris Bennett for reading semifinal and penultimate versions of the present paper and making several helpful comments, including pointing out a couple of erroneous numbers.
2. Depuydt (1996), p. 42; Depuydt (1997b), p. 127 (at line 18, for "last crescent visibility" read ". . . invisibility"); Depuydt (2007), 76 note 43; and Depuydt (forthcoming), section C.2.
3. Kugler (1922), p. 2 ("The first reappearance of the slim sickle in the western evening sky is the sign for the beginning of the month").
4. See, for example, Beaulieu (1993), Britton (2007), p. 115–119, Huber (1982), and Steele (2011), p. 478–481.
5. Hannah (2005), p. 43.
6. Pritchett (1959), p. 152.
7. Pritchett and Neugebauer (1947).
8. Jones (2005), p. 165 note 2.
9. The five principal contributions are Pritchett (1959), (1970), pp. 39–89, (1982), (1999), and (2001).
10. Parker (1950), pp. 9–23 ("Chapter I: The Beginning of the Egyptian Lunar Month"); Parker (1957), p. 39.
11. Pritchett (1959), p. 154.
12. Pritchett (1959), p. 152.
13. Pritchett (1959), p. 153.
14. Pritchett (1970), p. 73.
15. Pritchett (1959), p. 154.
16. Hammond and Scullard (1970), pp. 192–193.
17. Hammond and Scullard (1970), p. v.
18. Hammond and Scullard (1970), p. 192.
19. Pritchett (1970), pp. 72–73.
20. Pritchett (1970), p. 73.
21. The diagram at Pritchett (1970), p. 72 clearly shows daylight preceding nighttime in the same day. The expression "at dusk following the completion of the 28th day" makes it seem as if the day ends in the evening; but morning as the completion of the 28th day also seems possible as an interpretation of this phrase.
22. As noted at the outset of this paper, following in the footsteps of others, I used to consider the problem of when the day begins important. It is therefore slightly unfortunate that the only time Pritchett cites me, as far as know, it is to adduce my support of the significance of precisely that problem (Pritchett (2001), p. 95, note 2, referring to Depuydt (1997a), p. 27).
23. Pritchett (1970), p. 73.
24. Pritchett (1982), pp. 260–266.

25. Pritchett (1970), p. 72.
26. Pritchett (1982), p. 72.
27. Pritchett (1982), pp. 260–266.
28. When Pritchett ((1999), p. 88) writes that Pritchett (1982) "correct[s]" Pritchett (1959), pp. 151–154 and Pritchett (1970), pp. 72–73, which are "based on a theory of a day of evening visibility", two matters seem to have been confused, namely the beginning of the day and the beginning of the month. Pritchett (1982) "corrects" Pritchett (1959) in that it is assumed in Pritchett (1982), as it apparently is in Pritchett (1970), that the day began in the morning whereas it is assumed in Pritchett (1959) that it began in the evening. Pritchett (1982) "corrects" Pritchett (1970) in that it is assumed in Pritchett (1982), as it is in Pritchett (1959), that the beginning of the month was determined by watching the old crescent in the morning whereas it is assumed in Pritchett (1970) that it was determined by watching the new crescent in the evening.
29. Depuydt (1998).
30. Pritchett (1982), p. 266.
31. Pritchett (1982), p. 266.
32. Goldstine (1973).
33. Goldstine (1973), p. 37.
34. Thus Lange and Swerdlow (2006). Whenever dates of visibility and invisibility phenomena are mentioned in what follows without explicit reference to a source, the source is Lange and Swerdlow (2006).
35. After reading a semifinal version of this paper, Chris Bennett—who has a monograph in press on the history of the Macedonian calendar in Egypt for the series Studia Hellenistica—reported that there are only a handful more, dating to Roman times (personal e-mail communication of October 5, 2010). They have not been included in this paper. We agreed that they do not appear essential to this paper's line of argument. For information about them, I refer to Bennett's forthcoming study. [This study has now appeared as Bennett (2011).]
36. Grzybek (1990), pp. 135–137.
37. Grzybek (1990), pp. 135 note 42, 151–155.
38. See Grzybek (1990), p. 137 note 47.
39. See footnote ** in table 2.
40. For example, as part of his comments on a semifinal version of the present paper, Chris Bennett suggested using "Alexandria-only" dates, 15 according to his count (personal e-mail communication of October 5, 2010). The results do not deviate in any significant way from those obtained in the other selections.
41. See, for example, Pritchett (1959), p. 152.
42. Lange and Swerdlow (2006).
43. Personal e-mail communication of February 11, 2009.
44. See, for example, Parker (1950), p. 12 §§38–39.
45. Parker (1950), p. 13 §47.
46. Peter J. Huber, personal e-mail communication of February 11, 2009.
47. Pritchett (1982), pp. 260–261.
48. Bennett (2008). In Depuydt (1997a), pp. 184–186, I styled two temple service dates (nos. 5 and 24 in the tables in Bennett (2008), pp. 547–551) as "double dates". In a certain regard, they are. But in the present paper, I restrict the term "double date" to cases in which the lunar date is a day in the Egyptian lunar month and not a certain day in a temple service term.
49. Bennett (2008).
50. See also Depuydt (1998).
51. Luft (1992), pp. 189–195, 205–208.
52. See the tables in Bennett (2008), cols. 547–552.
53. Bennett (2008), cols. 551–552 note 82.
54. Personal e-mail communication of October 21, 2010.
55. An important early discussion of the dates is Bilfinger (1888), pp. 77–88 (dates with Macedonian month names), 147–154 (dates with Athenian month names).

56. Hannah (2008), p. 94 somehow obtains 18 November, 30 October, and 1 March as well as 14 November, 16 or 17 October, and 25 February as lunar Day 1, but it is not fully clear to me how.
57. van der Waerden (1960), p. 71; cf. Depuydt (1996), pp. 31–32.
58. Depuydt (1996).
59. Depuydt (1996), pp. 40–42. At p. 41, with note 30, conjunction for Babylon should have been conjunction for Alexandria. But in this case, the difference does not affect the conclusions drawn.
60. Sachs and Hunger (1988), p. 195.
61. Personal e-mail communication of February 11, 2009.
62. Epping (1889), p. 179.
63. Ideler (1825–1826), vol. 1, p. 279.
64. See, for example, Pritchett (1982), pp. 260–266 for the Greek calendar, and Depuydt (1998) for the Egyptian lunar calendar.
65. Pritchett (1982), p. 261 note 53.
66. Bennett (2008), col. 535 note 43 and col. 548, repeating a personal communication by Quack cited elsewhere.
67. For the related concept of the "straddle" month, a month encompassing a point in time that serves as a herald of the beginning of the year, in the sense that the first new moon following it is the beginning of lunar Month 1, see Depuydt (1997), pp. 43, 213–215.
68. Westermann and Kraemer (1926), pp. 1–59 (no. 1).
69. Westermann and Kraemer (1926), p. 22 note 1.
70. Pritchett (1982), p. 256.
71. Lange and Swerdlow (2006). I chose modern dates because, in the version of the program that I downloaded, Δt had been disregarded in computing the age of the moon. However, around 1900 CE, Δt is zero or close to zero.
72. See, for example, Hunger (1992), passim and Parpola (1993), passim.
73. Depuydt (1997b) (for errata, see Depuydt (2008), p. 74); Depuydt (2008), pp. 47–51.
74. Depuydt (1997b), pp. 121–124.
75. Lange and Swerdlow (2006).
76. See the section on the moon in the text entitled "Computation visibility phenomena" accompanying the program Lange and Swerdlow (2006).
77. In a personal e-mail communication of October 9, 2010, Peter J. Huber urged me to consider that the probability that the first crescent was seen in the evening of 11 June 323 BCE is not entirely negligible. Naturally, additional assessments of the probability in question remain desirable and would surely be welcomed.
78. Brack-Bernsen (2010), p. 278.
79. Beaulieu (1993); Steele (2011), p. 478.
80. See especially Brack-Bernsen (1997), (1999), and (2002), and Brack-Bernsen and Hunger (2002). Brack-Bernsen (2010) is a survey for a wider audience, with additional observations.
81. Steele (2007). The quote is from Brack-Bernsen (2010), p. 296.
82. Steele (2007), p. 143.
83. Stern (2008).
84. Huber and Steele (2008).
85. Britton (2008).
86. Steele (2011).
87. Such instances are gathered in Stern (2008). The author does otherwise assume that lunar months began with first crescent visibility in Babylonian astronomical texts.
88. For a summary with references See Stern (2008), p. 37 note 11.
89. See, for example, Sachs and Hunger (1996), pp. 210–211 (tablet BM 45830).
90. Parpola (1993).
91. Hunger (1992).
92. See, for example, the "list of text headings" in Hunger (1992), pp. 369–373.

93. Hunger (1992), p. 32 no. 53.
94. Parpola (1993), p. 177 no. 225.
95. Depuydt (2002), p. 473.
96. Stern (2000), p. 164.
97. Depuydt (2002), p. 475.
98. Hunger (1992).
99. Parpola (1993).
100. The following list of instances found in Hunger (1992) should be complete or very nearly complete, but the difference hardly matters statistically speaking: p. 10 nos. 11–12, p. 11 no. 13, p. 32 no. 53, pp. 35–36 no. 60 (twice), p. 36, no. 61, p. 51 no. 85 (twice), p. 66 no. 107, p. 75 nos. 120–121, p. 90 no. 150, p. 101 no. 171, p. 103 no. 176, p. 105 no. 183, p. 111 nos. 191–193, p. 136 no. 244 (twice), p. 137 no. 246, p. 146 no. 262 (Hunger: "on the 1st day"), p. 146 no. 263, p. 147 nos. 246–265, p. 161 no. 292, p. 170 no. 304, p. 171 no. 305, p. 180 no. 319, p. 188 no. 331, p. 196 no. 344–345, p. 197 no. 346, p. 204 no. 358, p. 213 no. 375, p. 224 no. 390, p. 225 nos. 391–393, p. 241 no. 424, p. 249 nos. 440–442, p. 264 no. 472, p. 266 no. 478, p. 268 no. 485, p. 274 no. 500, p. 284 no. 512, p. 285 no. 514.
101. The following list of instances found in Hunger (1992) should be complete or very nearly complete, but the difference hardly matters statistically speaking: p. 8 no. 7, p. 9 no. 9, p. 10 no. 10, p. 20 no. 37, p. 35 no. 57, p. 51 no. 86 (twice), p. 52 no. 87, p. 72 no. 113, p. 74 nos. 116–119, p. 90 nos. 148–149, 98 no. 165, p. 109 no. 188, p. 110 no. 189, p. 140 no. 251, pp. 140–141 no. 252, p. 143 nos. 256–257, p. 144 no. 258, p. 145 no. 259, p. 161 nos. 290–291, p. 180 no. 318, p. 187 nos. 329–330, p. 195 no. 342, p. 211 no. 372 ("It becomes visible on Day 1" as a hypothetical phenomenon), p. 212 no. 373, p. 232 no. 409, p. 238 no. 418, p. 239 no. 420, p. 240 nos. 421–422, p. 241 no. 423, p. 249 no. 439, p. 251 no. 446, p. 254 no. 449, p. 256 no. 456, p. 260 no. 463, p. 271 no. 492, p. 274 no. 498, p. 281 no. 505, p. 282 no. 506 (twice), p. 282 no. 507, p. 283 no. 508–510, p. 284 no. 511.
102. Hunger (1992), p. 141 no. 253; also p. 148 no. 267 (twice).
103. Hunger (1992), p. 257 no. 456.
104. In the following list, cases in which the new crescent was visible on Day 29 are marked by "crescent sighted" and cases in which the new crescent was not visible on Day 29 are marked by "crescent not sighted": Hunger (1992), p. 45 no. 79 (not sighted, according to Hunger's plausible reconstruction), p. 75 no. 120 (not sighted), pp. 80–81 nos. 126–129 (sighted), and p. 81 nos. 130–132 (not sighted); Parpola (1993), p. 105 no. 126 (sighted), p. 109 no. 138 (not sighted), p. 110 no. 139 (not sighted), p. 110 no. 140 (not sighted), p. 111 no. 141 (not sighted), p. 111 no. 142 (sighted), p. 112 no. 146 (not sighted; Parpola plausibly reconstructs Day 29 as day of the watch). In these 16 watches taking place in the evening of Day 29, the new crescent was sighted in 6 instances and not sighted in 10 instances.
105. Hunger (1992), p. 45 no. 79.
106. Hunger (1992), p. 75 no. 120.
107. Parpola (1993), p. 112 no. 146.
108. Hunger (1992), p. 82 no. 133.
109. Parpola (1993), p. 112 no. 145.
110. Peter J. Huber, personal e-mail communication of February 11, 2009.
111. Hunger (1992), p.197 no. 346.
112. Hunger (1992), p. 52 no. 88 (probably new crescent), p. 197 no. 346 (old crescent disappears on Day 27), and p. 219 no. 382 (old crescent disappears on Day 24).
113. Hunger (1992), p. 11 no. 14

References

Beaulieu, P.-A., 1993, "The Impact of Month-lengths on the Neo-Babylonian Cultic Calendar", *Zeitschrift für Assyriologie und vorderasiatische Archäologie* 83, 66–87.

Bennett, C., 2008, "Egyptian Lunar Dates and Temple Service Months", *Bibliotheca Orientalis* 65, 525–554.

——, 2011, *Alexandria and the Moon: An Investigation into the Lunar Macedonian Calendar of Ptolemaic*

Egypt, Studia Hellenistica 52 (Leuven: Peeters).

Bilfinger, G., 1888, *Der bürgerliche Tag: Untersuchungen über den Beginn des Kalendertages im Classischen Altertum und im Christlichen Mittelalter* (Stuttgart: Kohlhammer).

Brack-Bernsen, L., 1997, *Zur Entstehung der babylonischen Mondtheorie: Beobachtung und theoretische Berechnung von Mondphasen*, Boethius 40 (Stuttgart: Franz Steiner Verlag).

——, 1999, "Goal-Year Tablets: Lunar Data and Predictions", in N. M. Swerdlow (ed.), *Ancient Astronomy and Celestial Divination* (Cambridge, Mass.: MIT Press), 149–177.

——, 2002, "Predictions of Lunar Phenomena in Babylonian Astronomy", in J. M. Steele and A. Imhausen (eds.), *Under One Sky: Astronomy and Mathematics in the Ancient Near East*, Alter Orient und Altes Testament 297 (Münster: Ugarit-Verlag), 5–19.

——, 2010, "Methods for Understanding and Reconstructing Babylonian Predicting Rules", in A. Imhausen and T. Pommerening (eds.), *Writings of Early Scholars in the Ancient Near East, Egypt, Rome, and Greece: Translating Ancient Scientific Texts*, Beiträge zur Altertumskunde 286 (Berlin: De Gruyter), 277–297.

——, and Hunger, H., 2002, "TU 11, a Collection of Rules for the Prediction of Lunar Phases and of Month Lengths", *SCIAMVS* 3, 3–90.

Britton, J. P., 2007, "Calendars, Intercalations and Year-lengths in Mesopotamian Astronomy", in J. M. Steele (ed.), *Calendars and Years: Astronomy and Time in the Ancient Near East* (Oxford: Oxbow Books), 115–132.

——, 2008, "Remarks on Strassmaier 400", in M. Ross (ed.), *From the Banks of the Euphrates: Studies in Honor of Alice Louise Slotsky* (Winona Lake, Indiana: Eisenbrauns), 7–33.

Depuydt, L., 1996, "The Egyptian and Athenian Dates of Meton's Observation of the Summer Solstice (–431)", *Ancient Society* 27, 27–45. Three erroneous numbers in this article are "27,758¼" (for 27,758¾) at p. 37 line 20, "3 July" (for 15 July) at p. 43 line 20, and "27,759" (for 55,518) at p. 45 line 20.

——, 1997a, *Civil Calendar and Lunar Calendar in Ancient Egypt*, Orientalia Lovaniensia Analecta 77 (Leuven: Uitgeverij Peeters and Departement Oosterse Studies).

——, 1997b, "The Time of Death of Alexander the Great: 11 June 323 B.C. (–322), ca. 4:00–5:00 PM", *Die Welt des Orients* 28, 117–135.

——, 1998, "The Hieroglyphic Representation of the Moon's Absence (*Psḏntyw*)", in L. H. Lesko (ed.), *Ancient Egyptian and Mediterranean Studies in Memory of William A. Ward* (Providence, Rhode Island: Department of Egyptology), 71–89.

——, 2002, "The Date of Death of Jesus of Nazareth", *Journal of the American Oriental Society* 122, 466–480.

——, 2007. "Calendars and Years in Ancient Egypt: The Soundness of Egyptian and West Asian Chronology in 1500–500 BC and the Consistency of the Egyptian 365-Day Wandering Year", in J. M. Steele (ed.), *Calendars and Years: Astronomy and Time in the Ancient Near East* (Oxford: Oxbow Books), 35–81.

——, 2008, *From Xerxes' Murder (465) to Arridaios' Execution (317): Updates to Achaemenid Chronology (Including Errata in Past Reports)*, British Archaeological Reports S1887 (Oxford: Archaeopress). A sheet with addenda and errata is available from the author and may be published somewhere in the near future.

——, forthcoming, "Calendar, Ancient Near East", in R. S. Bagnall et al. (eds.), *Wiley-Blackwell's Encyclopedia of Ancient History*. The bibliography in this article could not exceed 15 items. I promised to list additional bibliography in the present article, but have refrained from doing so in light of space constraints.

Epping, J., 1889, *Astronomisches aus Babylon*, Ergänzungshefte zu den Stimmen aus Maria-Laach 44 (Freiburg: Herder'sche Verlagsbuchhandlung).

Goldstine, H. H., 1973, *New and Full Moons 1001 B.C. to A.D. 1651*, Memoirs of the American Philosophical Society 94 (Philadelphia: American Philosophical Society).

Hammond, N. G. L. and Scullard, H. H., 1970, *The Oxford Classical Dictionary* (Oxford: At the Clarendon Press).

Hannah, R., 2005, *Greek and Roman Calendars: Constructions of Time in the Classical World* (London: Duckworth).

Huber, P. J., 1982, *Astronomical Dating of Babylon I and Ur III*, Occasional Papers on the Near East 1/4 (Malibu: Undena).

——, and Steele, J. M. (2007), "Babylonian Lunar Six Tablets", *SCIAMVS* 8, 3–36.

Hunger, H., 1992, *Astrological Reports to Assyrian Kings*, State Archives of Assyria 8 (Helsinki: Helsinki University Press).

Ideler, L., 1825–26, *Handbuch der mathematischen und technischen Chronologie, aus den Quellen bearbeitet*, 2 vols. (Berlin: August Rücker).

Jones, A., 2006, "Ptolemy's Ancient Planetary Observations", *Annals of Science* 63, 255–290.

——, 2007, "On Greek Stellar and Zodiacal Date-Reckoning", in J. M. Steele (ed.), *Calendars and Years: Astronomy and Time in the Ancient Near East* (Oxford: Oxbow Books), 149–167.

Kugler, F. X., 1922, *Von Moses bis Paulus: Forschungen zur Geschichte Israels nach biblischen und profangeschichtlichen insbesondere neuen keilinschriftlichen Quellen* (Münster: Aschendorff).

Lange, R. and Swerdlow, N. M., 2006, Version 3.1.0 of the program "Planetary, Lunar and Stellar Visibility" dated November 20, 2006 and downloaded from www.alcyone.de in early 2009.

Luft, U., 1992, *Die chronologische Fixierung des ägyptischen Mittleren Reiches nach dem Tempelarchiv von Illahun*, Österreichische Akademie der Wissenschaften, Philosophisch-historische Klasse, Sitzungsberichte 598 (Vienna: Verlag der Österreichischen Akademie der Wissenschaften).

Parker, R. A., 1950, *The Calendars of Ancient Egypt*, Studies in Ancient Oriental Civilization 26 (Chicago: University Press).

——, 1957, "The Lunar Dates of Thutmose III and Ramesses II", *Journal of Near Eastern Studies* 16, 39–43.

Parpola, S., 1993, *Letters from Assyrian and Babylonian Scholars*, State Archives of Assyria 10 (Helsinki: Helsinki University Press).

Pritchett, W. K., 1959, "The Athenian Lunar Month", *Classical Philology* 54, 151–157.

——, 1970, *The Choiseul Marble* (Berkeley and Los Angeles: University of California Press).

——, 1982, "The Calendar of the Gibbous Moon", *Zeitschrift für Papyrologie und Epigraphik* 49, 243–266.

——, 1999, "Postscript: The Athenian Calendars", *Zeitschrift für Papyrologie und Epigraphik* 128, 79–93.

——, 2001, *Athenian Calendars and Ekklesias*, ΑΡΧΑΙΑ ΕΛΛΑΣ: Monographs on Ancient Greek History and Archaeology 8 (Amsterdam: J.C. Gieben).

——, and Neugebauer, O., 1947, *The Calendars of Athens* (Cambridge, Mass.: Harvard University Press).

Sachs, A. J. and Hunger, H., 1988, *Astronomical Diaries and Related Texts from Babylonia, I: Diaries from 652 B.C. to 262 B.C*, Österreichische Akademie der Wissenschaften, Philosophisch-historische Klasse, Denkschriften 195 (Vienna: Verlag der Akademie).

——, and Hunger, H., 1996, *Astronomical Diaries and Related Texts from Babylonia, III: Diaries from 164 B.C. to 61 B.C*, Österreichische Akademie der Wissenschaften, Philosophisch-historische Klasse, Denkschriften 247 (Vienna: Verlag der Akademie).

Steele, J. M., 2007, "The Length of the Month in Mesopotamian Calendars of the First Millennium BC", in J. M. Steele (ed.), *Calendars and Years: Astronomy and Time in the Ancient Near East* (Oxford: Oxbow Books), 133–148.

——, 2011, "Making Sense of Time: Observational and Theoretical Calendars", in K. Radner and E. Robson (eds.), *The Oxford Handbook of Cuneiform Culture* (Oxford: Oxford University Press), 470–485.

Stern, S., 2000, "The Babylonian Calendar at Elephantine", *Zeitschrift für Papyrologie und Epigraphik* 130, 159–171.

——, 2008, "The Babylonian Month and the New Moon: Sighting and Prediction", *Journal for the History of Astronomy* 39, 19–42.

van der Waerden, B. L., 1960, "Greek Astronomical Calendars and Their Relation to the Athenian Civil Calendar", *Journal of Hellenic Studies* 80, 168–180.

Westermann, W. L and Kraemer, Jr., C. J., 1926, *Greek Papyri in the Library of Cornell University* (New York: Columbia University Press).

Lunar Calendars at Qumran? A Comparative and Ideological Study

Jonathan Ben-Dov

This article revisits a question that has been discussed by generations of Qumran scholars: was there a role for a lunar calendar in the Yahad community? Despite the dominance of the 364-Day year in the calendar texts from Qumran, there is ample evidence for lunar counts within them too, both in the *mishmarot* corpus and outside it.[1] How, then, did these two time reckoning systems interact?

The present discussion assumes— without defending it in detail—a position that is still mainstream among scholars of the Dead Sea Scrolls, namely that most of the scrolls found in caves near the site of Qumran on the shore of the Dead Sea were either written or collected by a sectarian community that had lived in that area around the first centuries before and after the common era. This community (or communities) called itself 'the Yaḥad', and will thus be referred here. This is a rather minimal version of the scholarly consensus on the Qumran community. While more far-reaching insights on the identity of the community are defendable, they are not crucial for the discussion and thus not promulgated here.[2] It is important, however, to note that, while some of the Dead Sea Scrolls represent the literature written by Yahad members for their own use, not all scrolls collected in the caves comply with this kind of inner-circle literature. Some scrolls represent earlier or softer versions of sectarianism, possibly of other sects or of predecessors to the Yahad, while others may have originated with other, non-sectarian circles.

Had a question about the employment of various calendar systems been raised with regard to other ancient literary corpora, the answer would have come through the meticulous study of administrative texts, loan contracts and temple records, which can shed light on practical aspects of the question at hand.[3] However, the nature of the material from Qumran entails a different kind of discussion, since the calendars from Qumran concentrate more on ideal aspects of Time— either ritual or astronomical— rather than on administrative and civil usage. Even within the realm of ritual, the texts do not convey administrative scenes from the life of a real temple, but rather literary depictions of the cycle of priests in an ideal temple. The discussion of the lunar calendar in the Dead Sea Scrolls should thus focus on the ideological aspect of calendrical choices much more than we would allow this aspect to appear in other, more mundane calendrical studies.

In late Hellenistic and early Roman Judea, as well as in most of the Ancient Near East at that time, life was regulated according to a luni-solar calendar, which served for both sacral and administrative purposes. This calendar (or a very similar one) is reflected in later rabbinic sources, with the year beginning in the autumn and using the Aramaic month names Tishri, Nisan etc.[4] In later times Judea subscribed to the Julian calendar like the rest of the empire. On this background, the propagation of a 364-day calendar in the Dead Sea Scrolls and other related documents is a declaration of opposition to, if not direct polemic with the state calendar, and must have been fueled by strong ideological, religious and sociological reasons. What then should we make of the fact that the sectarian tradition does make note of lunar calendrical data alongside the 364-day year? A series of questions should be asked: Is the lunar calendar mentioned in Qumran texts the same as the civil luni-solar calendar practiced elsewhere in Judea? How did this calendar interact with the sectarian year of 364 days: was it kept alongside the sacred year? Was it used for special purposes? Was there any clash or tension between the two systems, or maybe the fact that they were recorded in the same document shows that they peacefully coexisted?

A comparative method may be helpful in answering the latter question. While studying some non-Jewish parallels to the lunar texts from Qumran, in which the luni-solar calendar is synchronized with other systems, it is possible to induce how the different systems interacted. When observing a society outside the hot debates of Dead Sea Scrolls' study, perhaps the ideological factors can be neutralized. Several exemplars of non-Jewish material are discussed below, some of them already published while the others not hitherto noted. The comparative discussion requires prudent methodology, which will do justice to the Jewish material as a culturally autonomous source while at the same time explaining it by the same set of tools which is used for non-Jewish contemporary texts. The comparison will illuminate the unique ideological factors which informed the calendrical discourse in the Yahad.

Previous Scholarship

A pioneering article by Shemaryahu Talmon from 1958 set the stage for the study of calendars in the Dead Sea Scrolls.[5] In this article he laid the foundations for future study, reconstructing the organizational principles of the year and the cycle of *mishmarot* based on the few fragments that were announced by J. T. Milik, as well as on the pioneering work by Annie Jaubert. This article—revised in 1989 while still retaining the original argument—remained for many years the proof text for all scholars in the field. Among the many insights of this article, its most dominant heritage for future scholars was the view that the calendrical dispute between the Yahad and 'mainstream Judaism' (a popular term in the 1950s) was nothing but a rivalry between the solar and lunar calendars. While the Qumran covenanters (a term often used by Talmon, referring to the prominence of covenantal themes in the worldview of the Yahad) promoted a solar calendar and aimed to diminish the importance of the moon, their opponents from 'mainstream Judaism' propagated a luni-solar calendar and aimed to achieve quite the opposite.[6] In this article as well in his later work, Talmon connected his notion of a lunar vs. solar calendar war with additional statements in Yahad literature:

> ... 1Q27 predicts the foreseen blissful future age, which signifies (אות) the concomitant evanescence of *wickedness* before *righteousness*, an event likened to the disappearance of *darkness*, over which the moon rules, before *light*, that is the sun's domain.[7]

Having embraced this extreme view, Talmon played down the presence of lunar data in Qumran calendars. Thus, when in the early 1990s he published the lunar text 4Q321 with

Israel Knohl their interpretation was based on the notion, even presupposition, that this lunar text is a heavily ideological document promoting an anti-lunar agenda. According to Talmon and Knohl, by recording distinct lunar phases in each month the Qumran authors sought to point out the moments of darkness of the moon in order to oppose it:

> The moon is portrayed as the source of dark days of evil... In contrast, the sun is seen as the fountainhead of holy and blessed days... The synchronization of the lunar nights (and days) of darkness with the parallel dates in the solar calendar is intended to enable the covenanters to beware of these negative periods of time in the lunar months.[8]

Other attestations of lunar data in the DSS, outside the circle of 4Q321 and related texts, were dismissed by Talmon as irrelevant for the sacred time-reckoning. This was his opinion not only with regard to the pre-Qumranic lunar data in the Book of Luminaries (1 Enoch 72–82), but also with regard to more clearly Qumranic texts like the calendrical list 4Q319, which is contained within a copy of *Serekh haYahad* and features numerous non-lunar calendrical lists alongside it.[9] Talmon thus denied that any role whatsoever was given to lunar data in the framework of sacred time at Qumran, and claimed, to the contrary, that various texts sought to oppose the moon.

In 1987 Joseph Baumgarten reached the opposite conclusion from Talmon on the basis of his study of the scroll 4Q503.[10] This document contains prayers to be recited twice daily at sunset and sunrise, with the content of the prayers mainly centered on praise to the creator for his maintenance of the regular motion of the luminaries. In the prayers the author employs a poetic count of various measures of light, corresponding to the days of the month. Baumgarten noticed that these measures cannot be other than the parts of light in the moon, as measured in the Book of Luminaries (1 Enoch 73–74 and 78). Having identified 4Q503 as a sectarian text, he thus concluded that the Qumran community did not altogether ignore the lunar calendar, but rather maintained a lunar count alongside the 364-day year.

As more and more was known about the scrolls 4Q320 and 4Q321 and about the non-*mishmarot* scroll 4Q317 scholars acknowledged that the lunar presence in these texts could no longer be seen as ideological statements opposing the moon. Rather, they were lunar texts par excellence.[11] Recent handbooks on the Dead Sea Scrolls usually claim that the Yahad used several types of calendars side by side. Thus for example in an authoritative semi-popular handbook, VanderKam and Flint write:

> The Qumran calendaric texts follow the Enochic pattern of accepting two schematic years—a solar one of 364 days and a lunar one of 354 days—but they align themselves with Jubilees in dating festivals by the 364-day system...
>
> Alongside the simple solar arrangement we find that the covenanters accepted a schematic lunar calendar, which they synchronized with the 364-day system.[12]

How then should we account for the co-existence of the 364-day year with a lunar reckoning? What is the normative force of each of the two? Was one of them a religious institution while the other a civil calendar, as was more or less the case in Pharaonic Egypt along many stages of its long history?

Most scholars naturally hold that the Yahad—in whatever form and place it existed—ran at least some part of its community life based on the 364-day year. This is intuitively accepted and remains the consensus until today because, put simply, this calendar is the prevalent one in the sources of the community. Recently, however, Stern propagated a different view, close to what had been suggested in the past by Wacholder.[13] According to this view the 364-day year was never meant to be a year to live by, nor to run any social mechanism by—neither religious nor civil. Rather it 'may still have been intended as the representation of some cosmological (or eschatological?) ideal where solar years, lunar months, priestly weeks, and liturgical days would combine in perfect harmony'.[14] In other words, the 364-year reflected an elevated ideal rather than any sort of reality. I concede that such calendars were indeed known, as for example the 360-day calendar in ancient Mesopotamia, which was purely ideal.[15] However, that year is never connected with routine temple work or religious festivals, which is quite often the case in the DSS. The fact the texts from Qumran repeatedly connect the 364-day year with religious festivals and liturgies draws a distinctly different picture than that of the Mesopotamian 360-day year. An innocent reading of the scrolls on the basis of a fairly moderate version of 'the Essene-Qumran hypothesis' would lead one to accept that the Yahad did practice its special calendar, at least during some stage of its history.

A frequent objection in this regard is the matter of intercalation. According to Stern, since there is no reliable sign that the 364-year was ever intercalated, it could not have been practiced over any significant period of time. The gap of 1.25 days between this year and the tropical year would soon accumulate to an inconceivable gap between the year and the actual seasons, until the spring festival would have to be celebrated earlier on, even in the middle of the winter. Thus the 364-year, according to this reasoning, was never practiced in reality by any living society, not even a sectarian community.[16] However, as I claimed elsewhere, this view is too far reaching and should be modified. Indeed, the proponents of an ideal 364-day calendar could not have maintained a regular mechanism for intercalation, because that would be tantamount to admitting that their year is essentially flawed. But the commitment to the ideal scheme was so ideologically powerful that its proponents could not have recognized the validity of any other calendar either. The only way out from this impasse would be *ad hoc* intercalations enacted every once in a while. The advantage of this method is that the year is occasionally corrected but without instituting a fixed mechanism for its correction. Alternatively, the 364-day year may have been used by the sectaries as a revolving year. Indeed it would be awkward to celebrate Passover in the winter, but sometimes the ideological zeal is so powerful as to override the constraints of reality. This way or another, the problem of intercalation does not preclude the possibility that a 364-day year has indeed been normative in sectarian circles, at least for a certain period of time.[17]

We now return to the question: what is the function of the synchronization with lunar data in *mishmarot* and other Qumran texts?

Lunar Data in Calendar Texts from Qumran

The calendrical corpus comprises a wide variety of lists, which combine various astronomical, cultic and calendrical entities into large and elaborate numerical structures.[18] While some of these elements purport to reflect real astronomical phenomena, others reflect mere numerical entities, making the entire calendrical clockwork a matter of arithmetic more than astronomy. Some understanding of this arithmetical system is required for the argument.

The 364-Day Year consists of 52 weeks, which are divided into four seasons (*tequfot*) of

13 weeks each. Within Yahad sources, the weekly service cycle of the twenty-four priestly families in the temple (*mishmarot*) is designed to fit the 364-year. A full rotation is achieved within 6 years, with each family serving exactly thirteen times in this cycle:[19]

$$6 \times 52 \text{ (weeks in a year)} = 6 \text{ years} = 312 \text{ weeks} = 24 \text{ (\textit{mishmarot})} \times 13$$

Some calendrical texts reflect this permutation of the 52 weeks with the *mishmarot* and key festivals of the year, for example 4Q325.[20] Yet other texts add lunar phases into the numerical scheme. The lunar months in this scheme are of 29 or 30 days, in alternating order. Thus, the lunar phenomena recorded in these lists are all schematic and entirely detached from reality. The staggering 29/30-day months produce a lunar year of 354 days, which, in turn, is correlated with the 364-day year every three years according to the following equation:

$$3 \times 364 = 3 \times 354 + 30$$

An 'intercalary' month of thirty days is added after three lunar years, and thus the cycle corresponds to three years of 364 days each. To be precise, three years of 364 days correspond to 37 lunations. It is important to clarify that this 'intercalation' does not do anything to correct the 364-day year; this year remains unaltered with no sign of it being corrected towards the true solar year (see above). Moreover, the addition of 30 days after every three lunar years does not align the 'lunar' year employed in the *mishmarot* scrolls with the real lunar year. It is a schematic move which is intended to smooth the ideal numbers rather than to actually correct the calendar.[21]

Several calendrical documents constitute elaborations on the basic theme of the triennial cycle. These documents expound the following components: days, months and weeks of the 364-day year; days of the week; the cycle of priestly families; lunar phenomena. The latter element appears in the *mishmarot* scrolls 3Q320, 4Q321 and 4Q321a.[22] The 3-year lunar cycle matches quite nicely with the 6-year cycle employed for the *mishmarot*. Thus, two lunar cycles fit into one *mishmarot* cycle, together celebrating the numerical harmony of the calendar. Furthermore, yet another calendrical document combines the 6-year cycle with the longer cycles of seven and forty-nine years of the Shemitah and Jubilee. The various cycles and figures noted so far coalesce into a grand structure of 294 years (= 6 × 49), represented in the scroll 4Q319, which brings the artistry of calendar arithmetic to its climax.

The opening list of the scroll 4Q320 represents the essential lunar 3-year cycle, as the list records each of the 37 lunations contained in it. A sample passage from this list reads:

4Q320 1 ii 4–8[23]

4. The second year
5. on the 2nd (day) in Malkiah for 29 (days) on the 20th (day) in month 1
6. on the 4th (day) in Jeshuʿa for 30 (days) on the 20th (day) in month 2
7. on the 5th (day) in Ḥuppah for 29 (days) on the 19th (day) in month 3
8. Sabbath in Happiṣṣeṣ for 30 (days) on the 18th (day) in month 4

In this passage, an unnamed monthly lunar phenomenon is recorded along each of the first four months of year II of the 3-year cycle. The first occurrence of this lunar phenomenon is recorded at the beginning of the list, not quoted here. It had originally occurred at the very beginning of the march of Time—the fourth day of creation week (4Q320 1 i 1–5)—which was necessarily also day 1 of the month. During subsequent months in the 3-year cycle this phenomenon floated through the days of the month, due to the gap between the 30/31 days of the schematic month and the 29/30-day length of the lunation. Thus it reached day 19 or 18 of the month, as in the passage quoted here.

Each line of the list contains the following components:

(a) Day of the week and priestly family in which the lunar phenomenon occurs
(b) Number of days in the previous lunation (29/30)
(c) Date of the lunar phenomenon in terms of the 364-day year

It is now time to ask again how the synchronization of the 364-day year with a lunar scheme should be understood. Let us have a closer look at VanderKam's statement quoted above: 'they... align festivals by the 364-day system'. The fact is, however, that not only festivals are dated by the 364-day year, which would be expectable in a priestly context, but also the lunar phases themselves are dated by this system (item (c) above)! This is the case not only in 4Q320 but also in 4Q321, where an additional lunar phenomenon is dated in the course of every month. For example 4Q321 I 6–8, describing the beginning of year II in the 3-year cycle, i.e. the same period of time like the passage quoted above from 4Q320:

> The] se[cond year:] The first (month). On the second (day) in (the week of) Malkiah (which falls) on the tw[entieth in it (the first month); and] its *dwq* (is) [on the third (day) in (the week of) Ḥarim (which falls) on] the seventh in it (the first month). On the fou[r]th (day) in (the week of) Jeshuʿa (which falls) [on] the twentieth in the second (month); and [its *dwq* (is) on the fifth (day) in (the week of) Ha]qqoṣ (which falls) on the seventh [in it (the second month).

The schematic, seven-based 364-day year thus functions as the measuring stick for astronomical phenomena which might have otherwise been considered to administer the lunar calendar. Furthermore, the lunar phenomena recoded in 4Q320 and 4Q321 are by no means real and observable, and do not purport to agree with the actual lunar calendar practiced in Judea at the time. Rather they are entirely schematic. This is easily discerned from the fact that these 'lunar phenomena' are placed in recurrent intervals precisely 13 days + 16 days (in a month of 29 days) or 13 + 17 days (in a month of 30 days) apart.[24] Whoever compiled these lists in 4Q320 and 4Q321 was not interested in recording *real* lunar phases. His motivation was somewhat awkward in the eyes of the modern reader: gearing the schematic 364-day year with an equally schematic 'lunar' plan.

Comparative Material

In order to point out the peculiarity of the lunar texts from Qumran, let us compare the lunar text 4Q320 with a Greco-Egyptian text, a parallel of some relevance, which was hitherto unnoticed. The document p.Ryl. IV 589, dated to the early second century BCE, contains various lists from the cultic and administrative life of an Egyptian-Greek gymnasium. At the very end of the extant papyrus, a fragmentary table correlates a list of new moons with the Egyptian dates in which they occur. A sample passage runs as follows (lines 125–130):[25]

> ... The lunar new moons (κατὰ σελήνην νουμηνίαι) in the first year are
> Thoth 20 days 29
> Phaophi 19 days 30
> Hathyr 19 days 30
> Choiak 19 days 29

This list dates the real new moons according to the Egyptian civil calendar. Each line gives the date of the new moon within the Egyptian month, followed by the number of days in that month, which is always 29 or 30 days. The important term κατὰ σελήνην νουμηνίαι indicates that the list is interested in the *real* new moon (*kata selene*), as opposed to calculated new moons. As known from other sources, a 25-year cycle was constructed in Egypt in order to synchronize lunar dates with civil dates; it is best known from the Demotic p.Carlsberg 9, while our p.Ryl IV 589 gives a somewhat different version of it.[26] The basic principle is that when the lunation lasts for 30 days like most schematic months, the new moon remains in the same day of the Egyptian month in the next month too; but when the lunation lasts for 29 days only, the date of the new moon moves backwards one day within the 30-day Egyptian month. Hence the shift from day 20 of Thoth to day 19 of the immediately following month Phaophi.

The inner structure of the lines in this list, as well as overall structure of the list, are essentially identical to 4Q320 1 i 6 – ii 14. For the sake of clarification we quoted here the passages from both documents which describe the shift from day 20 to day 19 of the month. The differences, of course, lie in the fact that each of the ancient documents synchronizes two different systems, as the following table shows. While 4Q320 correlates schematic lunar phases with the 364-day year, p.Ryl. IV 589 correlates real new moons with the Egyptian civil calendar.

	The dated phenomenon	The dating System
4Q320	Schematic lunar phenomenon	364-day year
p.Ryl. IV 589	Real new moon	365-day year

Quite a few Egyptian documents, many of them pre-Hellenistic, attest to the use of the lunar calendar for the purpose of regulating the service periods of priests in the temples. The various priestly *phylae* (either four *phylae* in the early period or five of them after the decree of Canopus in 238 BCE) entered service on the day of the beginning of the month. Temple records that show a series of such transition dates were conveniently collected and discussed by Depuydt.[27] The format of these texts, which give a list of new moons according to the civil calendar, is remarkably reminiscent of Jewish *mishmarot* rosters, especially 4Q320. They demonstrate that this type of calendrical speculation in Jewish texts was well-grounded in the practices of Hellenistic temples outside Jerusalem.

Returning to the lists of new moons, it is often assumed that in the Egyptian environment, synchronization was required because the real new moon maintained a certain normative dimension in the life of the pertinent society.[28] The comparison with 4Q320 thus could possibly be taken to show that the lunar count maintained a normative aspect also in the eyes of the Yahad authors of 4Q320, who did their best to synchronize it with the 364-day year. Possibly, as some scholars claimed, 4Q320 attests to a religious celebration of the new moon in Yahad circles.[29] In the Egyptian text there was practical importance for each of the two synchronized systems; apparently the dates with Egyptian month names were associated with the civil year, while the new moons were required for religious rituals. This agrees with the suggested ritual background for p.Ryl. IV 589 as part of the life of a gymnasium.[30] Could this be the case in the Qumran list too? Does 4Q320 note the schematic new moons because a specific religious or civil significance was attached to these days alongside the beginning of the months in the 364-day year? While some scholars believe that this is the case (notably Koch and Glessmer quoted above), a closer analysis of the find proves that the lunar phenomena are not counted in 4Q320 for religious-institutional reasons. The main proof comes not from 4Q320 but rather from the sister text 4Q321.

The extant fragments of 4Q321 contain two parts: a lunar text recording two lunar phenomena in the six-year *mishmarot* cycle, followed by a list of festivals in the same cycle. In the lunar section two phenomena are noted in each month of the triennial cycle, as noted above. The first phenomenon is the same one as in 4Q320, with the dates entirely overlapping, while the second one records a different phenomenon, 13 days before the former.[31] See for example the entry in 4Q321 I 6–7 for the month 2 in year II, which dates the first lunar phenomenon on day 20 of the month, corresponding to the passages cited above (italics mine):

> The second year: The first (month). On the second (day) in (the week of) Malkiah (which falls) *on the twentieth in it* (the first month); and its *dwq* (is) on the third (day) in (the week of) Ḥarim (which falls) *on the seventh in it* (the first month).

Here, the occurrence at the 20th day of the month is followed by the mention of another phenomenon, called *dwq*, which occurs on day 7 of the same month.[32] Going back to the question posed above, it is clear that no religious-cultic routine would require recording these two distinct phases in the course of each month. Why would a priestly or other administrative scribe require the routine record of a date which occurs thirteen days before the new moon or before the full moon? For the sake of the argument, let us concede to other scholars, who held that 4Q321 records the full moon (i.e., the unnamed phenomenon recorded also in 4Q320) alongside the day of first visibility (*dwq*).[33] Even if this is true, I fail to see how the lunar section of 4Q321 fulfills a religious function in the Yahad. No other Jewish text attests to a celebration of both the new and the full moon in the course of one month, nor is there any mention of such celebrations in the religious documents of the Yahad.[34] The attempt to read 4Q320 and 4Q321 as recording religious lunar celebrations is truly at odds with everything else we know about the life of the Yahad.

What then is the purpose of these texts? Other comparative material may be of help. I have previously claimed with Wayne Horowitz that the lunar data recorded in 4Q320 and 4Q321 are best understood in analogy with the recording of the Lunar Three in Babylonian astronomy.[35] Many cuneiform documents—from the horoscopes to the almanacs—give a basic description of lunar months by noting three key pieces of data about the lunation: the length of the previous month (29/30 days); the date on which, after full moon, the moon first sets in the morning after sunrise; and the date of last lunar visibility in the morning.[36] Taken together, the data given in 4Q320 and 4Q321 produce a similar list of three lunar items. We were able to point out a late Babylonian astronomical text whose structure resembles that of 4Q320–321.[37]

The analogy of the Qumran *mishmarot* texts with the Babylonian material suggests an altogether different motivation for the lunar lists in Qumran than the religious-cultic framework suggested earlier. The fact that two phases are recorded in each schematic month supports the idea that they were recorded there for astronomical purposes rather than for cultic ones. Just as in ancient Babylonia the Lunar Three were recorded as part of a scientific discipline, which enabled the use of these pieces of data for further calculations, so should a similar motivation be assumed for the lunar data in the scrolls 4Q320 and 4Q321. This should be of no surprise, as it is already known from elsewhere that Jewish-apocalyptic circles transmitted Babylonian-type astronomy and recast it in literary form in the Book of Luminaries of 1 Enoch.[38] Lunar data were thus recorded in the scrolls as part of a tradition that began already in ancient Mesopotamia and continued in the Book of Luminaries.

I would like to clarify the role of the comparative evidence in the present discussion. It is certainly not the case that the Jewish material must agree with one of the external paradigms, either Egypt or Babylonia. On the contrary, it is perfectly reasonable to expect Jews to produce their own astronomical and calendrical schemes which may or may not have agreed with those of other cultures.[39] The interpretation of the lunar texts from Qumran should be reached first of all 'from within', as was attempted here. Once an interpretation arises from the Jewish material, solving the difficulties of the text and following its lead, it may be buttressed with parallels and analogies from outside, especially if other similarities are known from closely related materials. The practice of some Yahad scribes to record three pieces of lunar data is not a reduplication of the Babylonian discipline, but rather an independent cultural phenomenon. It is related to the former but not identical with it. As in many cases of cultural borrowing, interaction, or as it is often called these days, hybridity, the common cultural item shares some central aspects of similarity between the two cultures—the source and the receiving culture. However, it is embedded in the receiving culture within a new cultural texture. Being aware of the original cultural context is important for scholarly study as a heuristic device, but does not necessarily entail that a similar context exists in the new culture.

A closer comparison of the lunar texts from Qumran and Mesopotamia highlights the differences alongside the similarities. The similarity lies in the existence of the three pieces of data, and in addition in the similar intervals of time that pass between each of the lunar phenomena. However, in the Jewish texts these lunar phenomena are separated by fixed, ever-iterating intervals, while in the Babylonian materials the length of the intervals varies. The fixed intervals in the Jewish material vis-à-vis the varying intervals in the Babylonian texts are explained by the fact that a) the Jewish texts refer to merely schematic lunar phenomena, as opposed to the real records in the Babylonian material (whether observed or calculated), and b) the Jewish dates work in a schematic 364-day year, while the Babylonian data are given according to a functioning luni-solar system, with an ever-changing length of months according to nature and to the needs of the kingdom.[40] For the present purposes it is important to note that the two lunar phases in the course of each month were not recorded in Babylonia for cultic reasons, at least not primarily for cultic reasons.[41] Rather they were recorded as part of the astronomical tradition and were meant for use in various calculations. As much as one can conceive of a separation of religion from science in antiquity, the Lunar Three lists in 4Q320–4Q321 do not convey any normative cultic measures.

To sum up this part, in Yahad documents cultic normativity is limited to the 364-day year exclusively. Lunar phenomena were included in the *mishmarot* scrolls because they were part of an astronomical tradition, and because they helped to construct the cosmological worldview of the covenanters, but without immediate cultic significance. In this specific sense, texts like 4Q320 and 4Q321 are more similar to the Babylonian lists of Lunar Three than to the Egyptian calendrical text p.Ryl. IV 589.[42]

Calendrical Ideology

The above discussion leads to a nuanced appreciation of the presence of lunar data in *mishmarot* calendars from Qumran. The writers of these documents were holding both ends of the rope: while constructing the normative calendar on the basis of the 364-day year, they also maintained a schematic lunar reckoning. Thus the lunar model is maintained while its sting is neutralized. Yahad writers did not acknowledge the lunar calendar as such, because they denied any normative force to the lunar data they collected. Nonetheless they continued recording schematic lunar data and integrating them in the calendrical lists of the community. This modus operandi reflects a complex attitude to calendrical diversity, one which did not exist in extra-Jewish calendrical texts. Such a complex attitude cannot be justified by practical needs and must be indebted in some way to ideological factors. Talmon was therefore correct in underscoring the role of ideology in calendrical convictions, but some of his explanations for the 364-day year were wrong. With an updated understanding of the calendrical tradition in the Yahad, the role of ideology can be properly illuminated.

Within the literary works of the Yahad, the concept of Time played a central role. As demonstrated in a series of studies by Devorah Dimant, the self-perception of the community depended greatly on its understanding of the march of Time and its effects for divine revelation and providence.[43] Time was thus not a circumstantial component in the world-view of the community, but rather an organic part of its self-perception, and found expression in the community's biblical interpretation, liturgy and historiography. This extraordinary importance of Time was partly inherited from earlier literature such as the books of Enoch and Jubilees, as well as from the general importance of Time in apocalyptic thought in general.[44] This earlier heritage found particular emphasis within the framework of the Yahad.

The centrality of time in the Yahad found expression in terms of the most explicit way of constructing it: the calendar. Within Yahad circles the calendar found primary cultural importance, a rather rare state of affairs in the ancient world.[45] Elsewhere it gained comparable centrality only under the political constellation at the time of Augustus.[46]

The calendrical documents from Qumran produced an elaborate system which synchronized a large number of various numerical units under one umbrella structure. In the words of Sacha Stern:

> Modern scholarship has not sufficiently recognized (if at all) that the complexity and sophistication of these cycles far exceeds whatever had been composed and designed until then throughout the ancient world...[47]

The main characteristic of this tradition is the synchronization of various temporal frameworks by means of the elaboration of numerical schemes. This system encompasses the numbers 3, 4, 7, 12, 13, and even the greater numbers 49 and 52, by producing cycles as long as 294 years in the document 4QOtot. The various numbers are orchestrated into a subtle, Bach-like Fugue, with every part fitting perfectly into the general scheme. At the nodes of this grand edifice stand the key points of transition in the march of Time: the beginnings of Jubilees, the years of Shemitah, and the beginning of each three-year cycle. The three-year lunar cycle was thus taken as further arithmetical challenge in this ongoing endeavor. This explains the special schematic and unrealistic character of the lunar count at Qumran, as opposed to the non-Jewish sources described above. While in those sources the moon was a real natural phenomenon, in the Yahad documents it functions as a numerical device within the dominant 364-day scheme, reinforcing the harmonic picture of peace and concordance in the world.

The Yahad descended from apocalyptic circles such as those that had produced the Book of 1 Enoch. However, the Yahad was not strictly an apocalyptic group, since it had other ideological characteristics.[48] Some elements of the apocalyptic outlook were retained in the Yahad, and the fascination of Yahad authors with the natural order is clearly one of them.[49] This interest was less pronounced than in the Book of Luminaries of 1 Enoch. Instead, much of the interest in Nature was expressed either through its halakhic implications (e.g. the calendar) or through other immediate needs of the Yahad.[50] A text like 4Q503 which contains prayers about the creation of the luminaries is part of this ideology too, as the count of lunar phases in 4Q503 corresponds quite nicely with it. Thus, for the author of 4Q503 the moon was an important natural object worthy of praise, but not a calendrical marker.

Finally, one more ideological aspect can be pointed out, although with less conviction: the distinction between pre-calculated schematic time and *ad hoc* calendrical convictions. This ideological tension was pointed out by Ron Feldman, in a paper included in the present volume,[51] and is included here with less conviction since it is not explicit enough in the scrolls. The thrust of the argument emerges from a comparison of the 364-day calendar tradition with the rabbinic calendar, as reflected in Mišnah Roš Haššanah. I shall draw it here based on Feldman's concepts with recourse also to Rachel Elior and to my own formulations.[52] The Tannaitic sages hallowed the actual observation of the moon and required the enactment of the new moon by human agency. The procedure they constructed for this purpose—whether historical or primarily literary—reflects a strong preference to domesticate the infinite stretch of time by means of the human court rulings. The choice of a lunar calendar by the rabbis is thus justified, since it is the lunar calendar in particular which requires the involvement of human observers and other agents, due to its irregularity. The Tannaim highlighted the unpredictable aspects of the calendar, those that cannot be calculated in advance and require an *ad hoc* observation of the moon. In contrast, the 364-day calendar tradition is essentially constructed not of astronomical observations but rather of numerical cycles. Although this tradition is indebted to the lunar theory transmitted by its Babylonian and Enochic predecessors, it found a way to integrate the lunar elements while at the same time subduing the moon's arbitrary and unpredictable power. In this tradition there is neither need of observation nor of other human involvement. The lunar phases are here more imaginary than real, more schematic than observable, and are in fact secondary to the dominant calendrical role of the priestly *mishmarot*.

The ideological elements depicted here created the peculiar situation encountered in the Dead Sea Scrolls. Although in some way paralleled by extra-Jewish calendars, the lunar texts from Qumran employ a unique attitude towards the lunar cycle: although integrating it within the core calendrical literature, it was not acknowledged as a lunar calendar per se. The lunar count played no normative role in the life of the community. This aspect was retained exclusively for the 364-day year. Instead, the lunar cycle was acknowledged as a further numerical component which can be integrated into the grand scheme of the calendar texts, demonstrating the perfect orchestration of various powers within the march of Time.

Notes

1. For the 364-Day year see Ben-Dov (2011). For the calendrical texts from Qumran see Talmon, Ben-Dov & Glessmer (2001). A group of texts from this corpus ties various aspects of the calendar with the rotating service of twenty-four priestly families (henceforth: *mishmarot*) in the Jerusalem Temple. This service cycle is similar to other cultic-administrative apparata from the ancient world, such as priteniai in the Athenian calendar or phylae in ancient Egypt. On *mishmarot* see Talmon (2000), Ben-Dov (2010).
2. For a standard formulation of the consensus see VanderKam and Flint (2002). For a recent, more nuanced, formulation see Collins (2010). I generally concur with Collins about the identity of the Yahad as multiple communities spread throughout contemporary Palestine, rather than a single community residing at Qumran. For current discussion of Qumran sectarianism see the articles collected in theme issue of Dead Sea Discoveries 9, 3 (2009), as well as Regev (2007).
3. See some prominent examples from the field of Hellenistic-Roman Egypt or Old Babylonian Sippar: Jones (1997), Lippert (2009), Greengus (1987, 2001).
4. For a short summary of the evidence for this reality see Stern (2011), pp. 42–43; also Stern (2001), pp. 1–38. That Judean society, temple, and administration were run according to a luni-solar calendar by the late Hellenistic period is accepted by most scholars. A debate still remains as to the situation in earlier stages of Judean history. While many scholars believe that a 364-day year was practiced in the temple prior to the decrees of Antiochus IV in 167 BCE, I cannot accept the historical credibility of this claim. The full debate lies outside the scope of the present study; See Ben-Dov and Saulnier (2008).
5. Talmon (1958), revised in Talmon (1989).
6. Talmon (1989), pp. 165–171, more recently Talmon (1999).
7. Talmon (1999), p. 38, (Talmon's italics). Talmon's understanding of the passage at hand, 1Q27 1 i 5–8 (parallel 4Q300 3 4–6) cannot be maintained, since the term מולד does not carry here the calendrical meaning that it has in the Mishnah and related literature. More feasibly the term מולדי עולה means 'the nativities of unrighteousness', possibly with reference to an astrological connotation of the term מולד (see below).
8. Talmon and Knohl (1995), p. 299. A somewhat similar point of view was taken by the French scholar Veronique Gillet-Didier. While agreeing with Talmon that 4Q321 aimed to oppose the moon, she thought it better to interpret the lunar phenomena mentioned in it differently from Talmon (Gillet-Didier 2001). See below for a discussion of the identity of the lunar phases in 4Q321 and related texts.
9. Talmon (2000).
10. Baumgarten (1987); compare a similar opinion in Schiffman (1994), pp. 304–306.
11. Albani (1992, 1994), Ben-Dov and Horowitz (2003).
12. VanderKam and Flint (2002), p. 257, later p. 260.
13. Wacholder and Wacholder (1995). Stern first raised this idea in Stern (2000), pp. 14–16 and recently in a different form: Stern (2011). The latter article is a full version of Stern's shorter article in the *Oxford Handbook of the Dead Sea Scrolls*.
14. Stern (2011).
15. Brown (2000), Brack-Bernsen (2007).

16. Stern (2000). For a survey of multiple previous proposals about methods for intercalating the 364-day year see: Ben-Dov and Saulnier (2008). I agree with Stern that none of these proposals is compelling, and that no regular mechanism existed for the alignment of the 364-day year with the true solar year.
17. See further Ben-Dov (2008), pp. 18–20 and (2011), pp. 84–86.
18. The various types (genres?) of calendrical lists were classified quite clearly in the edition by M.G. Abegg (in Parry and Tov 2005, volume 1). This edition is somewhat misleading, however, because it distributes the scrolls according to literary pattern, thus assigning parts of each scroll to separate ends of the edition. The reader may get the wrong impression that each type of list was represented on a separate scroll, while in reality most of the scrolls contain an anthology of various lists. Stephen Pfann claimed that the various types of calendrical lists in the DSS represent diverse calendrical systems rather than different rosters based on the same calendar (Pfann 2009). However, this hypothesis goes beyond the evidence. There is no need to assume that several varying calendars were practiced by the community when the textual variety can be explained more simply as mere literary variation.
19. Non-Qumranic Jewish traditions assigned each of the 24 courses a weekly service period within one single year. However, it is not entirely clear how this was achieved in a year of ca. 354 days, even more so in a leap year of ca. 384 days. See Ben-Dov (2010).
20. See the classification in Talmon, Ben-Dov & Glessmer (2001), pp. 7–14.
21. Similar measures can be detected in the schematic Mesopotamian 360-day year, as pointed out by Horowitz in the 'Second Intercalation Scheme' of the cuneiform compendium Mul.Apin; see Ben-Dov (2008), pp. 161–167. Thus the three-year cycle and its elaboration in the Otot list 4Q319 cannot be considered as intercalation in the full sense of the word. I thus withdraw my earlier opinion that 4Q319 was 'a manual for the maintenance of the lunar calendar' (Talmon, Ben-Dov & Glessmer 2001, p. 210); thus I cannot support the claims by Kugler (2010).
22. 4Q321a overlaps with the lunar section of 4Q321. Since 4Q321 is better preserved, it shall be mentioned alone in the discussion below.
23. Talmon, Ben-Dov & Glessmer (2001), pp. 42–43, brackets omitted. The translation differs from that in DJD XXI and reflects my current refined understanding of the text.
24. Compare for example the irregular time intervals between lunar phases in Babylonian lists of Lunar Three: Ben-Dov and Horowitz (2005), Ben-Dov (2008), pp. 234–235.
25. Turner and Neugebauer (1949), quotation from p. 92. On this papyrus see recently Depuydt (1997), pp. 151–152, Lippert (2009), p. 193. Brackets and parentheses omitted. About fourteen more lines of the same list are preserved at the end of the papyrus.
26. For p.Carlsberg 9, see Parker (1950), pp. 24–29, Jones (1997). The synchronizations in p.Carlsberg and in p.Rylands are not identical, and moreover it seems that some administrative records display yet another system of synchronization, inconsistent with any of the other two; see Lippert (2009), with extensive earlier bibliography. The book by Bennett (2011) appeared too late for me to consult it.
27. Depuydt (1997), pp. 177–186.
28. This is the standard view of students of the Egyptian 25-year cycle, who simply assume that since the cycle exists, it must have had some practical application. Depuydt opposed this opinion by claiming that 'There is no evidence that the 25-year cycle was ever used for dating religious festivals' (Depuydt 1997), p. 152. Rather, he thinks that this cycle was traced as 'The simple knowledge of a proportion in nature... a mere arithmetical game'. Not being an expert for Egyptian calendars myself, I am unable to judge this opinion. However, the presence of the 25-year cycle on the same papyrus with a collection of lists for gymnasium administration (p.Ryl. IV 589) does not support Depuydt's view.
29. This most extreme representation of this idea is Koch and Glessmer (1999). Koch and Glessmer not only claim that this list commemorates religious activity in the beginnings of lunar months, but also that these beginnings take place at the day of the full moon rather than at first visibility, contrary to the common practice in Israelite and other ancient near eastern lunar calendars. Other scholars would agree to the identification of the full moon as the lunar phenomenon counted in 4Q320, but are less sure about the cultic role of that day—see for example VanderKam (1994).
30. Turner and Neugebauer (1949). I thank Prof. Roger Bagnall for discussing this religious background with

me. Bagnall seems to confirm the conclusions by Turner and Neugebauer.

31. The identity of the lunar phases recorded in 4Q320 and 4Q321 remains debated. Since the full debate will divert the present argument I shall not describe it in detail. Most scholars would claim that 4Q320 records the full moon date in each lunation, and that 4Q321 adds to it the phase called here *dwq*, which corresponds with the first visibility (for example VanderKam 1994) and see below for more bibliography). Talmon and Knohl argued instead that 4Q320 records the day of last visibility and that the *dwq* of 4Q321 corresponds to the day after full moon, when the moon begins to wane: Talmon and Knohl (1995); Talmon, Ben-Dov & Glessmer (2001), p. 79 and elsewhere. My own research agrees with Talmon and Knohl, although based on different evidence: Ben-Dov (2008), pp. 197–244.
32. Curiously, although *dwq* occurs before the other lunar phenomenon, it is mentioned after it in the list. This state of affairs changes in other parts of the 3-year cycle; see Ben-Dov (2008), pp. 208–215.
33. See note 31 above.
34. Several earlier studies tried to correlate the lunar dates of 4Q320 and 4Q321 with the lunar data contained in the scrolls 4Q317 and 4Q503: Wise (1994), esp. pp. 111–118; Abegg (1999). These studies, however, were not successful—see Ben-Dov (2008), p. 139.
35. Ben-Dov and Horowitz (2005).
36. Some texts measure the period of lunar visibility in the latter two dates, while other texts only give the dates in which they occur without specific measurements. The latter is the practice in the DSS too.
37. For the comparison, including an analysis of the numerical data in the DSS and Babylonian texts, see Ben-Dov (2008), p. 234–235. For the Babylonian text BM 32327+ see also Steele (2007).
38. Albani (1994), Drawnel (2007).
39. I aim to comply with the severe limits posed for the comparative method by Talmon (1978); see also Malul (1990). The categories of inter-cultural dependence have been more daintily defined in the field of Judaism in the Greco-Roman period; see Barclay (1999); more recently Satlow (2008).
40. Although the Babylonian calendar in the late period was mostly calculated, there might have remained in it a factor of *ad hoc* fixing of calendrical rulings; see Stern (2008) but compare Steele (2007).
41. For the religious dimension of lunar observations in Babylonia see: Beaulieu (1993).
42. Alternatively, if one accepts the view of Depuydt quoted above (Depuydt 1997), p. 152 that the 25-year cycle was 'an arithmetical game' rather than a practical calendrical cycle, then 4Q320 is very similar to it.
43. Dimant (2000, 2009), Tzoref (2011).
44. Baumgarten (2000), Koch (1983), Scott (2005).
45. This point has been made by Talmon (1958) and remains generally valid despite the criticism voiced by Stern (2011). Although some of Talmon's convictions are based on over-interpretations of the scrolls, it cannot be denied that the calendar retained a central place in the conceptions of the Yahad. Thus, calendrical rosters are included in key places of the sectarian literature: at the start of the scroll 4Q394, a copy of the halakhic-historical treatise 4QMMT, as well as in the concluding columns of 4QSe, a copy of the foundation text Serekh ha-Yahad. The evidence for the centrality of the calendar is strong enough and hard to deny, as Stern himself seems to acknowledge on p. 60 of his article. The thrust of his argument is to deny that the calendar played a constitutive role in the definition of the Yahad, and to deny that calendar polemics were a crucial component of Jewish society in Second Temple times. While these two points may indeed be correct, they do not altogether negate that a 364-day year existed, and has been practiced and endorsed by the Yahad.
46. Wallace-Hadrill (2005), Rüpke (1995).
47. Stern (2011), p. 41.
48. Collins (1997), Boccaccini (2005).
49. On the apocalyptic interest in Nature and Science, see Stone (1976), Alexander (2002).
50. For example the interest in astrology and physiognomy, see Alexander (1996, 1997), Popovic (2007).
51. See in more detail Feldman (2004).
52. Elior (2004).

References

Abegg, M. G., 1999, "Does Anyone Really Know What Time It Is? A Reexamination of 4Q503 in Light of 4Q317" in D. W. Parry and E. Ulrich (eds.), *The Provo International Conference on the Dead Sea Scrolls* (Leiden: Brill), 396–406.

Albani, M., 1992, "Die lunaren Zyklen im 364-Tage-Festkalender von 4QMischmerot/4QS[e]", *Mitteilungen und Beiträge* 4, 3–47.

——, 1994, *Astronomie und Schöpfungsglaube: Untersuchungen zum astronomischen Henochbuch* (Neukirchen-Vluyn: Neukirchener Verlag).

Alexander, P. S., 1996, "Physiognomy, Initiation and Rank in the Qumran Community", in P. Schaefer (ed.), *Geschichte-Tradition-Reflexion: Festschrift für Martin Hengel zum 70. Geburtstag* (Tübingen: J.C.B. Mohr), I, 385–394.

——, 1997, "'Wrestling against Wickedness in High Places': Magic in the Worldview of the Qumran Community" in S. E. Porter and C. A. Evans (ed.), *The Scrolls and the Scriptures: Qumran Fifty Years After* (Sheffield: Academic Press), 318–337.

——, 2002, "Enoch and the Beginnings of Jewish Interest in Natural Science" in C. Hempel, A. Lange and H. Lichtenberger (eds.), *The Wisdom Texts from Qumran and the Development of Sapiential Thought* (Leuven: Peeters), 223–243.

Barclay, J. M. G., 1999, *Jews in the Mediterranean Diaspora: From Alexander to Trajan (323 BCE–117 CE)*, (Berkeley: University of California Press).

Baumgarten, A. I. (ed), 2000, *Apocalyptic Time* (Leiden: Brill).

Baumgarten, J. M., 1987, "4Q503 (Daily Prayers) and the Lunar Calendar", *Revue de Qumran* 12, 399–407.

Beaulieu, P.-A., 1993, "The Impact of Month-Lengths on the Neo-Babylonian Cultic Calendar", *Zeitschrift für Assyriologie* 83, 66-87.

Ben-Dov, J. 2008, *Head of all Years: Astronomy and Calendars at Qumran in their Ancient Context* (Leiden: Brill).

——, 2010, "Mishmarot", in J. J. Collins and D. C. Harlow (eds.), *The Eerdmans Dictionary of Early Judaism* (Grand Rapids, MI: Eerdmans), 958–960.

——, 2011, "The 364-Day Year in the Dead Sea Scrolls and Jewish Pseudepigrapha", in J. M. Steele (ed.), *Calendars and Years II: Astronomy and Time in the Ancient and Medieval World* (Oxford: Oxbow Books), 69–105.

——, and Horowitz, W., 2003, "The 364-Day Year in Mesopotamia and Qumran", *Meghillot* 1, 3–26 (in Hebrew).

——, and Horowitz, W., 2005, "The Babylonian Lunar Three in Calendrical Scrolls from Qumran", *Zeitschrift für Assyriologie* 95, 104–120.

——, and Saulnier, S., 2008, "Qumran Calendars: A Survey of Scholarship 1980–2007", *Currents in Biblical Research* 7, 124–168.

Bennett, C., 2011, *Alexandria and the Moon. An Investigation into the Lunar Macedonian Calendar of Ptolemaic Egypt*, Studia Hellenistica 52 (Leuven: Peeters).

Boccaccini, G. (ed), 2005, *Enoch and Qumran Origins: New Light on a Forgotten Connection* (Grand Rapids, MI: Eerdmans).

Brack-Bernsen, L., 2007, "The 360-Day Year in Mesopotamia" in J. M. Steele (ed.), *Calendars and Years: Astronomy and Time in the Ancient Near East* (Oxford: Oxbow), 83–100.

Brown, D., 2000, *Mesopotamian Planetary Astronomy-Astrology* (Groningen: Styx).

Collins, J. J., 1997, *Apocalypticism in the Dead Sea Scrolls* (London and New York: Routledge).

——, 2010, *Beyond the Qumran Community: The Sectarian Movement of the Dead Sea Scrolls* (Grand Rapids, MI: Eerdmans).

Depuydt, L., 1997, *Civil Calendar and Lunar Calendar in Ancient Egypt* (Leuven: Peeters).

Dimant, D., 2000, "Ressurection, Restoration, and Time-Curtailing in Qumran, Early Judaism, and Christianity", *Revue de Qumran* 19, 528–548.

——, 2009, "Time, Torah and Prophecy at Qumran" in R. Hirsch-Luipold, H. Goergemanns and M. von

Albrecht (eds.), *Religiöse Philosophie und philosophische Religion der frühen Kaiserzeit*, Studien und Texte zu Antike und Christentum / Studies and Texts in Antiquity and Christianity 51 (Tübingen: Mohr Siebeck) 147–161.

Drawnel, H., 2007, "Moon Computation in the Aramaic Astronomical Book", *Revue de Qumran* 23, 3–41.

Elior, R., 2004, *The Three Temples: On the Emergence of Jewish Mysticism* (Oxford and Portland: Littman Library of Jewish Civilization).

Feldman, R., 2004, *"New Moon After New Moon and Sabbath After Sabbath" The Tension Between Culture and Nature in the Cycles of Sabbath and Moon in Ancient Jewish Calendars*, Ph.D. Dissertation, Graduate Theological Union.

Gillet-Didier, V., 2001, "Calendrier lunaire, calendrier solaire et gardes sacerdotales: recherches sur 4Q321", *Revue de Qumran* 20, 171–205.

Greengus, S., 1987, "The Akkadian Calendar at Sippar", *Journal of the American Oriental Society* 107, 209–229.

——, 2001, "New Evidence on the Old Babylonian Calendar and Real Estate Documents from Sippar", *Journal of the American Oriental Society* 121, 257–267.

Jones, A., 1997, "On the Reconstructed Macedonian and Egyptian Lunar Calendars", *Zeitschrift für Papyrologie und Epigraphik* 119, 157–166.

Koch, K., 1983, "Sabbatstruktur der Geschichte. Die sogenannte Zehn-Wochen-Apokalypse (I Hen 93:1-10, 91: 11-17) und das Ringen um die alttestamentlichen Chronologien im späten Israelitentum", *Zeitschrift für die alttestamentliche Wissenschaft* 95, 403–430.

Koch, K. and Glessmer, U., 1999, "Neumonds-Neujahr oder Vollmonds-Neujahr? Zu spätisraelitischen Kalender-Theologien" in B. Kollmann, W. Reinhold and A. Steudel (eds.), *Antikes Judentum und Frühes Christentum. Festchrift für Hartmut Stegemann zum 65. Geburtstag* (Berlin; De Gruyter), 114–136.

Kugler, R., 2010, "Of Calendars, Community Rules, and Common Knowledge: Understanding 4QS[e]- 4QOtot, with Help from Ritual Studies" in M. Grosman (ed.), *Rediscovering the Dead Sea Scrolls: An Assessment of Old and New Approaches and Methods* (Grand Rapids, MI: Eerdmans), 215–228.

Lippert, S. L., 2009, "*Au clair de la lune* – The Organisation (sic!) of Cultic Service by Moon Calendar in Soknopaiou Nesos", in G. Widmer and D. Devauchelle (eds.), *Actes du IXe congrès international des etudes démotiques* (Paris: Institute francais d'archéologie orientale), 183–194.

Malul, M., 1990, *The Comparative Method in Ancient Near Eastern and Biblical Legal Studies* (Neukirchen-Vluyn: Kevelaer Verlag and Butzon & Bercker).

Parker, R. A., 1950, *The Calendars of Ancient Egypt* (Chicago: The Oriental Institute).

Parry, D. W. & Tov, E., (eds.), 2005, *The Dead Sea Scrolls Reader* (Leiden: Brill).

Pfann, S. J., 2009, "A Reassessment of Qumran's Calendars", *Henoch* 31, 104–110.

Popovic, M., 2007, *Reading the Human Body: Physiognomics and Astrology in the Dead Sea Scrolls and Hellenistic-Early Roman Period Judaism* (Leiden: Brill).

Regev, E., 2007, *Sectarianism in Qumran: A Cross-Cultural Perspective* (Berlin: De Gruyter).

Rüpke, J., 1995, *Kalender und Öffentlichkeit: die Geschichte der Repräsentation und religiösen Qualifikation von Zeit in Rom* (Berlin: De Gruyter).

Satlow, M., 2008, "Beyond Influence: Toward a New Historiographic Paradigm", in Y. Z. Eliav and A. Norich (eds.), *Jewish Literatures and Cultures: Context and Intercontext* (Providence: Brown Judaic Studies), 37–53.

Schiffman, L. H., 1994, *Reclaiming the Dead Sea Scrolls: The History of Judaism, the Background of Christianity, the Lost Library of Qumran* (Philadelphia: Jewish Publication Society).

Scott, J. M., 2005, *On Earth as in Heaven: The Restoration of Sacred Time and Sacred Space in the Book of Jubilees* (Leiden: Brill).

Steele, J. M., 2007, "The Length of the Month in Mesopotamian Calendars of the First Millennium BC", in J. M. Steele (ed.), *Calendars and Years: Astronomy and Time in the Ancient Near East* (Oxford: Oxbow), 133–148.

Stern, S., 2000, "Qumran Calendars: Theory and Practice", in T. H. Lim, L. W. Hurtado and A. G. Auld (eds.), *The Dead Sea Scrolls in their Historical Context* (Edinburgh: T & T Clark), 179–186.

——, 2001, *Calendar and Community: A History of the Jewish Calendar, Second Century BCE-Tenth Century CE* (Oxford: Oxford University Press).

——, 2008, "The Babylonian Month and the New Moon: Sighting and Prediction", *Journal for the History of Astronomy* 39, 1–24.

——, 2011, "The 'Sectarian' Calendar at Qumran", in S. Stern (ed.), *Sects and Sectarianism in Jewish History*, IJS Studies in Judaica 12. (Leiden: Brill), 39–62.

Stone, M. E., 1976, "Lists of Revealed Things in the Apocalyptic Literature", F. M. Cross, W. E. Lemke and P. D. Miller (eds.), *Magnalia Dei – The Mighty Acts of God; Essays on the Bible and Archaeology in Memory of G. Ernest Wright* (Garden City, NY: Doubleday), 414–452.

Talmon, S., 1958, "The Calendar Reckoning of the Sect from the Judaean Desert", in Y. Yadin and C. Rabin (eds.), *Aspects of the Dead Sea Scrolls* (Jerusalem: Magnes), 162–199.

——, 1978, "The 'Comparative Method' in Biblical Interpretation: Principles and Problems", in *Congress Volume: Göttingen 1977*, VTSup 29, 320–356.

——, 1989, *The World of Qumran from Within: Collected Studies* (Jerusalem and Leiden: Magnes and Brill).

——, 1999, "Anti-lunar-calendar Polemics in Covenanters' Writings", in M. Becker and W. Fenske (eds.), *Das Ende der Tage und die Gegenwart des Heils; Begegnungen mit dem Neuen Testament und seiner Umwelt; Festschrift für Heinz-Wolfgang Kuhn zum 65. Geburtstag* (Leiden: Brill), 29–40.

——, 2000, "Calendars and Mishmarot", in L. H. Schiffman & J. C. VanderKam (eds.), *Encyclopedia of the Dead Sea Scrolls* (Oxford: Oxford University Press), 108-117.

——, Ben-Dov, J. and Glessmer, U., 2001, *Qumran Cave 4 XVI. Calendrical Texts,* DJD XXI (Oxford: Clarendon Press).

——, and Knohl, I., 1995, "A Calendrical Scroll from a Qumran Cave 4: Mišmarot Ba, 4Q321", in D. P. Wright, D. N. Freedman & A. Hurvitz (eds.), *Pomegranates and Golden Bells: Studies in Biblical, Jewish, and Near Eastern Ritual, Law, and Literature in Honor of Jacob Milgrom* (Winona Lake: Eisenbrauns), 267–301.

Turner, E. G., and Neugebauer, O., 1949, "Gymnasium Debts and New Moons", *Bulletin of the John Rylands Library* 32, 80–96.

Tzoref, S., 2011, "Pesher and Periodization", *Dead Sea Discoveries* 18, 129–154.

VanderKam, J. C., 1994, "Calendrical Texts and the Origins of the Dead Sea Scroll Community" in M. O. Wise, N. Golb, J. J. Collins and D. Pardee (eds.), *Methods of Investigation of the Dead Sea Scrolls and the Khirbet Qumran Site; Present Realities and Future Prospects* (New York: New York Academy of Sciences), 371–388.

——, and Flint, P. W., 2002, *The Meaning of the Dead Sea Scrolls: Their Significance for Understanding the Bible, Judaism, Jesus, and Christianity* (San Francisco: Harper).

Wacholder, B .Z. and Wacholder, S., 1995, "Patterns of Biblical Dates and Qumran's Calendar: The Fallacy of Jaubert's Hypothesis", *Hebrew Union College Annual* 66, 1–40.

Wallace-Hadrill, A., 2005, "Mutatas Formas: The Augustan Transformation of Roman Knowledge", in K. Galinsky (ed.), *The Cambridge Companion to the Age of Augustus* (Cambridge: Cambridge University Press), 55–84.

Wise, M. O., 1994, "Second Thoughts on *dwq* and the Qumran Synchronistic Calendars" in J. C. Reeves and J. Kampen (eds.), *Pursuing the Text: Studies in Honor of Ben-Zion Wacholder on the Occasion of his Seventieth Birthday* (Sheffield: Academic Press), 98–120.

Tame and Wild Time in the Qumran and Rabbinic Calendars

Ron H. Feldman

It is well known that for about four centuries starting in the second century BCE the calendar was a focus of conflict between different Jewish groups: one tradition, which culminated in the rabbinic calendar still used by Jews today, favoured an observational lunisolar calendar while the other, extensively documented in the literature found at Qumran, favoured a strictly calculated 364-day year calendar.[1] While there are ongoing scholarly debates about when, where and by whom these different types of calendars were used, and especially their use by the priests officiating in the Second Temple, for purposes of this paper I want to set aside the arguments about historic usage.[2] Here I want to focus on the different ways time is constructed in these calendars, and what this tells us about how their proponents differently negotiated the relationship between humans, God and nature in the temporal realm.

The calculated Jewish calendar (properly called "the Hebrew calendar") in use today is the latest stage of a lunisolar calendar tradition known primarily from rabbinic texts; these begin with the Mishnah, which was redacted over a century after the destruction of the Temple in 70 CE.[3] The earliest texts of this tradition describe an observational lunisolar calendar that is focused on correctly marking the first crescent of the new moon, which is considered the beginning of a new lunar month. In contrast, the 364-day year calendar is based exclusively on calculation rather than observation, and was forgotten until modern times. Today it is known from the Qumran texts (also known, more popularly, as the Dead Sea Scrolls) and the related texts of Jubilees and the *Astronomical Book* of 1 Enoch.[4] The texts describing a 364-day year calendar were all composed prior to the destruction of the Second Temple, and are mostly dated to the second or first century BCE at the latest, thereby predating the Mishnah by at least 300 years. Most scholars agree that this calendar tradition reflects priestly views and traditions. In discussing both types of calendars we are really discussing not a single calendar but calendar *traditions* because there are various features that changed as they evolved over time.

While there are well-known conflicts between proponents of these calendars concerning when to observe certain holidays and what to do when annual holidays fall on a Sabbath,[5] in my view the underlying point of calendrical conflict between these traditions is that they gave different solutions concerning the problem of reconciling the 29.5 day cycle of the New Moon with the seven-day cycle of the Sabbath. We will see that the 364-day calendar favoured the Sabbath at the expense of the New Moon while the rabbinic lunisolar calendar favoured the New Moon at the expense of the Sabbath. Of course, these divergent solutions are associated with different ideological positions and world-views.

Central to my analysis is the qualitative difference between the temporal cycles of the lunation versus the Sabbath. The common distinction between sacred and profane time is of no use here because what is at stake are different ways of establishing sacred time.[6] Methodologically speaking, because I agree with those who argue that humans are embedded within nature rather than separate from it, I will characterize these different rhythms of time by borrowing a distinction made in environmental thought between *wild* versus *tame*, where wild describes something beyond human control while tame describes that which is subject to human domination.[7]

The *lunation* and the *Sabbath* exemplify two qualitatively different types of time, *wild* and *tame*. The rhythms of the sun, moon and stars are part of *wild* nature that cannot be controlled by humans and, according to contemporary astronomy, existed before humans came on the scene. The biblical creation account of Gen 1–2:4a seems to be in general agreement with this sequence: the sun, moon and stars are created on the fourth day of the primordial week for the joint purpose of illumination and time keeping (Gen 1:14–19), and therefore existed before the creation of humans on the sixth day. The precise visibility of the first crescent of the new moon is an especially *wild time*, not only because it is uncontrollable, but because it remains unpredictable for contemporary astronomy.[8]

In contrast, the seven-day week defined by the Sabbath is a rhythm of *tame time* based exclusively on human counting; it is a direct revelation with no natural intermediary. The Sabbath's seven day rhythm is not governed by the celestial lights created on the fourth day, but is a direct commandment from God. While the biblical editors assert that the Sabbath is integral to the world of creation ("For in six days the Lord made heaven and earth and sea, and all that is in them, and He rested on the seventh day" (Exod 20: 11)),[9] it seems they are simultaneously aware that the Sabbath is not known to other creatures or peoples, but is a revelation to Israel that is part of its covenant with God:

> The Israelite people shall keep the sabbath, observing the sabbath throughout the ages as a covenant for all time: it shall be a sign for all time between Me and the people of Israel. For in six days the Lord made heaven and earth, and on the seventh day He ceased from work and was refreshed. (Exod 31:16–17).

The commandment to rest on the Sabbath is unique to the Israelite settlements, including humans and domesticated animals (see the two versions of the Ten Commandments: Exod 20:10; Deut 5:14). The Sabbath is manifested in this world only through their observance of it.[10]

I now want to turn to the Qumran and rabbinic calendar traditions, which both base themselves on these biblical texts and each, in its own way, seeks to *tame the wild* cycle of the lunation and *wild the tame* cycle of the Sabbath.

Before focusing on their differences it is important to acknowledge their similarities. Both of them evolved out of the Mesopotamian lunisolar calendar tradition.[11] Both are concerned with the proper and accurate observance of sacred moments, for you do not want to "make a day of testimony something worthless and a profane day a festival" (Jub. 6:37), as succinctly stated by the second century BCE author of Jubilees. Both traditions use the Bible as a resource and prooftext, mapping the annual biblical holidays onto a matrix of time defined by days, months, years and the seventh-day Sabbath. Indeed, they seem to have no conflict over the actual day observed as the Sabbath,[12] which leads to the conclusion that the calendrical polemics we are exploring arose *after* the successful institutionalization of the seventh-day Sabbath.[13] It turns out, however, that biblical ambiguities provide the basis for differing interpretations over such things as the definition of "month" and "year". Indeed, I think of the Bible as an enabler of later calendrical diversity and conflict.

Turning now to their differences, I want to first take up the 364-day calendar. This is a strictly calculated calendar that neither requires nor allows for observation. If there was a method of intercalation to make up for the missing 1¼ days per year it is unclear from the existing literature, although there has been much scholarly debate about this.[14] Certainly the later documents of this tradition maintain the idea of a rigid, unchanging arrangement of holy times commanded by God for humans to follow. This view is expressed in Jubilees, a second century BCE retelling of Genesis and Exodus as told by the Angel of the Presence to Moses on Mt. Sinai: "Now you command the Israelites to keep the years in this number – 364 days. Then the year will be complete and it will not disturb its time from its days or from its festivals because everything will happen in harmony with their testimony. They will neither omit a day nor disturb a festival" (Jub. 6:32).[15] The Rule of the Community, a later composition that is a major sectarian text from Qumran, similarly states: "They shall not stray from any one of all God's orders concerning their appointed times; they shall not advance their appointed times nor shall they retard any one of their feasts" (1QS, 13–15).[16]

While many scholars have described this as a "solar" or "concordant" calendar because it mathematically synchronizes the rhythms of sun, moon, stars and Sabbath, I contend that this is best described ideologically as a *Sabbatarian* calendar because the Sabbath, observed by the weekly duty shifts of the priests in the Temple, is the central rhythm of time to which all others are subservient.[17] 364 days conveniently divides evenly into 52 weeks, including 4 seasons of 13 weeks. Because the year has an exact multiple of complete weeks, the dates of the annual holidays fall on the same day of the week every year and have been arranged so that they never fall on the Sabbath. A calculated cycle of lunations appears in important variants of this calendar tradition (the *Mishmarot* texts), yet the months used for setting the dates of the annual holidays are not lunar but purely calculated groups of 30 or 31 days. Over the course of a six-year cycle, the lunations, non-lunar months and years synchronize with the weekly duty cycle of the 24 priestly families.[18] Since its primary temporal cycle is the seventh-day Sabbath, a length of time that does not appear in wild nature, I characterize the 364-day calendar as a calendar of *tame time*.[19]

The importance of the Sabbath to the worldview of the 364-day calendar tradition can be seen in Jubilees. Among the important changes compared to the biblical account is that the Sabbath is the now the "first law" (Jub. 2:24), given to the angels at creation and later revealed to Moses at Sinai, not merely the conclusion of the first week of creation as in Gen 1–2:4a:[20]

> He [God] gave us the sabbath day as a great sign so that we should perform work for six days and that we should keep sabbath from all work on the seventh day. He told us – all the angels of the presence and all the angels of holiness (these two great kinds) – to keep sabbath with him in heaven and on earth. (Jub. 2:17-2:19)[21]

To get a sense of how the Sabbath was supposedly observed in heaven we can turn to one of the major Qumran finds, *Songs of the Sabbath Sacrifice*, a cycle of liturgical prayers referencing the heavenly temple and its angelic priesthood for a sequence of 13 Sabbaths in the 364-day calendar. The text is generally considered to reflect the mysticism of the priestly followers of the 364-day calendar, especially how their practices on earth are connected to those supposedly taking place in heaven.[22] Here we find a passage that presents an image of God in His heavenly sanctuary being praised for His judgments and justice by the assembled councils of angels on the Sabbath:

> Song of the sacrifice of the seventh Sabbath on the sixteenth of the month. Praise the God of the lofty heights, O you lofty ones among all the elim of knowledge... [For He is] God of all who rejoice... forever and Judge in His power of all the spirits of understanding....And they make acceptable their knowledge according to the judgments of His mouth and their confessions (do they make acceptable) at the return of His powerful hand for judgments of recompense.
> (4QShirShabb^{d} (4Q403) 1 I 30–39)[23]

Jubilees explains that the purpose of the people Israel is to manifest the Sabbath on earth:

> He said to us [the angels]: 'I will now separate a people for myself from among my nations. They, too, will keep sabbath. I will sanctify the people for myself and will bless them as I sanctified the sabbath day. ....Now you command the Israelites to observe this day so that they may sanctify it, not do any work on it, and not defile it for it is holier than all (other) days. (Jub. 2:19–2:26)

To summarize, the 364-day calendar constructs a temporal cosmos of *tame* time, in which the rhythms of the celestial lights are subservient to the rhythm of the Sabbath, a rhythm directly revealed by God to the Israelites without wild nature as an intermediary. These texts tell us that on the Sabbath the angels praise God, who dispenses justice, and while the Sabbath was observed by God and the angels in heaven from the time of the creation, it becomes the task of the Israelites to bring this revealed rhythm into the created world. Here we can see a dual ideological move of *taming the wild* and *wilding the tame*: the wild rhythms of the celestial lights (created on day four) are tamed by their subjugation to the rhythm of the Sabbath, while the tame Sabbath itself is asserted to be an original wild rhythm of time created along with "the heavens and the earth" even if not perceptible in any part of the visible creation.

In contrast to the strict structure of the calculated 364-day calendar tradition, the observational lunisolar calendar described in the Mishnah (redacted in about 200 CE) and later rabbinic texts is mutable. The Mishnah extensively describes the procedure by which the new lunar month is determined and declared by the rabbinic calendrical court after the observation of the first crescent of the new moon. The need for this (discussed more extensively below) is the uncertainty about whether the next first crescent of the new moon will be sighted on the 30th or 31st day following the previous one, which determines whether the month just completed was 29 or 30 days long. Other rabbinic texts discuss the uncertainty concerning the intercalation of an additional lunar month into the year, which is also determined by observation of natural phenomena.[24] Because the length of the months and year varies, the length of the year will not be evenly divisible by seven, with the result that the day of the week on which the annual holidays fall will vary year to year, as will the number of weeks in a year. Therefore, in the rabbinic calendar tradition, the seventh-day

Sabbath cycle runs in a parallel and uncoordinated fashion with the annual holidays, which will occasionally overlap the Sabbath.

Before exploring the construction of tame and wild time in rabbinic calendar texts, I want to draw a clear contrast between the rabbinic attitude toward sacred time versus those of the 364-day calendar tradition by considering rabbinic texts written about four to five hundred years after the 364-day calendar tradition texts referred to earlier, but whose settings and motifs are virtually identical. Just as in Jubilees and the *Songs of the Sabbath Sacrifice*, we find God in heaven surrounded by the angelic council dispensing justice on a holy day. The Babylonian Talmud (BT) preserves a *baraita* concerning Rosh Hashanah which says that "the heavenly Court does not assemble for judgment until the Court on earth has sanctified the month" (*b. Rosh HaShanah* 8b). Rosh Hashanah is the only annual holiday that falls on the first of a lunar month, and its timing will determine the observance of the rest of the year's annual holidays, including Yom Kippur; it is therefore the most important—and paradigmatic—new moon holiday of the year.[25]

A commentary (*gemara*) in the Palestinian Talmud (PT) colourfully elaborates on the image provided by the BT:

> The Court said: today is Rosh Hashanah. The Holy One Blessed be He says to the ministering angels: set up the platform, let the defenders rise and let the prosecutors rise; for my children have said, today is Rosh Hashanah. If the Court changed to make the month full [so that Rosh Hashanah will fall on the next day], the Holy One Blessed be He says to the ministering angels: remove the platform, remove the defenders, and remove the prosecutors; for my children have decided to make the month full. What is the proof? *Therefore it is a statue for Israel, an ordinance for the God of Jacob* (Psalms 81:5). If it is not *a statute for Israel*, as though it were possible, it is not *an ordinance for the God of Jacob*. (*y. Rosh HaShanah* 1:3 (57b)).[26]

Both the similarities and differences between the rabbinic and 364-day calendar tradition texts illuminate the way these different Jewish cultural formations constructed the relationship between humans, God and nature. To begin with, we see a shift in paradigmatic sacred time: while the 364-day calendar tradition texts are focused on the Sabbath the rabbinic texts are concerned with the new moon and the related observance of Rosh Hashanah. This signifies a shift in emphasis away from the Sabbath and toward the lunation, thereby valorizing the uncertainty that is inherent in the wild time of the new moon.

Of even greater importance is a reversal in the relations of power between humans and God. Rather than the reverential tone we find in the 364-day calendar tradition texts, in which the task of humans is to emulate the actions of the angels, the PT text is almost comical: God doesn't know what day it is! While the sabbatarian rhythm of the 364-day calendar is eternally set and repeatedly observed, in this PT story God must wait to hear from the rabbinical court to know when Rosh Hashanah will fall since the legal authority to declare the new moons, and therefore the holidays, has been delegated to the court.

We might say that this demonstrates an amazing amount of rabbinical *chutzpah*. In Jubilees the Israelites emulate God and the angels in their observance of Sabbath while in the Talmuds this is reversed: God and the angels *follow* the practice of Israel and must wait around for the human rabbinical Court to tell the heavenly Court when it can be in session to dispense divine justice! While God may know when the new moon occurs, the human sighting the new moon is an inherently uncertain proposition, so what we find here is that the uncertainty and wildness of the new moon extends, through the rabbinical Court, into heaven. Power over time is shifted from God to humans, i.e., the Rabbis.[27] These Talmudic stories about the Heaven's observing holidays as determined by the rabbinic court is consistent with a broader rabbinic arrogation of authority to interpret Torah as a whole.[28] In comparison to the doctrines of the Qumran documents and the 364-day calendar tradition, the relationship between humans, God and nature has been redefined by the rabbis, and in important ways both God and nature are tamed and subject to human control.

I want to now turn to a more detailed discussion of the evolution of rabbinic attitudes about sacred time as expressed in rabbinic calendar texts. While the basic operation of the observational rabbinic calendar seems similar throughout the period I will consider—indeed, the rabbinic tradition always asserts that later practice is consistent with former practice—I will identify three key shifts in attitude reflected in texts that are portrayed as taking place at different locations associated with different historical periods. While I do not take the characters of the rabbinic texts, or the texts themselves, as literal reports of historical events, the texts do place characters and events in a rough historical context—and for my purposes here, the sequence is more important than the actual dates. Supposedly there was a calendrical court in Jerusalem that met at *Beit Yazek* during the time of the Temple (*m. Rosh HaShanah* 2:5). After the fall of the Temple this was moved to *Yavneh*, which is the setting for the stories about the generation of Rabban Gamliel (as below in *m. Rosh HaShanah* 2:9); this period ended in the wake of Bar Kochba revolt in 136 CE. By the end of the second century CE the calendar court had moved to *Ein Tav* in the Galilee, which is the setting for the story of Rabbi Hiyya and Judah the Patriarch (as in *b. Rosh HaShanah* 25a discussed below).[29]

In the earliest Beit Yazek phase there is an agreement with the attitude of Jubilees, that aligning the human observance of sacred time with God's intent is the goal, although the technique is totally different. In the observational lunisolar calendar, the new moon is considered the key divinely ordained temporal marker and is *more important* than the Sabbath according to the Mishnah. We can see this priority through the rulings in which witnesses to the new moon's appearance are encouraged to *violate* the rules of Sabbath observance in order to testify about their sighting of the new moon to the calendrical court (see especially *m. Rosh HaShanah* 1:1–2:9). For example:

> If a person saw the New Moon and is unable to walk, they bring him on an ass, or even on a bed; and if any lie in wait for them, they take sticks in their hands; if the way is long, they take food in their hands, because for a journey lasting a night and a day they may *violate* the Shabbat, and they go forth to give testimony about the New Moon, as it is written, "These are the appointed seasons of the Lord...which *you* shall proclaim in *their* appointed season" (Lev 23:4). (*m. Rosh HaShanah* 1:9, emphasis added)[30]

The Mishnah is arguing that Lev 23:4 is referring to the sacred times marked by the heavenly lights which were assigned this task by God in Gen 1:14, which uses the same term, "appointed seasons – מועד." The human task ("you" in Lev 23:4) assumed by the Court is to "proclaim" the sacred times of Lord's appointed seasons in conformance with the times revealed by the heavenly lights—in the rabbinic view, this is the moon. This requires accurate observation and is construed to be a task so important that it outweighs the laws of the Sabbath, which "they may violate". This also justifies the Court's role in evaluating the veracity of witnesses and its power to make a final declaration. The ritualistic legal formula for proclaiming the month is described by the Mishnah as follows: "The head of the Court says, "It is sanctified!" And all the people answer after him, "It is sanctified, it is sanctified" (*m. Rosh HaShanah* 2:7).

The Beit Yazek phase portrays a calendar of *wild time* in which uncertainty is highlighted and made central. Accurately determining the time indicated by the wild cycle of the new moon is paramount, and the focus of uncertainty is whether the new month will start on the 30th day. This uncertainty about the appearance of the first crescent is the justification for the entire legal procedure described in *m. Rosh HaShanah* 1:3–3:1, a procedure which gives the rabbinical Court the power and authority to determine and proclaim sacred times. This focus on the 30th day may also explain the use of the odd term "in its proper time". Why is this day normative while the 31st is described as "not in its proper time"?[31] I suggest it is because the Court has something it would like to accomplish on the 30th to justify and project its authority: declare the new month. If the new moon is not seen on the 30th, there is no need for the Court.[32]

Historically, I believe that the extreme emphasis on observation in the Beit Yazek phase of the rabbinic observational lunisolar calendar described in *m. Rosh HaShanah* is best explained as a response to, and a polemic against, the dysfunction of the 364-day calendar due to its lack of intercalation, (at least at the latest stage of its development as reflected in Jubilees and Qumran).[33] The command to violate the Sabbath when necessary is in direct conflict with the attitude toward the Sabbath of the proponents of the 364-day calendar, for whom the Sabbath is sacrosanct, and whose calendar precludes the overlapping of holidays with the Sabbath. It may have been these opponents who engaged in "disruptive practices" (*m. Rosh HaShanah* 2:1) and were the ones who would "lie in wait" for them (not just robbers), for we know from other mishnaic passages that there were opponents who sought to disrupt these observational procedures.[34] This is a conflict between those who followed calendars that had different priorities when forced to choose between the Sabbath and the new moon. It was a conflict between those who preferred certainty versus unpredictability, of tame versus wild time.

The *Yavneh* stage of rabbinic calendrical thought builds on the rabbinical Court's authority to declare sacred time to reverse the view of Beit Yazek: rather than emphasizing the uncertainty of the wild rhythm of the new moon's appearance, the lunar cycle is tamed by making a distinction between the *appearance* of the new *moon* and the *declaration* of the new *month*. This lays the philosophical groundwork for the eventual development of a fully calculated lunisolar calendar of tame time—although in a different fashion from that of the 364-day calendar tradition's effort to tame time by entraining the moon and the annual holidays to the rhythm of the Sabbath. I think it is quite likely that this turn of thought and practice was part of changed historical circumstances—namely, that with the destruction of the Temple and Qumran the advocates of the 364-day calendar fade from history at the same time that the rabbis are struggling against growing groups of Jewish-Christians for leadership of a people defeated and diminished by the Romans.[35]

The Mishnah begins this turn by asking: What happens when the calendar court is mistaken and declares a new month on what is the wrong day according to observation? What we see is a claim by the rabbis that it is their right to declare sacred time, whether this is in concert with the heavenly lights or not. This key philosophical turning point in the rabbinic calendar tradition is demonstrated in a famous story recorded in *m. Rosh HaShanah* 2:8–9:

> Rabban Gamliel had illustrations of the shapes of the moon on a tablet and on the wall in his upper chamber, which he showed to the simple, and he said, "Did you see like this, or like this? It once happened that two came and said, "We saw it in the east in the morning and in the west in the evening." Rabbi Yohanan ben Nuri said, "They are false witnesses!" When they came to Yavneh, Rabban Gamliel accepted them. And other two came and said, "We saw it at its proper time, and on the night preceding the added day it was not seen"; and Rabban Gamliel accepted them. Rabbi Dosa ben Harkinas said, "They are false witnesses! Can they testify about a woman that she gave birth, and on the morrow her belly is between her teeth?" Rabbi Yehoshua said to him, "I approve your words". (*m. Rosh HaShanah* 2:8)
>
> Rabban Gamliel sent to him: "I order you to come before me, with your staff and with your money, on Yom Kippur that falls according to your calculation". Rabbi Akiva went and found him distressed; he said to him, "I have to teach, that all that Rabban Gamliel did is done, for it is written, 'These are the appointed seasons of the Lord, sacred occasions which you shall proclaim' (Lev. 23:4), whether at their proper time, or not at their proper time, I have no appointed seasons save these". He came to Rabbi Dosa ben Harkinas. He said to him, "If we were to investigate the Court of Rabban Gamliel, then we must investigate each and every Court that has arisen from the time of Moses until now; as it is written, 'Then went up Moses, and Aaron, Nadav, and Avihu, and seventy of the elders of Israel' (Ex. 24:9). And why are the names of the elders not mentioned? To teach us that each and every three that have risen up as a Court over Israel are as the Court of Moses." He took his staff and his money in his hand, and went to Yavneh to Rabban Gamliel on the day that Yom Kippur fell by his calculation. Rabban Gamliel stood up and kissed him on his head, and said to him, "Come in peace, my master and my disciple! My master in wisdom, and my disciple because you accepted my words." (*m. Rosh HaShanah* 2:9)

Rabbi Dosa argues that Rabban Gamliel, the head of the calendar court, has clearly declared a new month on an incorrect day (and from previous criticism by Rabbi Yohanan ben Nuri, this was not the first mistaken declaration). We can infer that this new month happens to be Rosh Hashanah because this mistake will cause Yom Kippur to be observed on a different day vis-à-vis the correct sighting of the moon (as we learn later in the story). Rabbi Yehoshua agrees with Rabbi Dosa's observation, but is ordered by Rabban Gamliel to violate Yom Kippur according to Rabbi Dosa's calculation. Rabbi Yehoshua is greatly vexed: should he observe the holy day as declared by the rabbinic court, or on the day indicated by the heavens? In the end, Rabbi Yehoshua conforms to the ruling of the court, which is the primary lesson of this story.[36]

Rabbi Yehoshua's surrender to Rabban Gamliel is justified by two arguments: one is that while courts may be fallible, their rulings still hold. The fact that the story is about the moon merely highlights the court's authority: this is not merely a story about the unclear facts of some case between individuals, but is blatantly public and apparent to anyone who can see the moon.

The second, made by Rabbi Akiva, is more important for the present discussion: Rabbi Akiva argues that God has handed over the authority to declare sacred time to the rabbinic court, an assertion based on his interpretation of the biblical verse, "These are the appointed seasons of the Lord, sacred occasions which *you* shall proclaim" (Lev 23:4, emphasis added). It is important to note that Rabbi Akiva is using the same prooftext that was used in *m. Rosh HaShanah* 1:9 (referred to above), where it was used to justify the importance of accurately observing the moon and reporting its appearance to the Court even if the Sabbath had to be violated, a Beit Yazek period view now represented by Rabbi Dosa and Rabbi Yehoshua. However, to justify his argument Rabbi Akiva conveniently leaves out precisely the words from Leviticus that were emphasized in m. Rosh HaShanah 1:9, "which you shall proclaim in their appointed season". By emphasizing human proclamation and eliminating the autonomy of the celestial lights to mark time, Rabbi Akiva is reversing the argument of *m. Rosh HaShanah* 1:9.[37] Rabbi Akiva further argues that the court's proclamations are what matter: whether declared "at their proper time, or not at their proper time, I have no appointed seasons save these". Rabbi Akiva's words may even be taken as ironic or satiric: "in its proper time" and "not in its proper time" are technical terms for when the moon is sighted on the 30th or 31st day after the prior sighting, but here Rabbi Akiva seems to also be using these terms in a more literal fashion, i.e., whether the time declared is correct or incorrect, the proclamation determines the "appointed season".

While we should not think that this story is historical, it nevertheless portrays the philosophical shift of rabbinic attitudes about sacred time, which seems to also correlate to the shift from Beit Yazek to Yavneh (at least as indicated by the change in characters and venue). The old view—held by the 364-day calendar tradition, the Beit Yazek rabbinic period (as in *m. Rosh HaShanah* 1:9) and Rabbi Dosa in this story—was that God determined sacred times and it was for humans to conform to them. The difference between the 364-day calendar tradition and this rabbinical view was one of technique, not attitude: the former held to a holiday calendar defined by the divinely commanded sabbatarian rhythm while the latter claimed the moon was the divinely created marker for the holidays, even when the result conflicted with the Sabbath.

The new view of Yavneh, put into the mouth of Rabbi Akiva and implicitly that of Rabban Gamliel, arrogates to humans the right to determine sacred times. There is no such thing as a mistaken proclamation because God has "no appointed seasons save these". The key innovation is the distinction between the actual *sighting* of the new moon and the *declaration* of the sighting of the new moon. This distinction lays the groundwork for shifting the purpose of the calendrical court from validating the sighting of the new *moon* to the declaration of the new *month* that is only tenuously connected to the actual appearance of the new moon; the moon and the month have been divorced. This is a radical and paradoxical move because—as discussed above—it is the hyper attention to the uncertain appearance of the moon that provides the Beit Yazek rabbis with the justification for the legal proceedings of the calendrical court. Yet, having attained that authority, the court now becomes the venue for the rabbinical assertion of power to determine not the *actual* appearance of the new moon (over which they have no control) but the *socially* determined—and *divinely* accepted, as we saw in the passages from the BT and PT—beginning of the new month.

This power to declare the new month, even when mistaken vis-à-vis the new moon, leads to the next logical step: not just to be accepted when wrong, but to willfully declare the month according to the rabbis' desires regardless of the appearance of the new moon. What I identify as this third, *Ein Tav* phase is illustrated by a story in the *gemara* of the BT:

> Rabbi Hiyya once saw the (old) moon in the heavens on the morning of the twenty-ninth day. He took a clod of earth and threw it at it, saying "Tonight we want to sanctify you, and are you still here! Go hide yourself." Rabbi thereupon said to Rabbi Hiyya, "Go to Ein Tav and sanctify the month, and send me the watchword, 'David king of Israel is alive and vigorous.'" (*b. Rosh HaShanah* 25a).[38]

This story presents the essence of the rabbis' conundrum: they want to declare the new month, but are constrained by the justifications they have used for legitimating their authority, i.e., correctly observing the new moon. While the court wants to declare the new month that evening,[39] Rabbi Hiyya is angered by the old moon's appearance in the early morning because this means the new moon will not be visible as desired when the evening comes. Throwing a dirt clod seems to express his frustration at the same time that it demonstrates that the moon is truly not subject to human control, since it is impossible to hit the moon with the dirt clod. The moon's undesirable appearance is *interfering* with the desire of the court, and Rabbi Hiyya is upset about it. At this point in the story, the wildness of the moon is a problem, and taming it remains a challenge.

To resolve the problem of the old moon's appearance, Rabbi Judah the Patriarch[40] sends Rabbi Hiyya to another location where he will declare the new month regardless of whether or not he sees the new moon: "Go to Ein Tav and sanctify the month".[41] The calendrical court has decided to *disregard* the physical evidence and do what it wants; the new month does not depend on the new moon but on the Court's declaration. Indeed, the BT debates the conditions for "intimidating" witnesses to provide testimony when the court wants to either shorten or lengthen the month (*b. Rosh HaShanah* 20a). Observation has become a legal fiction and the wildness of the new moon has been tamed for calendrical purposes. This attitude lays the groundwork for the eventual rise of the calculated rabbinic calendar, which further dissociates the month from the observed new moon by adding rules that adjust the length of some months by a day or two in order to satisfy the rabbinic desire to control the days of the week on which certain holidays fall.[42]

At this point we can see that the post-Beit Yazek stages of the rabbinic calendar texts carry out a *dual reversal*, of *taming the wild and wilding the tame*. While the 364-day calendar tradition performed a similar dual reversal by (theoretically) forcing the rhythms of the celestial lights into alignment with the Sabbath, the later rabbinic texts take a different route. While the Beit Yazek texts argued for the wildness of the new moon and the need to violate the Sabbath to testify to its precise arrival, the Yavneh and Ein Tav texts *tame the wild* by claiming that humans have the ability to recognize or withhold recognition of the natural cycle of the moon—a wild cycle which takes place whether or not humans mark its passage. This process of taming the new moon took place incrementally. In the Beit Yazek phase the rabbis claimed the authority to validate the moon's appearance; the Yavneh phase built on this authority to assert the power to declare the new month even if mistaken vis-à-vis the moon; the Ein Tav phase expanded this power to declare the new month when desired, which provided the philosophical basis for a fully calculated calendar that calculated a theoretical new moon and included various postponements due to other considerations.

On the other hand, the Rabbis *wild the tame* by agreeing with the doctrine of the 364-day calendar tradition that the seventh-day Sabbath rhythm is an aspect of the primordial creation that does not require human recognition. As we saw above, for Jubilees (Jub. 2:19) the Sabbath is established in heaven by God and merely echoed by humans on the earth. The rabbinic parallel is most concisely expressed by a passage in the BT that states, "The appointed seasons of the Lord need sanctification by the Court, but the Sabbath of creation does not need sanctification by the Court" (*b. Nedarim* 78b). That is, while the right to determine the new month was given to humans, the rabbinical court has no jurisdiction over the Sabbath, which is a part of creation whose timing is set by God without the intermediation of the heavenly lights.

Nevertheless, the rabbinic view is somewhat less strict than that of the 364-day calendar tradition, which is consistent with the less rigorous approach to the Sabbath that we saw above in the rabbinic preference for the new moon over the Sabbath. Despite their effort to "naturalize" the "Sabbath of creation", the rabbis are aware that neither the Sabbath day's rest nor its seven day rhythm are tied to the world of nature. This is expressed by a baraita in the BT that reads: "If one is travelling on a road or in a wilderness and does not know when it is the Sabbath, he must count six and observe one day" (*b. Shabbat* 69b). The issue being addressed here is a situation where it is impossible to know from nature when the counting begins. The uncertainty expressed here echoes the perspective about sacred time we find in the Beit Yazek period of the rabbinic calendar, where uncertainty concerning the appearance of the new moon is highlighted. But unlike the moon, there is no indicator of the Sabbath for the individual who is lost in the wilderness because wild nature knows nothing of the seventh-day Sabbath and does not rest on a seven-day rhythm.

The rabbinic solution is to focus on the seven day *rhythm* of the Sabbath since the exact day may be unknowable; this is in contrast to the attitude of Jubilees, for whom the exact *day* is important, since humans must be in alignment with heavenly observance. The rabbis seem to acknowledge that all we can know is how the Sabbath has been counted by a community; the individual who loses that social context has no choice but to begin to count for oneself a seven-day rhythm, which may not match the counting of society or God. The rabbis endorse the idea that the rhythm and ritual elements of Sabbath rest are still commanded, even if one is uncertain about the exact day. Even though wild nature does not observe the Sabbath, humans must remember to observe it. In comparison to Jubilees, whose emphasis was on manifesting on earth what is going on in Heaven, the Rabbis' focus is on fulfilling God's commandments as mediated through their interpretation of Torah.

The lunation and the Sabbath are examples of two types of time, wild and tame, expressing two ways of temporally relating to the more-than-human cosmos in which we find ourselves.[43] In examining and contrasting two different Jewish calendar systems—the 364-day tradition and the rabbinic tradition—we have seen how different tactics were used to make sense of, and assert human control over, temporal experience. The seventh-day Sabbath, a rhythm of time that is not part of wild nature, is construed as part of nature, a process I have called *wilding the tame*. Simultaneously, I have described a process of *taming the wild* in which the uncertain appearance of the new moon is managed, either by subsuming it to the seventh-day Sabbath as in the 364-day calendar or, in the case of the Rabbis, subjecting it to a legal procedure that asserts the human right to declare a new lunar month regardless of the actual appearance of the new moon.

Both taming the wild and wilding the tame are complimentary tactics for *taming time* engaged in by elite groups aiming to gain authority, legitimacy and power. Yet, time is a dimension of wild nature that is even less amenable to human control than space. Therefore, the words and actions of these groups are unable to tame time per se—which is untouched by human effort—but tame the human experience of time by asserting the right to structure the human interface with time in the context of a particular cultural formation. Calendars are models *of* and *for* social practices of time; a dominant calendar projects the ideology, practice and authority of the dominant group.[44] Every calendar is a map of time, a narrative of time that mediates the phenomenological experience of time itself; every calendar projects its own ideological position and world-view and can be "read" as a "text" that negotiates the relationship between humans and nature.

While calendars inevitably mediate the human experience of time, they vary in the degree to which they align cultural rhythms to those of wild nature. The lunar element of the calculated rabbinic calendar preserves a practice of sacred time that in principle adheres closely, if not precisely, to the lunation, while justifying this practice by remembering a past in which time was uncertain and less controlled.

On the other hand, like the later invention of the mechanical clock, in the ancient world the establishment of the seventh-day Sabbath was an important step in the process of taming nature by taming time, which in turn contributed to creating an experienced separation of human culture from wild nature.[45] But this shift was neither simple nor straightforward, involving transformed understandings of the interrelationship of humans, God and nature in what Rabbi Abraham Joshua Heschel described as Judaism's "*architecture of time*".[46] The calculated Hebrew calendar in use today preserves echoes of these ancient controversies and temporal transformations in the unruly lack of fit between the rhythms of the Sabbath and the lunation.

Notes

1. For discussion of the conflict between the proponents of these different calendars see Talmon (1951); Talmon (1958); Talmon (1999a); Talmon (1999b); Schiffman (1995), pp. 304–305; Rubenstein (1999), pp. 175–177; Baumgarten (1985), pp. 395–397.
2. The main debate is between two views. First are those who contend that the Second Temple's calendar was lunisolar, similar to the one described in rabbinic texts and has clear antecedents in Mesopotamian calendar practices. Second are those who contend that the Temple used a 364-day year calendar similar to that found in the Qumran (and related) texts, although there are various hypotheses concerning when and for how long such a calendar may have been used. For a brief discussion of academic positions concerning the possible historical use of the 364-day calendar and its priestly orientation, along with further references, see Feldman (2009), esp. pp. 343–350.
3. For standard descriptions of the history and functioning of the rabbinic calendar derived from examination of rabbinic texts (and assuming the use of this calendar in the Second Temple, as described in these texts) see Tabory (2000), pp. 19–34; Alon (1984), pp. 237–248. For a more critical evaluation of the historical use and evolution of the rabbinic calendar see Stern (2001).
4. For a concise explanation of the 364-day calendar along with a discussion of the texts, scholarship and debates surrounding it, see VanderKam (1998). For discussion of the variety among the texts of this calendar tradition, see n. 18.
5. This question of correct Sabbath observance was a key source of conflict: the rabbis decided that when there was an overlap between a holiday and the Sabbath, the holiday (and its sacrifices) took precedence, while the advocates of the 364-day calendar were completely opposed to this practice. The Bible never discusses what to do when there is such a conflict, and the 364-day calendar never has such a conflict. For further discussion, see references in n. 1.
6. Durkheim argued that cultures classified all things, including time, into two categories, the sacred and the profane; see Durkheim (1961), pp. 52, 347. Eliade greatly developed and popularized this distinction as characteristic of time and place; see for example Eliade (1959), p. 70. The Jewish calendars under

discussion agree on the idea of distinguishing between sacred and profane, but use different criteria to establish these differences.

7. This terminology is used widely in environmental thought, but derives specifically from the writings of Thoreau (Thoreau (1998), p. 31). While the distinction between tame vs. wild time may seem close to other more familiar dichotomies, such as culture vs. nature or artificial vs. natural, I prefer to use tame vs. wild because I am sympathetic to Merchant's critique of the common nature vs. culture dualism and am attempting to emphasize the embeddedness of humans in the natural environment, where nature and culture are both actors and subjects in systemic interactions; see Merchant (1990), pp. 143–144; Merchant (1989), pp. 1–26.
8. On the uncertainty of sighting the new moon see Stern (2001), pp. 110–111; Long (1998), pp. 43–47.
9. See also Gen 2:1–4a. Unless otherwise noted, Biblical translations are from the JPS *Tanakh* (1985).
10. It seems to me that the biblical editors awareness of the non-natural essence of the seventh-day Sabbath rhythm is revealed by the terms used to describe it and command its observance: one must "remember-זכור" (Exod 20:8) it, "preserve-שמור" (Deut 5:12) it, and "make-לעשות" (Exod 31:16) it. In the Ten Commandments it is the only time cycle mentioned and the only element termed "holy-קדוש", which means something that is "set apart". All of these terms imply a positive action (including restful inaction) that must be taken by humans to manifest the Sabbath in this world. In addition, the Sabbath's establishment on day seven, after the creation of humans on day six, could be seen to imply the need of humans in order to manifest the Sabbath, in contrast to the times determined by the luminaries created on day four, which do not require human participation.
11. There has been no question about the origins of rabbinic calendar practice, especially considering its use of Babylonian month names. As for the 364-day calendar tradition, I am convinced by the recent work of Horowitz and Ben-Dov; Horowitz (1996); Horowitz (1998); Ben-Dov and Horowitz (2003); Ben-Dov and Horowitz (2005); Ben-Dov (2008).
12. In contrast, there were conflicts over the observance of other holidays, especially the well-known dispute over the proper count of the Omer and the related observance of Shavuot reported in *m. Menahot* 10:3 and *b. Menachoth* 65a–b. For recent discussions about the rabbis' Boethusian opponents as supporters of Qumran calendar practice, see Sussmann (1989), p. 54; Schiffman (1995), p. 304; Tabory (2000), pp. 136–138.
13. There has been substantial scholarly debate for over a century about the origins of the Sabbath and the associated seven day week. One school credits the continuous seventh-day Sabbath with being a uniquely Jewish idea, while others attempt to derive it from Mesopotamian sources. While I am sympathetic to the first school, this is not a critical issue for the discussion in this paper. There is also a scholarly disagreement about when the seventh-day Sabbath was institutionalized, with the main positions being that it goes back to the dawn of Israelite practice, or that it was established during the Persian period of the Second Temple. For a further discussion of these positions and the process by which the seventh-day Sabbath was institutionalized, see Feldman (2009), especially pp. 351–356.
14. Among the suggestions for various intercalation schemes: Zeitlin suggested that the "Jubilee" year was not a year, but a 49 day block inserted every 49 years as an intercalation device; Zeitlin (1973), p. 186. Glessmer develops a possible intercalation scheme based on the sabbatical cycle; Glessmer (1996). Vanderkam suggests that 35 days (5 weeks) could have been intercalated every 28 years; VanderKam (1979), p. 406. Wise suggests an intercalary scheme based on adding four weeks to the solar calendar periodically, thereby preserving the coordination with the lunar cycle; Wise (1994), p. 109. Gardner hypothesizes that a week was intercalated every six years, with an additional week intercalated every 84 years; Gardner (2001), pp. 266–271. Among those who discount the possible use of this calendar are Herr (1976); Segal (1957), pp. 251–253; Davies (1983); Beckwith (1996), pp. 126–127; Ben-Dov (in this volume).
15. Translations from Jubilees are from VanderKam ed. (1989).
16. García Martínez and Tigchelaar (2000), p. 71.
17. For further discussion of the characterization of the 364-day calendar see Feldman (2009), pp. 347–350; Ben-Dov and Saulnier (2008).

18. While this brief summary of major features common to the 364-day calendar tradition is sufficient for purposes of this paper, there is a fair amount of diversity in the textual evidence itself from which commonalities must be inferred and differences respected. While the Sabbath is not even mentioned in the *Astronomical Book* of 1 Enoch, and the lunation is not part of the Jubilees calendar, I am highlighting the shared importance of the Sabbath in the Qumran texts and Jubilees. Jaubert's widely accepted reconstruction of the calendar underlying the text of Jubilees is identical to the calendrical texts found at Qumran; see Jaubert (1957), published in English as Jaubert (1965). Ravid critiques Jaubert's reconstruction, but does not diminish the importance of the Sabbath in Jubilees; Ravid (2003). For further analysis of the variety of 364-day calendars see Snyder (1997); Glessmer (1999); Feldman (2004), pp. 73–79; Ben-Dov (2008), pp. 1-67; Talmon, et al. eds (2001); VanderKam (1998). My view of the differing 364-day calendars is that they are more likely to express diachronic developments rather than synchronic competing calendars.
19. Elior similarly emphasizes the central importance of the "sevenfold cycles of Sabbaths and festivals, sabbatical years and jubilees" that "have no visible testimony in nature or any revealed expression other than the audible divine decree taught by the angels and kept by the priests". Elior (2004), p. 83.
20. As Doering points out, Sabbath observance even supersedes the commandment of Gen. 1:28 to procreate, the first commandment spoken to humans in the Bible, a commandment that is omitted from Jubilees account of the sixth day of creation in Jub. 2:14. Doering (1997), p. 187 n. 39.
21. For discussions of the importance of the Sabbath to the author of Jubilees, see VanderKam (1998), pp. 27–33; Doering (1997).
22. Newsom (1985), esp. pp. 5–21; Schiffman (1995), pp. 355–360; Elior (2004).
23. Newsom (1985), pp. 211–212. Elim, which means divine beings, is probably a reference to angels.
24. For example, a *baraita* (a mishnaic era text) in the BT explains: "Our Rabbis taught: A year may be intercalated on three grounds: on account of the premature state of the corn-crops; or that of the fruit-trees; or on account of the lateness of the Tekufah [season]. Any two of these reasons can justify intercalation, but not one alone" (*b. Sanhedrin* 11b); Talmudic translations are based on Epstein ed. (1983). All three criteria are observational; the first two are agricultural, while the third is the astronomical observation of the spring equinox (while "Tekufah" can refer to either equinox or solstice, intercalation was always a second month of Adar which means that the spring equinox was the event being referred to). There are also multiple versions of a story telling of Rabban Gamliel sending missives to the diaspora to intercalate the year; see *t. Sanhedrin* 2:6, *y. Sanhedrin* 1 (18d) and *b. Sanhedrin* 11b.
25. This is also the same day over which Rabbi Yehoshua and Rabban Gamliel have their dispute *m. Rosh HaShanah* 2:8–9 (see discussion below).
26. Translation from Goldman ed. (1988), p. 44. An even more elaborate version appears in *Pesiqta de-Rab Kahana* 5:20; see Stern (2001), p. 230.
27. The rabbinic assertion of power over time has been recognized for some time (for recent discussions see Rubenstein ed. (2002), p. 86; Stern (2001), p. 231). My point here is that this is a *reversal* of earlier attitudes.
28. As expressed in the much discussed text known as "The Oven of Akhnai" (*b. Bava Metzi'a* 59a–b), the rabbis reject revelation and miracles, asserting that because the Torah was given at Mt. Sinai "it is not in heaven" and its laws are exclusively the domain of human interpretation. For a discussion and references to interpretations of this story, see Rubenstein (1999), pp. 34–63, esp. n. 1–3.
29. For a history of these eras (which largely assumes the rabbinical texts are reporting actual events), see Alon (1984).
30. Translations of the Mishnah are based on Kehati ed. (1992–1994). Similar descriptions of Sabbath violations are in *m. Rosh Hashanah* 1:4, 5.
31. Virtually all translators use the word "proper" even though the Hebrew is literally "in its time" or "not in its time." I am using "proper" because this reflects the rabbinical attitude about these temporal moments (see Jastrow (1903)). Gandz discusses the various and changing terms used to describe the 30th and 31st days, but he understands the terminology as merely technical; I am suggesting a socio-political motivation; Gandz (1949). "In its proper time" is the earliest term, reflecting this early stage. A later pair of terms is

"not intercalated" and "intercalated" for the 30th and 31st day, indicating a reversal between which day is construed normative.

32. The Mishnah addresses this issue by relating a disagreement: "Whether it was seen at its proper time, or whether it was not seen at its proper time, they sanctify it. Rabbi Eleazar ben Zadok says, If it was not seen at its proper time, they do not sanctify it, for Heaven has already sanctified it" (*m. Rosh HaShanah* 2:7). While the Mishnah does not decide between these two views, the gemara in the BT debates the issue and decides in favour of R. Eleazar (*b. Rosh HaShanah* 24a), i.e., the court has nothing to do unless the new moon is sighted on the 30th day.
33. The reasons why the early rabbinic calendar used an observational method deserve a larger discussion than can take place here (see also Stern and Schiffman, in this volume). While assumed to be a practice from long tradition by much scholarship, the reason for using a calendar dependent on observation is not obvious, particularly since a calculated Babylonian calendar seems to have become standard in the Persian Empire by the fifth or fourth century BCE (on the calculated Babylonian calendar, see Wacholder and Weisberg (1976), p, 72. See also Rochberg (1995), p, 1938). Although their answers are inadequate, some scholars have posed this question; see Gardner (2001), pp, 6–7, 123; Schürer (1973), p. 594; Geller (1995), pp. 46–47. Indeed, this is a problem whether one contends the rabbinic system was the "official" calendar throughout the Second Temple period, or that it was introduced into the Temple in place of a 364-day year calendar at some point between the second century BCE and first century CE, for in both cases calculation techniques were available by the fourth century BCE.
34. They are identified as "Boethusians" and "Cutheans," See *m. Rosh Hashanah* 2:1; 2:2; *m. Menahot* 10:3. Boethusians are thought to be a priestly group while Cuthians are Samaritans. For discussions of various groups of this era see Baumgarten (1997); Sanders (1992); Baumgarten (1985); Schwartz (2001), pp. 91–99; Sussmann (1989).
35. D. Schwartz uses a similar diachronic historical analysis of the changes in rabbinic competitors to discuss evolving rabbinic views regarding nature; see Schwartz (2004).
36. While not critical for my discussion of this story, it is worth noting that the mishanaic text "He said to him", is ambiguous about who is doing the speaking: Rabbi Akiva, Rabbi Dosa or Rabbi Yehoshua? Most Jewish commentators have assumed that Rabbi Dosa is doing the speaking, but there are good reasons for thinking it is Rabbi Yehoshua. For a thorough discussion of this point, as well as the story as a whole, see Schwartz (2004), esp. pp. 24–33. D. Schwartz's proposal also makes the Yehoshua in this story more consistent with the Yehoshua in the "Oven of Akhnai" story, the one who stands up for the authority of the Court (see above n. 28). Of more relevance to my discussion is D. Schwartz's proposal that Rabbi Dosa's view about nature and law is "characteristic of priestly trends in ancient Judaism—that when the law conflicts with nature (in this case: the moon), it is the latter that prevails" (Schwartz (2004), p. vi, see also pp. 29–30; cf. D. Schwartz's earlier essay, Schwartz (1992), pp, 234–235 and Rubenstein's response in Rubenstein (1999), pp, 175–177). I suggest a more nuanced discussion: as I argue below, Rabbi Dosa and the advocates of the 364-day calendar share the idea that God determines sacred time and it is for humans to recognize and adhere to them, but they differ about the calendar and procedures used to determine sacred time. Schwartz's "priestly trends" were advocates of the 364-day calendar who were in conflict with rabbinic groups that, like Rabbi Dosa, advocated an observational lunisolar calendar. What counts as "nature" was part of what Rabbi Dosa and the priestly advocates of the 364-day calendar were arguing about; Rabbi Dosa's view exemplifies what I have described as "wild" nature (and time), to which humans are subject, over which they have no control, and which cannot be changed by legal decree.
37. D. Schwartz cogently argues that the appearance of Rabbi Akiva in the story "is secondary, added in by a tradent who missed the presence of this most prominent Yavnean figure and who also wanted...rabbinic rules to derive from the biblical text"; Schwartz (2004). This, however, does not impact the thrust of my argument; at most it means that the change in thinking reflected in the inversion of the interpretation of Lev 23:4 took place at a slightly later date.
38. For a slightly different version, see *y. Rosh HaShanah* 2:5 (58a). A significant difference in the PT version is that after Rabbi Hiyya throws a rock at the moon it disappears; it seems that the BT is not comfortable

with this solution, which is consistent with its views concerning the inadmissibility of supernatural evidence (see above n. 28).

39. The reason why the Court wants to declare the new month is not stated. The medieval commentaries of Rashi and Tosafot to this passage infer that this is because the Court did not want the Day of Atonement adjacent to the Sabbath; this view already assumes the knowledge of at least some of what became the calculated rabbinic calendar's rules of postponement — דחיות that can shift the observance of the new month by one or two days away from the calculated occurrence of the new moon. For a basic discussion of these rules, see Wiesenberg (2007), pp. 354–356; for the historical development of these rules, see Stern (2001), pp. 155–210.
40. Judah the Patriarch was also the redactor of the Mishnah.
41. Indeed, the name of this location is possibly a pun: while "Ein Tav — עין טב" would commonly mean "good spring"—i.e., good drinking water—it could also mean "good eye", implying that Rabbi Hiyya will need a *very* good eye indeed to see the new moon at the desired time. This is the only appearance of "Ein Tav" in the BT, enhancing the possibility of its use as a pun in this context; the term does appear in the PT as a place name, and the later BT commentators (Rashi and Tosafot) assume it refers to a location where the calendrical Court would meet. According to Safrai's analysis of the literature, this was the last location of the calendrical court mentioned in the Amoraic sources, which would also fit my suggestion that this reflects a later stage of calendrical ideology; Safrai (1965), p. 37.
42. For references on the postponements, see n. 39. While the shift to calculation has traditionally been dated to the fourth century CE, Stern has shown that this was a gradual process that did not reach the current from of the calculated Hebrew calendar until the Tenth century; see Stern (2001), pp. 211–275.
43. I have adapted this from Abram's use of "more-than-human world" in preference to "nature"; while I agree with the embeddedness this terminology indicates, I find the locution unfortunately awkward, and have therefore stayed with the overused term "nature". See Abram (1996).
44. I have drawn these dual aspects of calendars as models of and for reality from Geertz's discussion of cultural patterns in general, of which I take calendars as a particular case. Geertz (1973), p. 93. Durkheim was one of the pioneers concerning the social and political function of calendars: "A calendar expresses the rhythm of the collective activities, while at the same time its function is to assure their regularity" (Durkheim (1961), p. 23; see also pp. 488–491). Although I agree with Durkheim that the natural aspect of calendars serves to legitimize them, I contend that the origins of calendars lie in the effort of humans to make sense of found temporal rhythms. For a more recent sociological treatment see, Levine (1997).
45. On the similar significance of the clock and the Sabbath, and their function of distancing humans from nature, see Zerubavel, who writes: "The invention of the continuous week was therefore one of the most significant breakthroughs in human beings' attempts to break away from being prisoners of nature and create a social world of their own". Zerubavel (1985), p. 11. Zerubavel lauds this liberation of humans from nature, but this also helps set up the context for the subjugation of nature in the modern period; on this see Merchant (1990).
46. Heschel (1951), p. 8.

References

1985, *Tanakh: A New Translation of The Holy Scriptures according to the Traditional Hebrew Text* (Philadelphia: The Jewish Publication Society).

Abram, D., 1996, *The Spell of the Sensuous: Perception and Language in a More-Than-Human World* (New York: Vintage Books).

Alon, G., 1984, *The Jews in Their Land in the Talmudic Age (70-640 C.E.)*, Translated by G. Levi (Cambridge, Mass.: Harvard University Press).

Baumgarten, A. I., 1997, *The Flourishing of Jewish Sects in the Maccabean Era: An Interpretation* (New York: Brill).

Baumgarten, J. M., 1985, "Polemics in New Fragments", in J. Amitai (ed.), *Biblical Archaeology Today* (Jerusalem: Israel Exploration Society), 395–397.

Beckwith, R. T., 1996, *Calendar and Chronology, Jewish and Christian* (New York: E. J. Brill).

Ben-Dov, J., 2008, *Head of All Years: Astronomy and Calendars at Qumran in their Ancient Context* (Boston: Brill).

——, and Horowitz, W., 2003, "The 364-Day Year in Mesopotamia and Qumran", *Meghillot* 1, 3–26.

——, and Horowitz, W., 2005, "The Babylonian Lunar Three in Calendrical Scrolls from Qumran", *Zeitschrift für Assyriologie* 95, 104–120.

——, and Saulnier, S., 2008, "Qumran Calendars: A Survey of Scholarship 1980–2007", *Currents in Biblical Research* 7, 124–168.

Davies, P. R., 1983, "Calendrical Change and Qumran Origins: An Assessment of VanderKam's Theory", *Catholic Biblical Quarterly* 45, 80–89.

Doering, L., 1997, "The Concept of the Sabbath in the Book of Jubilees", in M. Albani, J. Frey, and A. Lange (eds.), *Studies in the Book of Jubilees* (Tubingen: Mohr Siebeck), 179–205.

Durkheim, E., 1961, *The Elementary Forms of the Religious Life*, Translated by J. W. Swain (New York: Collier Books).

Eliade, M., 1959, *The Sacred and the Profane: The Nature of Religion*, Translated by W. R. Trask (San Diego: Harcourt Brace Jovanovich).

Elior, R., 2004, *The Three Temples: On the Emergence of Jewish Mysticism*, Translated by D. Louvish (Oxford: The Littman Library of Jewish Civilization).

Epstein, I. (ed.), 1983, *Hebrew-English Edition of the Babylonian Talmud* (London: Soncino Press).

Feldman, R. H., 2004, *New Moon after New Moon and Sabbath after Sabbath: The Tension between Culture and Nature in the Cycles of the Sabbath and Moon in Ancient Jewish Calendars*, Dissertation, Graduate Theological Union.

——, 2009, "The 364-Day "Qumran" Calendar and the Biblical Seventh-Day Sabbath: A Hypothesis Suggesting their Simultaneous Institutionalization by Nehemiah", *Henoch* 31, 342-365.

Gandz, S., 1949, "Studies in the Hebrew Calendar I: A Study in Terminology", *Jewish Quarterly Review* 39, 259–280.

García Martínez, F. and Tigchelaar, E. J. C., 2000, *The Dead Sea Scrolls Study Edition* (New York: E. J. Brill).

Gardner, B. K., 2001, *The Genesis Calendar* (New York: University Press of America).

Geertz, C., 1973, *The Interpretation of Cultures* (New York: Basic Books).

Geller, M. J., 1995, "The Influence of Ancient Mesopotamia on Hellenistic Judaism", in J. M. Sasson (ed.), *Civilizations of the Ancient Near East* (New York: Scribner), 43–53.

Glessmer, U., 1996, "The Otot-Texts (4Q319) and the Problem of Intercalations in the Context of the 364-Day Calendar", in H.-J. Fabry, A. Lange, and H. Lichtenberger (eds.), *Qumranstudien: Vorträge und Beiträge der Teilnehmer des Qumranseminars auf dem internationalen Treffen der Society of Biblical Literature, Münster, 25.–26. Juli 1993* (Göttingen: Vandenhoeck & Ruprecht), 125–164.

——, 1999, "Calendars in the Qumran Scrolls", in P. W. Flint and J. C. VanderKam (eds.), *The Dead Sea Scrolls After Fifty Years: A Comprehensive Assessment* (Leiden: Brill), 213–278.

Goldman, E. A. (ed.), 1988, *The Talmud of the Land of Israel: Volume 16 Rosh Hashanah*, Translated by E. A. Goldman (Chicago: University of Chicago Press).

Herr, M. D., 1976, "The Calendar", in S. Safrai and M. Stern (eds.), *The Jewish People in the First Century* (Assen/Amsterdam: Van Gorcum), 834–864.

Heschel, A. J., 1951, *The Sabbath* (New York: Farrar, Straus and Giroux).

Horowitz, W., 1996, "The 360 and 364 Day Year in Ancient Mesopotamia", *The Journal of the Ancient Near Eastern Society of Columbia University* 24, 35–44.

——, 1998, "The 364 Day Year in Mesopotamia, Again", *Nouvelles assyriologiques breves et Utilitaires*, 49–51.

Jastrow, M., 1903, "זמן.", in *A Dictionary of the Targumim, the Talmud Babli and Yerushalmi, and the Midrashic Literature* (New York: The Judaica Press), 405.

Jaubert, A., 1957, *La Date de la Cène: Calendrier Biblique et Liturgie Chrétienne* (Paris: J. Gabalda).

——, 1965, *The Date of the Last Supper*, Translated by I. Rafferty (Staten Island, N.Y.: Alba House).

Kehati, P. (ed.), 1992–1994, *The Mishnah: A New Translation with a Commentary by Pinhas Kehati*, Translated by E. Levin (Jerusalem: Eliner Library, Dept. for Torah Education and Culture in the Diaspora of the World Zionist Organization).

Levine, R., 1997, *A Geography of Time* (New York: Basic Books).

Long, K., 1998, *The Moon Book* (Boulder, Col.: Johnson Books).

Merchant, C., 1989, *Ecological Revolutions: Nature, Gender, and Science in New England* (Chapel Hill: The University of North Carolina Press).

——, 1990, *The Death of Nature: Women, Ecology and the Scientific Revolution* (San Francisco: Harper).

Newsom, C. A., 1985, *Songs of the Sabbath Sacrifice: A Critical Edition* (Atlanta: Scholars Press).

Ravid, L., 2003, "The Book of Jubilees and its Calendar—A Reexamination", *Dead Sea Discoveries* 10, 371–394.

Rochberg, F., 1995, "Astronomy and Calendars in Ancient Mesopotamia", in J. M. Sasson (ed.), *Civilizations of the Ancient Near East* (New York: Scribner), 1925–1940.

Rubenstein, J. L., 1999, "Nominalism and Realism in Qumranic and Rabbinic Law: A Reassessment", *Dead Sea Discoveries* 6, 157–183.

——, 1999, *Talmudic Stories: Narrative Art, Composition, and Culture* (Baltimore: Johns Hopkins University Press).

—— (ed.), 2002, *Rabbinic Stories* (New York: Paulist Press).

Safrai, S., 1965, "The Localities of the Sanctification of Lunar Months and the Intercalation of Years in Palestine After 70 C.E. [Hebrew]", *Tarbiz* 35, 27–38.

Sanders, E. P., 1992, *Judaism: Practice & Belief, 63 BCE – 66 CE* (Philadelphia: Trinity Press International).

Schiffman, L. H., 1995, *Reclaiming the Dead Sea Scrolls* (New York: Doubleday).

Schürer, E., 1973, *The History of the Jewish People in the Age of Jesus Christ (175 B.C. – A.D. 135)* (Edinburgh: T. & T. Clark).

Schwartz, D. R., 1992, "Law and Truth: On Qumran-Sadducean and Rabbinic Views of Law", in D. Dimant and U. Rappaport (eds.), *The Dead Sea Scrolls: Forty Years of Research* (Leiden: Brill), 229–240.

——, 2004, "From Priests at their Right to Christians at their Left? On the Interpretation and Development of a Mishnaic Story (M. Rosh Hashanah 2:8-9)", *Tarbiz* 74, v–vi, 21–41.

Schwartz, S., 2001, *Imperialism and Jewish Society, 200 B.C.E. to 640 C.E.* (Princeton: Princeton University Press).

Segal, J. B., 1957, "Intercalation and the Hebrew Calendar", *Vetus Testamentum* VII, 250–307.

Snyder, G., 1997, *Mishmarot Calendars from Qumran Cave 4: Congruence and Divergence*, Dissertation, Hebrew Union College – Jewish Institute of Religion.

Stern, S., 2001, *Calendar and Community: A History of the Jewish Calendar, Second Century BCE – Tenth Century CE* (New York: Oxford University Press).

Sussmann, Y., 1989, "The History of Halakha and the Dead Sea Scrolls: Preliminary Observations on Miqsat Ma'ase Ha-Torah (4QMMT) [Hebrew]", *Tarbiz* 59, 11–76.

Tabory, J., 2000, *Mo'ade Yisra'el bi-tekufat ha-Mishnah veha-Talmud [Jewish Festivals in the Time of the Mishnah and Talmud]* (Jerusalem: The Hebrew University Magnes Press).

Talmon, S., 1951, "Yom Hakippurim in the Habakkuk Scroll", *Biblica* 32, 549–563.

——, 1958, "The Calendar Reckoning of the Sect from the Judean Desert", in C. Rabin and Y. Yadin (eds.), *Aspects of the Dead Sea Scrolls* (Jerusalem: Magnes), 162–199.

——, 1999a, "Anti-Lunar-Calendar Polemics in Covenanters' Writings", in M. Becker and W. Fenske (eds.), *Das Ende Der Tage Und Die Gegenwart Des Heils* (Leiden: Brill), 29–40.

——, 1999b, "Calendar Controversy in Ancient Judaism: The Case of the 'Community of the Renewed Covenant'", in D. W. Parry and E. Ulrich (eds.), *The Provo International Conference on the Dead Sea Scrolls: Technological Innovation, New Texts, and Reformulated Issues* (Boston: Brill), 379–395.

——, Ben-Dov, J., and Glessmer, U. (eds.), 2001, *Qumran Cave 4, XVI: Calendrical Texts* (Oxford: Clarendon Press).

Thoreau, H. D., 1998, "Walking", in J. B. Callicott and M. P. Nelson (eds.), *The Great New Wilderness Debate* (Athens, Ga.: The University of Georgia Press), 31–41.

VanderKam, J. C., 1979, "The Origin, Character, and Early History of the 364-Day Calendar: A Reassessment of Jaubert's Hypotheses", *Catholic Biblical Quarterly* 41, 390–411.

—— (ed.), 1989, *The Book of Jubilees* (Lovanii: Peeters).

——, 1998, *Calendars in the Dead Sea Scrolls: Measuring Time* (New York: Routledge).

Wacholder, B. Z. and Weisberg, D. B., 1976, "Visibility of the New Moon in Cuneiform and Rabbinic Sources", in B. Z. Wacholder (ed.), *Essays on Jewish Chronology and Chronography* (New York: Ktav Publishing House), 59–74.

Wiesenberg, E. J., 2007, "Calendar", in M. Berenbaum and F. Skolnik (eds.), *Encyclopaedia Judaica* (Detroit: Macmillan Reference USA), 354–359.

Wise, M. O., 1994, "Second Thoughts on *duq* and the Qumran Synchronistic Calendars", in J. C. Reeves and J. Kampen (eds.), *Pursuing the Text: Studies in Honor of Ben Zion Wacholder on the Occasion of his Seventieth Birthday* (Sheffield: Sheffield Academic Press Ltd), 98–120.

Zeitlin, S., 1973, "Some Stages of the Jewish Calendar", in S. Zeitlin (ed.), *Solomon Zeitlin's Studies in the Early History of Judaism* (New York: Ktav), 183–192.

Zerubavel, E., 1985, *The Seven Day Circle: The History and Meaning of the Week* (New York: The Free Press).

The Rabbinic New Moon Procedure: Context and Significance

Sacha Stern

Rabbinic sources from late Antiquity describe an elaborate procedure for determining when the new moon became first visible, and thus when the new month began. This procedure is referred to in Hebrew as *qiddush ha-ḥodesh*, an ambiguous phrase that can be translated as 'sanctification of the month' or 'sanctification of the new moon'; either way, the notion of sanctity, and hence the religious dimension of the rabbinic calendar, is evident. The procedure is described in full in the Mishnah (early third century CE), one of the earliest works of rabbinic literature, in tractate *Rosh Hashanah* (henceforth *mRH*).[1] In brief, the determination of the new moon and hence of the new month is presented as the sole prerogative of a single body of rabbis, referred to as a 'court', which would have been located originally in Jerusalem (*mRH* 2:5); after the destruction of the Temple in 70 CE, the court is presented in the western Judaean town of Yavneh (Jamnia), at least until the early second century (*mRH* 2:8), after which the sources refer to various locations in western Judaea and in Galilee.[2] Decisions were based on new moon sightings by members of the public, who came to the rabbinic court to testify that they had seen the new moon. Witnesses from all over Judaea were expected to travel to the court, if necessary even on the Sabbath, to report their new moon sightings.[3] The witnesses were duly interrogated by the court; if their testimony was deemed credible, the head of the court declared the beginning of the new month (*mRH* 2:6–7). The decision of the court was then disseminated throughout Palestine and to the Jewish Diaspora in Babylonia, originally through a chain of beacons, later by sending special envoys (*mRH* 2:2).

The historicity of this account is doubtful on several counts, as has been discussed in more detail elsewhere. The extent to which rabbis exerted any social or political authority in Jewish Palestine of the early Roman period is highly contentious, and this casts doubt on whether the rabbis would have been able to monopolize control of the calendar and impose their decisions about the new month on the Jewish communities of Palestine and beyond. It is also questionable whether the procedure described in *mRH* could have been observed in all its details: on practical grounds, for example, the feasibility of a system of communications through a chain of beacons (as described in *mRH* 2:3–4) seems somewhat unlikely.[4] The trend in recent scholarship is to view not only this particular account, but indeed the Mishnah in general, as legal and prescriptive rather than historical and descriptive. The literary characteristics of legal narratives such as in *mRH* have been emphasized, in particular, by Moshe Simon-Shoshan, suggesting that this text should be read as literature rather than as history.[5]

But even if a somewhat fictitious narrative, the new moon procedure described in *mRH* must have had at least some relation to the social and cultural context in which it was composed; for it is only in relation to this context that the narrative would have made sense to the authors and recipients of the Mishnah, and that it would have drawn its legal authority—whether theory or practice—and its conceptual meaning. In this article, accordingly, I propose to interpret the rabbinic new moon procedure in relation to what is known of other lunar calendars in Antiquity. Comparison of the rabbinic new moon procedure to the Babylonian and other ancient lunar calendars will enable us, in the first part of this article, to appreciate the unique peculiarities in the rabbinic new moon procedure. In the second part of this article, an attempt will be made to explain these peculiarities in relation to the social, historical context within which the rabbis would have been operating. I shall argue that by devising this unique procedure, the Mishnah and other rabbinic sources were making implicit, socio-political claims that were critical to the rabbinic movement and went well beyond the narrow question of how the Jewish calendar was (or should have been) reckoned.

In historical terms, the evidence of ancient literary and epigraphic sources (outside rabbinic literature) suggests that all Jews in the first few centuries of the Common Era determined the dates of their festivals according to a lunar calendar, and that this calendar was based on empirical sightings of the new moon; but the procedures that were employed are largely unknown. There is evidence in Talmudic literature that rabbinic courts applied the Mishnaic new moon procedure, or at least some form of it, as late as the fourth century CE,[6] which suggests that to the rabbis this procedure was not merely theory. But evidence of other, non-rabbinic Jewish calendars suggests that rabbis were not in sole control of the Jewish calendar in Palestine, and certainly not in the Diaspora. Palestinian and Diaspora communities determined the calendar independently from one another, which could lead to considerable variety and to festivals being celebrated by different Jewish communities on different dates.[7]

The account of the new moon procedure in *mRH* is the only description we have of how a Jewish calendar, in Antiquity, might have been determined and reckoned; indeed, it is the only ancient source I know that describes, in any meaningful detail, the workings of an empirical lunar calendar. Cuneiform sources provide us with some information about the Babylonian calendar, as we shall presently see, but only sporadically and without any continuous narrative such as is found in *mRH*. The uniqueness of the Mishnaic account is perhaps an indication, in itself, of the peculiarity of the rabbinic new moon procedure. Rather than assuming this procedure as 'typical' or as representative of what would have been the norm for Jewish or other ancient lunar calendars, I would like to argue that the Mishnah was promoting a procedure that was quite different and quite original.

Babylonian and Rabbinic New Moon Procedures

A very different conclusion was reached by Ben-Zion Wacholder and David Weisberg, a Jewish historian and an assyriologist, in a seminal article of 1971 where they compared the calendar and new moon procedures of the Mishnah with those attested in ancient Mesopotamia.[8] They argued that the procedures apparent in cuneiform sources for sighting the new moon and declaring the beginning of the month were so similar to those of the rabbinic calendar that the latter must have been in direct, historical continuity with the much earlier calendars of the ancient Near East. This article has been very influential, but in this present paper I shall take issue with it. In actual fact, the only common features between the Babylonian calendar—which was dominant in Mesopotamia in the first millennium BCE—and the Jewish, rabbinic calendar is that both were lunar and based on new moon sightings, both allowed only 29 or 30-day months, and that the Jews, like all Aramaic-speakers in the late antique Near East, used Babylonian month names. But in all other respects—especially in matters of procedure—the calendars were very different, as we shall presently see. This will cast serious doubt on the extent to which the rabbinic calendar may be regarded in continuity with earlier Near Eastern calendars.

The Babylonian calendar, during the first millennium BCE, was the most widely used calendar in Mesopotamia and in the Near East, because the great Near Eastern empires of the Assyrians, Babylonians, Persians, and eventually Seleucids (Hellenistic), adopted and disseminated it as their official, imperial calendar. The structure of this calendar, such as the names and sequence of its months, is well-known; although month lengths were not fixed but variable, it is reasonably certain that months could only count either 29 or 30 days. The determination of the length of the month, and thus of when the new month should begin, was important not only for cultic and astrological reasons,[9] but also for economic purposes and for the imperial administration. Yet in spite of the great wealth of astronomical cuneiform sources that have survived, surprisingly little is known about how Babylonian months were set: more precisely, how and by whom the length of each month was determined, and who decided when the new month began.[10]

The evidence, which Wacholder and Weisberg partially presented and which I shall review in some detail, is restricted almost entirely to letters from astrologers to the Assyrian king of the neo-Assyrian period (eight-seventh centuries BCE), a whole millennium before the Mishnah was composed—which immediately highlights the difficulty of establishing historical continuity between them.[11] These letters reveal that in the neo-Assyrian period, the beginning of the Babylonian month depended to a large extent on experts variously called in modern scholarship 'scribes', 'scholars', or 'astrologers', who were either temple officials or directly associated with the imperial court.[12] In these letters, the astrologers reported to the king when they had sighted the new moon; the purpose of these letters was, it seems, to establish the beginning of the new month. If the new moon sighting was uncertain or problematic, the king was expected to decide whether or not to accept it, and hence, when the month should begin, whether after 29 or 30 days of the outgoing month. This is evident, for example, in a letter of the astrologer Bulluṭu to the Assyrian king:

> We watched on the 29th day;[13] the clouds were dense, we did not see the moon. We watched on the 30th day; we saw the moon, (but) it was (already) very high. The (weather) of the 29th day has to do with it. What is it that the king my lord says?[14]

Bulluṭu is saying that if not for bad weather, the new moon should have been sighted on the 29th, because when it was actually sighted on the next evening, it was clearly no longer new. His question to the king is whether the month should begin when the moon was actually sighted or rather when it should have been first visible.

Sometimes the king's decision would depend on comparing new moon reports from different cities. The following letter was probably addressed to King Assurbanipal (669-633 BCE) from one of his leading astrologers:

> To the king, [my lo]rd: your servant Adad-[šumu-uṣur]. Good health to the king, [my lord]! ... I observed the (crescent of the) moon on the 30th day; it was high, (too) high to be (the crescent) of the

> 30th. Its position was like that of the 2nd day. (So) if it suits the king, my lord, the king should wait for the report from Assur before fixing the date. Perhaps the king, my lord, will say: 'Why didn't you decide (about the matter)?' Am I [...]? The king should [ask] the scri[bes]: the days ...[15]

Another letter provides further evidence:

> ... (There were) clouds. We did not (see) the moon, probably because of the cl(ouds). The king, my lord, should send messeng(ers) to the cities of Assur (and) Arbela, to go (and) find out definitely about the m(atter), (and to inform) quickly the king, my lord. (The report) of (Calah) ...[16]

Other letters confirm that reports of new moon sightings were frequently sent from various Assyrian cities (Assur, Arbela, etc.) to the royal court at Nineveh, some coming from even as far as Babylonia. It is unclear whether reports from other cities were sent as a matter of course, or only on demand. The letter of Adad-šumu-uṣur implies that a report from Assur was anyway expected, but the last letter cited implies that reports from Assur and Arbela needed to be actively sought.

The decision-making process that led to the determination of the beginning of the month is not clearly or explicitly described in these letters, but the relationship between astrologers and the king was obviously crucial. The astrologers—many of whom, in the neo-Assyrian kingdom, were directly employed by the king—were apparently invested with the responsibility, which also means the authority, of observing the new moon and communicating their observations to the king (this is certainly the case of the astrologers based around the royal court in Nineveh, from whom we have a large number of letters informing the king whether or not the new moon was sighted). But the ultimate decision, especially in doubtful cases, was taken by the king, as the letter of Bulluṭu (cited above), for example, makes clear.

The end of the letter of Adad-šumu-uṣur suggests that sometimes the king would try to hand over the decision to his astrologers, a responsibility which astrologers would be reluctant to accept. But in some cases, astrologers were more assertive. Thus, the Babylonian astrologer Ašaredu the Younger wrote:

> On this 30th day [the moon became visible]. The lord of kings will say: "Is [the sign?] not affected?" The moon disappeared on the 27th; the 28th and the 29th it stayed inside the sky, and was seen on the 30th; when else should it have been seen? It should stay in the sky less than four days, it never stayed four days.[17]

Ašaredu is justifying his new moon sighting on the 30th (i.e. end of 29th) on the grounds that after the last sighting of the old moon, the moon could not have remained invisible ('in the sky') for as much as four days. The tone of his letter appears to be polemical: either against some other astrologer, or perhaps even against the king himself. In any case, Ašaredu asserts his astronomical expertise and expects the king to respect it; in this letter, the king is not invited to express his own opinion. The sources thus suggest that the relationship, in calendar matters, between the king and astrologers was complex and varied.

The involvement of both the king and his astrologers in setting the beginning of the month gave this process at once a 'political' and an astronomical or 'scientific' character. But as courtiers or temple officials, the astrologers were politically significant in their own right. Moreover, astronomical observation, astrological omens, royal policy, and royal administration, were all closely interrelated, in such a way that it would be misleading in this context to treat 'politics' and 'science' as neatly distinct. The political significance of the neo-Assyrian new moon procedure is particularly evident from the frequent dispatch of new moon reports from the cities of the empire to the royal court, which was executed by astrologers (as we have seen above) but also in some cases by imperial officials (e.g. Mar-ištar, King Esarhaddon's agent in Babylonia).[18] Inasmuch as the month was determined on the basis of new moon reports from all over Mesopotamia, the calendar which the Assyrian king controlled was intended to represent a common time-frame for all the cities of the empire. Not only did this procedure embody the centralizing force of the king over the cities of the empire, but it also turned the standard Babylonian calendar into a truly imperial calendar.

In contrast to all this, the new moon procedure described in Mishnah *Rosh Hashanah* and other rabbinic sources had neither the 'political' nor the 'scientific' character of its Babylonian counterpart. In terms of the 'political', to begin with, the Jewish calendar as described in rabbinic sources was not controlled by astrologers or a king, but by a rabbinic court. The rabbis who composed this court are presented in rabbinic literature as scholars and occasionally as judges, but never as kings or political rulers. Indeed, most present-day scholars are of the view that the Palestinian-based rabbinic movement of the Mishnaic period (late first – early third centuries CE) was of marginal political significance. As in any province of the Roman Empire, those who ruled over Palestine were the Roman imperial administration, the city councils, and the local aristocracies that manned them; the rabbis, in contrast, were hardly invested with the powers and functions of political rule.[19]

In this respect, the rabbinic new moon procedure differed not only from the Babylonian calendar, but also from most lunar and other flexible calendars in the ancient world. Most ancient calendars, indeed, were set and determined by politicians or political rulers. In the Greek city-states, the months and years of the Greek lunar calendars were determined by the city councils and their magistrates; in the Roman Republic, intercalations were initiated by the *pontifex maximus* and the pontifical college—a priestly body, but elected from the senatorial class and in a high position of political influence; and in monarchical states, of course, the calendar was controlled by the king (e.g. Alexander the Great, who is reported to have tampered with the calendar on several occasions).[20] Control of the calendar and of the length of months and years gave political rulers the means of controlling economic activity, state administration, religious cult, and in some political systems, their own tenures of office—often to their personal advantage. In most ancient societies, consequently, the calendar was a fundamentally *political* function. The rabbis' usurpation—according to the Mishnaic narrative—of a function that was normally reserved, in the ancient world, for political rulers demands an explanation, and will be discussed in more detail below.

Babylonian and rabbinic new moon procedures also differed in the evidence on the basis of which the calendrical decisions were reached. As we have seen, the Babylonian new moon was determined, at least in the neo-Assyrian period, on the basis of written reports of expert astrologers, which were regularly sent and submitted to the king. The rabbinic month, in contrast, was determined on the basis of oral reports by ordinary, lay members of the public. The quality of information provided by these witnesses was much inferior to that of Mesopotamian astrologers, who had sufficient expertise to advise, for example, on whether the moon was seen at the expected altitude or at the expected time (as we have seen above). The non-expert nature of new moon reporting in the rabbinic procedure needs to be emphasized, as it had an important impact on the rabbinic procedure itself. Unlike Mesopotamian astrologers, witnesses in rabbinic courts were treated as potentially unreliable and needed, therefore, to be verified through a procedure of interrogation, to which I shall now turn.[21]

The Interrogation of Witnesses: Astronomy and Integrity

In this section, I shall argue that the interrogation of witnesses in the rabbinic new moon procedure—as presented in rabbinic sources—was designed to verify their personal integrity rather than the astronomical plausibility of their new moon reports. This suggests that the rabbinic new moon procedure was not 'scientific' (just as it was not 'political') but rather, as I shall discuss further on, 'judicial'.

This is not to say that astronomical plausibility counted for nothing in the rabbinic new moon procedure: quite on the contrary, reports of new moon sightings could be rejected if they were deemed astronomically implausible. For example, the famous conflict between Rabban Gamaliel and Rabbi Joshua about the date of the Day of Atonement arose when witnesses had reported seeing the new moon on the 30th but not on the 31st; the rabbis' initial disagreement was only about the astronomical plausibility of this report (*mRH* 2:8–9). It is true that their conflict was eventually resolved with the argument that a rabbinic court—R.Gamaliel's—had the authority to decide when the month began, even on the wrong time. But this argument, which is formulated elsewhere as a self-standing ruling,[22] does not mean that the rabbinic court has a free hand to decide what it wants: for the very same sources emphasize that new moon sightings must determine, in first instance, the court's decisions.[23] The point I am making, therefore, is not that the rabbinic calendar ignores empirical reality—as has been recently argued[24]—but only that its procedure for verifying the empirical reality is focused on the integrity of the witnesses rather than on strictly astronomical criteria.

The interrogation of the witnesses, in the Mishnaic narrative, consists of the following questions:

> Say, how did you see the moon: before the sun or after the sun? To its north or to its south? How high was it? Where did it incline? How wide was it? (*mRH* 2:6)

These questions relate to the shape and position of the new moon; but in astronomical terms, they are unclear and open to several interpretations. This lack of clarity suggests, in itself, a rudimentary level of astronomy;[25] and as I shall now explain, it is unlikely—partly because of their unclarity—that these questions could have helped to identify erroneous or false new moon reports. The correct answer to the first question, as the Mishnah goes on to state explicitly (*mRH* 2:6), is 'after the sun', because the new moon is always behind the sun and never ahead of it (and thus, the new moon sets typically one hour after sunset); but this is likely to have been common knowledge, and therefore, the question could only have served to expose very grossly erroneous moon sightings.[26] The next two questions might have required some astronomical knowledge on the part of the examiner: thus, the examiner would have needed to have known whether, at the time of the sighting, the moon was to the north or the south of the sun, and at what 'height' or altitude above the horizon.

However, 'north' and 'south' of the sun can be defined, astronomically, in several ways, and the answer to this question, just as the answer to the question of the moon's altitude, is likely to have depended very critically on the exact time when the moon was sighted—which, in normal circumstances, witnesses would not have been able to measure.[27] In astronomical terms, therefore, the second and third questions are unclear and would have been difficult for anyone to answer in a precise or meaningful way; it is difficult to know what they could have achieved by way of astronomical verification.[28] The fourth question, 'where did it incline', probably means quite simply whether the convex (or outer) side of the new moon crescent was to the left or to the right; but in the northern hemisphere, the latter is always the correct answer—so again, like the first question, this question is rather elementary and would only have exposed very grossly erroneous moon sightings. Finally, the question 'how wide was it' most probably refers to the size of the new moon crescent;[29] but in the absence of precise measuring instruments, the witnesses could hardly have been expected to return an accurate and verifiable measurement of the crescent's size, and hence one wonders how effective this question could have been—just like the first four questions—for purposes of astronomical verification.

The continuation of this same passage (*mRH* 2:6) reveals, in fact, that these questions were not designed, primarily, to expose astronomical implausibilities in the witnesses' new moon reports, but rather to verify that their testimonies were mutually consistent:[30]

> And after that, they would bring in the second (witness) and interrogate him. If their words were in agreement, their testimony was accepted. (*mRH* 2:6)

In astronomical terms, the agreement of the witnesses would not have proved much about the truth of their testimony. This is because the answers to the first and fourth questions are so obvious, as we have seen, that witnesses are hardly likely to have disagreed; whereas for the second and third questions (north or south, and altitude), it would not have been possible for witnesses to corroborate each other unless they had sighted the new moon at exactly the same time (which means, in practical terms, together).[31] However, this passage makes it clear that the agreement of the witnesses itself—regardless of astronomical considerations—is considered sufficient to establish the witnesses' integrity.[32]

The same can be inferred from a parallel passage in the roughly contemporary Tosefta (early third century CE), where the third question (altitude) and a variation on the fourth question are presented in a little more detail:

> If one (of the witnesses) says 'I saw it two goads high', and the other says 'three', their testimony is valid. If one says 'three' and the other says 'five', they cannot be joined together, but one (of them) can be joined to others.
> If one says 'I saw it leaning' and the other says 'I saw it upright', they cannot be joined together, but one (of them) can be joined to others. (*tRH* 2:2, Zuckermandel ed. p. 210)

It is unclear how the goad, which is attested in later sources as a standard linear measurement, is meant here to be used for the measurement of celestial angles.[33] However, the astronomical accuracy of the witnesses' report is not what matters in this passage. The veracity of their report is not established on the basis of known astronomical facts—e.g. the rabbinic court's knowledge of the moon's altitude at the time of the reported sighting, or their knowledge of the crescent's inclination—but only on the basis of whether the witnesses are in agreement with one another. Indeed, if they are in disagreement, it is sufficient for one of the witnesses to find a third person (or 'others') who agree(s) with him. All that counts, therefore, is the witnesses' agreement.

Particularly interesting, in the Tosefta passage, is the ruling that a disagreement of two and three goads (in the witnesses' report of the moon's altitude) is negligeable and insignificant, whereas a disagreement of three and five goads is significant and invalidates the testimonies. This finds an exact parallel in the context of criminal judicial procedure, where witnesses are interrogated about the date and time when the incident was witnessed:

> If one says 'on the second of the month' and the other says 'on the third', their testimony is valid, because one may know that the month was intercalated and the other not[34] ...; but if one says 'on the third' and the other 'on the fifth', their testimony is invalid.
> If one says 'at two hours' and the other says 'at three hours', their testimony is valid; but if one says 'at three' and the other 'at five', their testimony is invalid. (*mSanhedrin* 5:3)

This whole method of verification, indeed—which consists in interrogating the two witnesses separately, and then checking whether their testimonies corroborate each other—is not specific to the new moon procedure, but standard, according to rabbinic law, in all civil and criminal litigation.[35] This is precisely what valorizes the interrogation of witnesses in the new moon procedure, in spite of its astronomical inadequacy. The purpose of the interrogation is not to verify the astronomical plausibility of the witnesses' reports—which, as we have seen, would not have been very effective—but rather to establish the witnesses' integrity through a standard judicial procedure. In this respect, the verification procedure of the new moon witnesses is not astronomical, but primarily judicial.

A Quasi-Judicial Procedure

The 'judicial' nature of the interrogation of new moon witnesses leads us to consider, more generally, the strikingly judicial character of the rabbinic new moon procedure as a whole—a peculiarity of the rabbinic calendar which has been largely ignored in modern scholarship.

The entire new moon procedure in *mRH* has the likeness of a trial. It is closely modelled on rabbinic judicial procedures that appear elsewhere in the Mishnah, especially in tractate *Sanhedrin*. The rabbinic body that determines the new month is called, in the Mishnah, a *beit din* (lit. 'house of law', i.e. judicial court); its leader is *rosh beit din* ('head of the court'—*mRH* 2:7). This court must be composed of three judges, as for all civil cases and for some criminal trials (those that do not involve capital punishment: *mSanhedrin* 1:2). As in all trials, the verdict must be based on the testimony of two witnesses; even if the court itself saw the new moon, two of its members must detach themselves from the court (and if necessary, more judges must be co-opted to make up the quorum of three), and they present themselves to the court as witnesses (*mRH* 3:1). As in all trials, the witnesses need to be valid: this excludes relatives of one another, people disqualified on various grounds of immorality, slaves, and women (*mRH* 1:7–8).[36] As in all trials, the witnesses are interrogated separately, to verify that they corroborate each other (*mRH* 2:5–6—as we have seen above). As in all trials, the procedure must be completed in daytime (*mRH* 3:1), and there are rules in the Palestinian Talmud about which judges—junior or senior—express their opinion first.[37] As in all trials, the Mishnaic procedure concludes with the head of the court declaring his verdict: 'It [the month] is sanctified!', and all the people responding: 'It is sanctified, it is sanctified!'[38]

In spite of all this, rabbinic sources seem to recognize that the determination of the new month is not a real trial, and that the judicial character of its procedure is, in some ways, contrived. This explains why exceptions to normal judicial rules, some quite radical, are in certain cases allowed. Thus according to Rabbi Shimon, the testimonies of two relatives (such as father and son) are acceptable for the new moon (*mRH* 1:7)—whereas in normal judicial procedure, relatives would be unanimously rejected as invalid (*mRH* 3:4). Even more striking is the unanimous ruling in several sources that if the new moon was 'sanctified' without witnesses, or if the witnesses were later found to be false, the new moon declaration remains retrospectively valid.[39] Such a concession, in a normal judicial setting, would be totally inconceivable, as it would make no sense for a court to rely on the evidence of false witnesses. But the acceptability (*ex post facto*) of false witnesses in the new moon procedure is not so surprising: for this is no real trial—it is a trial without litigants and without accused.

The make-believe, quasi-judicial character of the new moon procedure betrays more fundamental problems about its origins, justification, and rationale. The origins of this procedure, whether Biblical or from some other source, are unclear even in rabbinic terms: the procedure is not supported, in the entire corpus of rabbinic literature, by any halakhic (legal) explanation or *midrash* (scriptural exegesis). Although the procedure is distinctly rabbinic, in that it conforms largely to the judicial procedures that halakhic sources lay down elsewhere for civil and criminal trials, hardly any explanation is given in rabbinic literature as to why a judicial procedure should be adopted and followed for the determination of the new moon.[40]

The judicial pretences of the new moon procedure are strange not only in rabbinic terms, but also in relation to other calendars in the ancient world. In the context of other ancient calendars, judicial procedures are unattested and would have appeared completely out of place. In the neo-Assyrian kingdom, as we have seen, the beginning of the month was decided by the king on the basis of written advice from his astrologers; there was nothing judicial about this procedure. Less is known about other calendars, but an Athenian inscription from the fifth century BCE gives us a glimpse of how intercalations were decided in the Athenian *polis*: this inscription, a decree issued by the *Boule* (city council) and the Assembly, concludes with a motion proposed by Lampon, an influential politician of late fifth-century Athens, which includes, among other points, that the next archon (i.e. the magistrate or leader of the city, elected for the next year) should intercalate a second month of Hekatombeon.[41] The inclusion of this motion in the text of the decree signifies its ratification by the Boule and the Assembly. What would have followed, one presumes, was in due course a formal announcement of the intercalation by the incoming archon. This is all a purely political procedure—there is nothing judicial about it.

Another new moon procedure appears with reference to archaic Rome, but only much later in Macrobius' *Saturnalia* (late fourth – early fifth centuries CE). Referring to a period when the Roman calendar was still lunar,[42] Macrobius describes how the new moon would be sighted each month by one of the minor pontiffs. The minor pontiff reported his sighting to another, more prominent member of the pontifical college, the *rex sacrorum* ('king of sacrifices'); they both made a sacrifice to mark the beginning of the month, and announced to the assembled people how many days were to elapse between the kalends (beginning of the month) and the nones.[43] Whatever the historicity of this account, it says something of what was expected, in late Antiquity, of a lunar calendar. The procedure described by Macrobius is a mixture of political and cultic; again, there is nothing judicial about it.

In short, the judicial, or rather make-believe judicial character of the rabbinic new moon procedure stands out as unique in the ancient world, as well as being poorly explained or justified in rabbinic literature itself. It calls, therefore, for an explanation.

Social Context: Rabbinic Courts and City Councils

The uniquely judicial character of the rabbinic new moon procedure was clearly related to the socio-political status of the rabbis who controlled it. As mentioned above, nearly all other ancient calendars were controlled by politicians and political rulers, who naturally employed political procedures; but the rabbis of Palestine were not political rulers, and therefore, their calendar procedures could hardly have been 'political'. This raises the question, however, of how and why the rabbis came to be in control of the Jewish calendar in Palestine—if indeed they ever did. In this section I shall investigate, from a historical perspective, how and by whom the Jewish calendar was controlled in Palestinian society. It is in relation to this social context that we will appreciate how and why the rabbinic new moon procedure—whether practice or mere theory, and with all its judicial peculiarities—was conceived.

Let us go back in history, and consider who controlled the Jewish calendar in Judaea in the centuries leading up to the rabbinic period. Under the great Near Eastern empires of the first millennium BCE, especially of the Achaemenid (sixth-fourth centuries) and Seleucid dynasties (third-second centuries BCE), the Jews appear to have simply adopted and used the official imperial, Babylonian calendar, for their own purposes; this calendar would have been set, at least in principle, by the king or the imperial administration, and may not have needed the input or interference of any Judaean political authority. After the defeat and withdrawal of the Seleucids from Judaea in the mid second century BCE, the evidence is patchy, but we may assume that control of the calendar in Judaea passed into the hands of the local, Hasmonaean dynasty. The Hasmonean rulers were called at various times 'ethnarchs' and 'kings'[44] but also held the position of high priests in the Jerusalem Temple, which makes them all the more than likely to have been responsible for setting the calendar of Judaea and the Jews. When in 37 BCE Judaea came under the rule of Herod, a non-priestly king, control of the calendar may well have remained in the hands of the high priest; and this arrangement could have been maintained during the period of direct Roman rule, from 6 CE to 70 CE.[45] Alternatively, the calendar might have been controlled in this period by a Judaean council of elders or Sanhedrin (provided a permanent and well-defined institution of this kind was really in existence).[46]

After the destruction of the Temple in 70 CE and the collapse of the high priesthood, of the Sanhedrin (if it ever existed), and indeed, of any form of central Jewish political authority, control of the Jewish calendar is likely to have become an issue: who, from this point onwards, would have taken it over and determined the dates of Jewish new moons and festivals? Modern historians have assumed that it was the rabbinic court that took care, from this point onwards, of the Jewish calendar.[47] This is actually a traditional view, which is based entirely on an assumption, in rabbinic sources, that rabbis had controlled the calendar even before 70 CE. However, it is most unlikely that a function that had always been reserved to political rulers or to figures of high political influence—in ancient Israel, in the province of Judaea, as well as throughout the ancient world—should have been handed over after 70 CE to individuals who were neither politicians nor rulers.

The only local, political authorities in Palestine to have survived the debacle of 70 CE were the city councils, and it is they who should be expected to have taken over control of the Jewish calendar. The role of the city council and magistrates in setting and running the calendar(s) of the city and its territory was a well established tradition in the Greek and Hellenistic world, and Judaea/Palestine was no exception. In the first century CE, the largely 'pagan' or 'Greek' cities of the Palestinian coastline, Gaza, Ascalon, and Caesarea, reckoned their own, distinctive calendars independently from one another. Caesarea may have done so since its foundation under the reign of Herod, whilst Gaza and Ascalon had begun setting their own calendars much earlier, ever since the end of Seleucid rule.[48] It is natural to expect that after 70, the predominantly Jewish cities of inland Palestine (especially Lod, Tiberias, and Sepphoris) would have seen it as their prerogative, indeed responsibility, to take over control of a Jewish calendar that could no longer be set by the high priesthood in Jerusalem Temple now that it had been destroyed.

That city councils took charge of the Jewish calendar in Judaea/Palestine after 70 CE is difficult to prove, especially on the basis of rabbinic sources that seek to promote the contradictory view that the Jewish calendar was controlled entirely by rabbis. Nevertheless, some passages in the Palestinian Talmud suggest that the reality differed somewhat from the rabbinic ideal. The Palestinian Talmud was redacted somewhat later than the Mishnah and the other early third-century sources we have been considering, but the relevant passages in the Palestinian Talmud refer to events of the early second and early-mid third centuries CE, and even if their historical authenticity is questioned, city councils are highly unlikely to have been less powerful in the second-third centuries than in the late fourth, when the Palestinian Talmud was redacted. Nevertheless, the evidence remains overall limited.

The first possible source of evidence is a story told in the Palestinian Talmud of a Shazkar, head of the city of Gadara (south-east of the Sea of Galilee), intercepting the new moon witnesses on their way to R.Gamaliel's court; the latter responded by having him removed from his headship.[49] Whether 'head' (*rosh*), presumably equivalent to the Greek *archon*, means leader of the city council or perhaps only of the Jewish community of Gadara, this story suggests a conflict between rabbis and civic or communal authorities over control of the new moon procedure—even if, as we are told here, the rabbi eventually prevailed.[50]

In the following passage of the Palestinian Talmud, where city councils are explicitly mentioned, it is the city councils who appear to have had the upper hand:

> R.Yoḥanan said: If they mention you for (candidature to) the *boule* [city council], let the Jordan be your frontier.[51]
> R.Yoḥanan said: One may appeal to the authorities for exemption from the *boule*.
> R.Yoḥanan said: One may borrow at interest for ... the sanctification of the month.
> R.Yoḥanan used to go into the synagogue in the morning, collect the crumbs, eat, and say: 'May my lot be with he who sanctified the month here (last) evening!'[52]

This passage, a list of four traditions of or about Rabbi Yoḥanan (early-mid third century CE), occurs in the context of a reference in the Mishnah to a meal that was held 'at the intercalation of the month' (*mSanhedrin* 8:2). The tradition that concerns us most is the last. The first two traditions seem out of context, as they do not relate at all to the intercalation or the calendar; but I think that their incongruous inclusion in this passage is significant and has an explanation. The story at the end suggests that Rabbi Yoḥanan was somehow excluded from a meal that had taken place, the day or night before, in a synagogue; at this meal, they had 'sanctified the month'—the same phrase used, elsewhere, by the head of the rabbinic court to declare the beginning of the new month.[53] Yet it seems unlikely that R.Yoḥanan—probably the most important rabbinic sage of mid third-century Palestine—was excluded from the calendrical rabbinic court, especially as he is mentioned elsewhere in the Talmud as a member of it.[54] The insertion at the beginning of this passage of traditions about the *boule* suggests rather that this was an assembly of the city council, that had gathered in the synagogue to hold a festive meal and sanctify the month. R.Yoḥanan—who earlier in this passage, paradoxically, had recommended exemption from or evasion of the *boule*—regretted his exclusion from it. This story suggests that 'sanctification of the month' was not really the monopoly of the rabbis. Calendar decisions could be taken by other authorities, perhaps the city councils, from which even great rabbis like R.Yoḥanan were excluded.

The story also suggests that the Jewish, communal practice of 'new moon meals' (also known as 'sanctification of the moon/month' or 'intercalation of the month' meals), which are frequently referred to in early rabbinic sources (from the early third century and later), could involve in practice, at least in some cases, a decision-making procedure of determination and declaration of the beginning of the month, which was distinct from (but functionally similar to) the procedure ascribed to the rabbinic court.[55] I propose to differ from the dominant, modern scholarly interpretation of these 'new moon meals', which considers their purpose to have been only to celebrate and publicize the date of the new month that had been determined, beforehand, by the rabbinic court.[56] This passage is not the only evidence, in fact, that new moon meals were a forum in which the beginning of the month could be decided. The liturgy for new moon meals, as presented at least in later Palestinian sources, includes the declaration 'sanctified is the moon!'—exactly the same phrase as attributed to the rabbinic court, at the conclusion of the rabbinic new moon procedure, in *mRH* 2:7—suggesting that the new moon decision was taken at these meals.[57] The procedure implicit in new moon meals is quite different from the new moon procedure of *mRH*. They are meals, not trials, and witnesses are never mentioned; it is possible that the participants at the meals are expected to sight the new moon themselves. The location of the meals appears, in some passages, to be synagogues,[58] and in a few places there are references to city councillors, although rabbis also sometimes appear.[59]

The practice of using new month meals as the occasion for deciding the beginning of the month—in contradistinction to the rabbinic new moon procedure described in *mRH*—may have been widespread. In the story of R.Yoḥanan cited above, it was presumably the *boule* of Tiberias, R.Yoḥanan's home city, that gathered for the new month meal and 'sanctified the month'. But the same could have been occurring simultaneously in other large cities in Palestine with sizeable Jewish populations, such as Sepphoris and Lod, where common meals could equally have been held to determine the beginning of the new month. In this manner, the city councils of Palestine would have combined the Hellenistic traditions of archontic calendar control, bouleutic common dining,[60] and the Jewish, possibly Biblical tradition of festive, new month meals.[61]

The Rabbinic Procedure: From Dissidence to Mainstream

The picture that emerges is one of a rabbinic court in competition with city councils over control of the Jewish calendar. Even if, as the Mishnah and other rabbinic sources indicate, the rabbis laid claim on the Jewish calendar, Jewish city councils had a stronger claim over it, because in Judaea as elsewhere in the Graeco-Roman world the calendar had always been controlled by political rulers. Being teachers of Torah did not entitle the rabbis to set the dates of the Jewish festivals, because in the context of the ancient world, even cultic calendars were normally determined by political rulers.

The legitimacy of the rabbinic court is more than likely, therefore, to have been an issue. Rabbinic sources address this in several ways, for example by claiming that the rabbinic court's decisions to sanctify the month and intercalate the year were divinely sanctioned.[62] But I would like to argue that the peculiarly *judicial*, or quasi-judicial, character of the rabbinic procedure which I have described in this article, whether a literary/legal construct or a reflection of actual practice, was a socio-political statement that constituted, in the context of competition from the city councils, a powerful source of rabbinic legitimization. Rabbis were not city councillors or civic rulers—as the R.Yoḥanan passage (cited above) shows, they advocated evasion from the city councils and did not participate in bouleutic meals—but they did act as arbiters and judges. A 'court of justice' (as opposed to a city council) was a locus of social authority in which the rabbis regarded themselves, and may have been regarded by other Jews, as entitled to belong. By defining the determination of new months as a specifically judicial procedure, rabbis were able to draw the calendar into their own space. The judicial procedure was at once a way of appropriating the Jewish calendar and of delegitimizing the city councils; it was a way of demonstrating that only rabbis could control the calendar. In this respect, the rabbinic court with its peculiar new moon procedure was politically dissident and subversive; more than a method of controlling the calendar, it represented to the rabbis a novel, alternative, and source of authority in Jewish Palestinian society.

As is well known, and as I have discussed in detail elsewhere, the rabbinic calendar underwent considerable change in the course of late Antiquity and the early medieval period. The Mishnaic calendar, based on empirical new moon sightings and on the authority of a rabbinic court, was gradually abandoned in favour of a calculated, fixed and standard scheme that could be reckoned by anyone independently, without resort to any judicial procedure or to any rabbinic authority; this is the system still in use today. There are various reasons why this radical change took place, which are beyond the scope of this article to discuss in detail. As I have explained elsewhere, this change was the reflection of a much more general, macro-historical trend that affected in similar ways all calendars in Antiquity.[63] In the case of the rabbinic calendar, however, the switch to a fixed scheme may have been more specifically influenced by the Christians' invention and adoption of fixed Easter cycles in the course of the third-fourth centuries CE; furthermore, a fixed and standard calendar responded to the growing need, or rather to what the rabbis of late Antiquity perceived as a growing need, to enable the large but distantly separated Jewish communities in Palestine and in Babylonia to observe the festivals on the same dates.[64]

I shall only add, however, one general observation. As I have shown elsewhere, the most significant changes in the rabbinic calendar, that eventually led to the rise of a fixed calendar scheme, occurred in the third-fifth centuries CE. This was a period of decline of the city councils in the Roman Empire, and of the political ascendance, instead, of religious leaders such as Christian bishops and Jewish patriarchs.[65] The gradual abandonment of judicial new moon procedures, which as I have argued in this article had been designed by rabbis in competition to city councils, and the switch instead to a fixed calendar, reflects perhaps a change in the socio-political status and authority of Jewish patriarchs and rabbis in late antique Palestine. This normalization of the rabbinic calendar in late Antiquity reflects, in a certain way, the more general process of normalization of the rabbinic movement in this period, from dissidence and marginality to mainstream and authority in Judaism and Jewish society.[66]

Notes

1. *mRH* 1:3–3:1; in English translation, Danby (1933), pp. 188–191. An earlier version of this paper was presented at the AJS Conference 2007; I am grateful to Hindy Najman and Ishay Rosen-Zvi for their useful comments, as well as to Jonathan Ben-Dov for his comments on this present paper.
2. E.g. Usha (Tosefta *RH* 2:1, Palestinian Talmud *RH* 2:1 (57d)), Ein Tav (*pRH* 2:5 (58a), Babylonian Talmud *RH* 25a), and Lod (*pSanhedrin* 1:2 (18c), *bḤullin* 56b). See Safrai (1965).
3. *mRH* 1:6, 1:9, 2:5; also *tRH* 2:1.
4. Stern (2001), pp. 162–163, 245.
5. Simon-Shoshan (forthcoming). The new moon procedure described in *mRH* is also assumed in other early rabbinic works, although they do not provide a complete, continuous account of it: thus the Tosefta expands on some elements of the Mishnaic account, and the Palestinian and Babylonian Talmuds provide a commentary on *mRH*. Laws relating to the Mishnaic procedure appear also in Sifra *Emor pereq* 10. For this reason, I shall refer to the Mishnaic new moon procedure more generally as 'rabbinic', and interpret it in relation to the full range of early rabbinic literary works (which, it should be added, are no less problematic as historical evidence than is the Mishnah). It is only in the later, redactional layers of the Talmuds (i.e. late fourth century for the Palestinian Talmud, and sixth century for the Babylonian) that other procedures for determining the month (in particular, the use of fixed calendar schemes) begin to emerge (see end of this paper and Stern 2001).
6. A number of rabbinic personalities from the second – fourth centuries are named in the Tosefta (third century) and Palestinian Talmud (late fourth century) as appointed members of the calendar court: *tSanhedrin* 2:1, *pRH* 2:6 (58b), *pSanhedrin* 1:2 (18c), *pRH* 3:1 (58d), and *pAvodah Zarah* 3:1 (42c). One passage suggests the late third-century survival, in eastern Galilee, of a local beacon system to disseminate the court's decisions: *pRH* 2:1 (58a).
7. Stern (2001).
8. Wacholder and Weisberg (1971).
9. Beaulieu (1993).
10. See Steele (2007) and Stern (2008).
11. There is also some evidence, slightly less relevant to the new moon procedure itself, from the neo-Babylonian and early Achaemenid periods (sixth century BCE): see below, note 21.
12. 'Astrologers' are referred to in the sources as *ṭupšarru enūma anu enlil*, literally 'scribes of [the astrological compendium] Enūma Anu Enlil', though this term is relatively rare, and other, more general titles are also used. The term ṭupšar itself has a broader meaning than 'scribe', as it is also used for scholars and high officials (cf Jeremiah 51:27), and these astrologers were certainly far more than scribes. See further Parpola (1970–83), vol. ii, p. xiv, and Rochberg (2010).
13. This means at the end of the 29th day of the outgoing month.

14. Parpola (ibid.), p. 102 (with a judicious interpretation); Hunger (1992), no. 120; and Beaulieu (1993), pp. 66–67 n. 4, whose translation is cited here.
15. Parpola (ibid.), vol. i, no. 119 (also ii, pp. 101–102); id. (1993), no. 225; see also Beaulieu (ibid.). The remainder of the text is unfortunately lost. The translation of Wacholder and Weisberg (1971) is incoherent. Brown (2000, p. 276) rejects any precise dating.
16. Parpola (1970–83), vol. i, no. 323.
17. Hunger (1992), no. 346.
18. Parpola (1970–83), vol. i, no. 290: Mar-ištar informs the king that he sighted the moon on the first (i.e. 31st of the old month), thus determining the beginning of the month of Du'uzu.
19. Goodman (2000), pp. 93–111; Hezser (1997), esp. pp. 353–404; Schwartz (2001), pp. 103–128; and Satlow (2005), pp. 153–158; *pace* Goodblatt (1994). A figure such as Rabban Gamaliel (II) may be presented, in the context of the new moon procedure, as exerting authority over his rabbinic peers (*mRH* 2:8–9), but this authority seems not to extend beyond rabbinic circles, and it does not involve the functions normally associated, in ancient society, with political rule, such as control of the military, taxation, public services, etc.
20. Greece: Samuel (1972), Pritchett (2001), Hannah (2005). Rome: Rüpke (1995). Alexander: Plutarch, Alexander 16 and 25. The Roman calendar of the Republic, prior to the institution of the Julian calendar in 46 BCE, was not lunar but consisted of a 355-day year that was irregularly adjusted with the intercalation of a 27-day month.
21. There are further differences between the Babylonian and rabbinic new moon procedures, which cannot be surveyed in detail here. For the dissemination of new moon decisions, for example, rabbinic sources envisage the use of chains of beacons (see above and note 6), which is not attested for this purpose in Mesopotamian sources (Beaulieu 1993, p. 72, n. 19). Instead, sources from the neo-Babylonian and/or early Achaemenid periods suggest that official reports about month lengths circulated in Babylonia in a haphazard way: there is thus a letter from some temple official to a governor at Sippar asking for a report on whether the last month was 29 or 30 days (Beaulieu ibid., pp. 70–71), and another from an official of the Ebabbar temple at Larsa to an official of the Eanna temple at Uruk mentioning that he has received a report that the last month was of 29 days (ibid. pp. 76-7).
22. Sifra *Emor pereq* 10:3: 'if they sanctified it but were forced, mistaken, or misled – whence (do we know that) it is sanctified? ... if you declared them, they are my festivals, and if not, they are not my festivals' (the last phrase, which epitomizes the overriding authority of the rabbinic court, is repeated several times in the passage like a *leitmotiv*). See parallels in *tRH* 3(2):1 (Zuckermandel edn., p. 211), *bRH* 25a. The triad 'forced, mistaken, or misled' is common in early rabbinic literature, and most probably the original version of this particular tradition. In the Babylonian Talmud (*bRH* 25a), however, the term 'deliberate' was substituted for 'forced', and in some manuscripts of *tRH* 3(2):1 'deliberate' was added as a fourth (and excessive) term, most probably under the influence of the Babylonian Talmud: see Henshke (2007), p.95 (I am grateful to David Henshke for discussing these sources with me). This later, Babylonian Talmudic version (sixth century CE?), which suggests that the rabbinic court has the power to ignore *deliberately* the empirical evidence of new moon sightings, represents a radical change in comparison to the earlier sources, and may be related to the gradual fixation of the calendar in the Talmudic period (on which see Stern 2001 and further below).
23. Sifra *Emor pereq* 10:4-5, with extensive discussions in *pRH* 3:1 (58c) and *bRH* 20a.
24. I disagree with Elior (2004), who argues that the Qumran calendar conformed to the 'laws of nature' whereas the rabbinic calendar was arbitrary and based human judgement (if anything, it should be the opposite: see my review in Stern 2005), and with Hayes (2011, pp. 124–128) who argues, following a suggestion of Daniel Schwartz (and perhaps unduly relying on the Babylonian Talmud's version of the ruling: see above, note 22), that the Qumran calendar was 'realist' and the rabbinic calendar 'nominalist'. Hayes is more guarded than Elior, as she concedes that the rabbinic calendar also comprises a significant 'realist' element. But Elior's and Hayes' categorizations of both the Qumran and the rabbinic calendars are, in my view, reductionist and misrepresentative of the fundamentally abstract nature of the Qumran calendar on the one hand, and the fundamentally empirical nature of the rabbinic new moon procedure on the other.

25. I differ with Wiesenberg (1962), who goes as far as seeing in this Mishnaic passage the 'elements of a lunar theory' (see further elaboration in his conclusion, p. 195). In this otherwise excellent article, Wiesenberg discusses in detail and which much erudition the different interpretations that have been given to these questions by medieval and modern scholars, deciding in favour of those that he considers astronomically most plausible. He does not note, however, that the vagueness and unclarity of the Mishnaic questions is itself significant.
26. I am assuming only one possible interpretation of the first question; this and other options are discussed by Wiesenberg (1962), pp. 174–180. Talmudic sources imply a different interpretation, for which, as the Mishnah goes on to state, the answer to the question would be equally straightforward (Wiesenberg (1962), pp. 193–195).
27. I am inclined to accept Wiesenberg's preferred interpretation of the second question, based on the Palestinian Talmud (*pRH* 2:5 (58a)), that it refers to whether the moon is to the north or the south of the sun's azimuth at sunset (Wiesenberg (1962), pp. 165–170).
28. See discussion Wiesenberg (1962), pp. 159–170.
29. It does not refer to the moon's latitude (as some have argued), an astronomical coordinate which a lay, naked-eye observer would never have been able to gauge. For the fourth question, Wiesenberg (1962), 182-8) favours a more sophisticated interpretation which assumes, in my view, an excessive amount of astronomical knowledge. But as to the fifth question, I am in agreement with Wiesenberg, pp. 156–157.
30. In this respect, I disagree with Wiesenberg's main argument (see above, note 25).
31. As noted by Wiesenberg himself, pp.159–163.
32. Further on, the Mishnah mentions that Rabban Gamaliel had pictures of moon crescents in his attic, which he used to show to the witnesses and ask: 'did you see it like this, or like that?' (*mRH* 2:8). This passage does not clarify, however, whether the intention was to establish the astronomical plausibility of the testimonies or only their mutual agreement.
33. See *bShabbat* 12b and Wiesenberg (1962).
34. The witness who dates the event to the second of the month knew that the month had been intercalated, i.e. that the previous month counted 30 days; whereas the witness who dates the event to the third of the month did not (yet) have this information, and assumed the previous month to count 29 days. Both witnesses, however, agree on the day when the event took place.
35. Civil litigation: *mSanhedrin* 3:6; criminal: ibid. 5:1–4. In both, the verdict normally depends on the the testimony of two witnesses
36. Note the parallel in *mSanh*. 3:3–4, with reference to normal judicial procedure.
37. First speaker in new moon courts: *pRH* 2:6 (58a), *pSanh*. 1:2 (18c); normal judicial procedure in *mSanh*.4:2.
38. *mRH* 2:7. Cf the verdict declaration in *mSanh*. 3:7 (for monetary cases), and the verdict announcement in 6:1 (capital cases).
39. *tRH* 3(2):1; other versions in Sifra *Emor pereq* 10:2, *pRH* 2:8 (58b), 3:1 (58c), the latter explaining that 'we are not particular about the new moon testimony'.
40. The Babylonian Talmud suggests, but only implicitly and incidentally, Psalms 81:4-5 as a scriptural source for the new moon being determined in a judicial setting (*bRH* 22b, 25b; the inference is typically very tenuous). According to another rabbinic source of the same period (*Pesiqta deRav Kahana* 5:15: Braude and Kapstein 1975, p. 117), the first month of Israel (in Exodus 12:2) was sanctified when God joined Moses and Aaron to form a court of three judges, with the angels Michael and Gabriel as witnesses. But this myth of origins still does not explain why the new moon procedure had to be designed as a judicial trial.
41. This decree is commonly known, in modern scholarship, as the 'First Fruits Decree': *Inscriptiones Graecae* I.3, no. 78, in Meiggs and Lewis (1988), no. 73. Hekatombeon is one of the months of the Athenian calendar.
42. The switch from a lunar calendar to the more familiar, schematic calendar of the Roman Republic (on which see above, note 20) is thought to have occurred either in the mid-fifth or the late fourth centuries BCE, although both dates are rather speculative. For the earlier date, see Michels (1967, pp. 121–130) and Rüpke (1995, pp. 204–207); for the later date, Humm (2000) and Rüpke (2006, pp. 26–27).

43. Macrobius, *Saturnalia* 1:15:9–12, partly following Varro, *De Lingua Latina* 6:27–28 (mid first century BCE) who mentions only the announcement of the 'pontiffs', and on that basis derives the word kalendae from an archaic Latin verb *kalare* (to call, to announce). The kalends are the beginning of the Roman month; the nones originally corresponded, perhaps, to the moon's first quarter (see Rüpke 1995, pp. 210–214). According to Livy, the minor pontiffs were originally called 'pontifical scribes' (22:57:3: *scriba pontificius*), which says perhaps something about their original function within the pontifical college, but also implies that the title of 'minor pontif' in Macrobius' account is an anachronism, as may well be other of its details. On the *rex sacrorum*, see Rüpke (1995), p. 313 n. 77.
44. Schürer (1973–87) i, Nadav (2010).
45. Some evidence of priestly involvement in the Jewish calendar before 70 CE appears in *mRH* 1:7, albeit in competition with a rabbinic court.
46. On the elusive identity of the Sanhedrin before 70 CE, and its disappearance after 70 CE, see Schürer (1973-87) ii, pp. 199–226; Goodblatt (1994), pp. 77–130; and Goodman (2007), pp. 327, 378.
47. E.g. Safrai (1965), Goodblatt (1994), p. 207, and Goodman (2000), p. 108. The latter writes, in relation to the calendar: 'when [after 70 CE] the Temple Sanhedrin could no longer give them [the Jews] certainty they accepted the authority of the self-appointed rabbinic Sanhedrin [i.e. court] with relief. This was the only religious function performed by the rabbis that was genuinely needed by the rest of the Jewish nation'. This notion that the calendar was somehow unique and a religious function that only rabbis could perform after 70CE is a common misconception, refuted if anything by the existence of local Jewish calendars in Diaspora communities (on which see Stern 2001).
48. See Samuel (1972), Meimaris (1992), and Stern (forthcoming) ch. 5.
49. *pRH* 1:6 (57b), offering a different version of the story told in *mRH* 1:6 where the person who intercepted the witnesses was R.Aqiva. For a parallel see *bRH* 22a, where the name Shazkar (so in *pRH* according to ms. Leiden) varies considerably in the manuscripts. This R.Gamaliel belongs to the early second century CE. See Henshke (2007, pp. 97–98, n. 80).
50. However, it may be possible to read, in this story, that the witnesses were intercepted in Gadara because of some other conflict between R.Gamaliel and Shazkar, without Shazkar laying claim on the calendar and on the determination of the new moon. I am grateful to Jonathan Ben-Dov for this critical observation.
51. I interpret this as meaning 'cross over the river Jordan and make it a frontier between you and them'.
52. *pSanhedrin* 8:2 (26a), with parallel in *pMoed Qatan* 2:3 (81b).
53. *mRH* 2:7.
54. *pRH* 2:6 (58b), *pSanhedrin* 1:2 (18c).
55. See *tMegillah* 4:15 (Zuckermandel ed. 226), *pMegillah* 1:6 (70b), *pBerakhot* 6:1 (10a), ibid. 6:4 (10c), *pPesahim* 1:1 (27b), ibid. 2:5 (29c), *pMoed Qatan* 3:8 (83d), and *bSanhedrin* 70b, and note 58 below; the practice of new month meals may have had ancient biblical origins, as evident in 1Samuel 20:24-34. Another type of public meal, laid on for the witnesses that presented themselves to the rabbinic calendrical court, is mentioned in *mRH* 2:5 but probably not relevant to the present context.
56. So Lieberman (1934), pp. 102–4; but medieval commentaries on *bSanhedrin* 70b (e.g. R.Meir Abulafia, *Yad Ramah*, 13th century) express different views.
57. Liturgy for new month meals is found in later Palestinian sources from the Cairo Genizah and in tractate *Soferim* 19:7–8 (see Fleischer 1973). Although *Soferim* was redacted in the Middle Ages, possibly in Europe, reference in this passage to *bulevtin* (i.e. in Greek *bouleutai*, city councillors) suggests that this text may be an authentic, late antique tradition: for the term bouleutai would not have made sense to any Jew writing in Palestine or elsewhere after the sixth century CE, when the institution of *boule* and *bouleutai* fell rapidly into oblivion (see Whittow 1990, Liebeschuetz 2001, pp. 104–109, and Rapp 2005, pp. 286–287). Fleischer (1973), pp. 346–348 rightly senses that the Soferim text suggests a procedure of 'sanctification of the month', but similarly to Lieberman, refrains from this interpretation because of his traditionalist assumption that sanctification of the month could only be carried out by the rabbinic court.
58. Thus the same R.Yoḥanan instructed the 'synagogue of Kifra' (or Kufra) to begin the new month meal while it was still daylight and to declare whether the new month was on time (i.e. on the 30th of the old month) or postponed (to the 31st): *pTa'anit* 4:5 (68b), *pRH* 4:4 (59c) (see Lieberman 1934, p. 104). In

this case, the synagogue is unlikely to have taken any calendrical decision itself: Kifra was a small village outside Tiberias (see *pMegillah* 1:1 (70a), where Kifra is identified as the ancient site of Tiberias; see also the Midrash *Song Rabbah* 1:27), presumably under the jurisdiction of both the boule of Tiberias and the rabbinic court of R.Yoḥanan—whichever of the two controlled, on this occasion, the calendar.

59. The main players in the *Soferim* text (above n. 43) are not rabbis, but 'assemblies of elders and *bouleutai* (city councillors)' and 'the twelve magistrates of the city' (Fleischer 1973, p. 339). But in some other passages (e.g. *pBerakhot* 6:1 (10a), 6:4 (10c)), the new month meal is said to be attended by 'rabbis'.
60. On control of the calendar by the archons or city magistrates in Greek and Hellenistic cities, see above and Stern (forthcoming), ch.1. On bouleutic common dining, i.e. the well established Hellenistic tradition of *bouleutai* or select members of the *boule* (archons, *gerontes*, *prytaneis*, etc.) holding common meals at festivals or on other regular occasions, see Schmitt Pantel (1992), pp. 168–170, 175–157, and 303–326, who rightly underscores the political function of this practice.
61. This argument does not necessitate that the councils of cities such as Tiberias, Sepphoris, and Lod, in which the population was always mixed and not exclusively Jewish, were composed entirely of Jews. The ethnicity and religious identity of city councils in late antique Palestinian cities is a complex question that cannot be done justice here: see for example Schwartz (2001), pp. 132–142. Non-Jewish city councilors, however, are likely to have been happy to share a lunar calendar with their Jewish colleagues, whether they were Samaritans (who are known to have also used a lunar calendar) or pagans (whose traditional calendars, at least before the arrival of the Romans, had always been lunar).
62. See in particular *pKetubot* 1:2 (25b) with parallels in *pNedarim* 6:13 (40a) and *pSanhedrin* 1:2 (19a), and elucidation in Stern (2001), p. 231; and more explicitly in later sources, *Exodus Rabbah* 15:2 and 15:20, *Pirqei deRabbi Eliezer* 8.
63. Stern (forthcoming).
64. Stern (2001), ch. 5.
65. City councils, bishops: Whittow (1990), Liebeschuetz (2001), pp. 104–168, and Rapp (2005). Patriarchs: Goodblatt (1994), Jacobs (1995), and Hezser (1997), although Schwartz (2001) rightly emphasizes that even in late Antiquity, the Jewish Patriarchs never achieved political status on the same level as bishops.
66. The extent of the authority of rabbis in late antique, Jewish Palestinian communities remains debatable: see references in previous note, Miller (2006), and most recently Millar (2011).

References

Beaulieu, P.-A., 1993, "The Impact of Month-Lengths on the Neo-Babylonian Cultic Calendar", *Zeitschrift für Assyriologie und Vorderasiatische Archäologie* 83, 66–87.

Braude, W. G. and Kapstein, I. J., 1975, *Pesikta de-Rab Kahana* (London: Routledge).

Brown, D., 2000, *Mesopotamian Planetary Astronomy-Astrology*, Cuneiform Monographs 18 (Groningen: Styx).

Danby, H., 1933, *The Mishnah* (Oxford: Oxford University Press).

Fleischer, E., 1973, "Beirurim beva'ayat yi'udam haliturgi shel piyyutei qiddush yerahim", *Tarbiz* 42, 337–363.

Goodblatt, D., 1994, *The Monarchic Principle: Studies in Jewish Self-Government in Antiquity* (Tübingen: Mohr Siebeck).

Goodman, M., 2000, *State and Society in Roman Galilee, A.D. 132–212*, 2nd edn. (London: Vallentine Mitchell).

——, 2007, *Rome and Jerusalem* (London: Allen Lane).

Hannah, R., 2005, *Greek and Roman Calendars. Constructions of Time in the Classical World* (London: Duckworth).

Hayes, C., 2011, "Legal Realism and the Fashioning of Sectarians in Jewish Antiquity", in S. Stern (ed.), *Sects and Sectarianism in Jewish History*, IJS Studies in Judaica 12 (Leiden: Brill), 119–146.

Henshke, D., 2007, "R.Joshua's Acceptance of the Authority of Rabban Gamaliel II: A Study of Two Versions of the Same Event", *Tarbiz* 76, 81–104 [Hebrew].

Hezser, C., 1997, *The Social Structure of the Rabbinic Movement in Roman Palestine* (Tübingen: Mohr Siebeck).

Humm, M., 2000, "Spazio e tempo civici: riforma delle tribù e riforma del calendario alla fine del quarto secolo

a. C.", in C. Bruun (ed.), *The Roman Middle Republic: Politics, Religion, and Historiography c. 400–133 B.C. (papers from a conference at the Institutum Romanum Finlandiae, September 11–12, 1998)* (Rome: Institutum Romanum Finlandiae), 91–119.

Hunger, H., 1992, *Astrological Reports to Assyrian Kings*, State Archives of Assyria 8 (Helsinki: Helsinki University Press).

Jacobs, M., 1995, *Die Institution des jüdischen Patriarchen: Eine quellen- und traditionskritische Studie zur Geschichte der Juden in der Späntantike* (Tübingen: Mohr Siebeck).

Lieberman, S., 1934, "Tiqunei Yerushalmi 6", *Tarbiz* 5, pp. 97–110 (repr. *Mehqarim be-Torat Eretz Yisrael*, Jerusalem: Magnes Press, 1991, 200–213)

Liebeschuetz, J. H. W. G., 2001, *The Decline and Fall of the Roman City* (Oxford: Oxford University Press).

Meiggs, R. and Lewis, D., 1988, *A Selection of Greek Historical Inscriptions to the End of the Fifth Century B. C.*, revised edn. (Oxford: Clarendon Press).

Meimaris, Y. E., with Kritikakou, K., and Bougia, P., 1992, *Chronological Systems in Roman-Byzantine Palestine and Arabia—The Evidence of Dated Greek Inscriptions*, Meletemata 17 (Athens: Diffusion de Boccard).

Michels, A. K., 1967, *The Calendar of the Roman Republic* (Princeton, NJ: Princeton University Press).

Millar, F., 2011, "Inscriptions, Synagogues and Rabbis in Late Antique Palestine", *Journal for the Study of Judaism* 42, 253–277.

Miller, S. M., 2006, *Sages and Commoners in Late Antique 'Erez Israel. A Philological Inquiry into Local Traditions in Talmud Yerushalmi*, Texts and Studies in Ancient Judaism 111 (Tübingen: Mohr Siebeck).

Nadav, S., 2010, "The Title *ethnarch* in Second Temple Period Judea", *Journal for the Study of Judaism* 41, 472–493.

Parpola, S., 1970–83, *Letters from Assyrian Scholars to the Kings Esarhaddon and Assurbanipal*, 2 vols. (Kevelaer: Butzon and Bercker).

——, 1993, *Letters from Assyrian and Babylonian Scholars* (Helsinki: Helsinki University Press).

Pritchett, W. K., 2001, *Athenian Calendars and Ekklesias* (Amsterdam: Gieben).

Rapp, C., 2005, *Holy Bishops in Late Antiquity: The Nature of Christian Leadership in an Age of Transition* (Berkeley: University of California Press).

Rochberg, F., 2010, "Scribes and Scholars: the ṭupšar Enūma Anu Enlil", in F. Rochberg, *In the Path of the Moon. Babylonian Celestial Divination and Its Legacy* (Leiden: Brill), 237–256.

Rüpke, J., 1995, *Kalender und Öffentlichkeit. Die Geschichte der Repräsentation und religiösen Qualifikation von Zeit in Rom* (Berlin and New York: Walter de Gruyter).

Rüpke, J., 2006, *Zeit und Fest: eine Kulturgeschichte des Kalenders* (Munich: Beck).

Safrai, S., 1965, "The Localities of the Sanctification of Lunar Months and the Intercalations of Years in Palestine after 70CE", *Tarbiz* 35, 27–38 [Hebrew].

Samuel, A. E., 1972, *Greek and Roman Chronology. Calendars and Years in Classical Antiquity* (Munich: C. H. Beck).

Satlow, M. L., 2005, "A History of the Jews or Judaism? On Seth Schwartz's Imperialism and Jewish Society, 200 B.C.E. to 640 C.E.", *Jewish Quarterly Review* 95, 151–162.

Schmitt Pantel, P., 1992, *La cité au banquet. Histoire des repas publics dans les cités grecques* (Rome: Ecole Française de Rome).

Schürer, E. , 1973-87, *The History of the Jewish People in the Age of Jesus Christ (175 B.C.- A.D. 135)*, revised and edited by G. Vermes and F. G. B. Millar (with M. Black (vols. i–ii) and M. Goodman (vols. iii.1–2)), 4 vols. (Edinburgh: T.& T. Clark).

Schwartz, S., 2001, *Imperialism and Jewish Society, 200 BCE to 640 CE* (Princeton: Princeton University Press).

Simon-Shoshan, M., forthcoming, *Stories of the Law: Narrative Discourse and the Construction of Authority in the Mishnah* (Oxford: Oxford University Press).

Steele, J. M., 2007, "The Length of the Month in Mesopotamian Calendars of the First Millennium BC", in J. M. Steele (ed.), *Calendars and Years: Astronomy and Time in the Ancient Near East* (Oxford: Oxbow Books), 133–148.

Stern, S., 2001, *Calendar and Community: A History of the Jewish Calendar, 2nd century BCE – 10th century*

CE (Oxford: Oxford University Press).
——, 2005, "Rachel Elior on Ancient Jewish Calendars: A Critique", *Aleph – Historical Studies in Science and Judaism* 5, 287–292.
——, 2008, "The Babylonian Month and the New Moon: Sighting and Prediction", *Journal for the History of Astronomy* 39, 19–42.
——, forthcoming, *Calendars in Antiquity. Empires, States, and Societies* (Oxford: Oxford University Press).
Wacholder, B.-Z. and Weisberg, D. B., 1971, "Visibility of the New Moon in Cuneiform and Rabbinic Sources", *Hebrew Union College Annual* 42, 227–242.
Whittow, M., 1990, "Ruling the Late Roman and Early Byzantine City: A Continuous History", *Past and Present* 129, 3–29.
Wiesenberg, E., 1962, "Elements of a Lunar Theory in the Mishna, Rosh Hashana 2:6, and the Talmudic Complements Thereto", *Hebrew Union College Annual* 33, 153–196.

From Observation to Calculation: The Development of the Rabbinic Lunar Calendar*

Lawrence H. Schiffman

This paper will explore the evolution of the Jewish lunar calendar from a calendar based on observation of the moon to one based on astronomical calculation. We will argue that establishment of the current calculated 'Jewish Calendar' involved a complex series of developments that took place during Late Antiquity and the early Middle Ages. We will further argue that the calendar of ancient Israel was indeed lunar, and that sectarian calendars, mentioned in Second Temple and rabbinic literature, represent reformist proposals.[1]

Various methodological pitfalls plague the study of the Jewish calendar. It is no accident that the calculation of the Jewish calendar has been called *sod ha-ʿibbur*, 'the secret of intercalation'. Although not the original meaning of this phrase,[2] this translation, reflecting later usage, correctly describes one of the major obstacles in the study of the Jewish calendar. It is clear that much of the astronomical and calendrical knowledge of Jews was not reduced to writing until the Middle Ages and, therefore, its origins will not be recovered from any written source. Our proposed historical analysis can only be evaluated as the most likely reconstruction.

The ramifications of this study extend beyond calendrical research. Within the last twenty-five years, contentious debates have erupted over the role of rabbinic leadership in the affairs of the ancient Jews and in the molding of Judaism as a whole.[3] While we cannot assume a monolithic Judaism, as did our academic forebears, arguments advocating a complete absence of unity or rabbinic leadership are equally fallacious. An exploration of the calendar, an important test case for unity and political-religious control,[4] will further substantiate our position that the social fabric of ancient Jewish society must be seen as a complex mix of variegation and pluralism alongside certain elements of common Judaism and even unity.[5]

I must emphasize that without the thorough work and expertise of my colleagues, Sacha Stern and Jonathan Ben-Dov, this presentation would have been infinitely more difficult. Although I am indebted to them, I will at times take a different perspective to the data they gathered and extensively analyzed.

* I wish to thank Mehran Etessami for his careful reading and critique of the mathematical aspects of this paper. A.J. Berkovitz (Yeshiva University) served as my research assistant for the completion of this project, and his excellent work and critical judgment contributed greatly to the article. The comments of J. Ben-Dov were most helpful.

The Canaanite, Phoenician, and Israelite Calendars

An exploration of the evidence from the Canaanite, biblical, and Phoenician-Punic calendar lends credence to arguments in favor of Israelite reliance on a lunar calendar with some form of solar synchronization. The Gezer Calendar, a 10th century inscription, also bolsters our argument. This calendar lists twelve months of the year (with no mention of intercalation and month names) by the agricultural activities such as: *ʾsf*, ingathering, *zrʿ*, seed time, *qṣr śʿrm*, barley harvest, *qṣr wkl*, wheat harvest, *qṣ*, summer fruits.[6] These terms are in use in the biblical books from Genesis to the Prophets.[7] Similarly, in 1 Kings 4:7 each of Solomon's twelve districts had to provide food for the royal palace for one month each year; in 1 Chronicles 27, one of the stewards of King David was on duty each month of the year. Neither of these passages provides for a possible extra month in the year.

Referring to the period of Solomon's reign, the Hebrew Bible mentions four month names, probably alluding to the seasonal events in nature. They are: Ziv (possibly the month of flowers), Etanim (possibly the month in which only permanent water courses still flow), Bul (possibly month of the great rains), and possibly also Aviv (the month of the ears of grain).[8] A variety of Phoenician and Punic texts also refer to at least three out of these four month names.[9] Israel and the ancient Canaanites, therefore, shared a common calendar. A full version of the Canaanite calendar, including all of the months, can be reconstructed from Punic and Phoenician inscriptions. While it is difficult to be certain, one of the month names, *mpʿ*, occurs in the phrase *mpʿ lpny* (i.e., former *mpʿ*). Since this month name is found in two inscriptions,[10] it may be specifically labeled to indicate that it was an intercalary month.[11] We may reconstruct the Canaanite calendar as a twelve-month lunar calendar, with the periodic interpolation of a thirteenth month to maintain synchronization with the seasons.[12]

Additionally, because the Israelites counted the day from one evening to the next evening,[13] the months started with the appearance of the crescent new moon. The Israelites' indication of this event was the *yrḫ* (lunation),[14] after the old Phoenician term. Although early evidence indicates the equal use of *ḥdš* (new moon),[15] the simultaneous employment of *yrḫ* may establish another link between the Israelite, Canaanite and Phoenician calendars.

Although no explicit biblical evidence mentions intercalation, some method of equalization with the solar year was necessary to guarantee the celebration of Passover in the spring.[16] Although three major festivals—Passover, Shavuot, and Sukkot—punctuated the Israelite year, the beginnings the year vacillated between autumn and spring.[17] The start of the year in the spring was apparently adopted when the kingdom of Judah became a vassal state of Babylonia.[18]

The Babylonian exile also affected the Israelite calendar. Exposure to the Babylonians allowed for a rich infusion of astronomical and calendrical knowledge into the Jewish elite, even in Judea. The Israelites also adopted the Babylonian month names. Initially, the Bible referred to months in ordinal procession.[19] Some evidence from the Ancient Near East possibly attests to this system too.[20] The post-exilic books of Zechariah, Ezra and Nehemiah, Chronicles, Esther, and the books of the Maccabees often identify numbered months with Babylonian names.[21] The book of Jubilees and many of the texts from Qumran did not adopt these names.[22] Eventually, the Babylonian month names became normative among the Jewish people.[23] Further, the Babylonian calendar was also used by the Jews of

Elephantine, since it was the administrative calendar of the Persian empire. They also used Egypt's 365-day civil calendar alongside it.[24]

We may understand the Israelite calendar as an adoption of the Canaanite calendar, comprised of lunar months, beginning with the earliest appearance of the new moon and extending for twenty-nine or thirty days. This calendar was synchronized with the solar year, although we must recognize that there is no conclusive evidence as to how this was done in biblical times.

The evidence amassed above vitiates claims that biblical Israel relied on a solar calendar and that the Pharisaic-rabbinic tradition introduced a new calendar at some time during the Second Temple period.[25] Rather, it is the solar calendar of Second Temple times, in all its sectarian manifestations, that represents reformist policies.

The Calendars of Second Temple Israel

Two types of solar based calendars circulated during the Second Temple period, alongside the luni-solar calendar.[26] A presumed *Vorlage* of the Astronomical Book of Enoch posits a 360-day calendar, [27] a type of schematic calendar found in ancient Babylonian literature.[28] Within the same text, however, the author puts forward a strong argument for a 364-day calendar, which he regards as the correct version of the calendar.[29] The 360-day calendar consisted of twelve months of thirty days each, and the 364-day calendar, advocated as well by the book of Jubilees,[30] operating on the pattern 30, 30, 31 days for each quarter.

Portions of the texts mentioned above are found in the Qumran corpus, even if the calendrical sections are not always preserved.[31] Indeed, it is the 364-day schema that serves as the basis for the Qumran solar calendrical texts.[32] Material designated as 4QEnastr (4Q208–211) specifically includes calculations of the synchronization of solar and lunar calendars according to a three-year cycle.[33] Although Milik considered this material to be part of the text of 1 Enoch,[34] scholars have contested his view.[35] Regardless, the material's presence indicates that in the late Persian period, when the Qumran Aramaic texts were by and large composed,[36] this form of calendrical calculation was known in the Land of Israel.[37]

Scholars discussing calendars in the Qumran manuscript collection have given primary attention to the 364-day calendar, because it is used to set out the festivals in various calendrical documents and also in the Temple Scroll.[38] The Astronomical Book included in 1 Enoch makes careful comparisons between the lunar cycle and the proposed solar calendar, even without mention of festivals. However, later Qumran documents also give the lunar calendar an important role.[39] Not only are most dates of the proposed solar calendar provided with equivalent lunar dating, but also some prayer cycles are clearly constructed according to a lunar count of the days of the month.[40] These solar-based calendars are reformist proposals and do not reflect the calendar used either by the society as a whole or by the Jerusalem Temple.[41]

By the early Hellenistic period, knowledge of Babylonian lunar astronomy and calendation was available in Jewish society in the Land of Israel, as evidenced by the material collected at Qumran. Following the Talmudic dating of the adoption of the Babylonian month names to the exilic period,[42] we argue that much of this astronomic and calendrical knowledge came into the Jewish community during the Babylonian exile, more or less at the very same time that the mathematically calculated lunar, 19-year calendar cycle became known in Babylonia.[43] It began to be consistently applied in Achaemenid times.[44] The presence of lunar calendrical material in the Qumran corpus, therefore, indicates that such information was available to elements of Second Temple period Jewish society in the Land of Israel, even if the Pharisees and the later rabbis continued to use lunar observation for the fixing of the months.[45] Because of this evidence for advanced knowledge of lunar calendation and astronomy, the claim of the Tannaim (mishnaic rabbis) to have had the ability to determine the lunation and position of the nascent moon, such as was done in ancient Babylonia, in order to check the reports of witnesses and to influence the beginning of months that needed to be adjusted, is completely plausible.[46]

It is worth noting that neither the Qumran 364-day calendar nor their lunar calendar, assumed by the lunar tables, preserves any indication of a system for intercalation. Absence of intercalation, however, is impossible in a Jewish milieu where the festivals must match the agricultural seasons. The expanded festival calendar preserved in some of the Qumran calendrical texts and in the Temple Scroll underscores the requirement of matching festivals to the agricultural seasons.[47] The Torah requires that the Passover festival take place around the time of the vernal equinox.[48] Repeated use of a calendar of twelve months, whether in a 364-day solar context or in a 354-day lunar context, with no system of intercalation, would displace the holidays from their respective seasons. Regardless of which calendar we consider normative, we must assume a system of intercalation, whether precalculated or determined by observation.

The nineteen-year cycle, known as the Metonic cycle, was widely known in the Greco-Roman world from the fifth century BCE. This system, still in use in the current Jewish calendar, requires the addition of a thirteenth month in seven out of nineteen years, and also entails specific alternations between twenty-nine and thirty-day months. This cycle was part of the entry of Babylonian astronomy and mathematics into the Greco-Roman world that took place as cuneiform culture was declining. In fact, the Greeks did not use calculated calendars before the first century BCE, relying mostly on observation. The process of applying astronomical knowledge to calendar procedures happened to the Jews long before the Greeks, most probably through the medium of Aramaic, the linguistic and cultural *lingua franca* of the later ancient Near East. Thus, in the Second Temple period, the Jewish society most likely relied on a lunar-based calendar with some form of intercalation, originally Canaanite in origin, but with additional technical knowledge drawn from cultural interchange with Babylon.

Indeed, the ancient Babylonians developed the intercalated luni-solar calendar early in their history. The Sumerians already used intercalation to maintain the correspondence between certain holidays and specific seasons of the solar year.[49] We continue to hear of such practice throughout ancient Babylonian history. For example, the Babylonian King Hammurabi (c. 1700 BCE) also intercalated the calendar.[50] By the time the sun had set on the Neo-Babylonian Empire, this calendar was well rooted. This tradition passed on to the Persian Empire. By 503 BCE, under Darius I the Great, the Persian Empire was already using a fully calculated calendar with an intercalary cycle of seven out of nineteen leap years, with an added month.[51] Unlike the later Jewish year, six of the years added second Addaru, and one an extra second Ululu. This ensured that the first day of the month of Nisanu never strayed far from the equinox. It is perhaps ironic that this is often called the Metonic cycle, after the Athenian astronomer who introduced it to the West in 432 BCE.[52]

Evidence of Hellenistic and Roman Period Sources

Numerous sources from the Greco-Roman period indicate that the Jews followed a lunar-based calendar. No doubt may be cast on this reality during the Seleucid period, since Judea, like the rest of the Seleucid Empire, followed the lunar calendar of the Babylonians.[53] By this time, the Babylonians followed a 19-year intercalation cycle, a feature that typifies the calculated lunar calendar of the Jews already in the fourth century CE. In addition to political necessity, the Jewish adaptation of the Seleucid lunar calendar eased the process of fixing festivals to their proper seasons.

Earlier, however, the Ptolemaic rulers employed a version of the Egyptian civil calendar based on a 365-day year. This left the Jews on their own to calculate the festivals with some version of the lunar calendar that had earlier been replaced by the Macedonian version of the lunar calendar.[54] We have no way of knowing how similar the calendar fixing process in Jewish Egypt and elsewhere was to what took place in the Land of Israel. It is safe to assume, however, that Jewish calendation took place in Egypt and other such locales independently. In fact, lack of central control must have typified the far-flung Jewish communities spread throughout the Hellenistic world and even further away.

When the Roman Empire switched to the Julian calendar in the first century CE, the Jews once again had to engage in their own lunar calendation, and fixed their own calendar. As Jews continued to use the lunar calendar, they effectively disconnected themselves from the calendation of the general societies in which they lived, except in a few places where the autochthonous populations maintained their archaic lunar ways.[55] The Sassanian Iranian rulers, under whom the Jews of the Babylonian Jewish community lived in Iraq, switched over to a solar calendar in the early third century. Knowledge of the ancient Babylonian calendar, however, continued.[56]

Little information regarding the maintenance of the rule of the equinox, the requirement that Passover take place on or after the vernal equinox, survives. Some date formulas from the talmudic period, however, appear to follow exactly the calendar of the ancient Babylonians.[57] In fact, Galen (131–201 CE) alludes to the Jews' having a system of intercalation in their calendar.[58]

Sources from the fourth century CE on, however, especially relating to the dating of Easter, assert that Jews ignored the rule of the equinox and celebrated Passover too early.[59] Willingness of Jews to diverge from the rule of the equinox is best explained as resulting from the destruction of the Temple. When it was no longer necessary to key the festivals to the agricultural seasons, since agricultural offerings could not be performed after the destruction of the Temple, Passover was allowed to migrate to an earlier date. Another factor may have been a loosening of the influence of the ancient Babylonian calendar as Jews developed that calendar into the specific Jewish version and administered it independently. We should note that communities that were located elsewhere or those not affiliated directly with the rabbinic movement seem to have continued to conduct their affairs independently, determining the dates of festivals and their own calendars. These communities would have followed the lunar calendar of their cities where it was maintained despite the reforms of Caesar and Augustus. This, of course, would have been their only option, given the nature of ancient communication.[60] On the other hand, it is clear that Jews in Palestine and Babylonia knew how to calculate the calendar according to the Babylonian nineteen-year cycle. Therefore, they would have maintained their calendar, even if determined by observation of the moon, to keep it as closely as possible in accord with the mathematical model.

We can assume that agricultural and other considerations,[61] when the Temple still stood, compelled the rabbis to diverge more from the standard ancient Babylonian calendrical model than would have later rabbinic authorities after the destruction of the Temple. Climatic and agricultural conditions varied from year to year, and deeply affected Temple worship. An important illustration of this diversity comes from the Zoar tombstone inscriptions, which date from the fourth and fifth centuries. These texts, found just a short distance to the south of the Dead Sea, towards the Arabian side, indicate that some Jews seem to have done their own intercalations based on observation of the moon. There is no evidence of conformity to the nineteen-year cycle or to any other cycle. Further, the rule of the equinox was not observed, as can be calculated from the dates on the tombstones.[62]

Rabbinic Calendation

Since all Jewish calendars were determined previous to the fourth century CE by lunar observation, the month must have began with the first appearance of the new moon. The rabbinic court in Palestine proclaimed the new month, a practice that continued until the calculated calendar became the norm, beginning in the fourth century. In fact, the current Jewish calendar uses the term *molad* to describe the calculated mean conjunction—the time when none of the moon shows because the earth, moon and sun, in that order, are approximately in a straight line.[63]

According to our sources,[64] although they do not necessarily constitute sufficient historical evidence, the claim is made that the Jews of Babylonia followed the lead of those of the Land of Israel, and that the calendar was regularly established in Israel by lunar observations. In fact, some sources indicate that Babylonians used mathematical techniques in order to determine the likely days on which their colleagues in the Land of Israel would declare the new moons, in order to deal with the fact that despite the descriptions of the Mishnah, communication was not always effective.[65]

Evidence from the Byzantine environment indicates that the change to a calculated calendar encompassed the entire Jewish world at about the same time. The documents in question revolve around the Easter debate and give specific dates for the observance of Passover during the period. These documents come from Sardica in Bulgaria and Catania in Sicily. In addition, the *ketubah* of Antinopolis, dating to 417 CE, provides indirect evidence for a calculated calendar and for the widespread nature of this change.[66] It seems that whatever the authority of the rabbis may have been regarding distributing information based on lunar observation, they either caused or shared in a total transformation of Jewish calendation in the fourth century.

The institution of the nineteen-year cycle is often ascribed to the Patriarch Hillel II, but his role is first mentioned only in a responsum of Hai Gaon (early 11th century), as transmitted by Abraham bar Ḥiyya (b. 1065). Hai's report is the earliest mention that Hillel II instituted this calendar reform in 358/9 CE. This tradition, however, is not the only dating given for these developments in traditional Jewish sources. Caution must be exercised in accepting this tradition, since Hillel II plays a very limited role in rabbinic literature.[67] It may be that the desire to standardize the date of Easter celebration after the Christianization of the Roman Empire in 325 CE was also a stimulus leading Jews to standardize their calendar and the dating of Passover.[68]

Medieval Developments

The history of the rabbinic calendar did not end with the shift to the Metonic cycle. The calendar, as practiced in rabbinic Judaism, has certain specific exceptions regarding the occurrence of the New Year festival and the observance of Passover.[69] These exceptions are designed to avoid such things as the Day of Atonement and the Sabbath occurring consecutively. These restrictions led to specific rules for delaying the onset of the new month for up to 48 hours after the mean conjunction. These are termed *deḥiyyot*, best translated as 'postponements'. The system of postponements assumes that the beginning of the month takes place at the mean conjunction rather than at the first crescent. This shows that the roles for 'postponement' developed after the transition from a calendar based on observation to one based on calculation.

In order to determine the system of intercalation by which a thirteenth month is added in seven out of nineteen years, one must assume the length of a solar year. The Babylonian Talmud provides a measurement, attributed to the Babylonian scholar Samuel (bar Abba; c. 165 – c. 257), which sets the length of the solar year at exactly 365 1/4 days.[70] This figure mirrors the measurements of the Julian calendar, already several hundred years old. Since this is the only proposed length of the solar year provided by the Talmudic rabbis of Babylonia or Palestine, it makes sense to understand that such an approximation was used in the calendrical calculations that ensued after the changeover in the fourth century to the calculated calendar. Modern traditional sources claim that a more accurate figure, 365 days, five hours and 55 and 27/57 seconds, was an alternative calculation that was known to the Talmudic rabbis. This opinion, even if attributed to the Babylonian rabbi Rav Adda in medieval sources, is first mentioned by Maimonides (1135–1204), who does not attribute it to Rav Adda.[71] This more accurate solar year was established by multiplying the *molad* interval of 29 days, 12 hours and 793/1080 minutes[72] by 235, the number of lunations in the 19-year Metonic cycle, and dividing by 19. By the time of Maimonides, this new calculated solar year length, attributed to Rav Adda, became normative for Jews. We must assume that this change in the assumed length of the solar year, based upon which the entire calendar was calculated, was a major step forward, even though it is not recorded anywhere in our sources.[73]

Numerous sources demonstrate that in the Geonic period various details of the calendar as we know it were either not yet in effect or not consistently applied. For example, certain holidays fell on days that would be impossible according to the rules incorporated in our calendrical system.[74] Apparently, the rule of *molad zaqen* (lit. 'overaged', i.e., late conjunction) was not observed. This rule states that if the conjunction of Tishri occurs on or after noon, the first of the month will be postponed. In fact, this rule sparked the controversy between Rav Saadyah Gaon and the Palestinian Gaon Ben Meir regarding the date of Passover, 922 CE. This conflict was the final gasp of Palestinian power over the calendar.[75]

Another example is a letter written in 835/6 CE by the Exilarch in Babylonia. According to this letter, Passover was to occur two days earlier than it would have under the present calendar. This document indicates without question that certain details of the present-day Jewish calendar were not instituted until after 835/6.[76] Any attempt to fill in the dates for the final fixing of the calendar must acknowledge that until Maimonides' Code and its full presentation of 'The Laws of the Sanctification of the New Moon', we do not have any authoritative document that presents our system in its entirety.[77] Furthermore, Maimonides' treatise shows influence in its third part, in which he discusses how to calculate the position of the sun and the moon at any particular time, of the Arabic astronomer and mathematician Muhammad ibn Jābir al-Harrānī al-Battānī (c. 858–929). Among other things, al- Battānī determined a more accurate length of the solar year of 365 days, 5 hours, 46 minutes and 24 seconds.[78]

In addition to determining the length of a solar year, the correct calculation of the Jewish calendar also requires a figure for the lunation, the actual length of time between conjunctions (*molad*). Indications of approximate knowledge of this information do occur in rabbinic literature, although some of this data may be a result of later interpolation. The lunation used by the present-day calendar, 29 days, 12 hours and 793/1080, was first attested by the early ninth century in the work of the Muslim astronomer al-Khwarizmi (c. 780 – c. 850 CE), and its use for calendar calculation is confirmed obliquely in the letter from the Exilarch of 835/6 CE.[79] This figure most likely comes from the second-century classic work of astronomy by Ptolemy (c. 90 – c. 168 CE), the *Almagest*, who himself got it from the second-century BCE Greek Astronomer Hipparchos (c. 190 – 120 BCE). The *Almagest* was translated into Arabic in the ninth century; the Jews could have absorbed it then, or from earlier Babylonian or Greek sources.

In general terms, we may say that the development of the calendar continued even after the decision to change over to a calculated calendar, and refinements of the system continued regarding the avoidance of certain 'scheduling conflicts' in the Jewish ritual year. The Jewish calendar's history seems to have made a full swing. Initially borrowed from Babylonian sources and practiced in the Land of Israel, its final touches seem to have been put on in Babylonia. The place of its origin was the place of its completion.

Conclusion

We have followed the journey of the Jewish calendar from its earliest stages until its completion. In the process, we discussed the ancient Near Eastern background, Greco-Roman contributions, and the competing solar calendars put forward by Jewish sectarians. There can be no question, however, that much of the story of the calendar remains shrouded by the absence of sources, and that only some reconstruction will bring about a realistic understanding. Nonetheless, we have seen without question that the history of the calendar reflects in many ways the various cultural interchanges that molded Judaism throughout the ages, even as it continued to build upon its own internal core.

The study requires a reflection upon issues pertaining to the cohesiveness of the Jewish people and the role of rabbinic leadership in shaping the community as a whole. The evolution of the calendar, specifically the consensus that eventually developed around it, conflicts along the way notwithstanding, must be seen as a result of the emergence of a consensus within the Jewish community to set up a variety of innovations and improvements that led from observation to calculation, and at the same time from local authority to centralized authority. Whatever may have been the power of the rabbinic establishment over the daily lives of individual Jews, it seems that the rabbinic class either took the lead or participated in the changeover from calendation by observation to calculation starting in the fourth century CE. It is impossible to understand the unified transference to this method by Jewish communities except by assuming either collective rabbinic authority or collective consensus among rabbis and other political and religious leaders. Whatever the case, this changeover is a result of unified action on the part of the Jewish people, even if that action was gradual, instinctive or less-than-perfect.

The Islamic conquest created a situation in which the Jewish community was now essentially unified under the leadership of the Babylonian rabbis who were connected with the court of the Caliphate during the early Islamic period. While we cannot attach specific dates to the various developments, it seems clear that the same centralization of authority that brought about the dominance of the Babylonian Talmud within the Jewish community as a whole, when connected with the decline of Hellenistic Judaism and other forms of non-rabbinic practice, made much easier the eventual consensus around the fully developed calendar that we know today. Nonetheless, the Karaite schism[80] and the Saadyah-Ben Meir controversy showed that aspects remained of the Babylonian versus Palestinian conflict. Ultimately, however, the Judaism of the Babylonian Talmud as codified by Maimonides became the normative approach to the Jewish calendar. Therefore, the calendar, in its completed form, making use of so much of the scientific developments of Late Antiquity, remained the dominant unifier of the Jewish people.

Notes

1. For a brief survey of the history of the calendar see Carlebach (2011), pp. 11–23.
2. See *pRosh Hashana* (henceforth *pRH*) 2:5 (58b), where it refers to the council that declares the intercalation of the month or year. See the article by S. Stern in this volume. Tractate Rosh Hashana is henceforth designated RH.
3. See Neusner (1993a), as well as his earlier critique of Sanders (1977) in Neusner (1978), p. 179. Neusner brings together his responses to Sanders in Neusner (1993b), especially pp. 1–10 for a methodological discussion. For a response to this methodology see Miller (2006), pp. 21–28.
4. Cf. Stern (2001), pp. 210, 234–235.
5. Cf. Sanders (1992), pp. 47–76.
6. See VanderKam (1998), p. 11, Talmon (1986).
7. The later Qumran sectarians (1QS 10:7) also cite four seasons: *qaṣir*, 'harvest', *qayiṣ*, 'summer fruits', *zeraʿ*, 'seed time' and *dešeʾ*, 'tender shoots'.
8. Aviv is in Exod 13:4; 23:15. Ziv is in 1 Kings 6:1, 37. Etanim is in 1 Kings 8:2. Bul is in 1 Kings 6:38. See also VanderKam (1998), p. 8.
9. De Vaux (1961), p. 183. See also VanderKam (1992). For Bul see KAI 14, 32, 38. For Ziv see RES 336 (Lidzbarski 99 = Slouschz 228). For *ʾtnm* see KAI 37A, 41. For Aviv, which is not mentioned in Phoenician menologies, see Licht (1958), pp. 37–38.
10. KAI 110:3; 137:5.
11. Stieglitz (1998), pp. 215–216. Cf. also figure 2, p. 221.
12. Stieglitz (1998). The Ugaritic ritual calendar was also lunar, perhaps following the official Mesopotamian one. Cf. Fleming (2000), p. 213. If the interpretation of this particular month name, *mpʿ lpny*, is incorrect, then it equally would be possible to assume that some form of a solar calendar similar to that used in Egypt might have obtained in Canaan. See Robbins (1995), pp. 1811–1812.
13. Contra Talmon, who built his argument on emending the medieval manuscript of the Damascus Document, a text later found among the Qumran scrolls: Talmon (1994). For further discussion see Ben-Dov and Saulnier (2008).
14. Licht (1958), pp. 37–38.
15. On these terms see Gandz (1949), pp. 259–263.
16. See the sources in Stern (2011), p. 48.
17. Clines (1974) believes that throughout pre-exilic times the New Year was in the spring; see pp. 23–26 for evidence for an autumnal New Year, and pp. 26–40 for Clines' rejection.

18. De Vaux (1961), pp. 189–193. See Jer 36:22 where the king is sitting in his winter house in the ninth month. The rabbis later tried to make sense of this confusion by pointing out that there were four different new years: Nissan for kings and festivals; Elul for cattle: Tishri the beginning of the count of the civil year and the year for sabbaticals and Jubilees; and Shevat, the new year of the trees (*mRH* 1:1). 11QT 14:9–18 has two new years, following Ezekiel 45:18-25. See discussion in Yadin (1983), v. 1, pp. 89–90. Note as well that the Book of Jubilees (6:23) mentions four divisions of the year and four memorial days to accompany them. See VanderKam (1998), p. 29. The Seleucids began their year in the fall. It is no wonder that the books of the Maccabees alternate between citing the beginning of the year in the spring and in the fall.
19. See VanderKam (1998), p. 9.
20. De Vaux (1961), pp. 184–185.
21. Nisan (Esth 3:7; Neh 2:1), Sivan (Esth 8:9), Elul (Neh 6:15), Kislev (Zech 7:1, Neh 1:1), Tebeth (Esth 2:16), Shebat (Zech 1:7), Adar (Esth 3:7, 13; 8:12, 9:1, 15, 17, 19, 21; Ezra 6:15). See VanderKam (1998), pp. 9–10.
22. For the use of these names in Judean Desert texts see Jacobus (2010), pp. 369–371.
23. De Vaux (1961), pp. 183–186.
24. Porten, et al. (1996), p. 152, n. 1.
25. Cf. Beckwith (1996), pp. 98–104, who considers the relationship between the biblical and sectarian calendars. For the opposite opinion, that the ancient Israelite calendar was the 364-day system, see also VanderKam (1998), p. 116, Jaubert (1953), p. 52, Guillaume (2004–2005), Guillaume (2009), pp. 33–121.
26. Cf. VanderKam (1998), pp. 17–33.
27. Ben-Dov (2008), pp. 34–37.
28. Ben-Dov (2008), pp. 119–125.
29. Ben Dov (2008), pp. 15–18, Stern (2001), pp. 5–9.
30. VanderKam (1998), pp. 27–33.
31. This includes astronomical sections from Enoch and Jubilees. For Enoch see Milik (1976), pp. 273–297. For Jubilees see VanderKam and Milik (1994).
32. Cf. Stern (2001), pp. 14–18, Talmon, Ben-Dov and Glessmer (2001), pp. 37–166.
33. See 4Q208 and 4Q209. For further detail see Drawnel (2011), pp. 71–133. See also Ben-Dov (2008), pp. 151–169, Beckwith (1996), pp. 113–120.
34. *The Books of Enoch: Aramaic Fragments of Qumrân Cave* 4, 274–284.
35. Stern (2001), p. 6. On this debate see also García Martínez (1992), pp. 45–96, Ben-Dov (2008), pp. 153–196 and 245–278, Drawnel (2011), pp. 15–30.
36. Wacholder (1990), pp. 257–281.
37. Drawnel (2011), pp. 39–53.
38. Yadin (1983), v. 1, pp. 116–119.
39. Talmon, Ben-Dov and Glessmer (2001), pp. 30–36; See 4Q320, 4Q321 and 4Q321a, pp. 37–91.
40. Baumgarten (1987), pp. 399–407; see also the article by Ben-Dov in this volume.
41. Cf. Schiffman (1994), pp. 301–305.
42. *pRH* 1:2 (56d).
43. Bickerman (1980), p. 23.
44. Cohen (1993), pp. 5–6.
45. *mRH* 1:4–3:1.
46. While in the Hasmonean Period it is likely that the Pharisees were not in control of the calendar, after the destruction of the Temple the only possible calendrical authority would have been the rabbis. Cf. Stern (2001), p. 153 who questions rabbinic control of the calendar in the tannaitic period.
47. Cf. VanderKam (2000).
48. Exod 23:16 and 34:22, Deut 16:9–10. Stern (2001), pp. 50–53. See also page 161 for agricultural support of intercalation. We follow the view that *tekufat hashanah* (Exod 34:22) does refer to the equinox.
49. Cohen (1993), pp. 5–6.

50. On the Mesopotamian calendar, see Fleming (2000), pp. 211–218.
51. Bickerman (1980), p. 24 attributes this to Artaxerxes II. For dating under Darius I see Assar (2003), p. 171.
52. Bickerman (1980), p. 29.
53. Stern (2001), pp. 27–28.
54. Bickerman (1980), pp. 38–40. See also Stern (2001), p. 33. For the possibility that Josephus used a version of the Macedonian lunar calendar see Stern (2001), pp. 8–34. Stern also suggests the possibility that Josephus may have simply borrowed the month names and used them as equivalent to the Babylonian month system that had been adopted by the Jews.
55. Stern (2001), pp. 42–46.
56. Stern (2001), p. 46.
57. Stern (2001), p. 47–62.
58. In *Hippocratis Epidemiarum Libros Commentarius* 1.1 (ed. Kühn, xvii/1.23). For Galen's comment in translation see Stern (2001), p. 23.
59. Regarding one polemical text, Stern (2001), p. 75 comments 'the polemical credibility of this argument would have depended on the authenticity of these Jewish dates; they are unlikely, therefore, to have been merely invented or forged'.
60. Stern (1994), pp. 49–55.
61. Such as availability of animals and crops for sacrifice, and the requirement that Passover take place on or after the vernal equinox; Stern (2001), pp. 161–162.
62. Stern (2001), pp. 87–98.
63. Gandz (1951), pp. 235–264. For this exact alignment see p. 236.
64. Cf. Stern (2001), pp. 241–252. For sources see ibid. p. 244, nn. 118–119.
65. Stern (2001), pp. 249–252). The tradition that diaspora Jews observe two days of festivals may be evidence that Babylonian Jews accepted the authority of the Land of Israel in establishing Rosh Ḥodesh, the day of the new moon and the beginning of the lunar month. Since the day of Rosh Ḥodesh often was not communicated to them in time, they observed the festivals based on both possibilities, namely, that the prior lunar month was either 29 or 30 days long. In order to continue this tradition, when the fixed calendar was established, it was designed so that the months preceding the festivals, Adar and Elul, are always 29 days long. Because of the difficulties in reaching the court faced by witnesses seeking to testify, New Year was celebrated for two days even in the Land of Israel.
66. Stern (2001), pp. 124–154.
67. Stern (2001), pp. 175–179.
68. Cf. the texts quoted in Stern (2001), pp. 81–85. See also p. 236.
69. For example, the first day of the New Year cannot occur on Sunday, in addition to Wednesday and Friday already prohibited by the Palestinian Talmud (*ySukkah* 4:1 [54b]). See Stern (2001), pp. 194–195. Another rule states that Passover cannot occur on Monday, Wednesday, or Friday. See Stern (2001), p. 192. See also Wiesenberg (2007), pp. 354–358.
70. *bEruvin* 56a. See Stern (2001), p. 199.
71. See Maimonides, *Code: Law of Sanctification of the New Moon* 9:1, and cf. 6:3. He gives 364 and ¼ (exactly) as the solar year in 6:4. For the earliest attribution of this statement to Rav Adda see Avraham b. Ḥiyya (1070–1136), *Sefer ha-ʿIbbur* 3:3-5; Stern (1996), pp. 10–105. On the esoteric understanding of this tradition, see Halbertal (2007), p. 42.
72. Despite the fact that this number is found in *bRH* 25a, Maimonides would not have derived it from here because it is an interpolation. It was most likely derived from Ptolemy's *Almagest*. See Stern (2001), pp. 201–202 and 207–209.
73. For a third possible rabbinic view on the length of the solar year, see Beller (1998), pp. 91–98.
74. See note 69 above.
75. Stern (2001), pp. 264–275.
76. Stern (2001), pp. 182–186.
77. Avraham b. Ḥiyya's *Sefer ha-ʿIbbur* did not attain authoritative status in Jewish tradition. The Four Parts

Table [*Luaḥ ʾArbaʿah Sheʿarim*] also seems never to have gained real authoritative status. See Stern (2001), pp. 268–270.

78. Beller (1988), pp. 51–66.
79. Stern (2001), p. 206.
80. One of the arguments of the Karaites against the Rabbinites was that the Karaite calendar was still observational and fixed according to Aviv (Stern 2001), p. 235.

References

Assar, G. R. F., 2003, "Parthian Calendars at Babylon and Selucia on the Tigris", *Iran* 41, 171–191.
Baumgarten, J. M., 1987. "4Q503 (Daily Prayers) and the Lunar Calendar", *Revue de Qumran* 12, 399–407.
Beckwith, R. T., 1996, *Calendar and Chronology, Jewish and Christian* (Leiden: Brill).
Beller E., 1988, "Ancient Jewish Mathematical Astronomy", *Archive for History of Exact Sciences* 38, 51–66.
——, 1998, "A Newly-Discovered Ancient Value for the Length of the Year", *Archive for History of Exact Sciences* 52, 91–98.
Ben-Dov, J., 2008, *Head of All Years: Astronomy and Calendars at Qumran in Their Ancient Context* (Leiden: Brill).
—— and S. Saulnier, 2008, "Qumran Calendars: A Survey of Scholarship: 1980–2007", *Currents in Biblical Research* 7, 124–168.
Bickerman, E. J., 1980, *Chronology of the Ancient World* (2nd ed.; Ithaca, NY: Cornell University Press).
Carlebach, E., 2011, *Palaces of Time: Jewish Calendar and Modern Europe* (Cambridge: Belknap).
Clines, D. J. A., 1974, "The Evidence for an Autumnal New Year in Pre-Exilic Israel Reconsidered", *Journal of Biblical Literature* 93, 22–40.
Cohen M. E., 1993, *The Cultic Calendars of the Ancient Near East* (Bethesda; CDL Press).
de Vaux, R., 1961, *Ancient Israel: Its Life and Institutions* (trans. J. McHugh; London: Darton, Longman & Todd).
Drawnel, H., 2011, *The Aramaic Astronomical Book (4Q208-4Q211) From Qumran: Text, Translation, and Commentary* (Oxford and New York: Oxford University Press, 2011).
Fleming, D. E., 2000, *Time at Emar: The Cultic Calendar and the Rituals from the Diviner's House* (Winona Lake, IN: Eisenbrauns).
Gandz, S., 1949, "Studies in the Hebrew Calendar I: A Study in Terminology", *Jewish Quarterly Review* 39, 259–263.
——, 1951, "The Problem of the Molad", *Proceedings of the American Academy of Jewish Research* 20, 235-64.
García Martínez, F.,1992, *Qumran and Apocalyptic: Studies on the Aramaic Texts from Qumran* (Leiden: Brill).
Guillaume, P., 2004–05, "Tracing the Origin of the Sabbatical Calendar in the Priestly Narrative (Genesis 1 to Joshua 5)", *Journal of Hebrew Scriptures* 5[Internet Journal: http://www.arts.ualberta.ca/JHS/Articles/article_43.htm].
——, 2009, *Land and Calendar: Priestly Calendar from Genesis 1 to Joshua 18* (New York: T & T Clark).
Halbertal, M., 2007, *Concealment and Revelation: Esotericism in Jewish Thought and Its Philosophical Implications* (trans. J. Feldman; Princeton: Princeton University Press).
Jacobus, H., 2010, "4Q318: A Jewish Zodiac Calendar at Qumran", in C. Hempel (ed.), *The Dead Sea Scrolls: Text and Context* (Leiden: Brill), 365–395.
Jaubert, A., 1953, "Le calendrier des Jubilés et de la secte de Qumrân: ses origins bibliques", *Vetus Testamentum* 3, 250–264.
Licht, J. S., 1958, "ḥodesh", *Encyclopedia Miqraʾit* 3, 36–41 (Hebrew).
Miller, S., 2006, *Sages and Commoners in Late Antique ʿEreẓ Israel: A Philological Inquiry into Local Traditions in Talmud Yerushalmi* (Tübingen: Mohr Siebeck).
Milik, J. T., 1976, with the collaboration of M. Black, *The Books of Enoch: Aramaic Fragments of Qumrân Cave 4* (Oxford: Clarendon Press).
Neusner, J., 1978, "Comparing Judaisms", *History of Religions* 18, 177–191.

——, 1993a, "The Mishnah in Philosophical Context and Out of Canonical Bounds", *Journal of Biblical Literature* 112, 296–301.

——, 1993b, *Judaic Law from Jesus to the Mishnah: A Systematic Reply to Professor E.P. Sanders* (Atlanta: Scholars Press).

Porten, B., et al., 1996, *The Elephantine Papyri in English, Three Millennia of Cross-Cultural Continuity and Change* (Leiden: Brill).

Robins, G., 1995, "Mathematics, Astronomy, and Calendars in Pharaonic Egypt", in J. M. Sasson (ed.), *Civilizations of the Ancient Near East* (New York: Charles Scribner's Sons), 3, 1799–1813.

Sanders, E. P., 1977, *Paul and Palestinian Judaism: A Comparison of Patterns of Religion* (London: SCM).

——, 1992, *Judaism: Practice and Belief 63 BCE–66 CE* (London: SCM).

Schiffman, L. H., 1994, *Reclaiming the Dead Sea Scrolls: The History of Judaism, the Background of Christianity, the Lost Library of Qumran* (Philadelphia and Jerusalem: The Jewish Publication Society).

Stern, S., 1994, "The Second Day of Yom Tov in Talmudic and Geonic Literature", *Proceedings of the World Congress of Jewish Studies* 11, Vol. C. 1, 49–55.

——, 1996, "Fictitious Calendars: Early Rabbinic Notions of Time, Astronomy, and Reality", *Jewish Quarterly Review* 87, 105–110.

——, 2001, *Calendar and Community: A History of the Jewish Calendar Second Century BCE-Tenth Century CE* (Oxford: Oxford University Press).

Stieglitz, R. R., 1998, "The Phoenician-Punic Menology", in M. Lubetski, C. Gottlieb, and S. Keller (eds.), *Boundaries of the Ancient Near Eastern World: A Tribute to Cyrus Gordon* (Sheffield: Sheffield Academic Press), 211–221.

Talmon, S., 1986, "The Gezer Calendar and the Seasonal Cycle of Ancient Canaan", in *King, Cult and Calendar in Ancient Israel: Collected Studies* (Jerusalem: Magnes), 89–112.

——, 2004, "The Reckoning of the Day in Biblical and Early Post-Biblical Periods: From Morning or From Evening?", in S. Japhet (ed.), *The Bible in the Light of its Interpreters: Sarah Kamin Memorial Volume* (Jerusalem: Magnes), 109–129 (Hebrew).

——, Ben-Dov, J., and Glessmer, U., 2001, *Qumran Cave 4. XVI: Calendrical Texts* (Oxford: Clarendon Press).

VanderKam, J. C., 1992, "Calendars", *Anchor Bible Dictionary* 1, 814–820.

——, 1998, *Calendars in the Dead Sea Scrolls: Measuring Time* (London and New-York: Routledge).

——, 2000, "Festivals", in L. H. Schiffman and J. C. VanderKam (eds.), *Encyclopedia of the Dead Sea Scrolls* (Oxford and New York: Oxford University Press), 1, 290–292.

—— and Milik, J. T., 1994, "Jubilees", in H. Attridge et. al. in consultation with J.C. VanderKam (eds.), *Qumran Cave 4. VIII: Parabiblical Texts Part 1* (Oxford: Clarendon, 1994), 1–140.

Wacholder, B. Z., 1990, "The Ancient Judeo-Aramaic Literature (500–165 BCE), a Classification of a Pre-Qumranic Text", in L. Schiffman (ed.), *Archeology and History in the Dead Sea Scrolls: The New York University Conference in Memory of Yigael Yadin* (Sheffield: Sheffield Academic Press), 257–281.

Wiesenberg, E. I., 2007, "Calendar", in F. Skolnik and M. Berenbaum (eds.), *Encyclopedia Judaica* (2nd ed.; New York: Macmillan), 4, 354–358.

Yadin, Y., 1983, *The Temple Scroll* (3 vols.; Jerusalem: Israel Exploration Society, The Institute of Archaeology of the Hebrew University of Jerusalem, The Shrine of the Book).

The Harmonization of the Lunar Year with the Julian Calendar by Anatolius, Bishop of Laodicea

Daniel P. Mc Carthy

Introduction – The Development of the Christian Computus

The medieval Christian computus was the science of the reckoning of the days of the week, of the sun, and of the moon, and of identifying the festival days for celebration within their separate calendars, especially that of the celebration of the Pasch. This developed directly from Christian beliefs arising from the life and death of Jesus of Nazareth. Specifically, because they believed that Jesus had resurrected three days after he had been crucified, Christians wished to celebrate these two events which they saw as demonstrating Jesus' triumph over death, and so they hoped for their own personal resurrection after death. As Christianity spread from its Judaic origin the recipient communities had first to resolve their interpretation of the received account of the chronology of Jesus' crucifixion and resurrection. They had then to appropriately place their celebration of these events within their own calendrical framework, typically the Julian calendar in the Latin-literate western empire, while in the Greek-literate eastern empire generally months with Macedonian names were arranged to maintain synchronization with the Julian calendar. Unfortunately for Christians, however, there existed no authoritative account of either the chronology of Jesus' crucifixion or resurrection, or indeed of how these events should be interpreted. In consequence what happened was a slow evolution over about eight centuries during which leading Christian philosophers contributed their own rationalizations, until by about the end of the ninth century stability of practice, but not uniformity, had been achieved. Examples of such contributions are: the Paschal table of Hippolytus † c. 236, the *octaëteris* of Dionysius † c. 247, the Paschal tract of Anatolius † c. 282, the Paschal letters of Athanasius † 373, the Paschal letter and 19-year table of Theophilus † 412, the *Latercus* of Sulpicius Severus † c. 420, the 532-year Paschal table of Victorius of Aquitaine † post 457, the Paschal letters and tract of Dionysius Exiguus † c. 550.[1]

These contributions had necessarily to commence with the four Gospels which were considered the most authoritative accounts of Jesus' life. These agreed that Jesus had been crucified on the sixth feria of the Jewish week, and that he had risen again three days later, counted inclusively. They also agreed that Jesus was crucified at the time of the Jewish celebration of Passover, but they conflicted on their relativity; John's Gospel placed it on Nisan 14, while the three Synoptic Gospels placed it on Nisan 15.[2] Because the crucifixion was located by the Gospels with reference to the Jewish celebration of Passover, and indeed early Christians identified Jesus' crucifixion with the sacrifice of the Paschal lamb, the Biblical authority for the Passover consequently represented a crucial authority for the

Christian Pasch. Thus essentially Jewish calendrical precepts such as the seven-day week, the Jewish nineteen-year lunar calendar, and Nisan, its first month, provided the elements of the early Christian computus. To these the Gospels added the details that the crucifixion had occurred on sixth feria, a Friday, on either Nisan 14 or 15, and the resurrection on first feria, a Sunday, on either Nisan 16 or 17. Thus it was these cryptic accounts, already incorporating some conflict, that had to provide the starting point for the rationalizations of the medieval Christian philosophers.

As Christianity spread to non-Jewish communities some of these Jewish elements were fairly quickly relinquished, specifically the Jewish nineteen-year lunar calendar and its determination of the first month, and these were replaced with alternative lunar calendars of cycle lengths variously 8-year, 19-year, and 84-year. Because Nisan falls around the time of the spring equinox, so the location of the first month in these alternative cycles was determined from the spring equinox. This equinox was located variously on 21, 22, or 25 March in the Julian calendar, or on 25 or 26 Phamenoth in the Alexandrian calendar which are equivalent to Julian 21 or 22 March. The 22 March date accorded with that given by Ptolemy in his *Almagest* where he estimated the spring equinox for 140 CE to be on Pachon 7 [i.e. 22 March], at about 1 pm. But since he also estimated a one day advance in the equinox in approximately 300 years, by the fourth century this 22 March date had necessarily to be adjusted to 21 March.[3] On the other hand, the 25 March date was the traditional date of the Roman equinox on the eighth kalends of April in the Julian calendar. Thus the initial ambiguity concerning the lunar age at the crucifixion was substantially expanded with alternative lunar cycles and a multiplicity of equinoctial dates. Further, after the Council of Nicaea in 325 it was generally accepted that the crucifixion should always be celebrated on a Friday, thereby requiring a seven-day range of lunar dates for both it and the celebration of resurrection on the ensuing Sunday. These lunar ranges for Easter Sunday varied from luna 14–20, to luna 15–21, to luna 16–20, according to the interpretation of the crucifixion-resurrection that was chosen. These diverse modifications of the initial Biblical-Gospel principles were introduced incrementally, often in a context of severe controversy, and each one added further complexity to the problem of reckoning on just which day the Pasch should be celebrated. It was this expanding corpus of Paschal tracts and calendrical computational theory that became known as the 'computus', and it grew to a formidable scale and complexity. For example, the eleventh-century Sirmond manuscript, Oxford University, Bodleian MS 309, is considered by scholarship to be the best index to the sources used by Bede † 735, the scholar monk of Northumbria whose computistical works came by the ninth century to dominate the monastic schools of all of western Europe.[4] The computus section of the Sirmond manuscript comprises 162 folios which contain copies of around twenty tracts written by earlier Christian philosophers, as well as accounts of the theory underlying the seven-day week, the Julian calendar, and the nineteen-year lunar calendar.[5] In 1943, Charles Jones, the editor of Bede's computistical works, pointed out that this knowledge of computus had come to Bede from Ireland, writing, 'Bede's works indicate that Northumbrian schools largely depended upon Irish knowledge and methods'.[6] Since then the research published by Ó Cróinín, Mc Carthy, Holford-Strevens, Warntjes and Bisagni has confirmed this, and greatly clarified the role of Irish computists in the transmission and development of this corpus.

Thus it was the task of the Christian computists to use the theory of this corpus to compute a date for the Pasch that satisfied their chosen interpretation in all three calendars;

namely, celebration of the resurrection falling on a Sunday in the ferial calendar, falling after the date of the equinox in the solar calendar, and falling between the designated lunar limits in the lunar calendar. Regarding the relationships between these various calendars, since none of them share a common factor, these relationships must change from year to year. For example, common solar years contain 365 days and since $365 = 52 \times 7 + 1$ then successive common years must commence one day later in the week than the preceding year. But every fourth year is bissextile and contains 366 days, so that the year following a bissextile year must commence two week days later. In consequence the complete relationship between the ferial and solar calendars extends over twenty-eight years, and this is known as the solar cycle. However, the most complex relationship is that found between the solar and the lunar calendars. This complexity arises because one solar year exceeds twelve lunar months by approximately eleven days, so that the relationship between solar and lunar months must shift by about eleven days from year to year. The Christian philosophers made various attempts to simplify the computation of this relationship, and it is that made by Anatolius, bishop of Laodicea, that forms the subject of this paper.

The Julian Year Versus a Lunar Year of Alternating Full and Hollow Months

The structure of the Julian year is familiar to everyone today who employs its Gregorian successor to schedule their affairs, for this retained the same Julian months. The lengths of these months are often memorised by a mnemonic verse such as, 'Thirty days hath September, April, June and November, All the rest have thirty-one, except for February alone, Which has twenty-eight rain or shine, but in a leap year twenty-nine'. We may conveniently represent this scheme graphically for a common year of 365 days, leaving the 31-day months clear, the 30-day months shaded, and cross-hatching the exceptional February, as illustrated in figure 1.

Jan	Feb	Mar	Apr	May	Jun	Jul	Aug	Sep	Oct	Nov	Dec
31	28	31	30	31	30	31	31	30	31	30	31

FIGURE 1. The twelve Julian months and their lengths in a common year, with 31-day months shown clear, the 30-day months shown shaded, and the exceptional February shown cross-hatched.

Turning to the lunar year, any calendar must approximate the lunar synodic month of 29.531 days, and in medieval lunar calendars the first approximation to this interval was achieved by alternating a full lunar month of thirty days with a hollow lunar month of twenty-nine days, so that the average lunar month of this bimenstrual period is 29.5 days. Six of these bimenstrual periods were then assembled to make a lunar year of 354 days with six alternating full and hollow months as illustrated in figure 2, showing the full months plain and the hollow months shaded.

30	29	30	29	30	29	30	29	30	29	30	29

FIGURE 2. The standard lunar year composed of six 30-day full months shown clear, alternating with six 29-day hollow months shown shaded.

This is, of course, the basic structural arrangement of the Metonic Cycle, and of the Christian, calculated Hebrew, and calculated Islamic lunar calendars, and so it effectively represents the standard computed lunar year for the entire Mediterranean basin.[7] Next we consider a Julian year and a lunar year which begin together on the same day, and these are illustrated in figure 3.

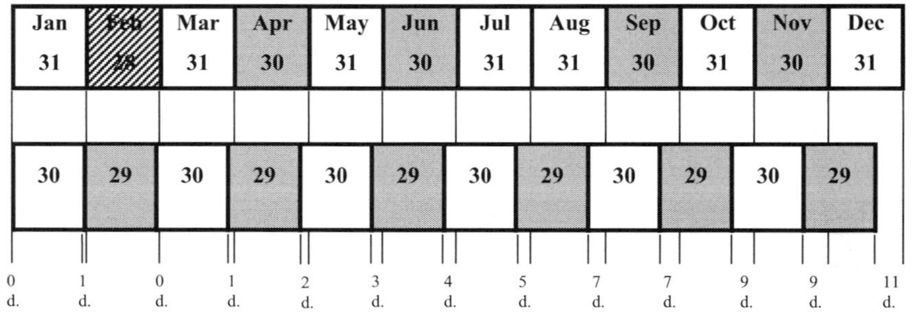

FIGURE 3. The Julian and standard lunar years are shown starting together, while at the bottom are shown the difference in days between the start of each Julian and lunar month.

Here, because Julian January has one day *more* than a full lunar month, there must be a one day difference between the start of February and the start of the second lunar month. Consequently the age of the moon on the kalends of February, i.e. 1 February, must be one day *older* than at the kalends of January, and hence a one day difference is shown below. Next, because Julian February has one day *less* than a hollow month, then both February and the second lunar month must finish on the same day, so that the age of the moon on the kalends of March must equal that at the kalends of January, which datum is known as the epact. Hence a difference of zero is shown below. Next, because Julian March has one day *more* than a full lunar month, March must finish one day later than the third lunar month, so that the age of the moon on the kalends of April must be one day *older* than at the kalends of January. Hence a difference of one day is shown below. Then, because the Julian months of April, May, June, and July each have one day more than their corresponding lunar month, then the age of the moon from the kalends of May through to the kalends of August must increase linearly from two through three through four to five as shown below. However, because Julian August has *two* days more than a hollow month, then August must finish seven days after the eighth lunar month, so that the age of the moon on the kalends of September must be seven days older than that at the kalends of January as shown below. Thus between the kalends of August and September there is a jump of *two* days in the age of the moon. After September, the increments to the age of the moon at the kalends of October through to January must increment non-linearly so that the differences go from seven to seven, to nine, to nine, to eleven, as shown below. In summary, while the age of the kalends moon varies linearly from Julian March to August, it is non-linear from September to December. In graphical terms, as long as a Julian clear month is aligned with a lunar clear month, or a Julian shadowed month is aligned with lunar shadowed month there is a unit increment in the kalends moon. On the other hand, whenever a Julian clear month is aligned with a shadowed lunar month, or vice-versa, then the increment must be either two or zero, and consequently linearity is lost.

The inevitable consequence of this is that for *any* lunar calendar with alternating full and hollow months there can be no extensive linear relationship between the age of the kalends moon and the Julian month. The resultant erratic sequence is an inescapable consequence of the Julian placing of the thirty-one day months after July, and this inevitably complicated the computation of the age of the moon on the kalends of the Julian months. For example, Bede in chapter twenty of his *De temporum ratione*, having presented a procedure for computing the age of the kalends moon for a year with any epact, added, 'Should anyone object that the order of either this or the preceding formula is unsound at any point, then let him teach a more accurate and handy formula for investigating questions of this kind, and we will gladly and gratefully accept it'.[8] However, about four and half centuries before Bede, Anatolius of Laodicea had in fact provided for a simpler computation, and I now turn to examine this.

Anatolius and his Works

Our knowledge of Anatolius derives entirely from the account given by Eusebius † c. 340, bishop of Caesarea, in his *Ecclesiastical History* (EH) completed in c. 326. In EH vii, 32.6 Eusebius first described Anatolius' secular achievements as follows:

> [Anatolius] was by race an Alexandrian who for his learning, secular education and philosophy had attained the first place among our most illustrious contemporaries; inasmuch as in arithmetic and geometry, in astronomy and other sciences, whether logic or physics and in the arts of rhetoric as well, he has reached the pinnacle. It is recorded that because of these attainments the citizens there deemed him worthy to establish the school of Aristotelian succession at Alexandria.[9]

Following this Eusebius gave a protracted account of Anatolius' role in securing the release of many Christians and others from Pyrucheum, the quarter of Alexandria besieged by a Roman army in c. 264. Not long afterwards Anatolius was ordained bishop by Theotecnus, bishop of Caesarea, with whom he served for a time as his co-adjutor, and then he became bishop of Laodicea in Syria when he passed through there enroute for the synod held in Antioch in 268 to deal with Paul of Samosate, and he died in c. 282.[10]

Regarding Anatolius' literary output, Eusebius EH vii, 32.13 refers to 'the Canons of Anatolius on the Pascha', and EH vii, 32.20 to his 'Introductions to Arithmetic also in ten complete treatises, and, as well, evidences of his study and deep knowledge of divine things'.[11] However, in the Latin translation of the *Ecclesiastical History* compiled by Rufinus of Aquiliea † c. 410, he rendered EH vii, 32.13 as 'Many distinguished treatises, also, composed by the said Anatolius have come down to us', and EH vii, 32.20 as, 'Anatolius left behind many other writings which were admired not only by religious men but also by philosophers', so it seems clear that a substantial corpus of Anatolius' work survived until Rufinus' time. Of these only fragments of his Introduction to Arithmetic have survived, but his Paschal tract has been preserved in the Latin translation of *De ratione paschali* (DRP), and it is this work that is relevant to the present discussion.[12]

DRP was cited and referenced by medieval Insular scholars such as Columbanus † 615, founder of Bobbio, and Colmán † 676 at the Synod of Whitby, when defending their adherence to the *Latercus*. This was the 84-year Paschal table compiled by Sulpicius Severus

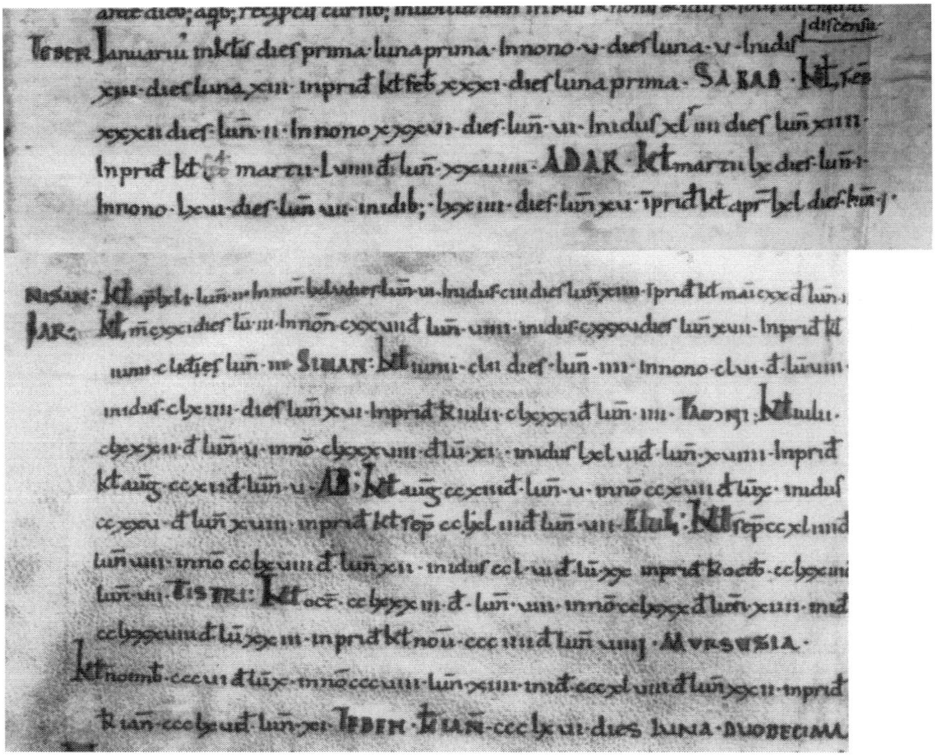

FIGURE 4. Padua Bibl. Antoniana MS I.27 ff. 74r–v, showing Anatolius' tabulation of the Hebrew months against the day count and lunar ages of the Julian kalends, nones, ides, and pridies, cf. Mc Carthy and Breen (2003), pp. 59–60. (Reproduced with the permission of the Biblioteca Antoniana, Padua.)

in c. 410, which was the Paschal tradition followed by some Insular churches from the fifth to the eighth centuries. It was also cited by those who were opposed to the *Latercus* such as Cummian † post 632, and Bede in his computistical works.[13] However, in 1733 Van der Hagen, as part of his treatise on 84-year Paschal cycles, dismissed DRP with the hypothesis that it was a forgery and he designated its author as 'pseudo-Anatolius'.[14] In 1736 he further asserted that DRP had been written in the first part of the seventh century by an author who was most likely Scottish or Irish.[15] In the subsequent 250 years most scholars effectively just repeated Van der Hagen's hypothesis as if it were fact; for example, Ideler (1825–6), Krusch (1880), Anscombe (1895), Turner (1895), Mac Carthy (1901), Nicklin (1903), Schwartz (1905), O'Connell (1936), Jones (1943), Cordoliani (1945–6), and Van de Vyver (1957) all referred to its author as 'pseudo-Anatolius'.[16]

However, in 1985 Dáibhí Ó Cróinín discovered in the manuscript Padua, Biblioteca Antoniana I. 27 ff. 76r–77v a full copy of the *Latercus*, the 84-year Paschal table compiled by Sulpicius Severus. Ó Cróinín's discovery was the first recorded sighting of this *Latercus* for over 1200 years, and the subsequent analysis of its structure and the comparison of this with DRP demonstrated that in fact DRP was an authentic translation of the Paschal tract of Anatolius.[17] In consequence, in 2003 Dr Aidan Breen and I published a critical edition of DRP.[18] Whilst only one section of DRP is relevant to this paper, a brief survey of its

Months		Kalends		Nones		Ides		Pridies	
Hebrew	Julian	day	luna	day	luna	day	luna	day	luna
TEBER	Jan.	1	1	5	5	13	13	31	1
SABAD	Feb.	32	2	36	6	44	14	59	29
ADAR	Mar.	60	1	66	7	74	15	90	2
NISAN	Apr.	91	3	95	7	103	15	120	3
IAR	May	121	4	127	10	135	18	151	4
SIUAN	Jun.	152	5	156	9	164	17	181	5
TAMNI	July	182	6	188	12	196	20	212	6
AB	Aug.	213	7	217	11	225	19	243	7
ELUL	Sep.	244	8	248	12	256	20	273	8
TISTRI	Oct.	274	9	280	15	288	23	304	9
MURSUSIA	Nov.	305	10	309	14	317	22	334	10
–	Dec.	335	11	339	15	347	23	365	11
TEBER	Jan.	366	12	–	–	–	–	–	–

Figure 5. Anatolius' tabulation of the Hebrew months, the Julian kalends, nones, ides, and pridies, giving for each their day count from January 1 followed by his age of the moon, cf. Mc Carthy and Breen (2003), p. 68.

content may help the reader to appreciate the typical content of the tracts included in the medieval computus discussed in the introduction.

DRP may be subdivided into six sections of which the following represents a brief summary:

1. A short review of Greek and Hebrew writing on the Pasch
2. An account of the principles determining the Jewish Pasch
3. An exegesis of the Christian Pasch and Anatolius' rejection of Roman and Gaulish Christian Paschal principles, while endorsing Asian Johannine principles
4. A tabulation of Anatolius' lunar year with the Julian calendar
5. A nineteen-year Paschal table
6. Theory of the annual ascent and descent of the sun

Of these it is section four, Anatolius' tabulation of his lunar year with the Julian calendar, that is relevant to this paper, and so in figure 4 I reproduce this table from the manuscript of *De ratione paschali* in Padua Biblioteca Antoniana I.27, where it immediately precedes the copy of the Latercus discovered by Ó Cróinín.

As may be seen, Anatolius first cited the Jewish month commencing at the tenth month, Teber, which he synchronised with the Julian month of January. He followed this by tabulating the day count and the lunar age at the Julian kalends, nones, ides, and pridies of the Julian months from January through to the January of the following year. Since it is not easy to follow these details in the manuscript in figure 5 is given the tabulation of this data from the edition of Mc Carthy and Breen.[19]

From this we can see that Anatolius gave the inclusive day count from the kalends of January, and his age for the moon on the day of the kalends, nones, ides, and pridies for each of the twelve Julian months, and he then concluded at the kalends of January for the following year. Now if we examine the ages of the moon, month by month, we find that the series for the kalends of January substantially represents a linear series, except for the single discontinuity in the month of March. In particular, these kalends moons show a linear increase from January to February, and then from April through to the following January, that is it goes from one to two, then from three to four to five etcetera through to twelve. Now this linear sequence for the age of the moon on the kalends of January has immediate implications for the structure of the lunar year assumed by Anatolius, and if we abstract this lunar year and represent it graphically we find that it is as illustrated in figure 6.

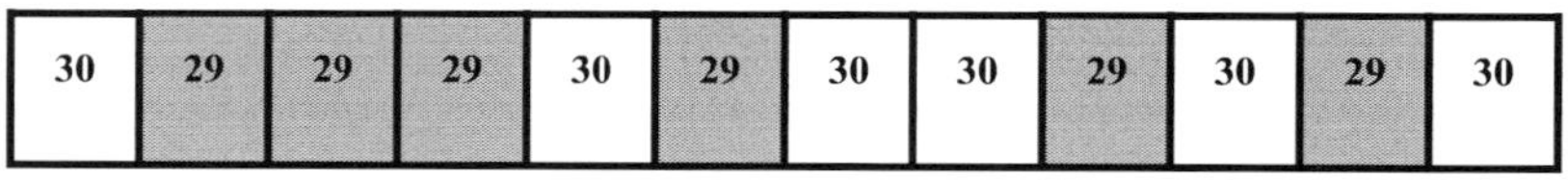

FIGURE 6. The twelve months of the Anatolian lunar year, comprising six 30-day full months and six 29-day hollow months not arranged in alternating order.

Thus Anatolius' lunar year comprised six full and six hollow months and so extended for exactly 354 days, but the distribution of these months varied very significantly from the standard alternating full and hollow sequence. In particular we find a sequence of three successive hollow months and a sequence of two successive full months, and at first sight this may seem to be a completely arbitrary arrangement. However, the basis for Anatolius' design becomes apparent when we set his lunar year alongside the Julian year, as is illustrated in figure 7.

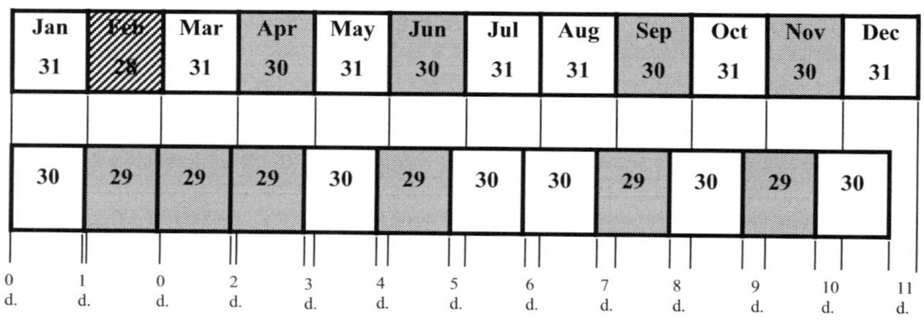

FIGURE 7. The Julian and Anatolian lunar years are shown starting together, while at the bottom are shown the difference in days between the start of each Julian and lunar month.

When we repeat the exercise of computing the difference in days between the end of each Julian month and the corresponding lunar month, we find a linear sequence from January to February, and from April through to the following January as shown below. When we examine the relationship between the Julian and the lunar months we find that, except for March, Anatolius aligned his full lunar months with the 31-day Julian months, and his hollow lunar months with the short, 28 or 30-day Julian months, as may be seen from the alignment between the clear and shaded months in figure seven. In this way Anatolius achieved a linear relationship between the age of the moon on the kalends of January, that is the epact, and the age of the moon on the kalends of the months April through to the following February. Thus Anatolius harmonized, as far as is possible, the months of his lunar year with the Julian months. The first person to identify this unique relationship between Anatolius' lunar months and the Julian calendar was Eduard Schwartz in 1905 in his *Christliche und jüdische Ostertafeln*, but unfortunately, having done this, he dismissed the work as fraudulent.[20] However, Ó Cróinín's discovery of a copy of the *Latercus* has shown that Schwartz was mistaken in his judgement, for analysis revealed that the Latercus was constructed using Anatolius' lunar year. In fact this linear relationship is only one of a number of instances of Anatolius employing linear sequences in DRP, for he also used these to represent the advances of the hours, to summarise the lunar days of the Paschal week, to represent the change in the ratio of day-to-night hours, and the change in solar elevation over the course of a solar year. It is clear from the extent of this usage, therefore, that Anatolius considered such linear progressions an essential tool of representation.[21]

The Purpose of Anatolius Harmonization

Anatolius' harmonization of his lunar year with the Julian year resulted in a linear relationship between the epact, the age of the moon on the kalends of January, and its age on the kalends of the months April through to the following February. This relationship thereby provided a straightforward means of computing the age of the moon on any day of the year. That this relationship was both known and used is confirmed by the survival of rules exploiting this linear relationship which have continued in use into modern times. One of these was published in c. 1922 by Canon Peter O'Leary in a book dealing with Irish numerals in which he cited a verse that he had evidently learned when he was a child in south-western Ireland around 1850. This provides a formula with which to calculate the age of the moon on any day in any month after March as follows:[22]

Cómhrímh síos ón Márta mbán,	Count from the beginning of March,
Go dtí an mí n-a mbeidh tú ann.	Down to the month in which you are.
Cuir aon fé n-a gceann, lá an mhí,	Put one less than the day of the month,
Agus an t-epacht.	And then the epact.
Aon nidh fé bhun nó os cionn trí dheich,	Then anything below or over thirty,
Sin agat aois na rae.	There you have the age of the moon.

This verse introduces the following four quantities: C, the inclusive count of months from March to the month of interest;[23] D, the day of that month; E, the epact or the moon's age on the kalends of March; A, the age of the moon on the date of interest. From the first three, the moon's age is to be calculated by taking the remainder on thirty of the sum of C, D−1, and E as follows:

$$A = [C + (D - 1) + E] \text{ remainder on } 30$$

We note first of all that the date of the epact is implicit from the fact that on the kalends of March, D = 1 and C = 0, so that A = 0 + (1 – 1) + E = E, that is, the epact is the age of the moon on the kalends of March, which, as we have seen, equals the age on the kalends of January. Secondly, it can be seen that the age of the moon on the kalends of the months following March is linearly dependent upon C, the month count, just as in Anatolius' tabulation. For example, in a year with epact luna 27 then the age of the moon on the subsequent kalends will be computed as:

$$A = [C + (1 - 1) + E] \text{ remainder on } 30 = [C + E] \text{ remainder on } 30.$$

Tabulation of this for E=27 yields the values:[24]

Kalends of	Mar	Apr	May	Jun	Jul	Aug	Sep	Oct	Nov	Dec
Count C =	0	2	3	4	5	6	7	8	9	10
Age A =	27	29	30	1	2	3	4	5	6	7

The increase by two from March to April is a consequence of the inclusive counting from March, but thereafter the increments are all by a unit except in June when the remainder on 30 results in a decrement of twenty-nine. To compute the age of the moon on a specific date, say 11 October, in such a year with epact luna 27, we have C=8 and D=11 and compute the age A as follows:[25]

$$A = [8 + (11 - 1) + 27] \text{ remainder on } 30 = 45 \text{ remainder on } 30 = \text{luna } 15$$

That these principles were known in other parts of Europe is demonstrated by the survival to the present day of a somewhat similar verse in the Italian Alps in the Bergamasque language as follows:[26]

Per sai' la luna de 'nko	In order to know the moon of today
ciapa ol de de 'nko,	take the day of today,
pio ol numer del mis,	add the number of the month,
pio la pata de st'an;	add the epact of this year;
se le pio de trenta,	if [the result] is more than thirty,
tira 'ndre trenta.	subtract thirty.

Here the 'number of the month' is implicitly reckoned from March, and in the computation the age of the moon is again made to depend linearly upon this number.

Conclusions

One might have expected, given Anatolius' highly-regarded mathematical and calendrical skills, that his lunar calendar and Paschal table would have found Christian followers in his own area of the eastern Mediterranean, but I know of no survival in these parts. The likely reason for this is that the Council of Nicaea in 325 rejected the Johannine Paschal principles which employed luna 14 on which Anatolius had based his tract. However, we do know that Anatolius' Paschal tract was welcomed in southern Gaul, where, in the early fifth century, Sulpicius Severus incorporated Anatolius' lunar calendar and Paschal principles into his 84-year *Latercus*. Then, with the collapse of the western Roman empire in the early fifth century, Sulpicius' Latercus was driven into the remote western and northern parts of Britain and to Ireland, and it was here that Anatolius' lunar calendar and Paschal principles found their most fervent followers. It is an interesting example of the unexpected quirks of transmission of ideas that techniques formulated on the shores of the eastern Mediterranean should have had to travel for over four thousand kilometres to find fertile soil in which to flourish.

Thus, to summarise:

1. Anatolius, by skilfully aligning his full months with six of the Julian 31-day months, achieved a linear relationship between the epact on the kalends of January, and the age of the moon on the kalends of April through to the following February. In this way he harmonized, as far as is possible, his lunar calendar with the Julian calendar.
2. Anatolius' lunar calendar and Paschal principles travelled by way of Sulpicius Severus in southern Gaul to the remoter parts of the British Isles, and here they flourished during the fifth, sixth, and seventh centuries.[27]

Notes

1. Jones (1943), pp. 3–77 provides the most comprehensive survey in English of the evolution of computus, but his account of Insular computus in pp. 78–113 is frequently obsolete and must be corrected by reference to the publications of Ó Cróinín (1982a, 1982b, 1983, 1997, 2003a, 2003b), Walsh and Ó Cróinín (1988), Mc Carthy and Ó Cróinín (1987–8), Mc Carthy (1993, 1994, 1996, 2008, 2010, 2012), Mc Carthy and Breen (2003), Warntjes (2005, 2007, 2009, 2010b), Bisagni and Warntjes (2007, 2008), Warntjes and Ó Cróinín (2010), Blackburn and Holford-Strevens (1999), and Holford-Strevens (2008, 2010). Mosshammer (2008), pp. 109–316 has provided an excellent update to Jones on the earlier period, but has not dealt with the Insular developments.
2. Jones (1943), pp. 7–9.
3. Toomer (1998), p. 138.
4. Jones (1937), p. 208.
5. Jones, (1937), pp. 213–219; Ó Cróinín, (2003a), pp. 202–203.
6. Jones (1943), p. 111.
7. Blackburn and Holford-Strevens (1999), pp. 723–725 (Hebrew), p. 732 (Islamic).
8. Wallis, (1999), p. 66.
9. Lawlor and Oulton, (1927), p. 247.
10. Cross and Livingstone (1983), p. 50.
11. Lawlor and Oulton, (1927), p. 249 (citations); Mc Carthy and Breen (2003) pp. 18–19.
12. Harnack (1904), pp. 52, 75–79; Migne (1857) X, cc. 231–236.
13. Mc Carthy and Breen (2003), pp. 39–41.
14. Van der Hagen (1733), pp. 157–376 (84-year Paschal cycles), pp. 336–355 (Latercus), pp. 332, 339 ('pseudo Anatolius').
15. Van der Hagen (1736), pp. 115–141 (Anatolius' Paschal table), p. 125 ('pseudo-Anatolius'), p. 140 (place and time of author).
16. Mc Carthy and Breen, (2003), p. 20 n.23 (all references except Cordoliani); Cordoliani (1945–6), p. 20.
17. Mc Carthy and Ó Cróinín (1987–8), pp. 227–228 (discovery of Latercus); Mc Carthy (1993), pp. 213–217 (Latercus structure); Mc Carthy (1994), pp. 39–40 (origin of the Latercus); Mc Carthy (1996), p. 308 (Paschal table); Mc Carthy and Breen (2003) , pp. 129–139 (authenticity of DRP).

18. Mc Carthy and Breen, (2003), pp. 45–53 (edition), pp. 54–62 (facsimile of Padua MS I.27 ff. 71v–75v), pp. 63–7 (translation).
19. Mc Carthy and Breen (2003), p. 68. While the ninth Hebrew month, Chislev, has been omitted from this tabulation, in the Irish computus, De ratione conputandi §41, the full list is given and this month is uniquely named as 'CORECLATH'; cf Walsh and Ó Cróinín (1988), p. 150, and Mc Carthy and Breen (2003), p. 96.
20. Schwartz (1905), pp 100–1.
21. Mc Carthy and Breen (2003), p. 119.
22. Mc Carthy (1993), p. 212; O'Leary (c. 1922), p. 1.
23. In the case of March itself C must be taken as zero.
24. Holford-Strevens (2008), p. 185 (Table 9, epact 27).
25. Mc Carthy (1993), p. 216, Table 1.
26. I gratefully acknowledge the kindness of Adriano Gaspani, Osservatorio Astronomico di Brera, Milan, in supplying me with both the Bergamasque text and English translation of this verse.
27. I wish to gratefully acknowledge the very constructive comments made by the anonymous referee upon the first draft of this article, and the invaluable assistance with the bibliography given by Mary Higgins, Research Librarian, Trinity College Dublin.

References

Anscombe, A., 1895, "The Paschal Canon attributed to Anatolius of Laodicea", *English Historical Review* 19, 515–535.

Bisagni, J., and Warntjes, I., 2007, "Latin and Old Irish in the Munich Computus: a reassessment and further evidence", *Ériu* 57, 1–33.

——, 2008, "The Early Old Irish material in the newly discovered Computus Einsidlensis (c. AD 700)", *Ériu* 58, 77–105.

Blackburn, B., and Holford-Strevens, L., 1999, *The Oxford companion to the year – An exploration of calendar customs and time-reckoning* (Oxford: Oxford University Press).

Cordoliani, A., 1945–6, "Les computistes insulaires et les écrits pseudo-Alexandrins", *Bibliothèque de l'école des Chartes* 106, 5–34.

Cross, F. L. and Livingstone, E. A. (eds.), 1983, *The Oxford Dictionary of the Christian Church* (Oxford: Oxford University Press).

Harnack, A., 1904, *Die Chronologie der Altechristlichen Litteratur bis Eusebius* (Leipzig: J.C. Hinrichs) II, 2.

Holford-Strevens, L., 2008, "Paschal lunar calendars up to Bede", *Peritia* 20, 165–208.

——, 2010, "Marital discord in Northumbria: Lent and Easter, his and hers", in I. Warntjes and D. Ó Cróinín (eds.), *Computus and its Cultural Context in the Latin West, AD 300–1200* (Turnhout: Brepols), 143–158.

Ideler, L., 1825–6, *Handbuch der mathematischen und technischen Chronologie* (Berlin: A. Rücker), 2 vols.

Jones, C. W., 1937, "The 'Lost' Sirmond Manuscript of Bede's 'Computus'", *English Historical Review* 52, 204–19.

——, 1943, *Bedae opera de temporibus* (Cambridge, MA: Mediaeval Academy of America).

Krusch, B., 1880, *Studien zur christlich-mittelalterlichen Chronologie, Der 84-jährige Ostercyclus und seine Quellen* (Leipzig: Verlag von Veit & Comp.).

Lawlor, H. J. and Oulton, J. E. L, 1927, *Eusebius Bishop of Caesarea – The Ecclesiastical History and the Martyrs of Palestine* (London: Society for promoting Christian knowledge), vol. I: translation; vol. II: introduction, notes and index.

Mac Carthy, B., 1901, *The Annals of Ulster vol. iv* (Dublin: HMSO).

Mc Carthy, D. P., 1993, "Easter Principles and a Fifth-Century Lunar Cycle used in the British Isles", *Journal for the History of Astronomy* 24, 204–224.

——, 1994, "The origin of the latercus Paschal cycle of Insular Celtic churches" *Cambrian Medieval Celtic Studies* 28, 25–49.

——, 1996, "The Lunar and Paschal Tables of De ratione paschali Attributed to Anatolius of Laodicea",

Archive for History of Exact Sciences 49/4, 285–320.
——, 2008, *The Irish Annals – Their genesis, evolution and history* (Dublin: Four Courts Press).
——, 2010, "Bede's primary source for the Vulgate chronology of his chronicles in De temporibus and De temporum ratione", in I. Warntjes and D. Ó Cróinín (eds.), *Computus and its Cultural Context in the Latin West, AD 300–1200* (Turnhout: Brepols), 159–189.
——, 2012, "On the arrival of the Latercus in Ireland", in I. Warntjes and D. Ó Cróinín (eds.), *Proceedings of the 2nd International conference on the Science of Computus in Ireland and Europe*, forthcoming.
——, and Breen, A., 2003, *The ante-Nicene Christian Pasch: De ratione paschali The Paschal tract of Anatolius, bishop of Laodicea* (Dublin: Four Courts Press).
——, and Ó Cróinín, D., 1987–8, "The "lost" Irish 84-year Easter Table rediscovered", *Peritia* 6–7, 227–242.
Migne, J.-P., 1844–64, *Patrologia Graeca* (Paris: J.-P. Migne), vol. 1–161.
Mosshammer, A. A., 2008, *The Easter Computus and the origins of the Christian era* (Oxford: Oxford University Press).
Nicklin, T., 1903, "The date and origin of Pseudo-Anatolius De Ratione Paschali", *Journal. of Philology* 28 , 137–151.
O'Connell, D. J., 1936, "Easter Cycles in the Early Irish Church", *Journal of the Royal Society of Antiquaries of Ireland* 66, 67–106.
Ó Cróinín, D., 1982a, "Mo Sinu maccu Min and the computus at Bangor", *Peritia* 1, 281–295, reprinted in Ó Cróinín (2003b), 35–47.
——, 1982b, "A seventh-century Irish computus from the circle of Cummianus", *Proceedings of the Royal Irish Academy* 82 C 11, 405–30; reprinted in Ó Cróinín (2003b), 99–130.
——, 1983, "The Irish provenance of Bede's computus", *Peritia* 2, 229–47; reprinted in Ó Cróinín (2003b), 173–90.
——, 1997, "The computistical works of Columbanus", in M. Lapidge (ed), *Columbanus, studies on the Latin writings*, Studies in Celtic History 17 (Woodbridge: Boydell Press) 264–270. Reprinted in Ó Cróinín (2003b), 47–55.
——, 2003a, "Bede's Irish computus", in Ó Cróinín (2003b), 201–212.
——, 2003b, *Early Irish History and Chronology* (Dublin: Four Courts Press).
O'Leary, P., c. 1922, *Irish numerals and how to use them* (Dublin: Browne and Nolan).
Schwartz, E., 1905, *Christliche und jüdische Ostertafeln*, Abhandlungen der Königlichen Gesellschaft der Wissenschaften zu Göttingen, phil.-hist. Klasse, Neue Folge, 8/6 (Berlin:Weidmannsche Buchhandlung).
Toomer, G. J., 1998, *Ptolemy's Almagest* (Princeton, NJ: Princeton University Press).
Turner, C. H., 1895, "The Paschal canon Attributed to Anatolius of Laodicea", *English Historical Review* 10, 699–710.
Van de Vyver, A., 1957, "L'Évolution du Comput Alexandrin et Romain du IIIe au Ve Siècle", *Revue d'Histoire Ecclesiastique* 52, 5–25.
Van der Hagen, J., 1733, *Observationes in Prosperi Aquitani chronicon* (Amsterdam: Johannes Boom).
——, 1736, *Dissertationes de cyclis Paschalibus* (Amsterdam: Johannes Boom).
Wallis, F., 1999, *Bede: The Reckoning of Time* (Liverpool: Liverpool University Press).
Walsh, M., and Ó Cróinín, D. (eds.), 1988, *Cummian's letter De controversia paschali and the De ratione conputandi* (Toronto: Pontifical Institute of Mediaeval Studies).
Warntjes, I., 2005, "A newly discovered Irish computus:Computus Einsidlensis", *Peritia* 19, 61–64.
——, 2007, "The Munich Computus and the 84 (14)-year Easter reckoning", *Proceedings of the Royal Irish Academy* 107 C, 31–85.
——, 2009, "The earliest occurrence of Old English gerīm and its Anglo-Irish computistical context", *Anglia* 127, 91–105.
——, 2010a, "The argumenta of Dionysius Exiguus and their early recensions" in I. Warntjes and D. Ó Cróinín (eds.), *Computus and its Cultural Context in the Latin West, AD 300–1200* (Turnhout: Brepols), 40–111.
——, 2010b, *The Munich Computus: Text and Translation* (Stuttgart: Franz Steiner Verlag).
——, and Ó Cróinín, D. (eds.), 2010, *Computus and its Cultural Context in the Latin West, AD 300-1200* (Turnhout: Brepols).

Between Crucifixion and Calendar Reform: Medieval Christian Perceptions of the Jewish Lunisolar Calendar

Carl Philipp Emanuel Nothaft

The moon would have played a marginal role in Christian religion, had it not been for the fact that Jesus Christ is reported by the evangelists to have been crucified on a Friday at the time of Passover. According to the Synoptic Gospels (Matthew 26:17–19; Mark 14:12–16; Luke 22:7–15), the Last Supper was conducted like a regular Passover meal, which took place on the evening that marked the transition from the 14th to the 15th day of the month of Nisan in the Jewish calendar. The Gospel of John (19:14, 31), by contrast, puts the Passion of Jesus on the 'day of preparation' for Passover and hence one day earlier, on 14 Nisan—a scenario that is seemingly confirmed by the Jews' refusal to enter the Roman praetorium on the day of the crucifixion, 'lest they should be defiled; but that they might eat the Passover' (John 18:28), implying that the meal lay still ahead. This striking discrepancy aside, the basic idea of a temporal connection between the Passion and Resurrection of Christ and the Jewish Passover emerges clearly from all four accounts.[1]

For pagan converts to Christianity who were eager to celebrate the anniversary of these events at the right time, this created a problem. After the introduction of the Julian calendar in 45 BCE, most inhabitants of the Roman empire had become used to reckoning with a solar year of 365.25 days, which did not take account of the phases of the moon, unlike the Jewish calendar, which, in Jerusalem at the time of Jesus, was in all likelihood still regulated by empirical principles such as the first sighting of the new moon crescent in the evening sky. The most straightforward way of solving the chronological problem thus posed was to make the date of Easter dependent on the Jewish calendar, either by always letting it coincide with the Passover on 14/15 Nisan (a custom later decried as the heretical doctrine of 'quartodecimanism') or by celebrating on the Sunday that followed it. This practice of 'following the Jews', common as it may have been during the first three centuries of Christianity, was eventually abandoned in favor of Easter cycles that projected lunar dates onto the grid of the Julian calendar, making it thereby possible to calculate the movable feast days long in advance and independently of the Jewish calendar. A leading role in this development was played by the church of Alexandria, which was the first to employ the 19-year lunisolar cycle that eventually became the sole basis for medieval Easter reckoning, also known as the science of *computus*.[2]

The adoption of Easter cycles fostered an increasing separation of the Christian from the Jewish *pascha*, which, owing to diverging standards of calculation and intercalation, were frequently celebrated a month apart. This is particularly palpable in the case of the Alexandrian style of Easter reckoning, which stressed the 'rule of the equinox', according to which the Easter full moon (the Christian version of 14 Nisan) could not fall earlier than

21 March, the assumed date of the vernal equinox, making 22 March the earliest permissible date for Easter Sunday. Scholars have occasionally come to regard this rule of the equinox, which had no direct counterpart in the Jewish calendar, as having been motivated by a deliberate attempt of the early church to sever its ties to the Synagogue, i.e. to emancipate Easter from Passover not just theologically, but also calendrically.[3]

While this interpretation is far from wrong, it only provides part of the picture. When Peter, who served as the patriarch of Alexandria during the first decade of the fourth century, defended his church's method of Easter reckoning, he appealed to the divine Law that had been revealed to Moses and laid down in the book of Exodus. In Peter's view, the Alexandrian rule of the equinox closely corresponded to the practice of the ancient Hebrews. If the Jews of his own age regularly violated the limit of the equinox, celebrating Passover a month earlier than the Christians, this was only a sign that they had abandoned the customs of their fathers—a clear sign of the degeneration that Judaism had supposedly gone through in the wake of the destruction of the Second Temple. According to Peter, one had to discern two 'Jewish' calendars: that of the Old Testament and that of contemporary Judaism. In this perspective, the Christian Easter calculation was not developed in order to abandon its roots, but it actually surpassed the contemporary Jewish calendar when it came to implementing Mosaic Law. The Christians, or so it seemed, were the better Hebrews.[4]

Most of the context of Peter's fragmentarily preserved statements is lost, but traces of his rhetoric can be detected in a number of other late antique sources. It was still present in the fifth century, when the episcopal sees of Rome and Alexandria repeatedly quarreled over the correct dating of Easter Sunday. In 444, Paschasinus, bishop of Lilybaeum (today Marsala, Sicily), wrote a letter to Pope Leo I, in which he advertised the Alexandrian 19-year Easter cycle as the 'calculation of the Hebrews, that is the calculation according to the Law' (*Hebraeorum, hoc est legalem supputationem*).[5] As is well known, the Roman pontiff eventually bowed to this kind of pressure from Egypt, which resulted in the Alexandrian 19-year cycle being adopted as the standard of reckoning in the medieval Latin church. The subtleties of the Alexandrian construal of the Hebrew vs. Jewish calendar, however, got evidently lost in transmission from Greek to Latin. A general lack of available sources on the Jewish calendar other than the Old Testament soon led to a situation in which early medieval scholars ceased to properly differentiate between the biblical calendar of the Hebrews and the calendar of the Jews that they may or may not have encountered in their own time. Instead, they went on to by and large identify Jewish lunar months with the theoretical lunations of the Alexandrian 19-year cycle, thereby providing their own method of Easter reckoning with a venerable pedigree that could be tracked back to the days of Moses or even Noah.[6]

This anachronistic view made it possible to apply the Easter computus to biblical exegesis. One noteworthy example is an obscure chronicle of the world, written in about 814 by Claudius of Turin, a Carolingian exegete and bishop, who is otherwise better known for his leanings towards iconoclasm. In order to highlight pivotal points on the timeline of salvation, Claudius invested a lot of effort in translating into Julian dates the 'lunar' calendar data that could be occasionally found in the Old Testament. Basing himself on an assumed equivalence between ancient Hebrew lunar months and Alexandrian computistical months, he furnished his chronicle with weekdays and calendar dates for events such as the beginning of Noah's flood (Saturday, 5 April), the reception of the Law at Sinai (Monday, 31 May), and the Babylonian capture of Jerusalem (Saturday, 23 June).[7]

Claudius of Turin's method is also present in the works of a number of eleventh and twelfth century monastic writers, who can be grouped together under the label 'critical computists'. One of their common goals was to employ the Easter cycle in order to solve a chronological problem that had been puzzling Christian scholars for centuries: the exact date of Christ's Passion and Resurrection. Their strategy was to search for a year in the Easter cycle in which the Julian date of the crucifixion (believed to be 23 or 25 March) and the 15th day of the moon (as posited by the Synoptic Gospels) both coincided on a Friday. In doing so, they effectively assumed that their computistical *luna* 15 of first spring lunation would automatically correspond to 15 Nisan as it had been observed by the Jews in first-century Jerusalem.[8]

Needless to say, this hypothetical Jewish calendar of the computists was very different from the type of calendar that Jews—first in Babylonia and Palestine, but eventually also in Spain and the rest of Europe—had actually adopted by the tenth century. Although the medieval Jewish calendar retained a 19-year cycle of intercalation, its method of fixing the beginning of the year represented a clear departure from the simple 'epact'-based calendrical arithmetic that had characterized late antique Easter cycles. The technical foundation of the fixed Jewish calendar, as it is still in use today, is the time of the mean conjunction or *molad*, calculated on the basis of a mean lunation of 29d 12h 44m 3 1/3s = 29.530594 days. Each date of *Rosh Hashanah* (1 Tishri) is computed by adding multiples of this value to a fictitious calendar epoch known as *molad baharad*, whose time is equivalent to Sunday, 6 October 3761 BCE, at 23 hours, 11 minutes, and 20 seconds.[9]

For the loss of simplicity, the astronomical precision thereby achieved was considerable, especially if compared to the lunisolar calendar still in use among Christians. The old Alexandrian equation of 19 Julian years (6939.75d) with 235 months implied a mean lunation of 29.530851 days, which caused the calculated new and full moons to fall behind the observable ones at a rate of one day in roughly every three centuries. By the end of the twelfth century, Christian computists had not only noticed this discrepancy, but, aided by the influx of Arabic astronomy from the Iberian peninsula, had begun to look for new models on which to base an astronomically improved Easter computus. Under these circumstances, it could not take long until the conjunction-based Jewish calendar made its appearance on Christian scholarly radars. A meticulous comparison between the 'Hebrew' (i.e. Jewish), 'Latin' (i.e. Christian) and 'Chaldean' (i.e. Muslim) values for the mean synodic month can be found in a *Compotus*, written in 1176 by the English astronomer Roger of Hereford. In the calculated version of their lunar calendar, frequently encountered in astronomical tables, the Muslims distributed 11 intercalary days over 30 lunar years (30 × 354 + 11 = 10 631 days), thereby producing a mean month length of 29d 12h 44m = 29.530555 days. According to the divisions of time used by Roger (where there are 40 moments to an hour, 564 atoms to each moment), this was equivalent to 29 days, 12 hours, 29 moments, and 188 atoms. A slightly larger value was proposed by the Jews, who increased the number of atoms to 208 8/9. That both estimates were significantly closer to the truth than the inflated 'Latin' value of 29 days, 12 hours, 29 moments, and 348 atoms did not remain hidden to Roger, who admitted that the tabulated 'ecclesiastical' moon could differ by up to four days from the observable moon.[10]

The accuracy of the Muslim lunar calendar became an important source of inspiration for various medieval proposals to reform the ecclesiastical lunisolar calendar, as they were advanced in the thirteenth century by Robert Grosseteste (ca. 1170–1253), bishop of

Lincoln, and the famous Franciscan scholar Roger Bacon (ca. 1214/20–ca. 1292).[11] An even more suitable template, however, was provided by the Jewish calendar, whose lunisolar form resembled that of the Christian Easter cycle more strongly than the purely lunar calendar of the Arabic astronomers. In particular, both Jews and Christians used the same 19-year intercalation cycle, with the exception that the Jewish cycle began three years later than the Christian one—a fact that had already been pointed out by Abraham Ibn Ezra in his *Liber de rationibus tabularum*, a Latin work written especially for a Christian readership in 1154.[12]

That there was in his own time a lively interreligious exchange on calendrical matters is hinted at in Ibn Ezra's 'Letter of the Sabbath' (*Igeret ha-Shabbat*, 1158), in which he criticized fellow Jews for communicating the details of their calendar to Gentiles in an inexact manner.[13] The most impressive twelfth-century manifestation of such contacts is the *Compotus emendatus*, written in 1170 or 1171 by the Westphalian cathedral canon Reinher of Paderborn, who wanted his church to abandon its erroneous ways and return to what he regarded as the astronomically sound paschal reckoning that had once been prescribed by Moses himself. In order convince his readers of the great advantages that the Jewish conjunction-based reckoning held in store, Reinher composed a competent and elaborate exposition of its rules of operation (the first of its kind by a Christian author) along with innovative tables for easy conversion between dates calculated on the basis of the *molad* and the Julian calendar. These tables made ample use Hindu-Arabic numerals, including zero, which puts the *Compotus emendatus* among the earliest known Latin texts to employ such numerals for technical purposes.[14]

Both the Jewish calendar and the reform of the Christian Easter cycle continued to play a prominent role in the writings of Roger Bacon, who sent his famous *Opus majus* to Pope Clement IV in about 1267. Although he did not go as far as Reinher of Paderborn in demanding that the church should adopt some version of the Jewish calendar, Bacon spoke very highly of the Jewish astronomers and the accuracy of their value of the mean month. In fact, he was so determined to convince the pope of the importance of studying the Jewish calendar that he spared no expenses to acquire a Hebrew calendar manuscript and send it to the papal curia in Viterbo along with his pupil John as an interpreter. According to Bacon, this Hebrew calendar was 'a wonderful work of astronomical art and highly useful for the understanding of the Law and the feasts prescribed by it. A person ignorant of it can never hope to properly understand the Law, nor can he converse with the Jews about such things, let alone convince them in any useful way'.[15]

That Bacon himself maintained some contact to learned Jews (and that he conversed with them about the calendar) is quite evident from his own profound knowledge on the subject, part of which reflects passages in the Babylonian Talmud. Amongst other things, he referred to the practice of the Jews in ancient Palestine, mentioned in the tractate *Rosh Hashanah* (22b–23b), of lighting beacons on high mountains in order to make known the beginning of a new month.[16] Yet while Bacon was without a doubt a brilliant and original author, his interest in the Jewish calendar was not entirely exotic even in the thirteenth century. Another detailed and technical description of the Jewish calendar was composed in about 1294 by a certain Robert of Leicester, who is presumably identical with the scholar of the same name who later served as the 48th master of the Oxford Franciscans (1321/22) and who also penned a treatise *On the Poverty of Christ*.[17]

Aside from the reform of the calendar, biblical chronology was the other main factor that motivated Christian interest in the Jewish calendar. Indeed, it is difficult to find a medieval discussion of the subject that does not in some way include an application of this knowledge to determine the exact date and year of Christ's Passion. In order to do so, scholars had to be able to calculate the Passover dates of the first century CE, a task at which the 'critical computists' had failed because of their anachronistic use of the Alexandrian lunisolar cycle. Since Reinher of Paderborn rejected this cycle in favour of the medieval Jewish way of reckoning, it was only natural that he would also be the first to use the *molad*-calendar in an attempt to re-calculate the date of the crucifixion. This method was subsequently applied by several other medieval Christian writers, including the aforementioned Robert of Leicester, but not by Roger Bacon, who apparently understood that the Jewish lunar month of antiquity was not based on the *molad* or mean conjunction, but on the day of first visibility of the new moon crescent. Whereas Reinher of Paderborn had presented a static picture of the Jewish calendar, which assumed that the *molad*-based reckoning had already been instituted by Moses (a view presumably shared by his Jewish informants), Roger Bacon implicitly acknowledged that the Jewish calendar had a history that had to be studied carefully in order to understand biblical time reckoning. It is thus no surprise if Bacon, both in his methods and in his conclusions, came remarkably close to modern discussions of the Passion date. Using astronomical tables, he calculated that Jesus had been crucified on 3 April 33 CE, the 14th day of Nisan, a date that is still widely accepted among experts today.[18]

The problems posed by the chronology of the Passion also determined Christian interest in another important aspect of the Jewish calendar and the feast days attached to it: its postponement rules or *deḥiyyot*. In the contemporary Jewish calendar, they make sure that the 15th day of Nisan can never fall on a Friday. If this rule had already existed in the first century, Jesus could not have been crucified on the first day of the Feast of Unleavened Bread, as the Synoptic Gospels seemed to imply. That the Jewish calendar thus favored John's account of the Passion was already acknowledged in the early twelfth century by Rupert of Deutz (ca. 1075–1129), who regarded the *deḥiyyot* as ancient Jewish practice.[19] An opposing viewpoint was expressed by an anonymous English computist, who, in 1175, dryly noted that the present-day Jewish calendar cannot have been in use at the time of Jesus precisely because it conflicted with the synoptic version of events.[20]

An ingenious attempt to harmonize both positions was eventually made in about 1430 by bishop Paul of Burgos (ca. 1353–1435), who had been a famous Castilian Rabbi named Solomon ha-Levi before his conversion to Christianity in 1390/91. Like others before him, Paul assumed that the Jewish calendar he was acquainted with had already been in use during the first century. Basing himself on the reckoning rules of this calendar, he found that in the year of the crucifixion (which he took to be 33 CE) the 15th of Nisan had been postponed from Friday to Sabbath. Yet according to the Synoptic Gospels, the Last Supper, held on Thursday evening, had been a Passover meal. Paul found it hard to accept the idea that Jesus, as a Jew faithful to the law of his people (Matthew 5:17), would have ever violated the Mosaic rules by celebrating Passover one day early. Luckily, all chronological problems seemingly dissolved if one looked more closely at the *deḥiyyot* and the reasons for their existence. As a former Rabbi, Paul knew that the rule which prevented Passover from falling on a Friday was a mere corollary of another rule, according to which *Hoshana Rabbah* on 21 Tishri could never fall on a Sabbath. Since the interval between Nisan and Tishri was always constant in the fixed Jewish calendar, a shift of *Hoshana Rabbah* from

Sabbath to Sunday automatically implied that the preceding 15 Nisan had to move from Friday to Sabbath. Yet this meant that the postponement of Passover in the year of the crucifixion was contingent upon a feast (*Hoshana Rabbah*) that still lay several months ahead, belonging to a future in which, due to Christ's self-sacrifice, mankind had been redeemed and the law of the Old Covenant abolished. The calendrical rules attached to *Hoshana Rabbah*, Paul argued, were hence null and void in this near future and this was the reason why Jesus, knowing what would happen, refrained from postponing the Passover shortly before his death.[21]

Paul's fame as a commentator on the Bible ensured that his solution for the Passion chronology would remain the object of scholarly debate for at least the next two centuries. To name just one example, we find his views discussed at the University of Louvain in the late 1480s, where Paul of Middelburg (1446–1534), a Dutchman who later became bishop of Fossombrone, and the philosopher Peter de Rivo (ca. 1420–1499) fought a heated controversy over the date of Christ's crucifixion. Paul of Middelburg added further substance to the theory of his namesake Paul of Burgos by presenting an array of obscure medieval Hebrew texts, which to him proved that the present-day Jewish calendar, including its *deḥiyyot*, had been in existence since the days of the founding of the Second Temple. His Hebrew citations in this context represent the earliest known instance of Hebrew printing in the Netherlands.[22] Peter de Rivo, on the other hand, dismissed this evidence as well as Paul of Burgos's entire strand of reasoning. In his view, the present-day Jewish calendar, with its astronomical precision, showed clear signs of Arabic influence and could only have been conceived after the rise of Islam, maybe even only after the development of the Alfonsine tables in the thirteenth century.[23]

Discussions of this kind would continue well into the sixteenth century and beyond, during a period that saw important developments in the field of Christian Hebraism.[24] As we have seen, however, scholarly interest in the calendar as a particular aspect of Judaism clearly predates the Renaissance. Owing to the joint roots of both calendars, Christian intellectual engagement with the Jewish calendar was kept alive throughout the Middle Ages, ranging from the naive identification of Alexandrian Easter reckoning with the calendar of Moses that characterizes the earlier period to the refined and erudite treatment we encounter in the works of Roger Bacon. To give up such pursuits was to impede a proper understanding of Scripture, at least according to Bacon, who reminded the pope that 'our Lord and the Apostles were Jews [*Hebraei*], as were the Patriarchs and Prophets'.[25]

Notes

1. See Strobel (1977), pp. 17–64; Finegan (1998), pp. 353–358. The following abbreviations are used throughout: PL = Patrologia Latina; CCSL = Corpus Christianorum Series Latina; CCCM = Corpus Christianorum Continuatio Mediaevalis.
2. For a comprehensive overview, see Pedersen (1983) and Thornton (1989).
3. Zerubavel (1982), p. 286; Simon (1996), pp. 310–322; Steel (2000), pp. 98–101.
4. Peter's arguments are cited in the seventh century *Chronicon Paschale*: Dindorf (1832), pp. 4–10.
5. See Paschasinus's *Epistola ad Papam Leonem* in Krusch (1880), p. 248. Photius, in his Bibliotheca (cod. 115), mentions a book entitled 'Discourse against the Jews and the heretics who follow them and against those [...] who do not celebrate Easter in the first month according to the Hebrews'. See Henry (1960), p. 86. On the general background, see Gerlach (1998).
6. See Bede, *De temporum ratione*, CCSL 123B: 312–315, 326–327, 420–422; ps.-Bede, *De argumentis lunae*, PL 90: 723, and Wiesenbach (1986), pp. 119–122. See also Warntjes (2010), pp. 242–243, for

further source references.

7. Claudius of Turin, *Brevis Chronica*, PL 104: 917–926. See further Allen (1998) and Nothaft (forthcoming).
8. The standard monography is now Verbist (2010). See further the editions by Weikmann (2004) and Wiesenbach (1986), who coined the term 'kritische Komputisten'.
9. For a useful introduction into the present-day Jewish calendar, see Feldman (1978), pp. 185–210. On its medieval origins, see Stern (2001), pp. 180–210; Stern and Mancuso (2007); Carlebach (2011), pp. 11–24.
10. MS Cambridge, University Library, K.K.1.1, f. 238r; Moreton (1995), pp. 581–586.
11. See the texts edited in Steele (1926); Brewer (1859), pp. 274–295; Bridges (1897), pp. 269–285.
12. See the edition by Millás Vallicrosa (1947), pp. 99–100.
13. Friedländer (1894/95), p. 71; Sela (1996), pp. 216–217.
14. The *Compotus emendatus* was edited by van Wijk (1951). See also Honselmann (1962) and Herold (2005). On the reception of Hindu numerals in the West, see Burnett (2006).
15. Brewer (1859), p. 220: 'Et in hac tabula est mirum artificium astronomiae, et summa legis intelligendae utilitas, et omnium festorum legalium, quam qui nescit numquam potest scire intellectum legis, ut oportet, nec cum Judaeis conferre de talibus, nec eius persuadere utiliter'.
16. Bridges (1897), p. 196: 'Et ideo Hebraei antiquitus per astronomiam certificaverunt primationem lunae, et cum non fuerat in visione novae lunae, nec potuit per visum cognosci, accenderunt faces in Jerusalem in monte alto, ut sciretur quod tunc fuit tempus primationis, quatenus homines essent parati facere solemnitates et festa quae habebant expedire'.
17. See MS Oxford, Bodleian Library, Digby 212, ff. 2r–10r; MS Erfurt, Universitätsbibliothek, Amploniana, Quart. 361, ff. 80rb–85rb; Russell (1936), p. 139; Walmsley (1953); North (1992), pp. 132–133.
18. See Bridges (1897), pp. 195–198, 206–210. For modern estimates of the crucifixion date see Finegan (1998), pp. 359–369. A comprehensive account of pre-modern attempts at dating the Passion is provided in Nothaft (2012).
19. Rupert of Deutz, *De sancta trinitate*, CCCM 22: 900; Idem, *De gloria et honore filii hominis super Mattheum*, CCCM 29: 300–301. On the *deḥiyyot*, see Stern (2001), pp. 166–167, 194–195; Belenkiy (2002).
20. MS London, British Library, Cotton Vitellius A.XII, f. 96ra. On the *Compotus constabularii*, see Moreton (1999).
21. See the commentary on Matthew 26 in Paul of Burgos (1634), cols. 441–446.
22. Paul of Middelburg (1513), sigs. D6v–E6r; Offenberg (1974).
23. Peter de Rivo (1488), sigs. c4r–6r; (1492), sigs. e5r–6v.
24. Weinberg (2000). On Christian Hebraism in the early modern period, see Coudert and Shoulson (2004).
25. Brewer (1859), p. 213: 'Nam Dominus noster et apostoli fuerunt Hebraei, sicut patriarchae et prophetae'.

References

Allen, M. I., 1998, "The Chronicle of Claudius of Turin", in A. C. Murray (ed.), *After Rome's Fall: Narrators and Sources of Early Medieval History* (Toronto: University of Toronto Press), 288–319.

Belenkiy, A., 2002, "A Unique Feature of the Jewish Calendar – Deḥiyot", *Culture and Cosmos* 6, 3–22.

Brewer, J. S. (ed.), 1859, *Fr. Rogeri Bacon Opera quaedam hactenus inedita*, vol. 1, Rerum Britannicarum Medii Aevi Scriptores (Rolls Series) 15 (London: Longman, Green, Longman, and Roberts).

Bridges, J. H. (ed.), 1897, *The Opus Majus of Roger Bacon* 1 (Oxford: Clarendon Press).

Burnett, C., 2006, "The Semantics of Indian Numerals in Arabic, Greek and Latin", *Journal of Indian Philosophy* 34, 15–30.

Carlebach, E., 2011, *Palaces of Time: Jewish Calendar and Culture in Early Modern Europe* (Cambridge, MA: The Belknap Press of Harvard University Press).

Coudert, A. P., and Shoulson, J. S. (eds.), 2004, *Hebraica Veritas? Christian Hebraists and the Study of Judaism in Early Modern Europe* (Philadelphia: University of Pennsylvania Press).

Dindorf, L. (ed.), 1832, *Chronicon Paschale* 1 (Bonn: Weber).

Feldman, W. M., 1978, *Rabbinical Mathematics and Astronomy*, 3rd ed. (New York: Hermon Press).

Finegan, J., 1998, *Handbook of Biblical Chronology: Principles of Time Reckoning in the Ancient World and Problems of Chronology in the Bible*, revised ed. (Peabody, MA: Hendrickson Publishers).

Friedländer, M., 1894/95, "Ibn Ezra in England", *Transactions of the Jewish Historical Society of England* 2, 47–75.

Gerlach, K., 1998, *The Antenicene Pascha: A Rhetorical History*, Liturgia condenda 7 (Louvain: Peeters).

Henry, R. (ed.), 1960, *Photius: Bibliothèque* 2 (Paris: Société d'Édition «Les Belles Lettres»).

Herold, W., 2005, "Der computus emendatus des Reinher von Paderborn", in W. Chobrak and K. Hausberger (eds.), *Kulturarbeit und Kirche. Festschrift Msgr. Dr. Paul Mai zum 70. Geburtstag*, Beiträge zur Geschichte des Bistums Regensburg 39 (Regensburg: Verlag des Vereins für Regensburger Bistumsgeschichte), 39–47.

Honselmann, K., 1962, "Magister Reinher. Schrittmacher für die Kalenderreform und die moderne Rechenkunst", in K. Honselmann (ed.), *Von der Domschule zum Gymnasium Theodorianum in Paderborn*, Studen und Quellen zur Westfälischen Geschichte 3 (Paderborn: Verein für Geschichte und Altertumskunde Westfalens), 107–126.

Krusch, B., 1880, *Studien zur christlich-mittelalterlichen Chronologie: Der 84-jährige Osterzyklus und seine Quellen* (Leipzig: Von Veit & Comp).

Millás Vallicrosa, J. M., 1947, *El libro de los fundamentos de las Tablas astronómicas de R. Abraham ibn Ezra* (Madrid, Barcelona: Casa Provincial de Caridad).

Moreton, J., 1995, "Before Grosseteste. Roger of Hereford and Calendar Reform in Eleventh- and Twelfth-Century England", *Isis* 86, 562–586.

——, 1999, "The Compotus of «Constabularius» (1175): A Preliminary Study", in J. Biard (ed.), *Langage, Sciences, Philosophie au XIIe siècle* (Paris: Vrin), 61–82.

North, J. D., 1992, "Astronomy and Mathematics", in J. I. Catto and R. Evans (eds.), *The History of the University of Oxford. Vol. II: Late Medieval Oxford* (Oxford: Clarendon Press), 103–174.

Nothaft, C. P. E., 2012, *Dating the Passion: The Life of Jesus and the Emergence of Scientific Chronology (200–1600)*, Texts and Studies on Time, Astronomy, and Calendars 1 (Leiden: Brill).

——, forthcoming, "Chronologically Confused: Claudius of Turin and the Date of Christ's Passion," in I. Warntjes and D. Ó Cróinín (eds.), *Proceedings of the 3rd Galway Computus Conference* (Turnhout: Brepols).

Offenberg, A. K., 1974, "The First Use of Hebrew in a Book Printed in the Netherlands", *Quaerendo* 4, 44–54.

Paul of Burgos, 1634, "Additiones", in *Biblia Sacra cum Glossa Ordinaria* 5 (Antwerp: Meursius).

Paul of Middelburg, 1513, *Paulina de recta Paschae celebratione: et de die passionis domini nostri Iesu Christi* (Fossombrone: Petrutius).

Pedersen, O., 1983, "The Ecclesiastical Calendar and the Life of the Church", in G. V. Coyne et al (ed.), *Gregorian Reform of the Calendar* (Vatican City: Specola Vaticana), 17–74.

Peter de Rivo, 1488, *Opus responsivum ad Epistolam apologeticam M. Pauli de Middelburgo de anno, die et feria dominicae passionis* (Louvain: Ravescot).

——, 1492, *Tercius tractatus de anno, die et feria dominice passionis atque resurrectionis* (Louvain: Westfalia).

Russell, J. C., 1936, *Dictionary of Writers of Thirteenth Century England*, Bulletin of the Institute of Historical Research, Special Supplement 3 (London: Longmans, Green).

Sela, S., 1996, "Contactos científicos entre judíos y cristianos en el siglo XII: El caso del Libro de las Tablas astronómicas de Abraham Ibn Ezra en su versión latina y hebrea", *Miscelánea de Estudios Árabes y Hebraicos*, Sección de Hebreo 45, 185–222.

Simon, M., 1996, *Verus Israel: A Study of the Relations between Christians and Jews in the Roman Empire (AD 135–425)* (London: Littman Library of Jewish Civilization).

Steel, D., 2000, *Marking Time: The Epic Quest to Invent the Perfect Calendar* (New York: Wiley).

Steele, R. (ed.), 1926, *Compotus Fratris Rogeri*, Opera hactenus inedita Rogeri Baconi 6 (Oxford: Clarendon Press).

Stern, S., 2001, *Calendar and Community: A History of the Jewish Calendar, Second Century BCE–Tenth Century CE* (Oxford: Oxford University Press).

——, and Mancuso, P., 2007, "An Astronomical Table by Shabbetai Donnolo and the Jewish Calendar in Tenth-Century Italy", *Aleph* 7, 13–41.

Strobel, A., 1977, *Ursprung und Geschichte des frühchristlichen Osterkalenders*, Texte und Untersuchungen zur Geschichte der altchristlichen Literatur 121 (Berlin: Akademie-Verlag).

Thornton, T. C. G., 1989, "Problematical Passovers. Difficulties for Diaspora Jews and Early Christians in Determining Passover Dates During the first Three Centuries A.D.", *Studia Patristica* 20, 402–408.

Verbist, P., 2010, *Duelling with the Past: Medieval Authors and the Problem of the Christian Era, c. 990–1135*, Studies in the Early Middle Ages 21 (Turnhout: Brepols).

Walmsley, C., 1953, "Two Long Lost Works of William Woodford and Robert of Leicester", *Archiv Franciscanum Historicum* 46, 458–470.

Warntjes, I., 2010, *The Munich Computus: Text and Translation. Irish Computistics between Isidore of Seville and the Venerable Bede and its Reception in Carolingian Times*, Sudhoffs Archiv. Beihefte 59 (Stuttgart: Steiner).

Weikmann, H. M., 2004, *Heimo von Bamberg: De decursu temporum*, MGH. Quellen zur Geistesgeschichte des Mittelalters 19 (Hannover: Hahn).

Weinberg, J., 2000, "Invention and Convention: Jewish and Christian Critique of the Jewish Fixed Calendar", *Jewish History* 14, 317–330.

Wiesenbach, J., 1986, *Sigebert von Gembloux: Liber decennalis*, MGH. Quellen zur Geistesgeschichte des Mittelalters 12 (Weimar: Böhlau).

van Wijk, W. E., 1951, *Le comput emendé de Reinherus de Paderborn (1171)*, Verhandelingen der Koninklijke Nederlandse Akademie van Wetenschappen, Afd. Letterkunde, n.s., vol. 57.3 (Amsterdam: North-Holland).

Zerubavel, E., 1982, "Easter and Passover: On Calendars and Group Identity", *American Sociological Review* 47, 284–289.

Christian Calendrical Fragments from Turfan

Mark Dickens and Nicholas Sims-Williams
with contributions by Thomas A. Carlson and Christiane Reck[1]

Introduction

The Turfan Collection in Berlin, split between the Berlin-Brandenburg Academy of Sciences, the Oriental Department of the State Library of Berlin and the Museum for Asian Art, contains approximately 40,000 manuscript fragments in more than 20 languages and scripts which were brought back by four Prussian expeditions to Turfan, western China between 1902 and 1914.[2] Alongside the many Buddhist and Manichaean texts in the collection are somewhat over 1100 Christian fragments, mostly written in Syriac or Sogdian in Syriac script. There are also smaller numbers of Sogdian texts in Sogdian script, Uyghur Turkish texts in either Syriac or Uyghur script, Middle Persian texts in Pahlavi script and New Persian texts in Syriac script.[3] Most of the Christian fragments were found at Bulayïq, a small town located 10 km north of Turfan, although small numbers were found at other locations in the Turfan area.[4] These fragments, currently in the process of being catalogued for the first time,[5] present intriguing glimpses into the status of East Syrian Christianity in the Uyghur Kingdom of Qocho located in the Turfan Oasis, a major junction on the Silk Road. Based on the few texts which have been reliably dated, the Church of the East had a presence in this multi-religious environment from at least the ninth to the thirteenth or fourteenth centuries.[6]

Alongside Syriac, the liturgical language of the Church of the East, Sogdian also played a crucial role in the Christian community at Turfan, reflecting the importance of this eastern Middle Iranian tongue as a major *lingua franca* along the Silk Road. The linguistic importance of Uyghur, the Turkic language spoken by many in the Turfan area, is also evident in the few extant Uyghur Christian texts in the Turfan Collection, particularly those from the Mongol era, when the majority of the Turfan Christian community were probably Uyghur native speakers.[7]

Most Syriac texts from Turfan are liturgical or biblical, with a small number of prayer texts, hagiographies and miscellanea. By contrast, the Sogdian Christian texts largely reflect genres that would be read by members of a monastic community—hagiographical, homiletic, and ascetical texts—although there are a number of biblical and liturgical texts, many of which are bilingual Syriac-Sogdian.[8] Finally, the few Uyghur Christian texts reflect a wide variety of genres, including hagiographies, apocryphal sayings, prayers and a wedding blessing. Amongst the Syriac and Sogdian fragments are a dozen or so that are calendrical in nature, the focus of this article.

Calendrical tables are an important component of the Syrian approach to time and dating systems, although there is little discussion of them in scholarly literature.[9] The primary use of such tables is to determine when dates in the liturgical calendar which depend upon the movable date of Easter will fall in a given year. A classic example is the table for calculating the beginning of Lent found in the second part of the *Opus chronologicum* of Elias of Nisibis, written in 1019 and published with a Latin translation by J.-B. Chabot in 1909–1910.[10] This table is reproduced in a slightly adapted form in the Appendix to the present article, along with a table for calculating the date of Easter kindly prepared for us by Thomas Carlson. A similar table, described by its editor as a "perpetual calendar, intended to give the beginning of Lent for any possible year",[11] is included as part of a short treatise on the Syrian calendar in a manuscript formerly in the possession of Mar Severius, Archbishop of Syria and Lebanon, later Mar Ignatius Ephrem I Barsoum, Patriarch of the Syrian Orthodox Church (1933–1957). According to a colophon preceding the treatise on the calendar, the manuscript was written in 1003/4 CE in the Syrian Orthodox Monastery of the Forty Martyrs near Melitene, but the date of composition of the treatise and accompanying tables is not known.

A different use for Syriac calendrical tables is evident in "The Book of Medicines" published by E. A. Wallis Budge.[12] The text consists of lectures on human anatomy and pathology, astrological material dealing with "omens, portents, spells, divinations and planetary forecasts", and prescriptions for various ailments. Two calendrical tables are included in the astrological section, accompanied by statements such as "If you wish to know on how many days of a solar month the moon will rise"; "If you wish to know on what day of the week the beginning of the lunar month is born"; "That you may know in what hour the moon is born, by night or by day"; and "if you wish to know how long the moon will shine, and when it will set".[13] The reason for the calendrical tables in this text becomes clear from the material that follows, namely "calculations concerning those who are sick", essentially divination techniques, in which the moon often plays an important role.[14] This is a far cry from the liturgical focus of the calendrical tables in the texts edited by Chabot and Dean, as well as those found amongst the Turfan fragments.

Syriac Calendrical Texts

In addition to calendrical tables, discussed below, the Syriac texts from Turfan include several small fragments containing lists of months with associated information, in particular the number of days in each month.[15]

Text 1: SyrHT 291 (T III B 99 No. 1), recto: 8.9 x 5.8 cm; 8 lines; verso blank

1 [ܫܒܛ]ܠ ܀ ܐܝܬܘܗܝ
2 [ܐܕܪ]ܠܐ ܀ ܐܝܬܘܗܝ
3 [ܢܝܣܢ]ܠ ܀ ܐܝܬܘܗܝ
4 [ܐܝܪ]ܠܐ ܀ ܐܝܬܘܗܝ
5 [ܚܙܝ]ܪܢ ܀ ܐܝܬܘܗܝ
6 [ܚܠܘ ܐ]ܝܪ ܠܐ ܀ ܐܝܬܘܗܝ
7 [ܚܠܘ]ܡܕܡ ܠܐ ܀ ܐܝܬܘܗܝ
8 [ܬܫܪܝ]ܐܝܪ ܠ ܀ ܐܝܬܘܗܝ

1 [... Ḥaziran] <u>30</u>: its sign
2 [... Iyar] <u>31</u>: its sign
3 [... Nisan] <u>30</u>: its sign
4 [... Adar] <u>31</u>: its sign
5 [... Shebaṭ] <u>28</u>: its sign
6 ... Latter [Kanun] <u>31</u>: its sign
7 ... Former [Kanun] <u>31</u>: its sign
8 ... Latter [Teshri] <u>30</u>: its sign

As with other texts discussed below (notably **U 3858**), **SyrHT 291** lists the months in reverse order,[16] along with the number of days in each and a reference to the "sign" of each. A similar list can be found in the text published by Dean (1934), where a full listing of the months is given, with the first, *Former Teshri* (October), at the bottom and the last, Elul (September), at the top. The first line of this text will suffice to show what **SyrHT 291** might have originally looked like:

Elul (has) 30 (days); and its sign is 7, and its excess is ½ (day).[17]

Dean's text explains how to derive the signs for each month as follows:

Take the days of the year, which are 365, and divide them by 7. Then there remains 1, which is the sign of First Tishri. Add its sign to the days of its own month and divide the sum by 7, and there remains 4, which is the sign of Second Tishri. Then, in like manner, add again the sign of Second Tishri to its days, and when you have divided the sum by 7 there remains 6, which is the sign of First Canun. And proceed likewise with the rest of the months – add the sign (of a month) to its days and divide by 7, and you find the sign of the following month.[18]

As the editor notes, "If the year began on Sunday these signs would indicate the day of the week on which each month began. In all other cases they indicate the relative positions of these first days in the weekly scheme."[19] In the case of **SyrHT 291**, the actual signs for each month must have been on an adjacent folio, now missing, since there is no obvious text missing in the left-hand margin of the fragment.

Text 2: SyrHT 264 (T II B 69 No. 3), recto: 3.2 x 4.8 cm; 3 lines; verso blank

ܙ
ܐ ܢܘ[ ](ܠܫܥ)ܬܐ
ܒ ܬܫܪܩܕܝܡ ܠܐ
ܓ [ ]

7:
1: [ ] hour(?)
2: Former Teshri: 31
3: [ ]

There is not enough of this fragment extant to discern its full significance. It is included at this point because one line clearly contains the name of a month (Former Teshri, equivalent to October, the first month in the Syrian year) and the number of days in that month (31). The right hand column contains part of a series of numerals between 1 (ܐ) and 7 (ܙ), presumably referring to the days of the week. Since only four numerals survive, one cannot see whether they belong to a simple repeating series [... 1, 2, 3, 4, 5, 6,] 7; 1, 2, 3, [4, 5, 6, 7; ...], or, as is perhaps more likely, a series which skips one day after every four in line with the occurrence of a leap year: [... 2, 3, 4, 5;] 7, 1, 2, 3; [5, 6, 7, 1; ...]. A series of the latter kind would give the weekday in a series of years of any day with a fixed position in the calendar, such as New Year or Christmas.[20]

Text 3: SyrHT 101 (T II B 22), recto: 2.1 x 5.9 cm; 1 line; verso blank

ܐܕܪ..ܠܐ..

Adar: 31

This is obviously the remnants of a list of the months with the number of days in each, similar to **SyrHT 291** above, although there is not enough extant to know if the sign of each month was also included in this case.

Text 4: U 3858 ([T II] B): 8.0 x 6.8 cm
recto = 6 rows x 3 columns; cells drawn free-hand in red and black

	ܠ	[ܢܝܣܢ]	1
ܐ[ܬܘܬܗ؟]	ܠܐ	(ܐܕ)ܪ	2
ܐ[ܬܘܬܗ؟]	ܟܚ	[ܫܒ]ܛ	3
	ܠܐ	[ܟܢܘܢ] ܐܚܪܝ	4
		(ܟܢܘܢ)[21] ܩܕܡܝ	5
		(ܬܫܪܝ) ܐܚܪܝ	6

1	[Nisan]	30	
2	(Ada)r	31	[its sign (?)]
3	[Sheba]t	28	[its sign (?)]
4	Latter [Kanun]	31	
5	Former (Kanun)		
6	Latter (Teshri)		

This text is very similar to **SyrHT 291**, consisting of a list of the months in reverse order, with the number of days in each and probably the word ܐܬܘܬܗ, "its sign" at the end of each line, although only the first letter is visible. Although there are also similarities with

SyrHT 70, the two fragments do not come from the same original table, in view of differences in the paper.

verso = 4 lines in Sogdian script

/1/](m)[
/2/](..)[22] xwrmzt(ʾ)[
/3/](... x)yδm[yδ[23]
/4/](.) mr(ys)rky[s[24]

/1/
/2/]... xurmazda[
/3/]... on that d[ay
/4/] Mar Sergius[

The word *xwrmztʾ*, in origin the name of the Zoroastrian god Ahura Mazda, is also used as the name of the first day of the month in the traditional Sogdian calendar. If it is so used here, it is possible that the text on the verso may be connected with the calendrical table on the recto. The use of the traditional (Zoroastrian) day-names by the Sogdian Christians at Bulayïq is proved by two further texts published below (**n295** and **SyrHT 67**). The proposed reconstruction of *(x)yδm[yδ* is also consistent with terminology one might expect in a calendrical text. If the reading *mr(ys)rky[s* is accurate, it would suggest that the author of the Sogdian text on the verso was indeed a Christian. The reference here may be to the commemoration of this saint.[25]

Text 5: SyrHT 70 (T II B 22 No. 2), recto: 3.9 x 5.7 cm; 1 row x 3 columns + 2 lines; cells drawn free-hand in red; verso blank[26]

ܐ[ܬܘܬܗ؟]	ܠܐ	1] ܬܫܪ[ܝ ܩܕܝܡ

2] ܚܕܒܫܒܐ ܕܣ[ܘܒܪܐ؟[27] [
3] ܬܫܪܝ ܐ[ܚܪܝ [

1 ... Former [Teshri]	31	[its sign (?)]

2 ... Sunday of A[nnunciation?]...
3 ... Lat[ter] Teshri...

The top line of this fragment is evidently the bottom line of a table similar to **SyrHT 291** and **U 3858**. Despite some palaeographic similarities between **SyrHT 70** and **U 3858**, the difference in paper rules out the possibility that they are from the same table. Beneath it, apparently in a different hand, are two fragmentary lines which seem to deal with ܣܘܒܪܐ, *Subbārā*, "Annunciation", the first season in the liturgical year, which begins in either Latter Teshri or Former Kanun (cf. Text 6 below).

Text 6: SyrHT 69 (T II B 22 No. 2): 4.3 x 8.9 cm
recto = 4 rows x 3 columns; cells drawn free-hand in red

[28][ܩܕܡܝ] ܢ[ܒܟܢܘ]	[ܐ]	[ܒܬܫܪܝ ܐܚܪܝ]	[ܓ]	1
ܒܬܫܪܝ ܐܚܪܝ[ ]	ܠ	[ܒܬܫܪܝ ܐܚܪܝ]	[ܒ]	2
ܒܬܫܪܝ ܐܚܪ[ܝ]	ܟܚ	[29] ܩܕܡܝ ܐ ܝ[ܒܬܫܪ]	[ܠܐ]	3
ܒܬܫܪܝ ܐܚ[ܪܝ]	[30] ܟܙ	ܩܕܡܝ ܝ[ܒܬܫܪ]	[ܠ]	4

1	[3]	[in Latter Teshri]	[1]	[in Former Kanu]n
2	[2]	[in Latter Teshri]	30	in Latter Teshri
3	[31]	[in] Former [Teshri]	28	in Latter Teshri
4	[30]	[in] Former [Teshri]	27	in Latter Teshri

This table looks at first sight similar to those above, but the numbers in the second column do not refer to the number of days in either Former or Latter Teshri, which have 31 and 30 days respectively. Rather, they are from a table giving the seven possible dates for the beginning of ܩܘܕܫ ܥܕܬܐ, *Quddāsh ʿEdtā*, "Hallowing of the Church" and ܣܘܒܪܐ, *Subbārā*, "Annunciation" (roughly equivalent to Advent), the last and first seasons in the Syrian liturgical year respectively. The date of the former can range from the 30th day in Former Teshri (October) to the 5th day in Latter Teshri (November) and that of the latter from the 27th day in Latter Teshri to the 3rd day in Former Kanun (December). It follows that the numbers in the second visible column (27, 28 and 30, from the bottom up) must belong with the months in the third column and refer to the beginning of *Subbārā*.

By counting back four weeks, the length of *Quddāsh ʿEdtā*, we can then reconstruct the numbers which would have been in the column immediately preceding the first column, to find the beginning of that liturgical season. The reconstruction of the figures 30 and 31 in rows 4 and 3 is confirmed by the fact that these are the only possible dates for *Quddāsh ʿEdtā* which are within the month of Former Teshri named in the adjacent column. The difference of 2 between rows 2 and 3 indicates the position of the leap year and thus enables us to reconstruct the missing numbers and month-names in rows 1 and 2. The scribal alteration of ܐܚܪܝ, "Latter" to ܩܕܡܝ, "Former" in the first visible column of row 3 can also be explained by the fact that this is where the month changes from Latter Teshri to Former Teshri.

verso = 4 lines in Sogdian script[31]

/1/	(βwγy?c)[[32]	*βūγ(i)č*[(9th month in the Sogdian calendar)
/2/	myšβwγ[yc	*mišβūγ*[(*i*)*č*[33] (10th month)
/3/	(ẓ)-y(mt)[yc[34]	*žīmt*[(*ī*)*č* (11th month)
/4/	xw[šmyc	*xu*[*šmič*[35] (12th month)

The names of four months of the traditional Sogdian calendar (*βūγič*, *mišβūγič*, *žīmtīč* and *xušmič*) suggest that there may be a connection between the text on the verso and the calendrical table on the recto side.

Sogdian Calendrical Texts in Syriac Script

Text 7: n295 (T II B 46), recto and verso: 10.5 x 6.5 cm; 10 lines on recto + 1 line on verso

This is a page of a codex in very small format. Amongst the Christian Sogdian manuscripts, such a size is characteristic of liturgical booklets intended, presumably, for individual use.[36] The present text, which seems to contain an account of work carried out on various days of the month, or perhaps rather predictions regarding the advisability of working on various days of the month, may have been written on a blank page at the end of such a booklet.

R1 (................................)
R2 **(m)[y](θy)..**[37] **[•••]**n [ ]
R3 (nʾ)[? ••](r)[••• ʾ](r)q
R4 (qt)[? •••](w ....)
R5 stm (p)tnym (xʾ)n?
R6 sʾr tyny (xw?)[ ]
R7 frnx[w]nty (w)[bʾt][38]
R8 **dtšy r(w)c(y *m*)[yθy]**
R9 ʾ(*yc*? ʾ)r(q)[ ]
R10 ny [••](•• ••)[ ]
V1 **rwcy (m)yθy..**
rest blank

"On the day [...-*rōč*]: ... un-(?)... work(?), (R4)so that(?) ... he might(?) bring a coarse(?) implement(?) to the *khan*(?) (who) ... He/it(?) ... [will be](?) fortunate. (R8)On the day *dhatšī-rōč*: No work(?) ... (V1) On the day ...-*rōč*: *(here the text breaks off unfinished)*."

R3. The restoration of [ʾ](*r*)*q* "work" is supported by the occurrence of this word in R9.

R5. On *stm* see the note to Text 9 (**n354**), line 5. The following (*p*)*tnym* is unfortunately obscure. One or more words so spelled are attested (in the plural form *ptnymt*) as a translation of Syriac ܡܐܢܐ in the sense "gear, trappings, implements"[39] and as the second element of two Manichaean Sogdian compounds: *jnʾ-ptnym* "advisor"[40] and *ṭryyy-pṭnym*, an epithet of "hands" with a negative connotation.[41] Finally, (*xʾ*)*n* could represent the Turkish title *khan*, but the reading is quite uncertain.

R6–7. If *tyny* in R6 is 3 sg. imperfect "introduced, brought" the obvious restoration in R7 will be *frnx*[*w*]*nty* (*w*)[*mʾt*] "[he/it was] fortunate". But it seems equally possible to interpret *tyny* as 3 sg. optative (for **tynyy*) and to restore a 3 sg. subjunctive *frnx*[*w*]*nty* (*w*)[*bʾt*], giving the sense "[he/it will be] fortunate" or "[may he/it be] fortunate".

R8. For *rwc* "day", cf. Text 10 (**SyrHT 67**), line 6. The day specified here is *dhatšī rōč*, the 8th, 15th and 23rd day of the month in the traditional Sogdian calendar. Two other days of the month were presumably named in R1 and R10, but they are unfortunately illegible.

Text 8: n288 (T II B 62 + C93 = T III B 61), recto: 13.5 x 8.5 cm; 11 lines; verso blank

This text is written on one side of a sheet made by gluing together two, in parts even three, layers of paper. The other side is blank. Although poorly preserved, it has proved crucial to the understanding of the calendrical tables which follow (Texts 9–13), since it contains instructions for consulting such tables.

The key to the reconstruction and understanding of this text is the sequence of numerals mentioned in it: 28, 12 and 19. Lines 1–4 contain instructions for calculating the position of the year within the 28-year solar cycle. The procedure described is as follows: Start with the number of the Seleucid year, divide by 28, and take the remainder. Lines 5–8 contain instructions for calculating the position of the year in the 19-year lunar cycle. In this case, the procedure is to add 12 to the number of the Seleucid year,[42] to divide by 19, and to take the remainder.

Exactly the same procedure is described by Elias of Nisibis, in the text accompanying his table for calculating the beginning of Lent:

> Subtract the years of Alexander 28 (by) 28, and what remains, enter with it the line of 28 ... Then add 12 years to the years of Alexander and subtract the total 19 (by) 19, and what remains, enter with it the line of 19 ... Then proceed with the remainder of the 19 as far as the remainder of the 28, and the number which you find opposite both numbers, if it is in black ink, then the Monday of the Fast falls in Shebaṭ, and if it is in red, then the Monday of the Fast (falls) in Adar.[43]

The same calculations are described in the Introduction to the printed text of the Ḥudra, the primary liturgical text of the Church of the East;[44] there, however, the resulting figures are used to consult two separate tables rather than a single table of 28 × 19 (= 532) cells, such as the table of Elias. Our Sogdian text states that one should place a "line" (Sogdian *wytq*) first on the axis of 28 and then on the axis of 19. Evidently the point of intersection indicates the required date. The implication that the text describes the procedure for consulting a 28 × 19 matrix receives strong support from the associated fragments of calendrical tables (Texts 9–13 below), most of which can now be seen to belong to tables of exactly this type.

1 [q’](m)y qt γrb(y)[
2 qd’m-z’yy sty. ’[
3 ’štwystq(y)’štwyst(q)[y bys’ w’c c’f ....]
4 p(r*xs’t*)x’t wytq wdy ’[wst
5 nw(*ts* ••)qy wytq-’y pyd’[r
6 pcm’ry cwpr( )dw’ts p(c)[m’r .... nwtsqy]
7 nwtsqy bys’( )w’c c’[f .... prxs’t x’t]
8 (w)ytq wdy ’wst.. [45][
9 m(yw)[46] srdy(θ)šryq(d*ym*)[m’x
10 (pc)m’r m’t [
11 (..... *x*w)t’w(•)[

"[If you des]ire to know [the ...], in which position it is, [take the number of the year, subtract] 28 (and again) 28 (as many times as possible). [However many] ... (4)may be left (over), p[ut] the line there.

Concerning the line of 19 ...: To (lit. over) the number [of the year add the] nu[mber] 12 ... subtract [19] (and again) 19 (as many times as possible). However many [... may be left over] ... (8)put the line there.

"[The month] Former Teshri of the tiger year ... the number (of the year of the Greeks) was ... [Our] Lord ..."

Line 1. The verbs are probably 2 sg. optative: "[if you des]ire that you may know". Virtually identical phrases are used to introduce the explanations of the calendrical tables in the text translated by Dean, as well as in the Syriac "Book of Medicines".[47]

Line 2. The compound *qdʾm-zʾyy* probably consists of *qdʾm* "which? what?" and *zʾy* "earth, ground", plus the suffix *-y* which turns the compound into an adjective, literally: "having what ground? in which position?"

Line 3. Sogdian has two ways of expressing a distributive sense: by repetition, as in *nym nym* "half-and-half",[48] or by use of the suffix *-ky*, as in *zʾrkyy* "by the thousands".[49] Here in *ʾštwystqy ʾštwystqy* "28 (and again) 28" the two methods are combined. The same construction is probably attested in lines 6–7: [*nwtsqy*] *nwtsqy* "19 (and again) 19". Since both examples are followed at a distance of one line by the words *wytq wdy*, it does not seem too fanciful to assume that the two passages are parallel, hence the restorations [*bysʾ wʾc cʾf*] in line 3, ʾ[*wst*] in line 4 and [*prxsʾt xʾt*] in line 7. The literal meaning of *bysʾ wʾc* (2 sg. imperative) is "send away, let out". In a mathematical context this must mean "take away, subtract". The Sogdian text thus expresses division as a process of repeated subtraction, as does Elias, who uses the Syriac verb ܢܦܩ, lit. "go out", hence "cast out, subtract" with repeated numerals: "28 (by) 28", "19 (by) 19".

Line 4. Forms such as *prxsʾt xʾt* (subjunctive + conditional particle *xʾt*) have recently been discussed by Yoshida,[50] who points out that they are restricted to late Sogdian texts whose syntax shows the influence of Uyghur Turkish. The Syriac equivalent in the text of Elias is ܦܐܫ, "remains". The word *wytq*, which also occurs in lines 5 (with oblique case ending *-ʾy*) and 8, is otherwise unknown. Etymologically, it may be understood as a diminutive of *wītē* "cord, rope" (in Sogdian script *wytʾk*, Yaghnobi *wīta*),[51] hence a "string" or "line". It seems to apply both to a row or a column of figures ("line of 19 ..."), in which sense it corresponds to Elias' use of the Syriac term ܣܘܪܛܐ, "line", and to the horizontal and vertical lines or markers whose point of intersection determines the required date.

Lines 9–11. The reference to a "tiger year" (the third year in every cycle of twelve years according to the Chinese and Central Asian animal cycle) suggests that this part of the text applies the method of calculation described to a specific year.

Calendrical Tables with Sogdian on the Reverse

Five fragments from Turfan have remnants of calendrical tables in Syriac script on one side and (with one exception, where the reverse is blank) Sogdian texts in either Sogdian or Syriac script on the other side.[52]

Text 9: n354 (T II B 66 No. 48a): 14.5 x 12.3 cm
recto = 9 rows x 8 columns; cells drawn free-hand in red and black; one indistinct line in Manichaean script in the lower margin[53]

						ܝܗ		1
[55] ܠܐ	ܐ[ܟ]	[54] ܙ	[ܝܕ]	ܟܐ	ܙ	[ܝܕ]		2
ܠ	[ܟ]	ܘ	ܝܓ	ܟܙ				3
[56] ܗ	[ܝܛ]	[ܕ]	ܝܒ	ܟܘ	ܗ			4
	ܟ[ܕ]	ܓ	ܝ	ܟܕ	ܓ			5
ܒ	ܟܓ	ܒ	ܝܘ	ܟܓ	ܒ	ܝܘ		6
ܐ	ܟܒ	ܐ	ܝܗ	ܟܒ	ܚ	ܝܗ		7
[57] ܠ[ܐ]	ܟܐ	ܘ	ܝܕ	ܟܚ	[ܙ]	ܝܕ		8
ܗ	ܝܛ	ܗ	ܒ[ܝ]	ܟܘ	[ܗ]			9

1		15	[8]	[22]	[15]	[1]	[22]	[1]
2		14?	7	21	14?	7	21?	31
3		[13]	[6]	27	13	6	20?	30
4		[19]	5	26	12	4?	19?	5
5		[17]	3	24	10	3	24?	[3]
6		16	2	23	16	2	23	2
7		15	8	22	15	1	22	1
8		14	7?	28	14	6	21	31?
9		[19]	5?	26	12?	5	19	5

Most of this fragment can be identified as a section of the Lenten table reproduced in the Appendix (table 1, columns 14–19; rows 20–28). However, the last column of **n354** does not correspond with any part of that table; in fact it includes two numbers (30 and 31) which can never occur in a Lenten table. A series of numbers which agrees with the surviving figures in this column does occur in the corresponding Easter table reproduced in the Appendix (table 2, column 17, rows 12–19) and the same pattern would also occur in a table for determining certain other dates dependent on Easter, such as Pentecost, though the comparison requires one to assume that the scribe wrote several numbers in red instead of black or vice versa. However, it is difficult to see how this fragment of an Easter or Pentecost table can relate to the Lenten table which stands beside it, since the dates indicated in this last column, whether Easter or Pentecost, do not align with the Lenten dates to their right.

verso = 9 lines of Sogdian in Syriac script

On the reverse of the calendrical table is the following Sogdian text. The two texts may have been written at different times, but are presumably connected, in so far as both are concerned with the calculation of certain dates in the liturgical year.

1](my)θ(•)[• xš]p(ʾ p)r(w) yw
2 m(γ)[w]n [cn](m)ʾx
3 (wy)t(wr)qw (knw)n q[dy]m mʾx
4 nwts (prm x)[š]p(γʾ *xwr*)ty (..) ty(m)
5 stm p(ʾšy pr)c[y .](.)cn q(ym)θ(ʾ)[y]
6 ʾwsʾr (ctfr)s m(y)θ(pr)c(y)
7 (s*w*l)qʾ[ʾ](γ)ʾm bw(t ∴) ʾt
8 [d](*s*)m(y)q myθ(y) [ ]
9 [ʾ](γ)ʾm bwt .. [ ]

"... both day [and] night alike, [from] the month ... until the nineteenth of the month Former Kanun one eats(?) at night. Again, (5)after the Great(?) Fast, after forty days from the Resurrection is the feast of the Ascension, and (8)on the [ten]th day is the feast [of Pentecost]."

Lines 3–4. The date (19th of Former Kanon, equivalent to December) is clear but its significance is unknown. The Church of the East observes a fast prior to Christmas, beginning on the 1st of December, and one might perhaps speculate that there was some change in the rules of the fast on the 19th, seven days before Christmas itself. However, the reading and translation of the phrase (*x*)[*š*]*p*(*γʾ xwr*)*ty* "one eats(?) at night" are quite doubtful.

Line 5. The word *stm* also occurs in Text 7 (**n295**), R5, but is otherwise unattested in this spelling. It apparently corresponds to the form written ʾ*stm* in Sogdian script,[58] which seems to be a later form of Manichaean Sogdian *stmb* "stern(?)," in Sogdian script ʾ*stʾnp* or ʾ*stmp* "coarse".[59] Here one could understand *stm pʾš* as "Severe Fast" or merely "Great Fast", corresponding to Syriac ܨܘܡܐ ܪܒܐ , the regular term for Lent. The Manichaean form *stmb* translates Middle Persian ʿ*stbr* (= New Persian *sitabr*), a word whose senses include the neutral "big, strong, solid" as well as the more negative "coarse, stiff".

Lines 5–9. The references to *stm pʾš* "Lent", *qymθʾ* "the Resurrection" and *swlqʾ ʾγʾm* "the feast of the Ascension" are all quite clear, so there can be little doubt that "the feast [of Pentecost]" should be restored in lines 8–9. That the community at Bulayïq celebrated the feast of the Ascension forty days after Easter and ten days before Pentecost, as is of course to be expected, is also clear from the Sogdian version of the Apostolic Canons.[60]

Text 10: SyrHT 67 (T II B 22 No. 2): 7.5 x 10.9 cm
recto = 7 rows x 9 columns; cells ruled in red

ܠܐ	[ܟ]ܙ		[ܟܓ]	[ܟ]ܙ					1
ܟܐ	ܟܗ	ܙ	ܣ	ܟܗ	ܟܐ				2
ܝ	ܟܙ	ܠ	ܚ	ܟܙ	ܘ	ܚ	ܠ	ܝ	[ܟ]ܙ	3
[ܘ]ܝ	ܟܠ	ܒ	ܗܘ	ܒ	ܬ	ܗܘ	ܒ	ܬ	[ܟܠ]	4
	[ܟܒ]	[ܣ]	[ܝܗ]	ܟܬ	ܣ	ܟܒ	ܟܬ	[ܝ]ܗ		5
		[ܘ]	ܟܠ	ܚ	ܘ	ܟܓ				6
		[ܗ]	[ܝ]ܬ	ܟ[ܘ]						7

1	[20]	[13]	[5]	[20]	[6]	27?	20?	[6]	27?	13
2	[25]	[11]	[4]	[18]	11	25	18	4	25	11
3	24?	10	3	17	10	24	17	3	24	10
4	23?	9	2	16	9	2	16	2	23	16?
5	[22]	15?	29	22	8	29	15?	8?	22?	[15]
6	[20]	[13]	[27]	20	6	27	13	6?	[27]	[13]
7	[19]	[12]	[5]	[19]	[5]	26?	19?	5?	[26]	[12]

Again, this fragment can be located in the Lenten table in the Appendix (columns 8–17, rows 11–17). As noted below, **SyrHT 273** may come from the same original table.

verso = 8 lines of Sogdian in Syriac script

On the other side of the calendrical table is a Sogdian text which appears to be concerned with a calendrical calculation, though its purpose is not clear. Two occurrences of a verb meaning "(it was) written" and (probably) one of the word for "scribe" suggest that it may be a colophon, in which the dates when the scribe began and finished writing the manuscript might have been mentioned.

1](•*wy*)st([61] n)ʾwqt[ ]
2 b]žʾwt dwy(s)t ʾšt(c)[ ]
3 p]cmʾry prw (s)ʾq q(wn)ʾ [ ]
4 p](c)r(w d)pyd([62] xw) ⁖ npxšt(y) qty ʾ•[ ]
5](tn)q[•]nʾ(yʾ n)wwyst (s)γt(yʾ sw)ryq
6](*tw* ⁖)d(*yny*) rwc(y) ⁖ (mʾ)t
7]••st ⁖ npxšt(y)
8 py](d)ʾr

"... twenty-...(?) ... increases, two hundred and eighteen ... (3)with the reckoning he(?) put(?) the number ... he is the scribe(?) instead of ... It was written ... on the 29th day of the month (by) the Syrian (6)[calendar] ... (The day) *dhēnē-rōč*. There was ... [It was] written ... because of ..."

Line 1. The first word may be a compound ending in *-wyst* "20", i.e. a number between 21 and 29. It does not seem likely that it can be read *dwyst* "200". The following *(n)ʾwqt*[appears to be the beginning of an otherwise unknown word.

Line 2. The spelling *ʾštc* "18" stands for [əštats], cf. *pncc* [pāǰats] "15". Other attested spellings in Syriac script are *štʾts* [štats] (in **U 7252**, side A, col. 1, line 20),[63] and *šts* [*štas*].[64]

Line 3. The noun *pcmʾr* or *ptšmʾr*—cf. also Text 8 (**n288**), lines 6 and 10—derives from the verb *pcmr-*, *ptšmr-* "to consider, reckon" and can generally be translated as "reckoning" or "number". In the Sogdian version of the Parable of the Minas (Luke 19:12-27), *pcmʾr* translates Syriac *mnyʾ* as a unit of currency, no doubt as a calque based on the meaning of the underlying root *mnʾ* "to count". As for *sʾq*, this too can usually be translated as "number" and is virtually synonymous with *pcmʾr*, as is clear for instance from the dyadic expression *pww sʾk pww ptšmʾr* "countless, innumerable".[65] However, it is possible that one or both of these words may have a more specific sense in a mathematical context such as this. Finally, *qwn*ʾ may be either 2 sg. imperative "make, put" or 3 sg. imperfect "(he) made, put".

Line 4. Since *dpyd* or *rpyd* is unknown, it seems likely that it is a mispointed form for *dpyr* "scribe".

Line 5. The phrase *(n)wwyst (s)yt(y*ʾ *sw)ryq* ... "on the 29th day of the month (by) the Syrian [calendar]" suggests that the preceding [...]*(tn)q*[•]*n*ʾ*(y*ʾ*)* might be a contrasting ethnic adjective with the Syriac suffix *-āyā* (spelled *-*ʾ*y*ʾ in Sogdian fashion rather than *-y*ʾ as in Syriac). However, it is difficult to find a plausible restoration.

Line 6. In Sogdian the word *rwc* "day" is exclusively used with the traditional names of the 30 days of the month, to which it is suffixed. There is virtually no doubt that the preceding word here should be read *dyny*, the name of the 24th day, since no other day-name is compatible with the traces. However, it is not clear whether the text indicates a synchronism between the 24th day (of a Sogdian month) and the previously-mentioned 29th day (of a Syrian month).

Text 11: SyrHT 68 (T II B 22 No. 2): 4.1 x 3.1 cm

recto = 4 rows x 2 columns; cells drawn free-hand in red

ܝܙ		1
ܟܓ	[ܐ]	2
ܟܐ	ܟܚ	3
ܟ[ܝ]	ܟܙ	4

1	[3]	17
2	1?	23
3	28	21
4	27	20?

This fragment can also be located in the Lenten table in the Appendix (columns 10–11, rows 2–5).

verso = 3 lines in Sogdian script

/1/ [ ](. ...)[
/2/ c[t](β)ʾr s(.)[
/3/ [1](δ)[

/1/ [ ](...)[
/2/ (four) ...[
/3/ [

The Sogdian word *ctβ*ʾ*r* "four" on the verso could relate to many contexts, including that of a calendar, but it is not possible to draw any clearer conclusions.

Text 12: SyrHT 273 (T II D), recto: 5.4 x 2.3 cm; 4 rows x 1 column; cells ruled in red; verso illegible[66]

ܕ	1
ܓ	2
ܒ	3
ܐ	4

1	4
2	3
3	2
4	1

There are two different places where this fragment could fit into the Lenten table (either column 10 or 18, rows 23–26), as well as two places in the Easter table (either column 6 or 17, rows 3–6). Of interest is the fact that it was apparently not found at Bulayïq, the site where most of the Christian fragments were discovered, but rather at Qocho, elsewhere in the Turfan oasis. However, similarities in palaeography, along with the red ruling, suggest that it may have originally belonged to the same table as **SyrHT 67**, the only other fragment with cells ruled in red, rather than drawn free-hand. In this case, it belongs in the Lenten table.

Text 13: So 15850 (T III T.V. B): 3.3 x 4.9 cm
recto = 2 rows x 5 columns; cells drawn free-hand in red and black

		ܘ			1
ܕ	ܝܐ	ܘ	ܝܚ [67]		2

1			6		
2		18	6	11	4

Due to the anomalous occurrence of the number 6 in two adjacent cells, which should not happen in Syriac calendrical tables, and the absence of a number in the upper right-hand cell, this fragment may represent a table which was left unfinished due to a scribal error.

verso = 3 lines in Sogdian script

/1/](....)[4](.)[
/2/](.)δy xw (ʾm/rδ)[[68]
/3/](..)[

Concerning the text on the verso, other than the possibility that the language is Sogdian, nothing more can be assumed.

As noted above, it is likely that **SyrHT 67** and **SyrHT 273** come from the same original Lenten table. Especially given the unique text on the reverse, **n354** seems to be the sole remnant of its original table, which apparently contained information for determining the dates of both Lent and Easter. Of the remaining calendrical tables, there are similarities in the scribal hand on **SyrHT 68** and **So 15850**, but given the anomalous nature of the latter, it is questionable whether they originally belonged together.

Conclusions

The Christian calendrical fragments from Turfan are few in number, but they offer interesting insights into the Christian community that existed in this outpost of the Church of the East for several centuries. The remnants of calendrical tables, which can be located in reconstructed Lenten and Easter tables show that the Turfan Christians were as concerned with determining the correct dates for these important events in the liturgical calendar as Christians elsewhere in Europe, the Middle East or Asia at that time.

The fact that Sogdian texts in either Syriac or Sogdian script which mention Sogdian months or provide calendrical calculations can be found on the reverse sides of most of the fragments containing calendrical tables further shows that these texts were not just being recycled (as is the case with many texts from Turfan which contain different languages on the recto and the verso). Rather, it seems that the Sogdian texts on the verso of the calendrical tables were intended to help members of the community in calculating important dates in the church calendar. In contrast, there seems no evidence that these tables were used for divination purposes, as is the case in the Syriac "Book of Medicines".

Finally, the use of three calendrical systems in these fragments—Syrian, Sogdian and the 12-year animal cycle used by the Chinese and subsequently the Turks—further attests to the multilingual and multicultural nature of the Christian community in Turfan

Appendix 1: Lenten and Easter Calendrical Tables

Two tables are reproduced below. The Lenten Calendrical Table is an adaptation of the table in the work of Elias of Nisibis.[69] The parallel Easter Calendrical Table has kindly been prepared by Thomas Carlson, to whom the authors are very grateful. The numbers in each cell give the dates of, respectively, Lent and Easter in any year. Figures in bold type in the Lenten table (red ink in the original) stand for dates in Adar while non-bold figures are dates in Shebaṭ; in the Easter table bold figures represent dates in Nisan, non-bold figures dates in Adar. The significance of the rows and columns shaded light grey is the same in both tables: the top row is the place in the 19-year lunar cycle (the remainder after adding 12 to the Seleucid year and dividing by 19); the left column is the place in the 28-year solar cycle (the remainder after dividing the Seleucid year by 28); the right column is the "foundation of the year",[70] while the bottom row indicates the day of the month on which Passover[71] falls, either in Adar (March) or Nisan (April). Darker grey shading indicates portions of the table which are paralleled in the Turfan fragments. In the Lenten table: **n354** = columns 14–19; rows 20–28; **SyrHT 67** = columns 8–17, rows 11–17; **SyrHT 68** = columns 10–11, rows 2–5; **SyrHT 273** = either column 10 or 18, rows 23–26. In the Easter table: **n354**, last column = column 17, rows 12–19; **SyrHT 273** = either column 6 or 17, rows 3–6.

TABLE 1. Lenten Calendrical Table, after Chabot (1909), pp. 124–125; (1910), p. 139.

	1	2	3	4	5	6	7	8	9	10	11	12	13	14	15	16	17	18	19	
1	11	25	18	4	25	11	4	25	11	4	18	11	25	18	4	25	11	4	18	1
2	10	3	17	3	24	17	3	24	10	3	17	10	24	17	3	24	10	3	24	2
3	9	1	16	9	23	16	8	23	9	1	23	9	1	16	9	23	16	1	23	3
4	7	28	14	7	28	14	7	21	14	28	21	7	28	14	7	21	14	7	21	5
5	6	27	20	6	27	13	6	20	13	27	20	6	27	13	6	27	13	6	20	6
6	12	26	19	5	26	12	5	19	12	5	19	5	26	19	5	26	12	5	19	7
7	11	3	18	4	25	18	3	25	11	3	18	11	25	18	4	25	11	3	25	1
8	9	2	16	9	23	16	2	23	9	2	16	9	2	16	2	23	16	2	23	3
9	8	1	15	8	22	15	8	22	8	1	22	8	1	15	8	22	15	1	22	4
10	7	28	14	7	28	14	7	21	14	28	21	7	28	14	7	21	14	7	21	5
11	13	27	20	6	27	13	5	20	13	5	20	6	27	20	6	27	13	5	20	6
12	11	25	18	4	25	11	4	25	11	4	18	11	25	18	4	25	11	4	18	1
13	10	3	17	3	24	17	3	24	10	3	17	10	24	17	3	24	10	3	24	2
14	9	2	16	9	23	16	2	23	9	2	16	9	2	16	2	23	16	2	23	3
15	8	29	15	8	29	15	7	22	15	29	22	8	29	15	8	22	15	7	22	4
16	6	27	20	6	27	13	6	20	13	27	20	6	27	13	6	27	13	6	20	6
17	12	26	19	5	26	12	5	19	12	5	19	5	26	19	5	26	12	5	19	7
18	11	25	18	4	25	11	4	25	11	4	18	11	25	18	4	25	11	4	18	1
19	10	2	17	10	24	17	2	24	10	2	17	10	2	17	3	24	17	2	24	2
20	8	1	15	8	22	15	8	22	8	1	22	8	1	15	8	22	15	1	22	4
21	7	28	14	7	28	14	7	21	14	28	21	7	28	14	7	21	14	7	21	5
22	6	27	20	6	27	13	6	20	13	27	20	6	27	13	6	27	13	6	20	6
23	12	26	19	5	26	12	4	26	12	4	19	12	26	19	5	26	12	4	19	7
24	10	3	17	3	24	17	3	24	10	3	17	10	24	17	3	24	10	3	24	2
25	9	2	16	9	23	16	2	23	9	2	16	9	2	16	2	23	16	2	23	3
26	8	1	15	8	22	15	8	22	8	1	22	8	1	15	8	22	15	1	22	4
27	7	28	21	7	28	14	6	21	14	28	21	7	28	14	7	28	14	6	21	5
28	12	26	19	5	26	12	5	19	12	5	19	5	26	19	5	26	12	5	19	7
	25	13	2	22	10	30	18	7	27	15	4	24	12	1	21	9	29	17	6	

TABLE 2: Easter Calendrical Table (by Thomas Carlson).

	1	2	3	4	5	6	7	8	9	10	11	12	13	14	15	16	17	18	19	
1	31	14	7	24	14	31	21	14	31	21	7	31	14	7	24	14	31	21	7	1
2	30	20	6	23	13	6	20	13	30	20	6	30	13	6	23	13	30	20	13	2
3	28	18	4	28	11	4	25	11	28	18	11	28	18	4	28	11	4	18	11	4
4	27	17	3	27	17	3	24	10	3	17	10	27	17	3	27	10	3	24	10	5
5	26	16	9	26	16	2	23	9	2	16	9	26	16	2	26	16	2	23	9	6
6	1	15	8	25	15	1	22	8	1	22	8	25	15	8	25	15	1	22	8	7
7	30	20	6	23	13	6	20	13	30	20	6	30	13	6	23	13	30	20	13	2
8	29	19	5	29	12	5	19	12	29	19	5	29	19	5	22	12	5	19	12	3
9	28	18	4	28	11	4	25	11	28	18	11	28	18	4	28	11	4	18	11	4
10	27	17	3	27	17	3	24	10	3	17	10	27	17	3	27	10	3	24	10	5
11	1	15	8	25	15	1	22	8	1	22	8	25	15	8	25	15	1	22	8	7
12	31	14	7	24	14	31	21	14	31	21	7	31	14	7	24	14	31	21	7	1
13	30	20	6	23	13	6	20	13	30	20	6	30	13	6	23	13	30	20	13	2
14	29	19	5	29	12	5	19	12	29	19	5	29	19	5	22	12	5	19	12	3
15	27	17	3	27	17	3	24	10	3	17	10	27	17	3	27	10	3	24	10	5
16	26	16	9	26	16	2	23	9	2	16	9	26	16	2	26	16	2	23	9	6
17	1	15	8	25	15	1	22	8	1	22	8	25	15	8	25	15	1	22	8	7
18	31	14	7	24	14	31	21	14	31	21	7	31	14	7	24	14	31	21	7	1
19	29	19	5	29	12	5	19	12	29	19	5	29	19	5	22	12	5	19	12	3
20	28	18	4	28	11	4	25	11	28	18	11	28	18	4	28	11	4	18	11	4
21	27	17	3	27	17	3	24	10	3	17	10	27	17	3	27	10	3	24	10	5
22	26	16	9	26	16	2	23	9	2	16	9	26	16	2	26	16	2	23	9	6
23	31	14	7	24	14	31	21	14	31	21	7	31	14	7	24	14	31	21	7	1
24	30	20	6	23	13	6	20	13	30	20	6	30	13	6	23	13	30	20	13	2
25	29	19	5	29	12	5	19	12	29	19	5	29	19	5	22	12	5	19	12	3
26	28	18	4	28	11	4	25	11	28	18	11	28	18	4	28	11	4	18	11	4
27	26	16	9	26	16	2	23	9	2	16	9	26	16	2	26	16	2	23	9	6
28	1	15	8	25	15	1	22	8	1	22	8	25	15	8	25	15	1	22	8	7
	25	13	2	22	10	30	18	7	27	15	4	24	12	1	21	9	29	17	6	

Appendix 2: Plates

Text 1: SyrHT 291 recto.

Text 2: SyrHT 264 recto.

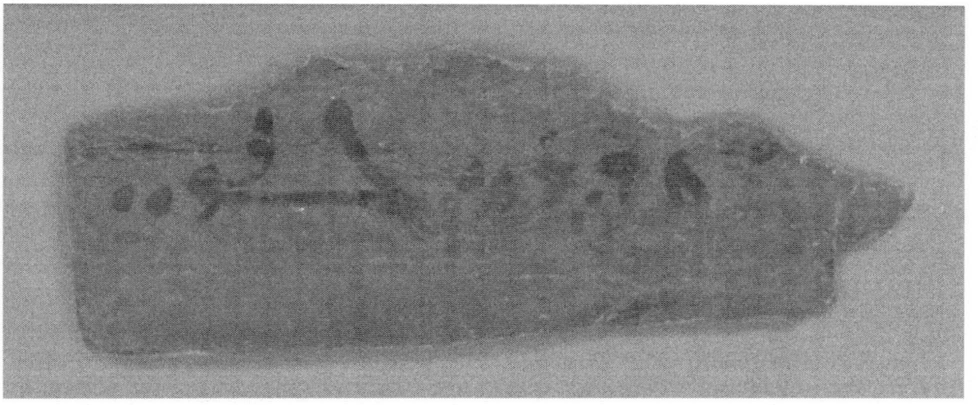

Text 3: SyrHT 101 recto.

Text 4: U 3858 recto (left) and verso (right)

Text 5: SyrHT 70 recto.

Text 6: SyrHT 69 recto (left) and verso (right).

TEXT 7: n295 recto (left) and verso (above).

TEXT 8: n288 recto.

Text 9: n354 recto (top) and verso (bottom).

Text 10: SyrHT 67 recto (top) and verso (bottom).

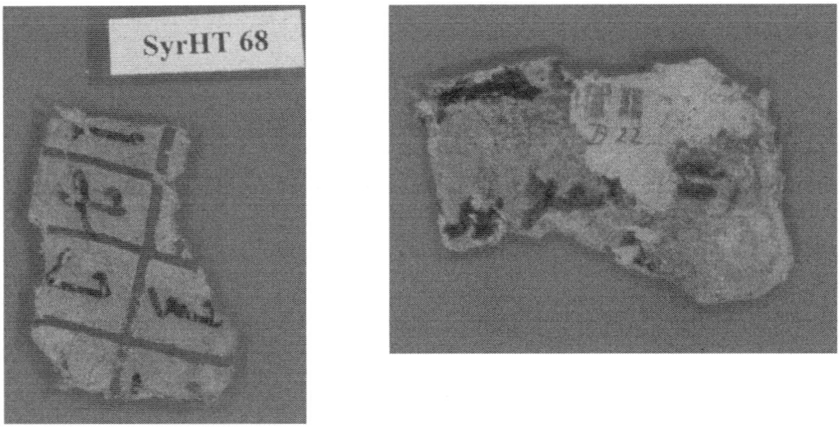

Text 11: SyrHT 68 recto (left) and verso (right).

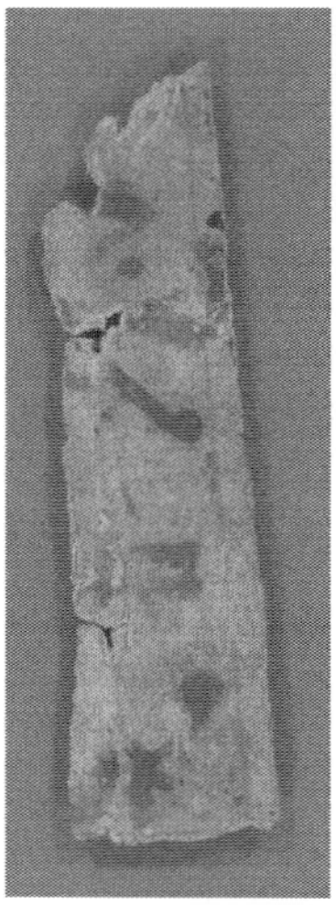

Text 12: SyrHT 273 recto.

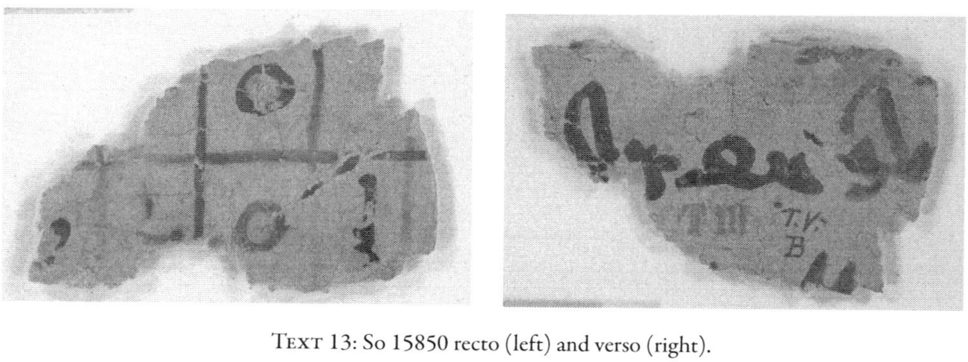

Text 13: So 15850 recto (left) and verso (right).

Notes

1. The Syriac texts are dealt with by MD and NSW, the Sogdian texts in Syriac script by NSW and the texts in Sogdian script by CR. The Easter table in the Appendix was contributed by TC. Since neither MD nor NSW is a specialist in calendrical matters, we owe an immense debt of gratitude to those who are and who have provided invaluable help in interpreting and reconstructing the calendrical texts and tables from Turfan presented in this article. François de Blois (FdB), Thomas Carlson, Gareth Hughes and Sacha Stern (SS) have all contributed important suggestions and insights, in particular with regard to what turned out to be a crucial point: the significance of the numbers 28, 12 and 19 in Text 8 (**n288**). Without their input this article would be much the poorer. Thanks are also due to the Staatsbibliothek zu Berlin - Preussischer Kulturbesitz and the Berlin-Brandenburgische Akademie der Wissenschaften for access to and permission to reproduce images of the relevant fragments. All images are copyright Depositum der BERLIN-BRANDENBURGISCHEN AKADEMIE DER WISSENSCHAFTEN in der STAATSBIBLIOTHEK ZU BERLIN - Preussischer Kulturbesitz, Orientabteilung.
2. On the Turfan expeditions and the resultant Turfan collection in Berlin, see Berlin-Brandenburg Academy of Sciences and Humanities (2007).
3. Although generally referred to as "fragments", the texts range from a 61-folio Syriac liturgical text to small scraps upon which only a few letters are visible. Several dozen of the Christian fragments (including some discussed in this article) are in fact bilingual, and a few are even trilingual.
4. On Bulayïq, see Sims-Williams (1989).
5. The fragments in Syriac script are being catalogued by The Christian Library from Turfan Project, funded by the Arts and Humanities Research Council of the United Kingdom and based in the School of Oriental and African Studies (SOAS), University of London. The project team consists of Erica C. D. Hunter (principal investigator), Nicholas Sims-Williams, Peter Zieme and Mark Dickens.
6. Rather than the problematic adjective "Nestorian", "East Syrian" is used throughout to refer to the Church of the East, as well as other churches (such as the Chaldean Catholic Church) which have subsequently split from it, all of which follow the same basic liturgical rite, albeit with certain theological differences.
7. The ethnic makeup of the Turfan community is discussed in Sims-Williams (1992), the multilingual nature of the Christian texts from Turfan in Dickens (2009).
8. Indeed, it is the nature of the literature found at Bulayïq which indicates that it was almost certainly home to a Christian monastery, as noted in Sims-Williams (1989).
9. On Syrian approaches to chronology, including the Syrian calendar and the Seleucid era, see Bernhard (1969); the work does not deal with calendrical tables.
10. Chabot (1909), pp. 124–125; Chabot (1910), p. 139. We are particularly grateful to François de Blois for drawing our attention to this important source.
11. Dean (1934), p. 129; the relevant table is Table I, ibid., p. 139. This West Syrian manuscript is currently located in Damascus, with the catalogue number Dam. Syr. 7/16. Thanks to Jonathan Loopstra (American University in Iraq) for this information.
12. Budge (1913). Page references will be given as follows: Syriac text/English translation.
13. Budge (1913), pp. 448/527, 451/529, 451/530, 452/531. Budge's rather archaic translation has been modified here. The phrase "if you wish to know" or "if you wish to understand" is also found throughout the text edited by Dean, cf. Dean (1934), pp. 130–138.
14. Budge (1913), pp. 453–468/531–549.
15. **SyrHT** signature numbers are kept in the Oriental Department of the State Library of Berlin, while **n**, **So** and **U** signature numbers are kept in the Berlin-Brandenburg Academy of Sciences. Signature numbers in parentheses (e.g. T II B 22) are those originally given to the fragments at the time of the Prussian Turfan expeditions. Fragment measurements are given in height × width order. The following conventions are used: [ ] for missing or illegible text, ( ) for letters which are only partially legible, italics within ( ) for traces compatible with the reading proposed, and ••• for unidentified letters. Grey shaded cells in the calendrical tables are either outside the fragment or have no visible letters in them. Rubrics are indicated by underlining for Syriac texts and their translations; for Sogdian texts, they are indicated by bold type. It should be

noted that one cannot always reliably distinguish red ink from black ink on these fragments, since the former can darken over time and the latter can fade.

16. We are grateful to Sacha Stern for noticing this point and thus making it possible to reconstruct the missing month-names in these texts. The reason for listing the months in reverse order remains unclear.
17. Dean (1934), p. 136. As noted here by Dean, "excess" refers to "the excess over the moon month of 29½ days."
18. Dean (1934), p. 130.
19. Dean (1934), p. 130, n. 4.
20. cf. the five right-hand columns in Dean's Table I (p. 139), which indicate the weekdays of the "Foundation of the Year", New Year, Christmas, Epiphany, and Holy Cross.
21. The first words on lines 5 and 6 are illegible due to the faded rubric.
22. The final letter of this word and the first letter of line 4 could be read as *p* in Syriac script, according to a proposal by Peter Zieme.
23. A less likely restoration is *(x)yδm*[*ʾx*, "in that m[onth]".
24. Peter Zieme proposes reading the word in this way, with the *s* in Syriac script, to be interpreted as Mar Sergius (Syriac ܡܪܝ ܣܪܓܝܣ).
25. Sergius features quite frequently in the Martyr's Anthems, which are an important part of the liturgy of the Church of the East; see Maclean (1894), pp. 28, 34, 39, 46, 116, 119. The current lectionary of the Church of the East also mentions a commemoration day for Sergius and his fellow martyr Bacchus, on the first Wednesday in October, although no special lessons are appointed in the lectionary; see Maclean (1894), p. 282. However, two Christian fragments from Turfan indicate that this commemoration was celebrated there, at least in the ninth century (based on provisional dating of one of the manuscripts concerned), even though it is not contained in modern liturgical texts of the Church of the East. Thus, a folio from a Sogdian lectionary with Syriac rubrics (**n164**), gives Matthew 16:24–17:6 (the text breaks off at this point) as the reading for the commemoration ܕܡܪܝ ܣܪܓ[ܝ]ܣ ܘܡܪܝ ܒܟܘܣ, "of Mar Sergius and Mar Bacchus"; see Müller (1913), pp. 12-16; Sundermann (1975), pp. 75-78. Significantly, Mar Awa, Bishop of the Assyrian Church of the East in California, has recently found the actual text of the commemoration of these saints in MIK III 45, a 61-folio liturgical manuscript which is part of the Turfan Collection: ܫܒܬܐ ܕܛܘܒܢܐ ܡܪܝ ܣܪܓܝܣ ܣܗܕܐ ܢܨܝܚܐ, "Saturday of the blessed Mar Sergius the victorious martyr" (fol. 13 recto – Bacchus is mentioned on fol. 13 verso – findings presented at the Christianity in Iraq VIII conference held at the School of Oriental and African Studies, London on 28 May, 2011).
26. **SyrHT 67–70** are glassed together.
27. The letter immediately following ܕ is partially torn off. Since it almost certainly refers to one of the seasons in the liturgical year, there are only a few options for the missing word. Based on the visible remnants of the letter and the reference to Latter Teshri in the next line, the most logical option is ܣܘܒܪܐ.
28. This is the most likely reconstruction, given what is visible of the bottoms of the missing letters, along with the final *nun*.
29. The last word in this cell has been altered from ܐܚܪܝ, "Latter" to ܩܕܡ, "Former", leaving the unexplained ܐ. Based on the final ܢ visible at the beginning of this cell and that immediately below, the first word must be ܬܫܪܝ, thus indicating the month of Former Teshri in both cells.
30. This number has a line over it.
31. Lower right corner of page, rounded and darkened; side margin = 1.4–1.6 cm; lower margin = 1.0 cm.
32. The reading of this word is very uncertain.
33. This is the form used in the Manichaean texts. The texts from Mt. Mugh call this month *tymych*, as discussed in Panaino (1990), p. 665.
34. There is probably a diacritic dot under *z*.
35. Presumably with metathesis for *xšwmyc*.
36. cf. Sims-Williams (1995), p. 258a.
37. Bold type in Sogdian texts indicates rubrics.
38. Or *(w)*[*mʾt*]?
39. Sims-Williams (1985), p. 179.

40. Gershevitch (1985), pp. 33–39.
41. "Sharply-inclined" according to Gershevitch (1985), pp. 35–36, "swiftly-grabbing" according to Sundermann (1992), p. 134.
42. "The purpose of adding 12 is to reconcile the Seleucid year count with the Alexandrian cycle" (SS).
43. Chabot (1909), p. 124; Chabot (1910), p. 139.
44. Darmo (1960), pp. 13–14, especially p. 13, r. col., bottom 12 ll; p. 14, r. col., ll. 1–11.
45. The rest of the line left blank?
46. Hardly *ms*.
47. Cited above at the end of the Introduction.
48. Gershevitch (1954), §1633.
49. Gershevitch (1985), pp. 27–29.
50. Yoshida (2009), pp. 281–282.
51. See Livšic (1962), p. 156.
52. For convenience, the calendrical tables will be considered the recto side of each fragment. Translations of each table with Arabic numerals are given in left-to-right format, as are the Easter and Lenten tables at the end of the article, in contrast to the Syriac text, where the original right-to-left format is obviously maintained.
53. Numbers in missing cells are supplied in the translation based on the Easter and Lenten tables reproduced in the Appendix.
54. Apparently in black ink, but should be red (according to expected sequence).
55. Apparently in red ink, but should be black.
56. In black ink, but should be red.
57. In red ink, but should be black.
58. Sundermann (1997), p. 105, line 4, where *ʾstm* occurs in a context parallel to *ʾstmp* in Yoshida (2000), p. 13, line 110. The proposal in Yoshida (2008), p. 59 to read *rxnʾm* in place of *ʾstm* is hardly justified.
59. Gershevitch (1954), §157; Yoshida (2000), pp. 13 (line 110), 83. On the late Sogdian development of [mb] to [m] see Gershevitch (1954), §453.
60. See Sims-Williams (1985), pp. 106–107, with n. 67.
61. Quite uncertain. Not *dwyst*.
62. Or *(r)pyd*. Wrongly pointed for *dpyr*?
63. An unpublished fragment discussed by Dickens (2009), pp. 30–32.
64. Thus correctly read in Müller (1913), p. 41, line 3, contra Sundermann (1974), p. 231.
65. Gershevitch (1954), §1164.
66. A label on the glass-plate links it with a now-lost fragment **T II B 66 No. 48b** and a typed hand-list of the Syriac fragments (prepared sometime between 1946 and 1972) further connects it with other calendrical texts discussed in this article: **T II B 22 No. 2** (**SyrHT 67–70**) and **T II B 66 No. 48a** (**n354**).
67. Written vertically in black ink in the cell (perhaps to save space).
68. This line seems to contain the Sogdian word *xw*, which functions as a demonstrative pronoun ("that"), an article ("the"), the nominative 3rd person singular pronoun ("he") or the 3rd person singular present of the verb "to be."
69. Chabot (1909), pp. 124–125; Chabot (1910), p. 139. The first three rows of Elias's table have been omitted, since they are not relevant to this article.
70. As Dean (1934), p. 131, n. 6 notes: "the day of the week on which the preceding year ended, expressed as a numeral, 1, 2, 3, 4, 5, 6, or 7". This is also referred to as the dominical letter.
71. Here and in what follows the term "Passover" refers "not to the Passover of the Jews but to the notional 'Passover' of Christian Easter computus, that is: the 14th day of the Paschal lunar month" (FdB).

References

Berlin-Brandenburg Academy of Sciences and Humanities, 2007, *Turfan Studies* (Berlin: Berlin-Brandenburg Academy of Sciences and Humanities): http://www.bbaw.de/bbaw/Forschung/Forschungsprojekte/turfanforschung/bilder/Turfan_engl_07.pdf.

Bernhard, P. L., 1969, *Die Chronologie der Syrer,* Sitzungsberichte der Österreichischen Akademie der Wissenschaften, Phil.-hist. Klasse 264 (Wien: Hermann Böhlau).

Budge, E. A. W. (ed.), 1913, *Syrian Anatomy, Pathology and Therapeutics or "The Book of Medicines", Vol. I: Introduction, Syriac Text* (Oxford: Oxford University Press).

—— (tr.), 1913, *Syrian Anatomy, Pathology and Therapeutics or "The Book of Medicines", Vol. II: English Translation, Index* (Oxford: Oxford University Press).

Chabot, J.-B. (ed.), 1909, *Eliae Metropolitae Nisibeni, Opus chronologicum, Pars posterior. CSCO, Scriptores Syri, III/8, Textus* (Parisiis: e Typographeo Reipublicae).

—— (tr.), 1910, *Eliae Metropolitae Nisibeni, Opus chronologicum, Pars posterior. CSCO, Scriptores Syri, III/8, Versio* (Romae: Karolus de Luigi).

Darmo, T. (ed.), 1960, *Ktaba da-Qdam wad-Batar wad-Hudra wad-Kashkol wad-Gazza w-Qala d-'Udrane 'am Ktaba d-Mazmure, Vol. I* (Trichur, Kerala: Church of the East).

Dean, J., 1934, "The Old Syriac Calendar", *Journal of the American Oriental Society* 54, 129–142.

Dickens, M., 2009, "Multilingual Christian manuscripts from Turfan", *Journal of the Canadian Society for Syriac Studies* 9, 22–42.

Gershevitch, I., 1954, *A Grammar of Manichean Sogdian* (Oxford: Blackwell).

——, 1985, *Philologia Iranica* (Wiesbaden: Reichert).

Livšic, V. A., 1962, *Juridičeskie dokumenty i pis'ma,* Sogdijskie dokumenty s gory Mug 2 (Moscow: Izd-vo vostočnoĭ lit-ry).

Maclean, A. J., 1894, *East Syrian Daily Offices* (London: Rivington, Percival; repr: Piscataway, NJ: Gorgias Press, 2003).

Müller, F. W. K., 1913, *Soghdische Texte I,* Abhandlungen der Preussischen Akademie der Wissenschaften 1912 No. 2 (Berlin: Preussische Akademie der Wissenschaften).

Panaino, A., 1990, "Pre-Islamic calendars", *Encyclopaedia Iranica* 4, 658–668.

Sims-Williams, N., 1985, *The Christian Sogdian Manuscript C2,* Berliner Turfantexte 12 (Berlin: Akademie-Verlag).

——, 1989. "Bulayïq", *Encyclopaedia Iranica* 4, 545.

——, 1992, "Sogdian and Turkish Christians in the Turfan and Tun-huang Manuscripts", in A. Cadonna (ed.), *Turfan and Tun-huang, the Texts: Encounter of Civilizations on the Silk Route,* Orientalia Venetiana IV (Firenze: Leo S. Olschki Editore), 43–61.

——, 1995, "A Sogdian version of the 'Gloria in excelsis Deo'", in R. Gyselen (ed.), *Au carrefour des religions. Mélanges offerts à Philippe Gignoux,* Res Orientales 7 (Leuven: Diffusion, Peeters Press), 257–262.

Sundermann, W., 1974, "Nachlese zu F. W. K. Müllers 'Soghdischen Texten I', 1. Teil", *Altorientalische Forschungen* 1, 217–255.

——, 1975, "Nachlese zu F. W. K. Müllers 'Soghdischen Texten I', 2. Teil", *Altorientalische Forschungen* 3, 55–90.

——, 1992, *Der Sermon vom Licht-Nous,* Berliner Turfantexte 17 (Berlin: Akademie-Verlag).

——, 1997, "Three fragments of Sogdian letters and documents", in *La Persia e l'Asia centrale da Alessandro al X secolo,* Atti dei convegni Lincei 127, 1996 (Rome: Accademia Nazionale dei Lincei), 99–111.

Yoshida, Y., 2000, "Sute wen kaoshi" [Studies of Sogdian texts], in Liu Hongliang (ed.), *Tulufan xinchu Moni jiao wenxian yanjiu* [Studies in the Manichaean texts recently discovered at Turfan] (Beijing: Wenwu chubanshe), 3–199 [in Chinese].

——, 2008, [Review of C. Reck, *Mitteliranische Handschriften,* Teil 1: *Berliner Turfanfragmente manichäischen Inhalts in soghdischer Schrift*], *Indo-Iranian Journal* 51, 51–61.

——, 2009, "Minor moods in Sogdian", in K. Yoshida and B. Vine (ed.), *East and West. Papers in Indo-European Studies* (Bremen: Hempen), 281–293.

Lunar Tables in Medieval Russia

Michael L. Gorodetsky

Introduction: Tables of 'Lunar Stream'

Celestial events attracted great interest in early Russia. Russian chronicles known as 'letopises' (annals) record hundreds of observations of solar and lunar eclipses, comets and meteor showers.[1] Special interest was also devoted to the movements of the Sun and the Moon as these luminaries determine calendrical cycles, in particular the 19-year lunar cycle which formed the basis of Christian Easter calculation. In addition, the cycle of lunar phases governed an earlier, traditional pagan lunar calendar; unfortunately, traces of this earlier calendar are very rare as all preserved written sources from medieval Russia date to after the Christianizing of Russia in AD 988 and are mostly written by ecclesiastical scribes using the Julian calendar. However, sometimes one can still find references to 'months of scribes' as opposed to 'celestial months in early texts.[2]

In various medieval Russian sources, usually of ecclesiastical content—books of hours, canons, psalters, etc—one can sometimes find tables of the so-called 'lunar stream'.[3] These tables give a schedule of new and full moons for a single 19-year lunar cycle. They generally include both the date and the time of the new and full moon. The role of such tables is not very clear; they are not directly related to ecclesiastical practice as the date of Easter was determined by another method and moreover Easter full moons are usually not synchronized with the dates in the lunar stream tables.

The origin and history of the development of the tables of lunar stream have not been investigated in detail. When these tables are compared to astronomical calculations it becomes evident that the numerical data of the tables are very imprecise and that there are a large number of scribal and numerical errors. Tables of this type are mostly based on very simple calendrical models or on some rounded value of the length of the synodic lunar month which is added repeatedly to some starting date (for example in a book of hours of the fifteenth century it is stated that the calculations are made from the length of '29 days and half a day and half an hour and 1/5 of an hour').[4] Some lunar stream tables are based on the length of the lunar month which is the basis of traditional Hebrew calendar ('Jewish lunar circle' as it is called in the text).[5] One such text defines the length of the month to be '29 and half a day and half an hour and 1/5th of an hour and 33 partial hourlets'. The number '33' is a scribal error for '37', as can be seen from the numerical values in the table. These partial 'hourlets' are the *heleqs* (1/1080 of an hour) of the Hebrew calendar as 1/2+1/5+37/1080=793/1080 gives the well-known value for the length of the synodic month of 24.5 days + 793 *heleqs*. These particular tables in Russian sources are calculated starting from the New Moon of Tishri 1 in year 5227 of the Hebrew calendar (Autumn of AD 1466).[6]

A particularly interesting lunar stream table is found in the compilation of St. Kirill of Belozersk.[7] These tables are placed on the lists 173–186v and are dated by watermarks to the end of the fourteenth or the beginning of the fifteenth century. These tables attracted my attention because of the unique pictures on the margins with phases of possible solar and lunar eclipses (see figures 1 and 2).[8] The tables are fixed to the 19-year Easter lunar circle, but the first year is stated as 6898 of the Byzantine Era (BE) which corresponds to AD 1390 (according to the Byzantine Era of the world started 1 September 5509 BC). In the following, I present an astronomical analysis of these Kirillo-Belozersky tables. However, before turning to this analysis it is instructive to look at the general level of astronomical knowledge in medieval Europe up to the end of the fourteenth century.

Medieval Astronomical Tables in Europe

During the Middle Ages, the main locus of astronomical activity was within the church. The need to explain growing divergences of Easter Tables with observed lunar phases led to a revival of interest in astronomical observations and calculations.[9] The renaissance of astronomy in Europe began on the Iberian Peninsula, on the junction of the Islamic and Christian worlds. In the second half of the eleventh century Arabic astronomers, gathered in the Caliphate of Cordoba under the leadership of al-Zarkali, compiled a set of astronomical tables that became known as the Toledan Tables. The Toledan Tables incorporated material from Arabic *zījes* (especially those of al-Khwarizmi and al-Battani) as well as Ptolemy's astronomy.[10] The error of eclipse times calculated using the Toledan Tables sometimes reached as much as seven hours, much worse than Ptolemy's own tables or contemporary Arabic *zījes*.[11]

In the twelfth century the Toledan Tables became widely known in the Latin World and were adapted to the Christian calendar (the Toulouse Tables). In 1252–1270 in then Christian Toledo under the patronage of the King of Castile and Leon Alfonso X the Wise, the Jewish astronomers Isaac ben Sid and Judah ben Moses ha-Cohen compiled a more accurate set of tables which became known as the Alfonsine Tables. Improvements to these tables were formulated by scholars in Paris not long before 1321, and the tables were finally published in 1485 as the Alfonsine Tables *editio princeps*.[12] Johannes de Muris, who participated in the development of the Tables in Paris, was able to predict the time of eclipses to an accuracy of about six minutes, compared to both his own observations and to modern calculations, much better than the accuracy of his observations (only about twelve minutes).[13] The Alfonsine Tables rapidly spread over the Europe and were widely used to predict eclipses. They were not the only astronomical tables in use at this time, however. In the first half of the fourteenth century Gersonides developed his own tables and was able to predict the times of three solar and seven lunar eclipses observed by him between 1321 to 1331 with an accuracy of a little over six minutes.[14]

An extensive study of European manuscripts containing astronomical tables of the fourteenth and fifteenth centuries was made by Lynn Thorndyke. In the Utrecht manuscript №317 he discovered tables with calculations of the circumstances of eight solar eclipses for the period 1366–1386.[15] Other tables, including calculations of solar and lunar eclipses for the period 1361–1392, were possibly compiled in France,[16] and tables of conjunctions

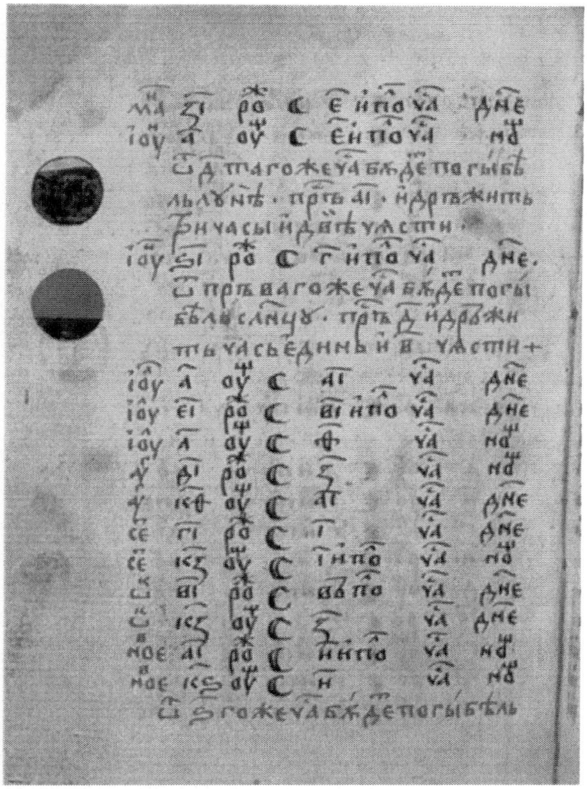

FIGURE 1. List 184v in the compilation of Kyrill of Belozersk syzygies for the year 1406.

May	17	birth	☾	5 and half an hour	day	
June	1	waning	☾	5 and half an hour	night	

From the 4 hour the perishing of the Moon will be 11 fingers and it holds three hours and 2 parts

June	16	birth	☾	4 and half an hour	day	

From the first hour the perishing of the Sun will be 4 fingers and holds one hour and 2 parts

July	1	waning	☾	11 hour	day	
July	15	birth	☾	12 and half an hour	day	
July	30	waning	☾	9	hour	night
Aug.	14	birth	☾	7	hour	night
Aug.	29	waning	☾	11	hour	day
Sep.	13	birth	☾	10	hour	day
Sep.	27	waning	☾	10 and half an hour	night	
Oct.	12	birth	☾	at half an	hour	day
Oct.	27	waning	☾	7	hour	day
Nov.	11	birth	☾	8 and half an hour	night	
Nov.	16	waning	☾	8	hour	night

From the 6 hour the perishing of ...

FIGURE 2. Translation of List 184v in the compilation of Kyrill of Belozersk syzygies for the year 1406.

and eclipses for the period 1327–1386 were compiled by Walter de Elveden in England. Further tables for the period from 1387–1462 were compiled in England by John Somer (Somur/Somour) for the meridian of Oxford. Like the Kirillo-Belozersky tables these latter tables include pictures of the phases of solar and lunar eclipses.[17]

Astronomical Analysis of the Kirillo-Belozersky Tables

Table 1 lists all the syzygies of the Kirillo-Belozersky tables for the period of 19 years from 6898 BE (AD 1390) to 6916 BE (AD 1408) in September style. The astronomical tables are aligned on a boundary of the 19-year Orthodox Easter lunar cycle and they begin from the first cycle of the moon on January 1. In addition, for each year the age of the moon on January 1, called 'Easter Base', is given calculated by (11×N+3) mod 30. In the first year of the table (AD 1390), January 1 is the 14th day moon which is indeed the day of astronomical full moon. According to N. V. Stepanov,[18] this synchronization was achieved thanks to the fact that in the beginning of the fourteenth century Nicephorus Gregoras increased lagging ancient bases by two days. However, the actual Paschal full moons appears 2–3 days later than dates in the table. For example, in 1390, which is the first year of the cycle of the moon, Paschal Full Moon ('Jews Pascha' in another text of Kyrill of Belozersk collection)[19] occurred on April 2, whereas the table indicates the waning of the moon on March 31. This discrepancy was the basis for the Gregorian reform in the West.

Table 1 lists for each new moon (indicated in the text by the Russian word for 'birth ☾') and full moon ('waning ☾') the date and the time of syzygy given in hours after sunrise (day hour) or sunset (night hour) with a precision of half an hour. Unusually, the times are given in equinoctial hours rather than the more common seasonal hours (1/12 part of a day or night),[20] as is evident from the fact that there are values given which are greater than 12. The third and fourth rows of each cell in table 1 give the difference in hours between the times found in the text and the time deduced from astronomical calculations taking into account the time of sunrise for Kirillov (59° 52' N, 38° 23' E) and Belgrade (44° 49' N, 20° 17' E). The reason for the latter choice will be explained below.

To analyze the accuracy of Kirillo-Belozersky tables, in figure 3 I have constructed a histogram of errors, breaking the whole range of errors into half-hour intervals. It may be clearly seen that the accuracy of the tables is much higher than one might expect from looking only at the numerical data. Most errors are concentrated around the central peak with a nearly normal distribution. The rms error is about 1.5 hours, several times smaller than would be expected if the Kirillo-Belozersky tables had been constructed using only the mean length of the synodic month. A significant number of errors corresponding to 1, −1 and 2 days is probably to be explained as either incorrect conversion of dates or scribal errors. Such obvious errors in subsequent calculations have been adjusted. Of the 66 errors of one day, 49 give the time of syzygy during the day. Of the two-day errors, all 7 examples concern daytime syzygies. By contrast, in case of an error on minus one day, 13 of the 15 cases are for nighttime syzygies. Errors in the range 6–18 hours probably have the same nature and are associated with a patchwork of daytime and nighttime hours.

Let us focus on the errors in the main peak of the distribution in the range from −6 to 6 hours. It may be seen that the peak of this distribution is shifted to the left, which may indicate that the tables were computed for a longitude to the west of Kirillov. The size of the shift to the left suggests a location about 20 degrees to the west of Kirillov. Because

	1.01fm	16.01nm	31.01fm	15.02nm	2.03fm	16.03nm	31.03fm	15.04nm	29.04fm	14.05nm	29.05fm	13.06nm	
	1.0n	2.0n	7.0d	4.0n	5.5n	6.5d	6.0d	5.0n	4.0n	6.0d	10.0d	15.0d	
Year 1	-9.8	-8.1	2.3	1.4	24.3	-22.8	0.9	1.0	1.8	-24.2	1.0	0.5	
Base 14	-9.9	-8.4	0.2	0.5	23.1	-24.1	-0.1	-1.0	-0.6	-24.3	1.2	0.7	
(1390)	27.06fm	12.07nm	27.07fm	11.08nm	26.08fm	10.09nm	25.09fm	9.10nm	24.10fm	8.11nm	23.11fm	7.12nm	22.12fm
	7.0n	5.0n	14.0d	3.0d	6.0d	10.0d	7.0n	10.0n	1.0n	0.5n	12.0n	1.0d	3.0n
	2.3	2.1	0.6	0.5	2.1	24.4	23.9	24.8	1.4	24.7	21.7	2.6	0.6
	-0.6	-0.6	0.3	-0.1	1.2	23.1	22.7	23.9	0.8	24.3	21.6	-0.1	0.6
	5.01nm	20.01fm	5.02nm	20.02fm	6.03nm	20.03fm	5.04nm	12.04fm	4.05nm	18.05fm	3.06nm	17.06fm	
	9.0n	13.0n	5.0n	1.5n	11.0d	5.5n	2.0d	4.5d	3.0n	1.0n	4.0n	5.0n	
Year 2	-0.7	0.8	25.8	28.7	3.0	0.4	-0.0	-168.0	2.8	3.1	17.9	20.8	
Base 25	-0.8	0.4	25.0	27.7	1.5	-1.1	-0.9	-168.7	0.4	0.4	15.0	17.8	
(1391)	2.07nm	16.07fm	31.07nm	15.08fm	30.08nm	14.09fm	28.09nm	13.10fm	27.10nm	12.11fm	26.11nm	11.12fm	25.12nm
	14.0d	11.0n	7.0n	5.0d	4.0d	9.0n	1.5n	6.0n	10.0n	7.5d	9.0d	12.0n	13.0n
	0.8	13.5	3.4	0.3	1.5	24.9	2.3	2.7	-0.8	2.5	2.8	-1.9	-0.6
	0.9	10.9	1.1	-0.4	0.5	23.5	1.2	1.8	-1.4	0.0	0.2	-1.8	-0.6
	10.01fm	24.01nm	9.02fm	23.02nm	9.03fm	24.03nm	7.04fm	22.04nm	7.05fm	22.05nm	5.06fm	20.06nm	
	5.0n	7.0n	1.0d	9.0d	11.0d	2.0d	6.0n	6.0n	4.5d	11.0d	13.5d	9.0n	
Year Base	2.2	2.5	2.4	1.8	1.0	-0.4	0.5	1.5	-0.7	-1.1	-0.2	3.1	
6	2.0	2.1	0.5	0.1	-0.4	-1.5	-1.4	-0.7	-0.9	-1.0	0.0	0.1	
(1392)	4.07fm	20.07nm	3.08fm	18.08nm	1.09fm	17.09nm	1.10fm	16.10nm	31.10fm	14.11nm	30.11fm	14.12nm	29.12fm
	5.0n	11.0d	10.0d	6.0n	9.0n	5.0d	4.0n	4.0n	2.0n	13.0n	2.5d	1.0d	11.0d
	0.9	0.6	1.5	1.7	0.9	1.9	1.5	1.9	3.5	-0.3	2.3	-5.5	-6.7
	-1.9	0.4	1.0	-0.2	-0.7	0.5	0.4	1.2	3.1	-0.5	-0.4	-8.3	-9.3
	12.01nm	28.01fm	11.02nm	27.02fm	13.03nm	28.03fm	11.04nm	27.04fm	11.05nm	26.05fm	10.06nm	24.06fm	
	12.0n	3.0n	2.0n	2.0n	8.0d	1.0n	8.0n	9.0n	11.0n	5.0d	6.5d	12.5d	
Year 4	-0.1	1.3	1.2	12.6	5.0	2.5	2.7	27.7	15.2	-0.5	1.6	0.1	
Base 17	-0.4	0.7	0.4	11.5	3.7	0.8	0.7	25.4	12.6	-0.4	1.9	0.3	
(1393)	9.07nm	23.07fm	8.08nm	22.08fm	6.09nm	20.09fm	6.10nm	20.10fm	4.11nm	19.11fm	4.12nm	18.12fm	
	4.0n	4.5n	7.0d	6.5d	5.0d	6.0n	5.5d	9.0d	4.5n	12.5n	1.0d	5.0n	
	3.4	1.9	0.5	2.0	-12.4	1.4	2.3	2.1	-0.2	19.4	3.2	-7.8	
	0.7	-0.5	-0.1	1.2	-13.5	0.1	0.5	0.1	-0.6	19.2	0.4	-7.7	
	2.01nm	17.01fm	31.01nm	16.02fm	2.03nm	18.03fm	31.03nm	16.04fm	30.04nm	15.05fm	30.05nm	14.06fm	
	2.0n	8.0n	12.0n	11.0d	11.0d	1.0d	11.0n	12.5d	1.0n	7.0n	10.0d	5.5d	
Year 5	-0.0	0.9	-0.1	2.0	1.0	0.6	1.1	-0.1	3.2	2.6	4.6	-0.2	
Base 28	-0.1	0.6	-0.8	0.2	-0.5	-0.7	-0.7	-0.8	0.8	-0.1	4.8	0.1	
(1394)	28.06nm	13.07fm	28.07nm	11.08fm	27.08nm	10.09fm	25.09nm	9.10fm	25.10nm	8.11fm	23.11nm	7.12fm	23.12nm
	3.5n	12.0n	2.0n	5.0n	1.0d	3.5d	5.0n	4.0n	5.0n	4.5d	6.5n	12.0n	4.0d
	1.6	17.7	7.6	1.8	-0.0	0.9	2.1	1.0	11.5	2.5	-0.1	-0.3	3.8
	-1.3	15.1	5.2	-0.2	-1.0	-0.3	0.9	0.1	10.9	0.1	-0.2	-0.2	1.1
	6.01fm	21.01nm	5.02fm	20.02nm	7.03fm	21.03nm	5.04fm	19.04nm	5.05fm	19.05nm	3.06fm	17.06nm	
	8.0n	5.5n	3.5n	13.0n	6.0d	11.5d	8.5n	7.5n	12.0d	9.5d	5.5n	6.0n	
Year 6	1.0	1.5	2.2	24.1	0.4	1.0	1.0	1.4	0.9	-0.1	2.7	2.4	
Base 9	0.9	1.1	1.5	23.1	-1.0	-0.1	-0.8	-0.7	0.6	-0.1	-0.3	-0.5	
(1395)	3.07fm	17.07nm	1.08fm	16.08nm	30.08fm	14.09nm	29.09fm	14.10nm	28.10fm	13.11nm	27.11fm	12.12nm	26.12fm
	5.5d	12.0d	13.0d	2.5d	6.0n	6.5d	3.5d	1.0n	4.5n	3.0d	2.0d	7.0n	8.0n
	0.1	0.0	1.1	-0.2	1.2	-11.6	1.5	1.2	1.5	3.1	2.6	-0.7	-0.7
	0.2	-0.1	0.7	-0.9	-0.5	-12.9	-0.1	0.4	0.9	0.7	-0.1	-0.6	-0.7
	11.01nm	25.01fm	9.02nm	24.02fm	10.03nm	25.03fm	8.04nm	23.04fm	7.05nm	23.05fm	6.06nm	21.06fm	
	6.0d	2.0n	7.5n	6.0n	2.5n	2.5d	12.0d	3.0n	5.5n	7.0n	6.0d	4.5n	
Year 7	3.2	0.6	1.9	11.7	12.9	2.5	0.6	2.0	1.9	17.6	-0.5	4.0	
Base 20	0.7	0.2	1.1	10.6	11.5	1.5	-0.2	-0.2	-0.6	14.8	-0.3	1.0	
(1396)	5.07nm	21.07fm	4.08nm	19.08fm	2.09nm	17.09fm	2.10nm	17.10fm	1.11nm	15.11fm	1.12nm	15.12fm	31.12nm
	2.0n	4.0d	5.0d	13.0d	6.0n	8.0n	1.5n	5.0d	8.0d	6.0n	14.0n	2.0d	8.0n
	3.3	-0.1	0.4	1.5	0.7	1.3	1.6	2.5	4.4	1.2	23.4	2.7	25.1
	0.5	-0.3	-0.1	0.7	-0.9	-0.0	0.6	0.6	2.2	1.0	23.4	-0.1	25.0

TABLE 1. Syzygies in the Tables of Kyrill of Belozersk: Date of full (fm) or new moon (nm), time in day (d) or night (n) in equal hours given in the tables and error for Kyrillov and for Belgrade.

	13.01fm	29.01nm	12.02fm	27.02nm	13.03fm	29.03nm	12.04fm	27.04nm	12.05fm	26.05nm	11.06fm	25.06nm	
	5.0n	7.0d	7.0d	8.5n	12.0d	4.5d	5.0n	13.0d	11.5d	5.0n	0.5d	4.0d	
Year 8	-1.5	2.7	1.0	2.6	-11.4	0.9	2.8	0.3	0.9	2.3	-1.6	-0.9	
Base 1	-1.7	0.5	-0.9	1.4	-12.8	-0.0	0.8	-0.1	0.8	-0.5	-1.3	-0.7	
(1397)	10.07fm	24.07nm	9.08fm	23.08nm	7.09fm	22.09nm	6.10fm	21.10nm	5.11fm	20.11nm	4.12fm	19.12nm	
	14.0d	13.0d	1.0d	10.0n	12.0n	2.0n	10.0n	5.5d	7.0d	14.0n	6.5n	9.5n	
	-0.7	0.0	-0.1	25.3	14.7	25.9	1.3	2.1	3.1	23.2	-1.2	-1.2	
	-0.8	-0.2	-0.7	23.5	13.2	24.7	0.4	0.0	0.8	23.0	-1.1	-1.1	
	3.01fm	18.01nm	2.02fm	17.02nm	3.03fm	18.03nm	2.04fm	17.04nm	1.05fm	16.05nm	31.05fm	14.06nm	
	3.5d	4.5n	6.0n	5.5d	5.0n	5.5n	6.5d	5.5d	11.0n	0.5n	1.5d	5.0d	
Year 9	2.4	0.7	25.0	1.6	12.7	0.7	11.7	0.6	16.8	4.2	-1.7	-16.6	
Base 12	-0.2	0.4	24.3	-0.1	11.5	-0.8	10.8	0.0	14.4	1.5	-1.5	-16.3	
(1398)	29.06fm	14.07nm	29.07fm	13.08nm	27.08fm	12.09nm	26.09fm	11.10nm	25.10fm	10.11nm	24.11fm	9.12nm	23.12fm
	4.0n	3.5d	7.0d	11.5d	9.0n	8.0n	10.5d	9.5d	11.5n	14.0n	0.5n	7.0n	9.5n
	4.1	-0.3	-0.9	25.0	2.4	49.2	2.4	26.5	0.9	47.4	1.2	21.1	-0.8
	1.3	-0.4	-1.3	24.3	0.7	47.8	0.8	24.7	0.3	47.1	1.2	21.2	-0.7
	7.01nm	22.01fm	6.02nm	21.02fm	8.03nm	22.03fm	6.04nm	21.04fm	6.05nm	20.05fm	4.06nm	18.06fm	
	6.0n	6.0d	9.5d	6.5n	2.0d	3.5d	5.0n	3.5n	4.5d	4.5d	14.5d	4.0n	
Year 10	1.1	2.5	2.1	26.0	0.6	0.6	2.8	27.6	-0.5	-0.8	0.0	3.4	
Base 23	1.0	0.2	0.1	25.0	-0.8	-0.5	0.9	25.4	-0.8	-0.8	0.2	0.5	
(1399)	3.07nm	18.07fm	2.08nm	16.08fm	31.08nm	15.09fm	29.09nm	14.10fm	29.10nm	13.11fm	28.11nm	12.12fm	27.12nm
	5.5n	8.5d	3.5d	10.0n	11.0d	6.0n	8.0n	7.0d	7.0d	11.0n	11.5n	2.0n	5.5n
	2.4	-0.8	0.4	1.1	1.4	3.5	1.0	-22.2	2.1	-0.6	23.3	-23.1	0.5
	-0.4	-1.0	-0.0	-0.9	0.4	2.2	-0.2	-24.1	-0.0	-0.8	23.3	-23.0	0.5
	11.01fm	26.01nm	10.02fm	25.02nm	12.03fm	26.03nm	10.04fm	25.04nm	9.05fm	24.05nm	8.06fm	23.06nm	
	12.0n	9.0d	8.5d	4.5d	5.0n	9.0n	3.5d	14.0d	14.0d	2.0d	9.0d	13.0d	
Year 11	0.3	2.6	2.4	2.2	48.2	25.3	23.9	24.3	23.9	-1.0	32.1	24.2	
Base 4	0.0	0.4	0.5	0.6	46.8	23.6	23.2	23.8	23.7	-0.9	32.4	24.4	
(1400)	6.07fm	21.07nm	5.08fm	20.08nm	4.09fm	19.09nm	3.10fm	18.10nm	2.11fm	16.11nm	1.12fm	15.12nm	30.12fm
	11.0d	5.5n	1.0d	4.5d	5.0n	1.0n	11.0d	10.0n	3.5d	7.0d	9.5n	10.0n	2.0n
	-1.0	2.4	0.3	1.7	26.5	26.7	2.8	24.6	2.6	2.5	-1.0	-0.9	-22.9
	-1.0	-0.1	-0.2	0.9	25.0	25.5	1.1	23.9	0.3	-0.1	-1.0	-0.8	-22.9
	14.01nm	29.01fm	13.02nm	28.02fm	14.03nm	29.03fm	13.04nm	27.04fm	12.05nm	27.05fm	11.06nm	25.06fm	
	1.5n	13.0n	3.5d	11.0d	9.0n	6.0n	3.0n	4.0d	7.0d	13.0d	5.0n	5.0n	
Year 12	0.7	0.7	1.7	2.1	0.7	0.3	2.8	-24.7	-25.1	-0.3	1.7	2.2	
Base 15	0.4	0.2	-0.2	0.6	-0.7	-1.4	0.8	-25.2	-25.2	-0.1	-1.3	-0.7	
(1401)	10.07nm	25.07fm	9.08nm	24.08fm	8.09nm	23.09fm	7.10nm	22.10fm	6.11nm	21.11fm	5.12nm	21.12fm	
	9.5d	6.5d	5.0n	5.0n	4.5d	10.0d	3.5n	3.0d	3.5d	13.5n	8.0n	9.0d	
	-23.6	0.1	2.2	25.2	1.7	25.5	2.5	1.7	7.7	24.9	8.6	19.9	
	-23.7	-0.2	0.1	23.4	0.5	24.1	1.5	-0.4	5.3	24.8	8.7	17.1	
	4.01nm	19.01fm	3.02nm	18.02fm	5.03nm	20.03fm	4.04nm	17.04fm	2.05nm	16.05fm	31.05nm	15.06fm	
	10.0n	9.0d	9.5d	12.5n	12.0n	1.5n	3.5n	5.5n	7.5n	5.0d	9.5n	13.0d	
Year 13	23.6	2.5	26.5	24.8	48.8	27.5	50.2	-0.3	15.8	-24.7	3.5	0.3	
Base 26	23.5	0.2	24.4	23.8	47.5	26.0	48.4	-2.4	13.4	-24.7	0.6	0.5	
(1402)	30.06nm	15.07fm	30.07nm	14.08fm	29.08nm	13.09fm	28.09nm	11.10fm	26.10nm	10.11fm	25.11nm	9.12fm	24.12nm
	15.0d	4.0n	4.0d	3.5d	4.5n	3.5n	4.0d	4.5d	5.5n	11.0n	0.5d	8.0n	2.5n
	-0.1	26.3	0.2	24.4	26.7	50.3	25.9	1.6	2.1	23.5	3.3	0.9	1.2
	0.1	23.7	-0.2	23.7	25.0	48.9	24.3	-0.2	1.6	23.2	0.6	1.0	1.2
	8.01fm	23.01nm	7.02fm	22.02nm	8.03fm	23.03nm	6.04fm	21.04nm	6.05fm	21.05nm	5.06fm	20.06nm	
	4.0n	11.0n	6.0d	9.5d	8.0n	9.5n	2.0d	11.0d	6.5d	0.5n	6.0d	14.0d	
Year 14	2.0	23.9	1.7	25.7	-1.4	25.7	-32.9	0.2	-15.1	17.2	0.1	21.9	
Base 7	1.8	23.4	-0.3	24.1	-2.7	24.1	-33.8	-0.4	-15.3	14.5	0.3	22.2	
(1403)	4.07fm	19.07nm	3.08fm	18.08nm	1.09fm	16.09nm	30.09fm	16.10nm	30.10fm	14.11nm	29.11fm	14.12nm	28.12fm
	13.0d	7.0d	4.5n	8.0n	3.0d	1.0n	2.0n	2.0d	1.0d	6.5n	6.5n	2.5d	2.5d
	0.6	0.3	25.9	25.4	1.0	1.8	1.6	1.9	2.1	1.8	23.1	3.6	-5.3
	0.7	0.1	23.7	23.5	-0.0	0.5	0.5	-0.0	-0.1	1.6	23.1	0.8	-8.0

TABLE 1 (cont).

	12.01nm	27.01fm	11.02nm	26.02fm	12.03nm	27.03fm	11.04nm	26.04fm	10.05nm	25.05fm	8.06nm	23.06fm	
	5.0n	6.0d	13.0n	1.0d	10.5d	5.5n	6.0n	5.0d	7.5d	4.0n	5.0n	5.5d	
Year 15	1.6	1.8	24.4	0.7	25.2	25.1	48.8	20.3	24.1	26.0	28.5	0.2	
Base 18	1.4	-0.4	23.6	-0.9	23.9	23.4	46.8	19.8	24.0	23.1	25.5	0.4	
(1404)	7.07nm	22.07fm	6.08nm	21.08fm	4.09nm	19.09fm	4.10nm	18.10fm	2.11nm	16.11fm	2.12nm	16.12fm	
	8.0d	13.5d	7.5n	5.5n	3.0n	2.5d	7.0d	2.5n	13.0n	14.5n	6.5n	6.0n	
	-0.1	1.3	24.9	25.1	2.5	0.6	1.6	1.3	0.3	0.3	1.7	1.1	
	-0.1	1.0	22.8	23.3	0.9	-0.8	-0.1	0.5	-0.1	0.1	1.7	1.2	
	1.01nm	15.01fm	30.01nm	14.02fm	1.03nm	15.03fm	30.03nm	14.04fm	28.04nm	13.05fm	27.05nm	12.06fm	
	4.0d	6.5d	7.0n	0.5d	2.0d	6.5n	12.0d	1.0n	5.0n	3.5d	4.5d	1.5n	
Year 16	3.3	2.4	2.0	1.2	1.0	0.5	1.1	3.7	1.3	-24.7	-24.9	3.0	
Base 29	0.6	-0.0	1.4	-0.7	-0.5	-1.0	0.1	1.7	-1.0	-24.8	-24.7	0.0	
(1405)	26.06nm	11.07fm	25.07nm	10.08fm	24.08nm	8.09fm	23.09nm	8.10fm	23.10nm	6.11fm	21.11nm	5.12fm	21.12nm
	15.0d	3.0d	1.0d	12.5d	2.5n	7.5n	4.5d	5.0d	1.5d	5.5n	8.5n	14.0n	3.5n
	0.0	-24.4	-24.7	1.1	2.0	1.5	-1.2	2.6	3.0	2.1	-1.0	-1.4	0.2
	0.2	-24.4	-25.0	0.5	0.3	-0.0	-2.7	0.9	0.9	1.8	-1.0	-1.4	0.3
	4.01fm	20.01nm	3.02fm	18.02nm	4.03fm	20.03nm	3.04fm	17.04nm	3.05fm	17.05nm	1.06fm	16.06nm	
	4.5n	3.5d	3.5d	7.0n	7.5n	7.0n	13.5d	13.5d	5.5d	5.5d	5.5n	3.5d	
Year 17	0.6	2.0	1.5	1.9	0.3	17.1	1.4	-22.9	-0.1	-15.4	2.1	-1.2	
Base 10	0.5	-0.3	-0.5	0.9	-0.9	15.6	0.5	-23.5	-0.4	-15.4	-0.8	-0.9	
(1406)	1.07fm	15.07nm	30.07fm	14.08nm	29.08fm	13.09nm	27.09fm	12.10nm	27.10fm	11.11nm	26.11fm	10.12nm	26.12fm
	11.0d	12.5d	9.0n	7.0n	11.0d	10.0d	10.5n	0.5d	7.0d	8.5n	8.0n	4.5n	2.0d
	-1.0	0.3	1.9	25.0	1.7	25.7	3.2	1.7	3.6	23.8	25.5	-0.5	26.4
	-0.8	0.2	-0.4	23.0	0.7	24.4	2.0	-0.1	1.5	23.6	25.5	-0.4	23.7
	9.01nm	23.01fm	8.02nm	22.02fm	9.03nm	23.03fm	8.04nm	22.04fm	7.05nm	21.05fm	5.06nm	20.06fm	
	8.5d	3.0n	1.5d	2.0d	6.0n	4.5n	4.5d	7.0d	1.5n	6.5n	5.5n	13.5d	
Year 18	2.1	-0.8	1.3	1.0	2.7	2.6	0.9	0.6	4.9	2.1	2.4	-0.6	
Base 21	-0.4	-1.3	-0.7	-0.6	1.3	1.0	0.1	0.1	2.4	-0.6	-0.6	-0.3	
(1407)	5.07nm	20.07fm	3.08nm	18.08fm	2.09nm	17.09fm	1.10nm	16.10fm	30.10nm	15.11fm	29.11nm	14.12fm	28.12nm
	4.0d	3.5d	0.5n	5.5n	6.0n	8.5d	6.5d	9.0d	10.0n	8.5d	5.0n	8.0n	8.5d
	-0.4	-0.8	5.9	2.6	25.2	2.6	2.1	-8.6	0.3	3.9	0.7	-1.4	-21.1
	-0.3	-1.0	3.7	0.7	23.6	1.2	0.4	-10.5	-0.2	1.4	0.7	-1.3	-23.8
	13.01fm	27.01nm	11.02fm	26.02nm	12.03fm	27.03nm	11.04fm	26.04nm	10.05fm	25.05nm	9.06fm	24.06nm	
	5.0d	13.0n	5.5n	10.5n	2.0d	2.0n	0.5n	3.0d	1.5d	14.0d	0.5n	5.5d	
Year 19	2.6	-4.5	1.4	0.6	0.9	2.2	26.3	-0.1	-0.6	0.1	27.6	7.7	
Base 2	0.1	-5.0	0.6	-0.5	-0.4	0.5	24.3	-0.6	-0.8	0.2	24.7	7.9	
(1408)	8.07fm	23.07nm	6.08fm	21.08nm	5.09fm	19.09nm	5.10fm	19.10nm	3.11fm	17.11nm	3.12fm	17.12nm	
	4.5d	4.0d	5.5n	11.5d	1.0n	6.5n	3.5d	5.5d	7.5n	7.5n	9.5d	1.0n	
	-0.9	0.2	1.6	1.7	3.5	0.9	1.9	2.5	-0.7	-1.0	4.4	0.4	
	-0.9	-0.0	-0.5	0.8	1.9	-0.3	0.2	0.5	-1.1	-1.2	1.7	0.5	

TABLE 1 (cont).

the times of the syzygies given in the Kirillo-Belozersky tables are given relative to sunrise and sunset it is possible to investigate the latitude of the place for which they were computed. The width of the peak curve depends upon the latitute chosen for the calculation of the moment of syzygy relative to sunset or sunrise using modern astronomical theory: the width of the peak will reach a minimum when we assume the same latitude in the modern calculations as was used in the computation of the original tables. In figure 4 I present the same calculation for two different latitudes, approximating the histogram with a Gaussian profile. From this, in figure 5 I have constructed a graph of the peak width as a function of the latitude from which it is possible to find a minimum error and hence the best fitting latitude for the data in the Kirillo-Belozersky tables. In this figure, black circles with error bars mark the results of treatment with a step of 5 degrees in latitude and the solid line is the approximating 5th order polynomial. The optimal latitude is about 45° N. For this latitude the tables significantly improve their accuracy down to 0.85 hours. As discussed above, the shift of the central peak for this latitude allows us to estimate that the optimal longitude is about 20° E. Formal error estimates of the obtained coordinates give about

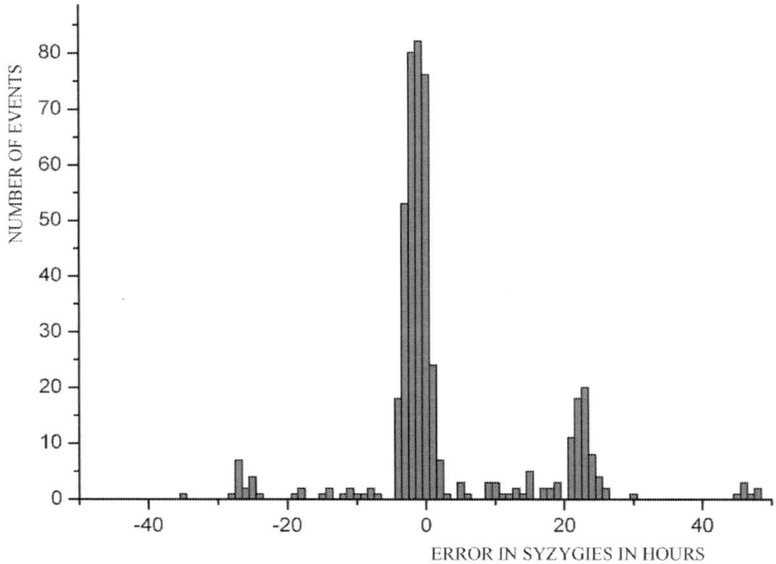

FIGURE 3. Histogram of errors in times of syzygies in the Kirillo-Belozersky tables.

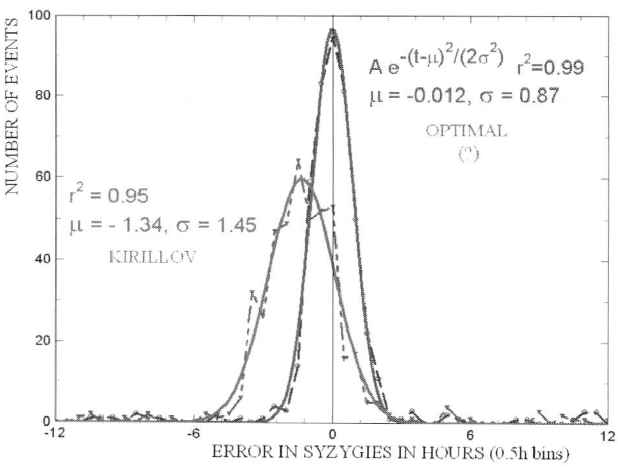

FIGURE 4. The determination of the latitude and the longitude of a place where the Kirillo-Belozersky tables were compiled using mean value and variance of errors. On the left is the histogram for the coordinates of Kirillov, approximated with normal distribution, on the right the histogram for the optimal coordinates 45° N, 20° E.

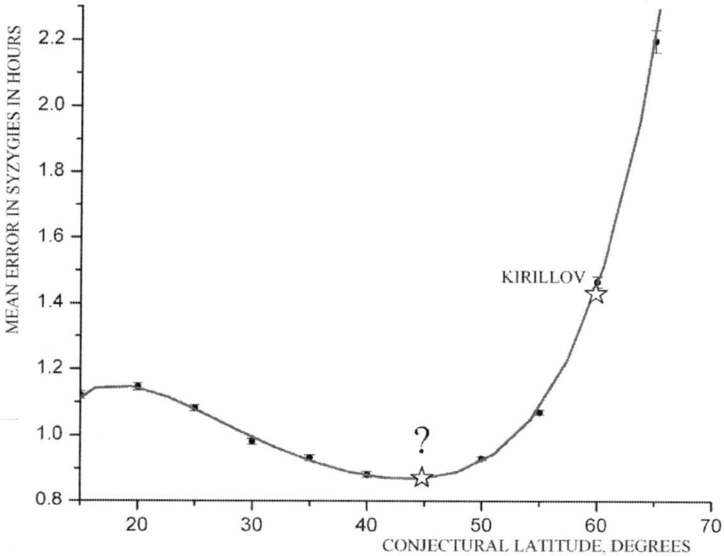

FIGURE 5. The determination of the latitude of a place where the Kirillo-Belozersky tables were compiled. Stars correspond to the latitude of Kirillov (on the right) and Belgrade (in the centre).

one degree (as approximately mean-square error divided by the square root of the number of values), but should be taken with caution since the method of calculation used by the originator of the tables and his probable biases are not known. Nevertheless, the coordinates suggest that the Kirillo-Belozersky tables were compiled for a site in the territory of Serbia, in the vicinity of its ancient capital Belgrade (see figure 5). The accuracy of the tables was obviously quite satisfactory for a Russian scholar, but still worse, even allowing for rounding errors, than one could, in principle, obtain using the Alfonsine tables. The originator, perhaps, used a simplified version of the calculations or the calculations were severely rounded.

In addition to the dates and time of syzygies, the Kirillo-Belozersky tables also list possible solar and lunar eclipses. In total sixteen lunar and five solar eclipses are mentioned. For example, in the table for the first lunar cycle after the full moon on April 29 the table states: 'In the third [hour] and half an hour of the night the perishing of the moon starts three fingers and holds one hour and two parts'. For each of the lunar or solar eclipses, in addition to the date the following information is given:

1) Start time of the eclipse to within half an hour. In some cases, instead of the start time the record is introduced by word meaning 'than', 'the same time'. One would think that in such cases the estimated start time coincides with the time specified in the table of syzygies but verification shows (see table 3) that it is not. In some cases it is not specified whether night or daytime was used. For example, in the case of the eclipse on 6 November 1307, although the new moon is given in hours from sunrise, calculations show that 'eighths same hour' should be taken from the previous sunset.

2) The maximum phase of the eclipse in 'fingers' apparently corresponds to a traditional method of phase measurement used since Ptolemy of 12th fractions of the diameter. Phase 12 fingers corresponds to a total eclipse, and it is appropriate specified in the text: '12 fingers, that means total'.

3) The duration of the eclipses in hours. Durations may be given in whole and half-integer hours and in other fractions called 'parts'. In the treatise of Kirik of Novgorod 'fractional hours' corresponding to a fifth of an hour are found, as we can assume the same use here. Note the disproportionately high frequency of use of durations of 'one hour and two parts', probably due to inclusion in this category of all eclipses below a certain magnitude.

Tables 2 and 3 summarize the eclipses in the Kirillo-Belozersky tables and the results of modern calculations. For each eclipse I give the calculated local time, the time of sunset or sunrise in Belgrade and the resulting difference in hours from the time of the eclipse given in the tables. The local circumstances of solar eclipses were determined using the program EmapWin by Sh. Takesako which uses the modern long numerical ephemerides JPL DE406.[22] Lunar eclipses are directly calculated from these ephemerides. The transformation from ephemeris to universal time (ΔT correction) to account for deceleration and fluctuations in the rotation of the Earth was also applied.[23]

The composer of the Kirillo-Belozersky tables missed three lunar eclipses on 20 March 1391 (2.4 fingers), 9 March 1392 (total) and a very small one on 26 October 1398 (0.5 fingers). The eclipses of 6 November 1397, 22 July 1404 and 15 November 1407 took place during the hours of daylight, while the eclipse of 12 April 1397 was only penumbral. For all the lunar eclipses the rms error in the time of the eclipse is 1.6 hours; however, if only those eclipses where the start time is explicitly indicated are considered, the rms is 0.88 hours, similar to the accuracy of general syzygy times.

During the 19-year interval covered by the tables, not five but eight solar eclipses could be observed in Serbia. Missing are three small eclipses: 24 March 1392 (4.5 fingers), 8 August 1393 (4.0 fingers), and 11 January 1396 (4.9 fingers). The eclipse of 6 May 1399, on the contrary, could not be seen in Europe—it was only visible in the South Atlantic. The accuracy of the predicted solar eclipses times is approximately 1.5 times worse than the accuracy of syzygies and lunar eclipses. The accuracy of the predictions of local circumstances is bad, which is not surprising given the level of development of astronomy at the time and complexity of the calculation of the local circumstances.

The Origin of the Tables

Astronomical analysis of the tables in the collection of Kirill of Belozersk has shown that they are a copy of some astronomical tables compiled in Serbia. This finding is in agreement with what is known about the other texts of this collection. In particular, the tables are preceded by a Serbian edition of 'Kormchaya kniga' ('Book of the Helmsman').[24] The astronomical tables were composed in a difficult time for South Slavs. In the year preceding the beginning of the table the fateful battle of Kosovo happened, and by the middle of fifteenth century, Ottoman Turks conquered almost the whole of Serbia. The entry into the Russian territory of intellectuals fleeing from Turkish conquest of the Balkans in the

Date	Text time (hours)	Text duration (hours; parts)	Text phase (fingers)	Calculated beginning of eclipse and error	Calculated Duration	Maximum phase (fingers)
29.04.(1390)	3.5 n	1;2	3	29.04 23:21 (+0.67)	0:17	0.1
1.09.(1392)	7.5 n	3.5	12 t	02.09 02:15 (+0.40)	3:41	>12
26.12.(1395)	6.0 n	3.5	12 t	26.12 23:27 (+0.90)	3:38	>12
21.06.(1396)	3.0 n	3.5	12 t	21.06 21:21 (-1.45)	3:43	>12
11.06.(1397)	8.0 (n?)	2	4	11.06 04:52 (+1.07)	1:50	2.6
4.12.(1397)	– (6.5 n)	1;2	2	—	—	0
21.04.(1399)	1.0 n	3.5	12 t	20.04 19:25 (-0.48)	3:27	>12
3.10.(1400)	– (11 d)	3.5	12 t	03.10 14:07 (-3.12)	3:35	>12
18.02.(1402)	11.5 n	1;2	2.5	18.02 05:39 (+0.42)	1:46	2.8
3.08.(1403)	2.5 n	3.5	12	02.08 22:13 (+0.52)	3:30	>12
22.07.(1404)	– (13.5 d)	3	11	22.07 15:32 (-2.68)	3:01	10.5
5.12.(1405)	– (14 n)	1;2	2	06.12 07:26 (+1.08)	0:46	0.5
1.06.(1406)	4 n	3;2	11	02.06 00:26 (+0.72)	3:13	9.2
26.11.(1406)	6 n	3.5	12 t	25.11 21:12 (-1.15)	3:22	>12
21.05.(1407)	4 n	3.5	12 t	22.05 00:39 (+1.08)	3:53	>12
15.11.(1407)	– (8.5 d)	3.5	12 t	15.11 12:26 (-3.22)	—	—

TABLE 2. Lunar eclipses in the Kirillo-Belozersky tables.

Date	Text time (hours)	Text duration (hours; parts)	Text phase (fingers)	Calculated beginning of eclipse and error	Calculated Duration	Maximum phase (fingers)
5.04.(1391)	10.5 d?	1;2	2	6:47		
(+1.6 -9.05)	1:43	3.0				
6.05 (1399)	- (4.5 d)	1;2	4	—	—	—
29.10.(1399)	- (7.0 d)	2	6 t	12:32 (-1.23)	2:42	9.9
16.06.(1406)	1 d	1;2	4	6:25 (+1.2)	2:00	10.6
19.10.(1408)	4 d	2	4	9:19 (-1.23)	2:12	4.9

TABLE 3. Solar eclipses in the Kirillo-Belozersky tables.

fourteenth and fifteenth century (the second South Slavonic influence) is well known to scholars.

I could not find any information about the level of development of computational astronomy in Serbia at this time, although the Kirillo-Belozersky tables indicate the existence of advanced astronomical scholarship there. It is known, however, that the famous historian and astronomer Nicephorus Gregoras visited Serbia[25] and that he knew both Alfonsine and Toledan Tables.[26] A possible indirect indication of the existence of such astronomical school in Serbia may be found in the figure of Planes Dalmatius (Dalmau Planes/Sesplanes/Ces-planes), who came to Barcelona with his teacher Petrus Engisberti de Rucherna on the invitation of Peter III of Aragon in 1360–1366. Following the death of his teacher, Dalmatius completed the compilation of a set of astronomical tables for the years 1361–1433.[27] We also know that Dalmatius was an active astronomer at the court and about 1379 wrote a book about solar and lunar eclipses.[28] His name indicates a possible origin or connection to Dalmatia, which was the territory of modern Croatia and Montenegro.

Further studies must obviously be related to the search of similar manuscripts in the collections of Russian and Serbian sources. Both directions resulted in some success. A large number of Easter and lunar tables are analyzed and described by Pentkovsky.[29] Among these tables he found one in a *Sledovannaya Psaltyr* (Psalter with the Order of Services) of the end of the fifteenth century which also contains references to eclipses.[30] Inspection of this source showed that this table is identical to the Kirillo-Belozersky tables, covering the same 19-year period 1390–1408 but without drawings of phases of eclipses. In this way these tables are, apparently, a secondary copy of Serbian tables. I found similar copies in a *Sledovannaya Psaltyr* of the fifteenth century in the collection of the *Troitse-Sergieva lavra*.[31]

Many astronomical and astrological texts are collected in the monograph of N. Jankovic.[32] In particular, this monograph describes and partly reproduces five Serbian manuscripts with calculated characteristics of eclipses. Although the sets of eclipses are somewhat different, in the manuscripts of Hilandar Monastery 651 and Belgrade Community Library 36/a they nearly coincide with the Kirillo-Belozersky tables, and the textual descriptions are almost identical. All these tables are written in the same Old Church Slavonic language as the Kirillo-Belozersky tables.

In the manuscript of National Library of Serbia 36/b the table of syzygies and eclipses is apparently intended for the 19 years following those in the Kirillo-Belozersky tables (1409 to 1427). In the manuscript Hodoshki zbornik 182 there are no descriptions of eclipses for the years 8 to 15 of the cycle but schematic drawings of maximum eclipse phases similar to those in the Kirillo-Belozersky tables are present on the margins.

Of greater interest is, apparently, the manuscript of the Serbian Academy of Arts and Sciences #147, in which many of the numerical data equivalent to Kirillo-Belozersky tables are presented with much greater accuracy. For example, for the first lunar eclipse of April 29 1390, this manuscript states: 'The same night a diminishing of the Moon will happen with 3 fingers and 2 parts. It will begin on the 3rd hour of the night with 25 lepts and ends before 5th hour and 11 fractions. From the start to the end it will be 1 and 46 partials' (ms. 147, p. 88). Both 'lepts' and 'fractions' apparently correspond to modern minutes. It seems likely that all the numerical data obtained in the other manuscripts, including the Kirillo-Belozersky tables, are rounded values of the data in this manuscript, which, if not an original source, is closest to it. In this way the Serbian origin of the tables defined previously by astronomical analysis is confirmed by Serbian sources.

Summing up the historical-astronomical analysis of the Tables in the compilation of Kyrill of Belozersk it is possible to deduce that these tables are at the best level of astronomical knowledge of this time and have parallels in European astronomical texts. These tables witness the existence of South Slavonic astronomical scholarship and show vivid interest of Russian medieval scribes to the latest astronomical achievements.

Notes

1. Svyatsky (2007).
2. Juravel (2005).
3. Troitse-Sergieva Lavra #17, 38, 46, 761, 762, 765, 795 (http://www.stsl.ru/manuscripts).
4. Ibid. #17 (2007), p. 336.
5. Ibid. #765 (1654), p. 88.
6. M.Gartsman, Private communication.
7. *Entsiklopediya* (2007); Prokhorov & Rozov (1981).
8. Pamyatniki Literatury Drevney Rusi. Vtoraya polovina XV veka (Monuments of Ancient Russian Literature, in Russian). Moscow. 1982. Illustrations. L. 184 v. In colour.
9. McCluskey (1998).
10. Toomer (1968).
11. Steele (2000), p. 135.
12. Chabas & Goldstein (2003).
13. Steele (2000), p. 138.
14. Steele (2000), p. 137.
15. Thorndike (1951).
16. Thorndike (1942).
17. Thorndike (1957).
18. Stepanov (1917), p. 8.
19. Entsiklopediya (2007), pp. 198–208.
20. Simonov (2004).
21. Kirik Novgorodets (1953).
22. Shinobu Takesako http://www2c.biglobe.ne.jp/~takesako/index_e.htm.
23. Morrison & Stephenson (2004).
24. Entsiklopediya (2007). pp. 298–299, 310.
25. Theodossiou et al. (2006).
26. Pingree (1976).
27. Thorndike (1937, 1950).
28. Thorndike (1946).
29. Pentkovsky (1990).
30. Uvarov collection.#670 (806), State Historical Museum, Psaltyr s vozsledovaniem, pp. 647–651v.
31. Troitse-Sergieva Lavra, #314 (847), pp.595–601.
32. Jankoviħ (1989), pp. 137–146.

References

Chabas, J. and Goldstein, B. R., 2003, *The Alfonsine Tables of Toledo* (Dordrecht/Boston/London: Kluwer Academic Publishers).

Entsiklopediya russkogo Igumena, 2001, (Encyclopedia of Russian father superior, Collection of St. Kyrill of Belozersk, in Russian), Rossiyskaya Natsionalnaya Biblioteka, Kirillo-Belozersloe sobranie, #XII, (Sankt-Petersburg), 113–125.

Jankoviћ, N.Ђ., 1989, *Astronoija u starim srpskim rukopisima* (Beograd).

Juravel, A.V., 2005, "Months 'of scribes' and 'celestial': their relation in letopises" (in Russian), *Historical-Astronomical Research* 30, 56–75.

Kirik Novgorodets, 1953, "Ucheniye imzhe vedati cheloveku chisla vsekh let" (facsimile, translation, comments by Zubov V. P.), *Istoriko-Matematicheskie issledovaniya* (Historical-mathematical researches, in Russian), 6, 174–212.

McCluskey, S. C., 1998, *Astronomies and Cultures in Early Medieval Europ*, (Cambridge: Cambridge University Press).

Morrison, L.V. and Stephenson, F. R., 2004, "Historical Values of the Earth's Clock Error ΔT and the Calculation of Eclipses", *Journal for the History of Astronomy* 35, 327–336.

Pentkovsky, A. M., 1990, "Calendar Tables in Russian Manuscripts of the XIV–XVI Centuries" (in Russian), *Metodicheskie rekomendatsii po opisaniyu slavyano-russkikh rukopisnykh knig* 3, 136–200.

Pingree, D., 1976, "The Byzantine Version of the Toledan Tables: The Work of George Lapothes", *Dumbarton Oaks Papers* 30, 85–132.

Pkokhorov, G. M. and Rozov, N. N., 1981, "Perechen knig Kirilla Belozerskogo", *TODRL* 36, 363–364 (with facsimile grayscale copies of L.184v and L.185r).

Simonov, R. A., 2004, "'Kosoy' chas i osmyslenie prirody bogosluzhebnogo vremeni", in *Drevnerusskaya kosmologiya* (Ansient Russian Cosmology, in Russian) (Saint Petersburg, Aleteya), 317–325.

Steele, J. M., 2000, *Observations and Predictions of Eclipse Times by Early Astronomers* (Dordrecht/Boston/London: Kluwer Academic Publishers).

Stepanov, N. V., 1917, *Kalendarno-khronologichesky spravochnik* (Calendar-chronological reference book, in Russian) (Moscow).

Svyatsky, D. O., 2007, *Astronomiya drevney Rusi* (Astronomy of ancient Russia), edition, comments, appendixes by M.L.Gorodetsky (Moscow: Russkaya Panorama).

Theodossiou, E. Th., Manimanis, V. N., Dimitrijevic, M S. and Danezis, E., 2006, "The Greatest Byzantine Astronomer Nicephoros Gregoras and Serbs", *Publications of the Astronomical Observatory Belgrade* 80, 269–274.

Thorndike, L., 1937, "Introduction and cCanon by Dalmatius to Tables of Barcelona for the years 1361–1433 A.D.", *Isis* 26, 310–320.

Thorndike, L., ,1942, "Other Astronomical Tables Beginning in the Year 1361", *Isis* 34, 6–7.

Thorndike, L., 1946, "Dalmatius Again", *Isis* 36, 158.

Thorndike, L., 1950, "The Tables of Barcelona of the XIVth Century", *Isis* 41, 283–285.

Thorndike, L., 1951, "Prediction of Eclipses in the Fourteenth Century", *Isis* 42, 301–302.

Thorndike, L., 1957, "Eclipses in the fourteenth and fifteenth centuries", *Isis* 48, 51–57.

Toomer, G. J., 1968, "A survey of the Toledan tables", *Osiris* 15, 5–174

Telling Time with the Moon: An American Overview

Stanisław Iwaniszewski

0. Introduction

As a phenomenon observable in the day and night sky, the moon has played an important role in human lifeworlds for thousands of years. The size, brightness and changing phases of the Moon made it the subject of many earthly projections, giving rise to a plethora of metaphorical and symbolic meanings. The relationship between the moon and events on earth can also be inferred from the human practices that use her changing phases to count time. Human societies throughout the world observed the moon in relation to seasonal-meteorological, natural-vegetation and social-cultural events linking the lunar cycles to the rhythms of seasons and human affairs. The rise and growth of the life on the earth was metaphorically connected to the rise, growth and decline of the moon. Drawing on these observations many ancient and non-modern peoples started to order their activities (political, ritual, and economic) according to lunar phases. Some of them counted and reckoned time developing what today we call lunar calendars. In those societies the moon became a powerful celestial instrument used to count time. Gazing at the moon is, therefore, a starting point leading to the study of the history of the development of human concepts of time.

Somewhere in between the first attempts to synchronize diverse sequences of events occurring in human lifeworlds and the development of precise lunar and luni-solar calendars fall the systems of time-reckoning based on changing lunar phases invented by diverse Native American groups.

The present paper is divided into two main parts. The first section examines the concept of time-reckoning based on the observation of lunar phases with a particular attention paid to 'time indicators'. The second part of the paper describes the types of 'lunar calendars' utilized by native Americans in prehispanic America.

1. Concepts of time indication and time reckoning

Ethnographers have shown that many ancient and non-modern societies used the moon for counting time. Examples of various moon-based systems adapted for time-reckoning are provided by Martin P. Nilsson in his important study on *Primitive Time-Reckoning*.[1] However, in providing multiple examples of time-reckoning based on the perception of various astronomical events, Nilsson insists that they do not necessarily refer to a continuous recording of time which we call a 'calendar'. Instead, Nilsson argues that all human societies ultimately develop two types of time perception called 'time indication' and 'time reckoning'. Put simply, time-indication events are those which mark the passage of time. Thus all temporal judgments based on references to concrete recurrent and natural events, either derived from astronomical, meteorological and seasonal environmental cycles (such as 'sunrise', 'first moon', first rain, the start of the warmer part of the year, the arrival of migratory birds, the flowering of certain plants, and so on) or from human social and cultural cycles (such as political, ritual, or economic sequences) are manifestations of time-counting systems called 'time indication'. Nilsson considers that regularly returning natural events constitute a recurring sequence of qualitatively distinct events of variable duration, with gaps between the end of one event and the start of the following one, so each stage may easily be anticipated, but not computed in advance because time indications are punctiform (point-like) and concrete rather than durative and numerical.[2] Therefore, the passage of time is conceptualized as the number of days, moons, rains, summers, bird- and flower-periods, etc., or as the number of years of a particular ruler, the number of generations of rulers that belong to the same dynasty, in terms of the time spans before or after wars, the founding of capitals, constructions of buildings, etc. The passage of time is counted since that time indicator event. In other words, the indicative time refers to all the relevant events upon which the social life of societies are built. For that reason the indicative time which is represented by a succession of activities important for peculiar lifestyles reveals the fundamental interests of native populations: rituals and feasts, public annual gatherings, changes in power of local authorities, agricultural horticultural, hunting and gathering activities, and so on. As various time indicators are unevenly distributed, they cannot represent formal arrangements of uniform or standardized units of time. Even if time indicators are ordered according to some fixed sequence, they do not represent regular divisions, nor can they stay for the idea of a homogenous and continuous time. Time indicators refer to culturally defined and socially meaningful 'guide-marks' which tell people when specific social activities ought to be initiated. In turn, they can be converted into 'dividing marks', structuring the sequence of recurring events.[3] At this moment it is still not necessary to ascribe a fixed duration to each of the recurring events, or to conceive of a total succession of events.

In addition to this Nilsson argues that since only some societies have invented calendars based on regular and continuous time units, the 'time-indication' system precedes that of time-reckoning.[4] Following Hallpike[5] and Lucas,[6] I consider that both systems of the counting of time may be operational among the same societies because they are activated at different levels of social interaction. A similar idea is expressed by Elias who notices that much of the use of peculiar time-reckoning systems depends on the development of the social requirements.[7] It may therefore be concluded that with all probability, and especially within more complex societies, individuals may operate within a variety of kinds of temporal orientation schemes which are thought of as appropriate to specific practices or discourses. As Lucas observes '... even clocks may largely have been used as time indication rather than time-reckoning'.[8]

Bearing this in mind, I shall seek to show that most of what is said about the methods of time-reckoning used by the historical Native Americans falls within the indicative time category. Even where the solstice moments were recognized as the turning points of a year and used as starting points for a series of differentiated lunations or 'moons', it is doubtful they were used to record the passage of greater units of time, that is, a series of years, initiated from a definite fixed point.

Nilsson suggests that the changing phases of the moon had for many non-modern societies the character of time indicators rather than that of measuring the passage of time, clearly indicating that this does not necessarily demonstrate the existence of a formal time-reckoning system, or a lunar calendar. It should be emphasized that for many non-modern societies the sky provides meanings through connections existing between the movements perceived in the heavens and diverse domains of human life in the context of changing climatic-meteorological processes. By this I want to say that the sighting of the new or full moon is usually made in the context of important social (economic, political, ritual) activities and in relation to some meteorological or earthly events rather than in the context of a purposeful intellectual search for some new perception of temporality. People have observed the moon not because they were interested in determining the length of a lunar month, but because the appearance of the thin narrow crescent in the western sky told them that they should (or should not) behave in a particular manner. In traditional non-Western societies, moon-sighting is always embedded in a social and cultural context.

1.1 The relation of lunar phases to the indicative time system

The indicative time method proposes that there are certain recurrent natural and non-human events that coincide with other important climatic-environmental changes and with social activities that are linked to them; and those events are selected to announce the arrival of important changes in the environment and to signal the need to start the adequate activities. As stated above, indicative time is irregular, discontinuous, puntiform, and aoristic.[9] The starting point for the moon-reckoning is attached to a concrete event perceived in the sky (the putative time indicators being for example 'first moon', 'full moon', 'no moon', or 'last moon') and their regular succession serves to construct longer chronological sequences.[10] However, even if the lunar month is called 'moon', it is often not a complete month in our sense; instead it is limited to the time when the moon is visible in the sky. Counting the moons represents gaps between the last visible waning moon and the first visible waxing moon and depending on visibility conditions such lunar months may be of varying length, implying that these time units are not uniform, homogenous or continuous.

When seasonal variations are involved encompassing longer periods of time, such as half-years, counted in winters and summers, or in rainy and dry seasons, a system of several lunations is required to synchronize environmental changes with prevailing or communal activities. In such a system various successive lunations are arranged into a series of recurrent moons that are named after the seasonal events they refer to. Time periods defined in such a way do not depend on assigning any definite or precise beginning or end to the seasonal change, it is sufficient to have named moons coinciding with the seasonal changes they allude to. From time to time, when the names of moon-months do not coincide with the season, additional thirteenth lunation is added. The time is reckoned in terms of prevailing seasonal activities rather than in terms of abstract lunations. Thus, for example the 'harvest' moon which coincides with the time adequate for ripening maize cobs only reaffirms the sense of the sequence of events. Obviously one can compute, utilizing a sequence of named moons, the number of moons that are placed between the 'planting' moon and the 'harvest' moon, but this does not mean there are formal, homogenous and bounded time units separating both events.

1.2 Earliest examples of lunar count records

We do not know when the people started to use the changing phases of the moon to mark the passage of time. Drawing on many well-documented ethnographical records of hundreds of peasant calendars usually made of incised notches on wooden tables or sticks, cultural astronomers working on the Upper Palaeolithic have interpreted a series of marks cut onto diverse bone and antler 'batons' as systems of notation or tallies. Several such items were described by Alexander Marshack.[11] From archaeological findings we know that the European Upper Palaeolithic environment was intensively exploited by diverse groups of hunters and gatherers, and the environment was highly seasonal, with a continuous change of times of year. So, it seems likely that hunters and gatherers observed the phases of the moon. If Marshack's interpretations are correct, we may assume that the oldest known object rendering the lunar count is incised on a bone plaque, an archaeological 'polisher', from the site of Blanchard (Dordogne, France) dated around 29,000 – 28,000 BCE. It displays a serpentine line consisting of 69 incised marks (2⅓ lunations) which is several times turned to model the periods corresponding to the waxing or the waning moon. Though we will never be sure if these bone objects really record a lunar count, they nevertheless seem to suggest that some kind of information was recorded as notches stored on Upper Palaeolithic artifacts, representing what d'Errico calls 'an artificial memory system'.[12] On the other hand, Palaeolithic notations may represent other events and numbers, for example, hunting tallies, as was already suggested by Eduard Lartet 150 years ago.

Marshack also attempted to link the seasonal images of animals from the cave at Lascaux and other sites to the meandering lines found on the Blanchard plaque.[13] Indeed, as Marshack proposes, the serpentine or zigzag-like lines of incised marks on the Blanchard plaque appear to subdivide the lunar cycle into parts corresponding either to the waning or waxing periods of the moon. The turns in lines may be associated with the shifts from one phase to another. Furthermore, observing that Upper Palaeolithic rock art images render clear seasonal and sexual representations and relations, Marshack proposed that the changes of the lunar phases were correlated with the seasonal changes in the environment. If his hypothesis is true, it reveals the aspect of time indication concept.

While discussing the existence of possible lunar calendars in America, one should also bear in mind that in the beginning the month may not necessarily be related to the annual or seasonal cycle.[14] However, the continuous observation of the phases of the moon may lead people to develop lunar calendars capable of regulating human relationships on a much bigger and longer scale. This is to say that uninterrupted observations of the phases of the moon inform people that a thin crescent-like object and a bulbous face-like object are one and the same thing.[15] Having realized this, the Moon seems to enable people to create a time-keeping device marking the flow of time through the sequence of differentiated lunations, which we call a 'lunar calendar'. In lunar calendars the lengths of the months are usually determined by actual lunations, each beginning with the observation of the first crescent of the moon in the western sky, but formal lunar calendars consist of a series of months alternating between twenty-nine and thirty days. Where greater periods of time are involved and the names for the lunar months are derived from climatic-meteorological and environmental events, a series of moon-months may be fitted into the solar year. Depending on the way the beginning of the year is defined, a series of lunar months are kept in phase with natural events by intercalating an additional moon-month. Many Old World lunar calendars consist of complex rules defining the intervals of time necessary to pass to reach the moments suitable for intercalations.

2. Moon-reckoning in North America

Most North American peoples did not record the passage of long intervals of time. However, they were aware of the changing seasons and observed the movements of migratory animals (birds, fish, big game animals), as well as the moments of the flowering of certain plants; so it is likely they also watched the phases of the moon. Many of those recurrent events could have been used to denote important social and environmental events by virtually all native groups. But this does not necessarily create lunar-based calendars capable of expressing the idea of a continuous flow of time. Therefore, one can rightly ask to what extent their observations of the lunar cycle can be seen as calendars.

2.1 Methodological statements

The aim of this paper is to describe the ways in which American indigenous peoples utilized the moon to tell the time. Each study of this kind must necessarily rely on the ethnographic record. Cultural material items are very scant and almost exclusively limited to several calendar sticks bearing notches representing tallies or to a very few rock art depictions (see below). While most calendar sticks were made in late nineteenth century, renditions of the lunar count made on rock panels and huge boulders are impossible to date with precision. So the examination of native lunar count systems is largely made through the study of native month names. In addition to months or moons often seasons are named.

Today it is almost impossible to know whether any Native American peoples ever had words for 'moon' and 'month'. In most cases Native American dictionaries inform us that the same terms are used to denote both things.[16] However, what I would suggest is that for most ethnographers educated within the Indo-European linguistic milieu the terms for 'month' and 'moon' are cognates.[17] Indeed in many Native American languages today the word 'month' is the same as the word for 'moon'.[18] Assuming that 'moons' had a much longer history as time indicators and that they had widely been used to mark such events as rituals and planting or sowing, it seems hard to demonstrate that time was counted precisely by months. The counting of 'moons' does not necessarily imply that there is a system of formal units based on a homogenized duration of a lunar month (synodic month, lunation). In this context I would like to highlight that many Native American peoples lack a word for 'year' employing words for 'summer', 'winter', 'rain', 'land', and 'world' instead.[19] This suggests that most Native Americans lacked formal systems of recording the passage of time consisting of minor units called 'months'.

In light of the above I would suggest that most societies living in the prehispanic Americas utilized the changing phases of the Moon not as part of a formal time-reckoning system but first and foremost as time indication guide marks. My paper is divided into three parts. I shall deal first with the types of so-called 'lunar calendars' used by the nomadic or horticultural indigenous population in North and South America. The next topic of my inquiry will be the lunar calendars developed by the peoples living in two cultural regions where higher social complexity appeared—the Mesoamerican and the Andean. Then I shall consider the possibility of the use of the lunar sidereal month. Finally, I will examine the uses of the lunar phases for agricultural purposes.

2.2 Rock art tallies

So far, there is no firm evidence for monthly calendars in prehistoric North America. However, there are a very few rock art sites from northern Mexico that seem to display lunar tallies. In his publications Breen Murray argued that patterns displayed on rock surfaces called 'tallies, or stroke marks' and 'dot grids' could have been used to mark the lunar cycles.[20] Tally counts found at Presa de la Mula and Boca de Potrerillos (Nuevo León, Mexico) register numbers 206 and 207, a good approximation of the length of 7 lunations.[21] Possibly this tally was related to the average gestation period of the white-tail deer. This relationship was based on some of the images placed on the same panel at Presa de la Mula representing the images of diverse projectiles, interpreted as hunting weapons, and on a petroglyph located near to a panel at Presa de la Mula featuring the tally of 28 marks (perhaps representing a lunar cycle). In addition, the structure of a petroglyphic count displayed at Presa de La Mula bears some similarity with the historical Pueblo calendar sticks; this analogy is not conclusive, however.

The rock art of northern Mexico provides numerous examples of the images made during the Archaic period and as Murray noticed many dot grids and tallies might have been created by a representatives (putative shamans) of hunting-gathering societies.[22] However, petroglyphic counts at Presa de la Mula or those found at a nearby site of Icamole (Nuevo León) seem to display numerical patterns known from Mesoamerica and registering eclipse intervals.[23] In Mesoamerica however, time was counted by formal calendars. Higher time periods were composed of smaller and standard intervals or units. On the other hand, North American examples suggest that in many cases no attention is paid to the exact number of days in the month and to the exact number of moons in the year. In the context of interpreting the past record, it is very useful to recognize that some tallies based on repetitive practices may have been invented independently by their practitioners. So while some tallies from northern Mexico appear to be inventions of local time-keepers, others may have been recorded using the schemes borrowed from the Mesoamerican calendar tradition.

2.3 Calendar sticks

Calendar notched wooden sticks acting as crude mnemonic devices to represent the passage of successive years were examined by Alexander Marshack.[24] Many such calendar records used by various Native American groups indicate that painted narrative buffalo skins, sticks made from interior of giant cactuses, even strings of berries or beads were used to mark the passage of time.[25] Only a few of them refer to lunar time-reckoning. The calendar stick owned by Tshi-zun-hau-kau, one of the chiefs of the Winnebago group represents a series of lunar months divided into three parts. The lunar month is divided into two parts of ten notches (days) each plus a third part of a variable length. According to Marshack, 'the calendar stick represents a non-arithmetic, observational, lunar-solar intercalary calendar ... made in the 1820s'.[26] Though some influences from Western societies may be expected, Marshack concludes that the Winnebago calendar stick stems from the Eastern Woodland hunter-gatherers cultural traditions and demonstrates a structural resemblance to certain Siberian calendars also executed on portable artifacts.[27]

2.4 Lunar and seasonal systems of reckoning time

Though the idea of the periodicity of the solar movements is generally known to Native Americans, it is the moon which plays a more important role in their life. The moon is often conceptualized as an animate being, sometimes it is even personified and its importance in social (tribal) life is both symbolic and entirely direct and functional.

A brief survey of the ethnohistorical bibliography on North American moon reckoning demonstrates enormous importance that the lunar phases played in the creation of indigenous time-reckoning systems.[28] The division of the seasons is generally associated with the changes in vegetation cycles, and with the arrival of cold or hot temperatures, of winds, rains, snows, droughts and other meteorological and climatic phenomena. In North America the moons often received their names from the principal natural events coinciding with them and the additional thirteenth moon was inserted, from time to time, to readjust the solar and lunar cycles.[29] Solar observations tended to be concerned with the solstices and involved the division of the year into two 6-month halves. In general Native American groups used descriptive, numerical, or astronomical (solstitial) moon series.[30]

2.4.1 Descriptive moon series

The division of the year into two basic seasons is based on the recurrence of specific climatic-environmental phenomena (winters, summers, cold and warm seasons) which cannot be defined in a very accurate way. The return of the seasons marks the passage of the year. The sequence of natural events is correlated with the moons which bear the appropriate names selected to suit the prevailing climatic-environmental conditions, basic subsistence activities or ceremonial feasts related to them. Over the whole continent, observations of moon phases have been used to mark the cycles of plant growth, and systems of time reckoning based on the moon served to schedule the periods suitable for planting and ripening. Examples of descriptive moon names are provided in table 1.

Keeping the lunar series in a proper sequence defines the order of those basic events, activities or feasts. It is evident that from time to time an extra month should have been added, usually at the moment suited for performing basic activities. Thus, if for example the 'salmon' moon already passed at the time adequate for salmon fishing, then an extra 'salmon' month was added. If the 'flying geese' moon passed but geese did not appear, an additional 'flying geese' month was added. Needless to say, this intercalation was irregular and nobody bothered to predict precisely the moments to do it. This is a good example of the time indication method: counting of time is based on sequences of qualitatively defined time points or time intervals that are synchronized with climatic-environmental phenomena, basic communal activities and ritual ceremonies. No one knew how many days were in a lunar month, or how many lunations were in a year.[31] Very often there is some confusion in the beginning and order of the moon names. Sometimes only some moons are named, some are unspecified and unnamed.

The names for the seasons were also descriptive and the number of the seasons varied from one native group to the other. The seasons constitute an independent count of the time, but often enter into the time reckoning, especially when the moons are grouped into a summer and winter series. Major time units were years. The year was the interval between two recurrent events or seasons. The beginning of the year depended either on astronomical

Month number	Descriptive Type Dakota (Plains) Cope (1919), p. 160	Numeral Type Yurok (Nortwest) Cope (1919), p. 154 Starts at the winter solstice	Astronomical Type Kwakiutl Cope (1919), p. 150 Begins at the winter solstice
1	Hard moon	First	Spawning season
2	Raccoon moon	Second	Elder brother
3	Sore eyes moon	Third	Under, that is under elder brother
4	Moon in which the geese lay eggs	Fourth	Next one under, that is next one under elder brother
5	Planting moon	Fifth	Trying oil moon
6	Moon strawberries are ripe	Sixth	Sockeye moon
7	Moon choke berries are ripe and geese shed feathers	Seventh	Between good and bad weather
8	Harvest moon	Eighth	Raspberry season
9	Moon rice is laid up to dry	Red berries	Eldest brother
10	Drying rice moon	Tenth	Right moon
11	Deer rutting moon	Beginning to camp out to gather acorns	Sweeping houses
12	Moon when deer shed horns	Acorns fall	Staying in dance house
13		Bad cold	Split both ways (winter solstice)

TABLE 1. Examples of moon names: the descriptive, numerical and astronomical groups (based upon Cope 1919).

(solstitial) or climatic-environmental phenomena. Springtime, winter, the rutting season of various animals, and the harvest time were the most popular moments to start a new year.[32]

Descriptive moon series appear to have been used by the majority of Native American groups: they are found among the groups inhabiting the prairies and woodlands of eastern North America.[33]

2.4.2 Astronomical and numerical moon series

In addition to descriptive lunar names, Cope provides arguments for the analytical division of the rest of the lunar time-reckoning systems into numerical and astronomical types.[34] Both types of time-reckoning systems may utilize astronomical references to fix the start of a year. Moons are often numbered in the Northwest coast region where they are also pivoted upon the winter solstice. This kind of calendar is based on the moon series which bears numerical designations, partially or totally replacing the descriptive terms (see table 1). The astronomical type retains the descriptive element. However, in contrast to the descriptive type, in defining time, Native American groups use observations of the sun or stars. For example they employ certain landmarks to observe the place on the horizon of the sun's rising or setting on days of solstices. In this way, the year is divided into two halves framed by the solstices. As a rule each half of a year comprises six moons, but the scheme is not to divide the year into a number of moons; rather it is a method of counting moons, especially new moons, standing for important meteorological, natural and social events. The solstice points may be used as starting points for numbered series of moons, but they do not coincide with activity cycles (hunting, gathering, agricultural) so they often do not mark the beginning of the year. So we have two parallel groups of six moons or of six and five moons. Astronomical knowledge used in time reckoning does not limit itself to the solstices. The heliacal rising and setting of the Pleiades, or of Orion's belt were also important astronomical markers.[35]

The division of the year into two groups of moons is common in Western North America. There can be twelve different moon names, repeated six-moon sets of names, or numbered moons. In the Northwest prevails a system of unnamed moons which are numbered and only occasionally named for public ceremonies. Most groups have an intercalary period pivoted upon solstices.

In the Southwest the lunar series are named and also ordered in two series pivoted upon solstices. Their names refer either to seasonal events or ritual activities. There is another group which consists of a lunar series with duplicating names.[36]

2.4.3 Lunar calendars in North America – Conclusions

Already in 1923 Kroeber noticed that relatively complex and advanced systems of moon reckoning which combine lunar, solar and sometimes stellar observations and develop mechanisms for a systematic intercalation or correction, are found in the area that was influenced by Mesoamerica.[37] The maps provided by Cope clearly denote the vast regions of Eastern North America dominated by descriptive lunar series, and limited areas where the moon series were attached to the solstices, arranged in two series of numbered or name repetitive moons (this information is presented in table 2).[38] This has been reaffirmed by a later regional study made by Spier.[39]

As has been remarked above, the native groups in the Southwest used lunar calendars to schedule ritual ceremonies and agricultural activities. On the other hand, many of their 'calendars' are symmetrical, showing 6 moons on each side of the half-year period. Their names are duplicated but most scholars agree that they have 13 named moons. For example,

Region	The role of solstices	Structure	Intercalation or correction	Type
Northwest	Solstices pivotal	Months in 2 series	Intercalation of non-lunar period	Months often numbered, occasionally named for rituals
Southwest	Solstices pivotal	Months in 2 series	Sometimes duplicating names	Months named after ceremonies or descriptive
General Inuit	Winter solstice		Sometimes lunar series corrected	Month names descriptive
Remainder of the continent	No use made of solstices	Rarely months in 2 series	No intercalation or system of correction	Month names descriptive, rarely numeral

TABLE 2. North American types of moon-reckoning systems according to Cope (1919).

the Zuñi and Hopi had seven named moons in their winter half of the year and only six in the summer.[40] So, it may be deduced that a set of 13 named moons reflects the need for intercalation.

3. Mesoamerica

The Mesoamericans recorded the passage of time by counting days. Their days were named, numbered and animated, possessing patron deities. The days had augural (negative or positive) values. Units of time, days, 20-day periods and greater were also sometimes conceived as bearing a sacred burden. The counting system was quasi-vigesimal (twenty-base system) and in consequence its units of time were primarily multiples of twenty. Two main calendars consisted of 260 days and of 365 days. The Mesoamerican 365-day year (presumably corresponding to a solar year) was composed of eighteen 20-day periods called months plus 5 additional days. Though in modern Mesoamerican languages the words for moon and month are the same, it is difficult to find traces of a lunar calendar, though many authors expressed opinions of its possible use in earlier times.[41] The origins and early phases of this calendar system remain unknown.

3.1 Luni-solar calendar in Mesoamerica?

Although historians of calendars have not been able to find evidence of the use of lunar calendars in Mesoamerica, Stewart proposed, on linguistic grounds, to 'suppress' 5 Mexica (Aztec) months thus reducing the total number of moons to 13 and suggesting the existence of a luni-solar calendar before the adoption of the system consisting of 18 months of

20 days each.[42] More complex lunar time-reckoning systems were developed by the Classic Maya skywatchers who grouped the moons into series of six and/or eighteen differentiated lunations. The system served to compute the phases of the moon, both backward and forward in time, though in several cases it stemmed from actual observations. Some scholars have claimed that this system was derived from a system operating in the American Southwest,[43] but others have suggested that the diffusion of the lunar knowledge was rather from Mesoamerica northward.[44]

3.2 The Maya Lunar Series

The so-called Lunar Series were attached to dates written in a complex quasi-vigesimal positional notation (the so-called Initial Series). The Lunar Series never occur alone, so they cannot be treated as an autonomous or separate counting system—they are always attached to the dates expressed by 260-day and 365-day calendars. This suggests that the Maya rulers, the authors and protagonists of the texts placed on stone monuments, were interested in timing their activities in accordance to lunar cycles. Despite the incompleteness of the record, the Lunar Series appears to represent a continuous lunar count, consisting of 29-day and 30-day periods, a rough correspondence with hollow and full lunar months known from the Old World.

It may be supposed that the observation of the moon led the Maya skywatchers to the development of the complex system of moon reckoning which is the Lunar Series. At present it is difficult to say whether this system derives from an earlier hypothetical lunar or luni-solar calendar or from the necessity to find periods enabling the prediction of eclipses. Within the Lunar Series system three types of the information about the lunar cycle were considered: the age of the current moon, the number of moons completed in a series 6 (or 18) differentiated lunations, and the alternating of 29 and 30-day formal lunations. This information is represented by hieroglyphic signs, called by epigraphers Glyphs A, B, C, D, E, and X. Glyphs D and E include information about the current phase, reporting on the number of days that elapsed from the moment when the moon 'arrived' (*huliiy*). The Moon's 'arrival' is interpreted as the appearance of a thin lunar sickle in the western sky after the period of her invisibility. Occasionally it may well refer to the day of the astronomical new moon. Glyph C is composed of two variable elements: numerical coefficients varying from 1 to 6 and three head variants. Although different scholars gave them diverse names, the consensus is that the variants followed a fixed order. Assuming, they are represented by a skull (s), young female (f) and young male or mythological male (m) the sequence of Glyph C head variants is represented by a following order: 1–6C skull (s) followed by 1–6C female (f) and 1–6C male (m) (see figure 1).

Glyphs C and X have long been believed to form a meaningful pair and attempts have been made to link forms of Glyph X to variable components of Glyph C. Though scholars have yet to identify all eighteen expressions of Glyph X,[45] it has been proposed that Glyph X, like Glyph C, was related to the sequence of eighteen lunar months, and that each of the 18 different lunations was named by a specific form of Glyph X. Glyph A defines the lunation as being either 29 or 30 days in length. Grammatically Glyph A is read as a noun, a number of twenty-nine or thirty, which is referred to other glyphs through Glyph B which is read as 'his sacred/holy/divine or youthful name' (*u ch'ul k'ab'a*, *u ch'ok k'ab'a*, respectively). Assuming there was a fixed correspondence between Glyph C and Glyph X variants

FIGURE 1. Variable head-variants of Glyph C: the skull, young feminine, and mythical jaguar heads.

the whole sentence may now be formulated as follows 'n days elapsed after (the moon) arrived, X (the name of the moon determined by the number and head patron of C) is/was the young/holy name of the twenty-nine/thirty'. We may now link 18 different lunations to either a 29-day or a 30-day month, generating, at least in theory, as many as 36 differentiated lunations.[46] Table 3 provides a series of Glyph Cs assembled in groups of 6 lunations each, patronized by one of the three head variants of Glyph C and associated with durations of lunations marked by alternating coefficients of Glyph A. For the sake of simplicity I composed one group of Glyphs C which starts with a 30-day month (called Group I) and another group which begins with a 29-day formal lunar month (called Group II).

The arbitrary nature of the alternating 29- and 30-day lunations does not relate them to observed lunations, so the Maya might have added an occasional 30-day lunation to keep track with the current moon. Teeple first noticed that values of Glyph A varied with coefficients of Glyph C suggesting that 'whenever Glyph C has an odd coefficient, 1, 3, 5, the chances are about three to one that Glyph A will show 30 days; whenever Glyph C has an even coefficient, 2, 4, 6, the chances are about three to one for a 29-day Glyph A'.[47] Though other scholars found these proportions biased, they all observed a strong correlation between a 30-day Glyph A and odd-numbered Glyphs C and between a 29-day Glyph A and even-numbered Glyphs C.

Closer inspection of the distribution of odd and even numbered Glyphs A shows that each addition of an extra 30-day lunar month breaks up with the symmetry displayed in table 3 since each such correction calls for two (or more) consecutive 30-day months.[48] Two consecutive 30-day months placed in the string of alternating 29- and 30-day lunations cause all following lunations to change from either even to odd or odd to even numbered ones; three consecutive 30-day month leave these relations intact.[49] However, given the tendency to display some regularity between odd- and even-numbered lunations and 30- and 29-day values of Glyph A mentioned above, any regular and long term reversal in the sequence of 29-day and 30-day lunations must be excluded. On most Maya inscriptions appear lunar notations belonging to Group I (table 3). This indicates that in most cases the series started with a 30-day lunation. However, from time to time on carved monuments appear the Series belonging to Group II, suggesting that 29- and 30-day lunations did not

Group I: Start with a 30-day lunar month			Group II: Start with a 29-day lunar month		
1Cs 30	1Cf 30	1Cm 30	1Cs 29	1Cf 29	1Cm 29
2Cs 29	2Cf 29	2Cm 29	2Cs 30	2Cf 30	2Cm 30
3Cs 30	3Cf 30	3Cm 30	3Cs 29	3Cf 29	3Cm 29
4Cs 29	4Cf 29	4Cm 29	4Cs 30	4Cf 30	4Cm 30
5Cs 30	5Cf 30	5Cm 30	5Cs 29	5Cf 29	5Cm 29
6Cs 29	6Cf 29	6Cm 29	6Cs 30	6Cf 30	6Cm 30
177 days	177 days	177 days	177 days	177 days	177 days
	531 days			531 days	
18 lunations x 29.53058562 (the average length of a lunar month in 700 CE) = 531.550541 days					

TABLE 3. Hypothetical semesters patronized by Glyph C head variants linked to an alternating Glyph A numerical coefficients. Two alternative sequences (starting either with a 30-day Glyph A or with a 29-day Glyph A) are displayed. Head variants: s – skull, f – female, m – mythological jaguar.

succeed each other in a regular sequence. Since in most cases the Lunar Series were not based on the first visibility of the 'first moon', it must be concluded that they depended on some previous computations, possibly indicating complex cyclic patterns. Some kind of predictions was made. However, bearing in mind that the dedicatory dates of monuments are much later than the Initial Series dates to which the Lunar series are attached, it is also possible that some lunar computations were made backward, or postdicted rather than predicted. To date, two basic methods of correcting lunar months have been proposed: short cycles of 886 days and longer periods of 11960 days.

Similar strings of groups consisting of 6 and 5 lunations (periods of 177–178 and 147–148 days, respectively) appear in the Dresden Codex, one of the few surviving codices from the prehispanic times. The lunar table of the Dresden Codex (pages 51–58) describes eclipse cycles involving the sun and the moon, and was probably applied to eclipse prediction. So the purpose of preparing the tables was predominantly related to the eclipses rather that to the problem of the length of a lunar month.

The whole system was not set up at once. The earliest known uses of Glyphs D and/or E (Moon Age) and Glyphs C (Moon Number) are from Uaxactún (Stela 18 at 8.16.0.0.0, or 357 CE), of Glyph A (Moon Duration) is from Río Azul (Tomb 1 at 8.19.1.9.12, or 417 CE), of Glyph X is from Copán (Stela 63 at 9.0.0.0.0, or 435 CE), and of Glyph B is from Balakbal (Stela 5 at 9.7.1.6.0, or 575 CE). The earliest known Maya lunar record is from the Seattle stela with a Glyph C variant (at 8.8.0.7.0, or 199 CE).

4. South America

4.1 The lunar sidereal calendar of the Inkas

The Inka culture was the last exponent of a long tradition of time reckoning in the Andes. Moreover they are the only native South American group about which there is detailed information preserved on the concepts of skywatching and calendars. Inka society developed a kind of complex state bureaucracy that needed to correlate various cycles of husbandry, irrigation, agriculture, trade and warfare. The society operated within the very complex of kinship, age classes and socio-political organization.

Cuzco, the capital of Tawantisuyu, the Inca empire, is located at the west end of the huge valley, and the Inkas used 41 directions radiating from the central Temple of the Sun in Cuzco, called Coricancha. These lines, called *ceques*, were based on sightlines toward the horizon. The directions of *ceques* are known with the help of sacred places or shrines called *huacas*, artificial markers placed along the *ceques*, in numbers varying from 3 to 15.[50] In general, the *ceque* lines formed a system of coordinates by which information of different orders (political, economic, and ritual) was encoded and ordered.

The elaboration of the *ceque* system is attributed to Pachacuti Inka Yupanqui who started the expansion of the Inka Empire outside the valley of Cuzco around 1440s. The lines radiating from Coricancha divided Cuzco and the rest of empire into four *suyus* (regions). The division into four quarters was involved in the organization of settlements, social groups and connected to the locations of many ceremonial sites.

Investigations made by Thom Zuidema established that the *ceque* system consisted of 41 lines encompassing 328 *huacas* in and around the city.[51] Observing that strings of *ceques* with *huacas* situated along them were similar to the strings or bundles of knotted cords which represented numbers called *quipus*, Zuidema proposed that the *ceque* system around Cuzco represented a system of numerical notation. As some of the *ceque* lines were astronomically aligned, Zuidema proposed they represented a complex lunar sidereal calendar. He observed that 328 divided into 41 yielded 8, the hypothetical number of days within an Andean week, and more importantly he discovered that the number of 328 days describes the cycle of 12 lunar sidereal months ($12 \times 27.33 = 328$ days).[52] The hypothetical yearly cycle consisted of 12 lunar sidereal months, giving a total of 328 days, and a period of 37 days to complete one 'lunar sidereal year'. To this, 37 days were added to arrive at 365 days, one solar year. The 37-day interval coincided with the invisibility period of the Pleiades (roughly between May 3 and June 9). With the heliacal rise of the Pleiades around June 9 the sidereal lunar count of 328 days (nights?) was initiated to reach the day of May 3, the night of the disappearance of the Pleiades from the sky. Zuidema's reconstruction of the lunar sidereal calendar was later modified.

It should be emphasized that the Inka solar calendar consisted of 12 months with the four most important Inka festivities, called Raymi, placed on solstices and equinoxes. The number of days in a single solar month should be that of 30 days or longer, no intercalary mechanism is known up to date.

The Aymara like the Inka had also a lunar calendar.[53]

4.2 Other South American lunar count systems

Unfortunately, much less is known of other South American calendars. As in North America, seasonal changes observed in the environment created conventional year divisions. South American groups often associated important ceremonies with the major turning points. However they all lacked the development of formalized divisions of the year. Many Native American groups, especially those from the Amazonian tropical rain forest, determined the annual cycle by observations of the stars, with a preeminent role played by the Pleiades.[54] In southern South America, a division of the year into four seasons is found, while people in the Amazonian region recognized only two seasons, a rainy and a dry one.

The division of the year affects the structure of lunar time-reckoning systems. For example, the Ona, who once inhabited the island of Tierra del Fuego (Patagonia), considered winter and summer as the major seasonal divisions, but also distinguished four different seasons. They divided the moons according to asymmetrical winter and summer series, assigning 6 moons to winter and only 5 moons to summer parts of the year. The names of the moon were descriptive and derived from environmental changes.[55]

In general, the system of descriptive names for moons prevailed in South America, though not all of the lunations were named or counted.

5. Lunar agriculture

In addition to the above, the division of the lunar phases is associated with the growth of vegetation. The lunar month is divided into two basic phases: waxing and waning. The waxing moon phase is generally considered as a good time to encourage plant growth and proliferation. The waning moon phase is associated with ripening or harvesting. Choosing the proper time for planting, ripening and woodcutting is widely evidenced by ethnographic reports.

Planting with the phases and positions of the moon is often found in agricultural societies in Mesoamerica.[56] The Mesoamerican lifeworld was largely structured or patterned around the planting cycle of maize, the major staple food. Homologies found between the cycles of human life and the growth of the maize plant provided the Maya with the conceptual means (symbols, metaphors, allegories, etc.) necessary to understand the lunar cycle. As a result, activities related to different phases of the agricultural cycle were often scheduled according to the moon. The farmer who sows seeds during the waxing phase is able to say how many moons have to pass before the time appropriate for harvesting. Timing agricultural activities involves some rudimentary knowledge of the lunar phases. Agricultural lunar calendars are, however, limited by the length of the agricultural season, so they cannot be used to count time continuously. They nevertheless, divide the lunar month into two basic periods, those of the waxing and waning moons, and this may be used as the basis for the further structuring of the lunar month. Such agricultural ad hoc time-recording methods rely on actual observations of the moon, but times appropriate for intermediary agricultural or horticultural activities and for final harvesting, are counted in 'moons'. Though the concept of planting by the lunar phases may be extended to other social activities, this system was predominantly produced in the service of agriculture and horticulture. Since this field of research lacks more intensive studies, it is not entirely known whether lunar agricultural practices were performed in all regions in the Americas before the European arrival. With all probability this system was developed in Mesoamerica long before the Europeans arrived, but in other regions native populations had a long-term contact with the European settlers who may have brought lunar agricultural concepts from the Old World.

6. Conclusions

With the exception of the Andean and Mesoamerican regions it is difficult to find systems of time-reckoning in the Americas based on standardized and formalized divisions. In spite of the interest in temporal events, very few groups were able to sufficiently homogenize their time units. It is difficult to say they had true calendrical systems. Most native groups in America had systems of naming moons according to their relationship with other seasonal and recurrent events, but they were not indented to represent a whole year, nor were regular corrections made with the passage of the sun. Corrections were made *ad hoc* and depended more on individual or collective decisions of political authorities than on well-defined regulatory mechanisms.

A few of native American groups developed calendars based on a series of differentiated lunations ('moons') pivoted upon other astronomical events, in most cases solstices, thus making possible the development of a formal mechanism of adjusting a lunar cycle to a solar one. These are predominantly the groups living in the North American Northwest and Southwest, and the north of Mesoamerica, and it is possible that their relatively complex calendrical systems evolved due to influences coming from the Mesoamerican calendar tradition. The same may eventually be said of a system from Columbia.

Contrary to Old World societies, the Mesoamericans invented a unique calendar system which apparently did not use lunar referents. The only systematic and continuous time-reckoning system comparable with the Old World lunar and luni-solar calendars is derived from the Maya Lunar Series. In contrast with Old World calendars, the Lunar Series were not correlated with Mesoamerican months, which consisted of 20 days only. The lunar sidereal calendar consisting of irregular periods, combining the sidereal and synodic periods and the invisibility period of the Pleiades was proposed by Zuidema for the Inka capital of Cuzco but it finds almost no support in ethnohistorical sources. So it must remain only as a hypothetical possibility.

At present, the only system which is widely evidenced through epigraphic record and which can serve as a type of a counterpart to all Old World lunar calendar systems appears to the Maya Lunar Series. While the general functioning of the Lunar Series has been successfully decoded, details concerning intercalations, borrowings from the nearby city-states, or computations or earlier dates are less known. For some time onwards, the Maya Lunar Series will remain a fascinating research field within a wider domain of Mesoamerican archaeoastronomy.

Acknowledgements

I would like to thank the organizers for their invitation to participate in the conference *Living the Lunar Calendar: Time, Text and Tradition*. The paper is part of the research project 'Lunar Theory in Maya Astronomy' (Project no. 296 of the Instituto Nacional de Antropología e Historia).

Notes

1. Nilsson (1920).
2. Nilsson (1920), pp. 8–9, 355–359.
3. Bourdieu (1977), pp. 103, 106–107.

4. Nilsson (1920), p. 361.
5. Hallpike (1979), p. 349.
6. Lucas (2005), pp. 68–69.
7. Elias (2007), p. 37.
8. Lucas (2005), p. 74.
9. Nilsson (1920), pp. 356–357.
10. Nilsson (1920), pp. 356–357.
11. Marshack (1972).
12. d'Errico (1998).
13. Marshack (1972, 1985a).
14. Cope (1919), p. 120; Nilsson (1920), p. 173.
15. Elias (2007), p. 55.
16. See Cope (1919), pp. 128–129.
17. Buck (1971).
18. Compare Nilsson (1920), p. 148.
19. Cope (1919), pp. 132–139; Kroeber (1923), pp. 320–322.
20. Murray (1982, 1986).
21. Murray (1982), p. 197 and (1986), p. 49.
22. Murray (1982), p. 202.
23. Murray (1985).
24. Marshack (1985, 1989).
25. Underhill (1938), pp. 8–11.
26. Marshack (1985), p. 43.
27. Marshack (1985), p. 47.
28. Swanton (1911); Cope (1919); Kroeber (1923); Spier (1955).
29. Swanton (1911), p. 109.
30. Cope (1919); Kroeber (1923), pp. 374–376; Spier (1955).
31. Cope (1919), p. 129.
32. Cope (1919), p. 136.
33. Cope (1919), p. 147.
34. Cope (1919).
35. Cope (1919), p. 141; Kroeber (1923), pp. 323–324.
36. Cope (1919), p. 147; Spier (1955), pp. 18, 25 fig. 3.
37. Kroeber (1923), p. 375.
38. Cope (1919).
39. Spier (1955).
40. Spier (1955), pp. 18–19; McCluskey (1982), pp. 44–47.
41. For example Caso (1967), pp. 34, 79.
42. Stewart (1984).
43. Satterthwaite (1947), p. 76.
44. Kroeber (1923), p. 375.
45. Schele et al. (1992), p. 5 and Rohark (1996), p. 71 produced lists of up to 14 possible variants; Justeson (1989), p. 90 founded only 12 forms.
46. The set of 36 differentiated lunations should be taken as collateral effect of my theoretical speculations.
47. Teeple (1928), p. 398 and (1931), p. 63.
48. Satterthwaite (1947), p. 92.
49. Lounsbury (1978), p. 800.
50. Zuidema (1982), p. 60.
51. Zuidema (1977, 1982).
52. Zuidema (1977), pp. 228–233.
53. Tschopik (1946).

54. Lévi-Strauss (2005a, 2005b).
55. Bennett (1949), p. 607.
56. Iwaniszewski (2006).

References

Bennett, W. C., 1949, "Numbers, Measures, Weights, and Calendars", in J. H. Steward (ed.), *Handbook of South American Indians. Volume 5. The Comparative Ethnology of South American Indians*, Bulletin 143 of the Bureau of American Ethnology (Washington: Smithsonian Institution), 601–619.
Bourdieu, P., 1977, *Towards a Theory of Practice* (Cambridge: Cambridge University Press).
Buck, C. D., 1971, *A Dictionary of Selected Synonyms in the Principal Indo-European Languages. A Contribution to the History of Ideas* (Chicago & London: University of Chicago Press).
Caso, A., 1967, *Los calendarios prehispánicos* (México, D.F.: Universidad Nacional Autónoma de México).
Cope, L., 1919, *Calendars of the Indians North of Mexico*. University of California Publications in American Archaeology and Ethnology, Vol. 16, No. 4, 119–176.
D'Errico, F. 1998, "Palaeolithic Origins of Artificial Memory Systems", in C. Renfrew & C. Scarre (eds.), *Cognition and Material Culture: The Archaeology of Symbolic Storage* (Cambridge: MacDonald Institute for Archaeological Research), 19–50.
Elias, N., 2007, *An Essay on Time* (Dublin: University College Dublin Press).
Hallpike, C. R., 1979, *The Foundations of Primitive Thought* (Oxford: Clarendon Press)
Iwaniszewski, S., 2006, "Lunar agriculture in Mesoamerica", *Mediterranean Archaeology and Archaeometry* 6/3, 67–75.
Justeson, J. S., 1989, "Ancient Maya Ethnoastronomy: An Overview of Hieroglyphic Sources", in A. F. Aveni (ed.), *World Archaeoastronomy* (Cambridge: Cambridge University Press), 76–129.
Lévi-Strauss, C., 2005ª, *Mitológicas 1. Lo crudo y lo cocido* (México, D.F.: Fondo de Cultura Económica). [*Mythologiques 1. Le cru et le cuit*, Paris 1964].
——, 2005b, *Mitológicas 2. De la miel a las cenizas* (México, D.F.: Fondo de Cultura Económica). [*Mythologiques 2. Du miel aux cendres*, Paris 1966].
Lounsbury, F. G., 1978, "Maya Numeration, Computation, and Calendrical Astronomy", in C. C. Gillespie (ed.), *Dictionary of Scientific Biography* 15 (New York: Charles Scribner's Sons), 757–818.
Lucas, G., 2005, *The Archaeology of Time* (Abingdon and New York: Routledge).
Marshack, A., 1972a, *The Roots of Civilization: The Cognitive Beginnings of Man's First Art, Symbol and Notation* (New York: McGraw-Hill).
——, 1972b, "Upper Paleolithic Notation and Symbol", Science 178, 817–828.
——, 1985, *Hierarchical Evolution of the Human Capacity: The Paleolithic Evidence* (New York: American Museum of Natural History).
——. 1985, "A Lunar-Solar Year Calendar Stick from North America", *American Antiquity* 50, 27–51.
McCluskey, S., 1982, "Historical Archaeoastronomy: The Hopi Example", in A. F. Aveni (ed.), *Archaeoastronomy in the New World* (Cambridge: Cambridge University Press), 31–57.
Murray, W. B., 1982, "Calendrical Petroglyphs of Northern Mexico", in A. F. Aveni (ed.), *Archaeoastronomy in the New World* (Cambridge: Cambridge University Press), 195–204.
——, 1985, "Petroglyphic Counts at Icamole, Nuevo Leon (Mexico)", *Current Anthropology* 26, 276–279.
——, 1986, "Numerical Representations in North American Rock Art", in M. P. Closs (ed.), *Native American Mathematics* (Austin: University of Texas Press), 45–70.
Nilsson, M. P., 1920, *Primitive Time-Reckoning*, Skrifter utgivna av humanistiska vetenskapssamfundet i Lund (Lund: Gleerup).
Rohark, J., 1996, "Die Supplementärserie der Maya", *Indiana* 14, 53–84.
Satterthwaite, L. Jr., 1947, *Concepts and Structures of Maya Calendrical Arithmetics*, Joint Publications of the Museum of the University of Pennsylvania and the Philadelphia Anthropological Society (Philadelphia: University Museum).

Schele, L., Grube, N., & Fahsen, F., 1992, "The Lunar Series in Classic Maya Inscriptions: New Observations and Interpretations", *Texas Notes on Precolumbian Art, Writing, and Culture* 29.

Spier, L., 1955, *Mohave Culture Items*, Museum of Northern Arizona, Bulletin no. 28 (Flagstaff).

Stewart, J. D., 1984, "Structural Evidence of a Luni-Solar Calendar in Ancient Mesoamerica", *Estudios de Cultura Nahuatl* 17, 171–191.

Teeple, J. E., 1928, "Maya Inscriptions VI: The Lunar Calendar and Its Relation to Maya History", *American Anthropologist* 30, 391–407.

——, 1931, *Maya Astronomy*. Contributions to American Archaeology, No. 2, 29-115, Publication 403 (Washington: Carnegie Institution of Washington).

Tschopik, H. Jr., 1946, "The Aymara", in J. H. Steward (ed.), *Handbook of South American Indians. Volume 2. The Andean Civilizations*, Bulletin 143 of the Bureau of American Ethnology (Washington: Smithsonian Institution), 501–573.

Zuidema, R. T., 1977, "The Inca Calendar", in A. F. Aveni (ed.), *Native American Astronomy* (Austin: University of Texas Press), 219–259.

——, 1982, "The Sidereal Lunar Calendar of the Incas", in A. F. Aveni (ed.), *Archaeoastronomy in the New World* (Cambridge: Cambridge University Press, Cambridge), 59–107.

Lunar Ceremonial Planning in the Ancient American Southwest

James Walton

Introduction

The formal lunar calendar of the peoples of the socio-complex Chaco Civilization of the American Southwest, ca. 800–1200 CE, is unknown, unlike the high civilizations of Mesoamerica or the Ancient Middle East. However, because there was exceptional opportunity to observe the moon in the clear air and dramatic skylines of their surroundings, investigators of archaeoastronomy and architectural planning have proposed that the ancient Chacoans incorporated lunar phenomena in their building orientation. This controversial assertion has yet to be accepted by the scholarly community.[1] However, maintaining a lunar calendar, in conjunction with the solar, is of particular importance in setting the ceremonies and dances that characterize the religious life of the Puebloan descendants of the builders of Chaco. Ethnographic studies of the Hopi and Zuni cultures, today residing in the western reaches of the Chacoan Regional System, document that these related societies have ceremonial calendars based on successive new crescent moons which must be intercalated to keep in synchronization with the solar year. Each moon brings a specific traditional ceremony and guidelines for the agricultural cycle.[2] Historical studies of post colonial Pueblo communities, descendants of the Chacoans, indicates that traditional Puebloan calendar keeping involves three points of reference: 'the passing of the solstices, the phases of the moon, and the seasonal positioning of recognizable constellations'.[3] The following discussion will attempt to apply the well documented ceremonial lunar calendar of the Hopi to the Chacoans and their immediate successors, often referred to as the 'Anasazi', with special attention to its implications in recovering unwritten calendar information.

The Chacoan World

Chaco Canyon in northwest New Mexico is an arid valley surrounded by spectacular cliffs which became the seat of one of the most important civilizations on the North American continent before the arrival of Europeans. The Chaco Wash is cut into the Colorado Plateau and drains into the San Juan River system, which meanders from the mountains of Southwestern Colorado to its confluence with the Colorado River near the Glen Canyon, Utah. Most of the sites discussed in this paper are found along tributaries of the Colorado River and are located in the Four Corners region of Colorado, New Mexico, Utah and Arizona (figure 1). By 1090 CE the residents of Chaco Canyon had created enormous

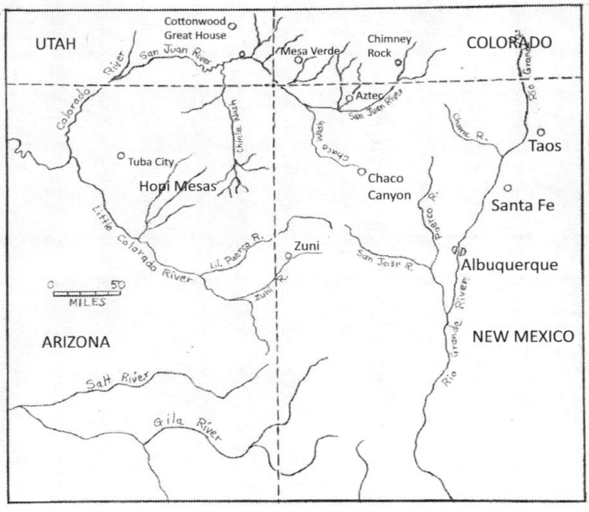

FIGURE 1. Four corners region and sites discussed.

FIGURE 2. Photograph of Pueblo Bonito.

planned city-like buildings called 'great houses', typified by Pueblo Bonito which covers more than 3 ½ acres and contains more than 650 rooms (figure 2).[4] The great houses in Chaco Canyon are the largest free standing masonry structures in the continental United States, rivaling even the palaces of the Aztecs in Mexico and the Inca of Peru. Archaeological studies have disclosed that outside the canyon itself, contemporaneous great houses, referred to as 'outliers', were associated with a system of roads and ceremonial structures that integrated a vast regional system without the benefit of domesticated beasts of burden or

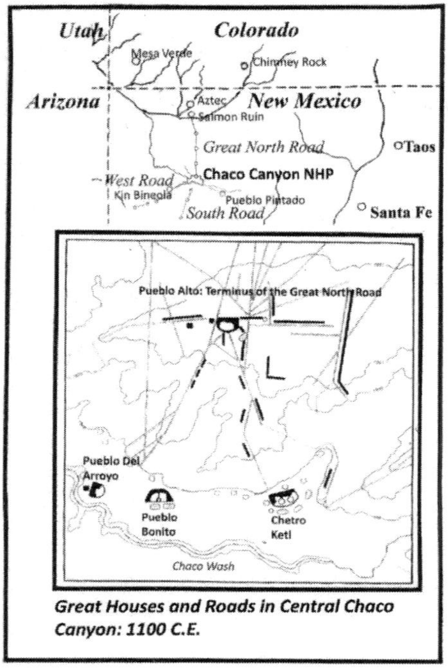

FIGURE 3. Chaco Regional System, ca. 1100 CE,

wheeled transportation (figure 3).[5] Geographically, the Chaco Regional System embraces four separate languages, and dozens of dialects and it is unlikely that the integration of their society could have taken place without a coordinated calendar that crosscut linguistic boundaries.[6]

By ca. 1125 CE great houses were no longer being built in Chaco canyon itself, but the great house tradition continued for another 100 years until drought precipitated abandonment of the region in the late 13th century. During the approximately two hundred years of great house construction many were established to the north and west—in the Totah region of the Animas and La Plata rivers; in Mesa Verde and the Montezuma plain of Southwestern Colorado; in the multiple drainages of the San Juan in southeastern Utah; and along the course of the little Colorado in northeastern Arizona.[7] These dwellings were also accompanied by ceremonial features such as great kivas, platform mounds and roads. Astronomical orientation of these later great houses is frequently aligned with the solar solstices, such as the group in Aztec, north of Chaco Canyon, built from ca. 1115–1135 CE.[8] Solar orientation relates to the known ethnographic practices of the Hopi and Zuni tribes which still reside in these territories. These tribes mark their year by the winter and summer solstice and carefully mark the position of the sun on the horizon on these dates from the perspective of shrines especially dedicated to solar observations.[9] However, evidence from the field is accumulating that indicates there was lunar orientation of ceremonial sites during the peak of Chacoan development. This is especially striking at the site of Chimney Rock in southwestern Colorado, built during the classic Bonito phase of Chacoan culture.

Chacoan Lunar Calendars and the Chimney Rock Great House

Prehistorically, the Chacoans may also have observed the extended calendar anniversaries created by the positions of the moon on the horizon which change over an 18.61 solar year period known by astronomers as the 'lunar standstill cycle'.[10] Nowhere is this more evident that at the Chimney Rock Great House, a Chacoan outlier dated archaeologically to the end of the 11th century CE when classic Chacoan civilization reached the peak of its development. The Chimney Rock Great House is positioned on a high mesa just to the west of two imposing spires of sandstone where the moon will rise within the gap between the rock pillars only during major northern lunar standstill as seen from the rooftop entrance of the east kiva of the structure. A series of lunar phases are visible rising between the spires would be anticipatory—beginning with the last crescent in early morning in July and leading up to full moon in December near winter solstice. These observable lunar events may have served the Chacoan Regional System in staging ceremonies to be attended by pilgrims streaming at great distance from the outliers to the central place. Over the past 40 years, archaeoastronomers studying architecture and rock art in the Chaco regional system, suggest the lunar standstill cycle may have created a multi-year ceremonial calendar where festivals in Chaco Canyon would be heightened every eighteen years by the phenomenal rise of full moon either in its northern most position near winter solstice, or in southern most position near summer solstice. From Chimney Rock information could have been sent to Chaco Canyon by runners, or by signaling between the Chimney Rock Mesa and Chaco Canyon with a relay station on Huerfano Mesa to the south.[11]

In this modeling of the ceremonial calendar, Chacoan 'New Year' close to winter solstice may have been significantly extended and enhanced during major northern standstills, every 18 years or so. Archaeological studies of the Great Houses indicate they are linked to a network of 'roads' and rock cut stairs that would direct pilgrims to the canyon's numerous ceremonial structures for feasting and celebrations.[12] Outlier studies have also proven that major lunar standstills in the southern mode after the demise of Chaco, are integrated in the ruins of Mesa Verde and other late Ancestral Puebloan sites.[13] Research is ongoing, in that putative lunar standstill observatories must be confirmed by actual photographic documentation, an opportunity that will not be presented for another 16 years.

My work has been to study and confirm sites in the Chimney Rock archaeological area where the lunar path on the horizon could be observed and photographed. During the past lunar 'standstill season', December 26th, 2004 – August 2007, there were many opportunities to confirm lunar standstill alignments proposed by various researchers. Because the moon 'rests' in standstill position, it is possible to have as many as three successive days when the moon is at the same approximate azimuth. During standstills, the moon reaches its maximum and minimum declinations, north and south, in every lunar synodic month but in different phases. Anticipatory moons before the full moons of winter or summer solstice will be in the second quarter and rise in darkness. Ironically, it is easier to photograph the moon in standstill position after the full moons of the solstices when it is visible in daylight as the third quarter begins. Photographically, I have documented three sites in addition to the Chimney Rock Great House, where the northern lunar standstill can be photographed between the spires. These are the guardhouse site, immediately to the west of the Chimney Rock Great House, the parking lot site below the high mesa, and from Peterson Mesa Site 5AA793, approximately three miles away from the spires. I have also photographed a 13th century site in Utah, near the Cottonwood Falls Great House, where the major southern lunar standstill moonrise is clearly marked by archaeological feature (Figures. 4a–c).

Chacoan Astronomy and Ethnography

As mentioned above, many ascribe sophisticated lunar alignments to the great houses of Chaco Canyon. But despite confirmation of lunar standstill alignments by myself and others, my search through the ethnographic literature has not discovered a direct connection between lunar standstills and historic calendar keeping. Both the Hopi and Zuni tribes observe a lunar calendar to stage their festivals and ceremonies. The observation of lunar extremes would have been inescapable within the context of the lunar ceremonial calendar as recorded in the Hopi villages in the late 19th century. Although the Hopi did not commit their lunar observations to public records, their procedures for observing the lunar month are well documented. The following discussion intends to provide guidelines by which the known Hopi lunar calendar might be coordinated with Chacoan archaeoastronomy to reconstruct the as yet speculative lunar standstill calendar, periodically repeating every 18.61 years. But first, a resume of solar horizon observations made by the Hopi will explain the tools at their disposal that could also establish a multi-year lunar calendar. Hopi and Puebloan solar horizon calendars are a practical method of timing agricultural work during the growing season, but also mark ceremonial occasions throughout the solar year.

Chacoan Solar Orientation and the Yearly Solar Calendar of the Hopi

The predecessors of the Hopi, also known as the Anasazi, may have been more sophisticated in combining their solar observations with their architecture. A major example is Aztec Ruins, built in a single episode in the early 12th century, where the northern base wall of the Aztec West Great House is aligned with summer solstice sunrise and winter solstice sunset, an axis observable in PIII sites such as Yellow Jacket 5MT-5 in the 13th century.[14] In addition there are several structures in Chaco Canyon that exhibit cardinal orientation in their ground plan or features, such as the Great Houses, Pueblo Bonito and Pueblo Alto, where the equinoxes rise along their boundary walls. Solstitial orientation is found in the internal windows in Pueblo Bonito and the great kiva, Casa Riconada, which admit light on the winter solstice and the summer solstice respectively.[15] When Old Oraibi, the oldest Hopi village, was established in ca. 1120 CE , it was roughly solstitial, though not planned. Before contact with Europeans the Hopi relied almost exclusively on dry land maize agriculture for subsistence and correct timing of plantings in their harsh semi-desert environment was critical for survival. The long baselines and numerous features of their landscape enabled a precise solar calendar for planting. Consequently, the Hopi define their cosmological directions by the solstitial points empirically perceptible on the local horizon, unlike the High Civilizations of Europe and Asia which define cosmological directions by north, south, east and west.[16] This pragmatic division used by the Hopi for

FIGURE 4a–c. Ancestral Puebloan sites positioned to view both major northern and major southern standstills.

Solar Position	Direction: Azimuth	Hopi Name	Color
Summer Solstice S Set	Northwest: Az. 300°	*Kwinini**	Yellow
Winter Solstice S Set	Southwest: Az. 240°	*Tevyuna*	Blue-green
Winter Solstice S Rise	Southeast: Az. 120°	*Tatyuka**	Red
Summer Solstice S Rise	Northeast: Az. 60°	*Hopoko*	White
(* = Sun's houses proper, where the sun is supposed to dwell for four days)			
"Above"	Zenith; North; Summer	*O'mi*	Black
"Below"	Nadir; South; Winter	*At'kyami*	Multi-Colored

TABLE 1. The Hopi solar directions.

agricultural purposes also highlights the ceremonial and cosmological use of the solstitial points in their religious life. Ceremonial observations are counterclockwise, commencing with the Summer Solstice Sunset in the northeast, proceeding to Winter Solstice Sunset in the southeast, etc. In Pueblo world view the sun is a spiritual entity who resides at his winter or summer 'houses' for a number of days each year. In the village of Walpi each of these four horizon points had a shrine at some distance from the village where prayer offerings, (Pahos), composed of sticks ornamented with feathers, were deposited. In addition to the solstial points, the abstract conceptual directions, 'Above' and 'Below' are added to create a framework of six directions, summarized in table 1). Lunar observation is anchored to the solar framework of the horizon calendar, but as yet no specific Hopi terms have been found for lunar horizon phenomena (see figure 5 and table 3 below).

The division between summer solstice sunset in the northeast and winter solstice sunrise in the southeast also created the division between sacred and secular time.[17] Observations of the sunset on the western horizon after summer solstice were used to set the great unmasked communal dances, such as Snake Dance in August, and observations of sunrise on the eastern horizon between winter and summer solstice set the planting calendar (figure 5). Playing into this duality is the fact that the full moon is opposite the sun in azimuth, so that when the sun is furthest south at winter solstice, the full moon rises furthest north in the same month. These complementary opposites are compared in figure 6.

The Hopi Lunar Calendar

Today's Hopi lunar calendar is intimately associated with the Kachina Cult which arrived in the southwest after the abandonment of the San Juan communities in the late 1200s CE.[18] It is possible that there are two interpenetrating calendars reflecting the original 'Chacoan' calendar and the more recent calendar of the kachinas which will be discussed below. The ceremonial Kachina year begins in November at the 'New Fire', when young men are initiated into ceremonial societies. This ceremony, the '*Wuwuchim*' has a longer

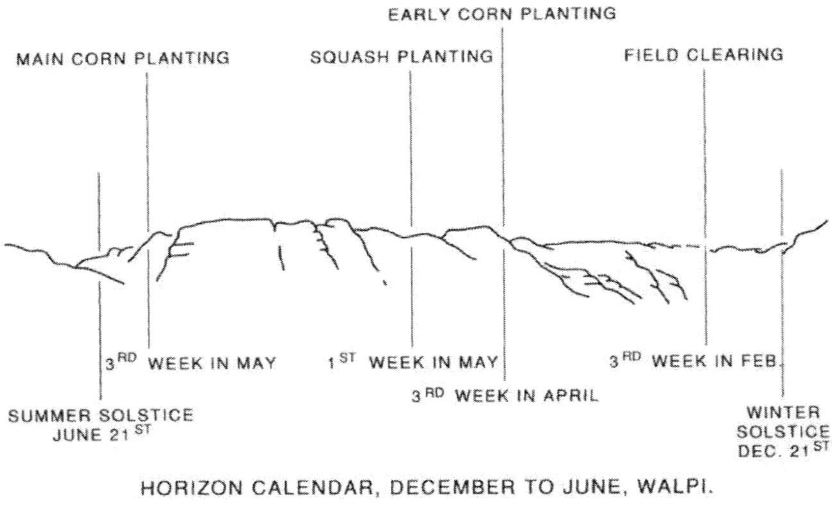

FIGURE 5. Hopi Solar Horizon Planting Calendar for Walpi, Arizona, late 19th century. (After Ford (1931).

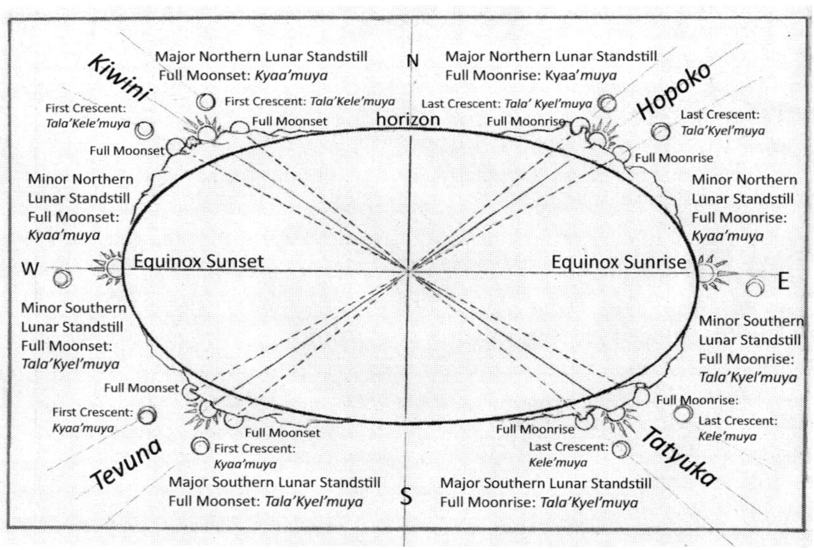

FIGURE 6. Schematic lunar ceremonial horizon calendar for major and minor standstills.

form every 4 years. The 'New Year', or Winter Solstice ceremony, '*Soyal*', fell in the second lunar month of the ceremonial calendar. In years when the *Wuwuchim* is long, the *Soyal* is doubled from eight days to sixteen. The lunar month is standardized to 28 days, its period invisibility near solar conjunction being ignored.[19] The Moon is called *Mu'uyawu* in Hopi language and the root, '*muya*', is suffixed to the name of each lunar month.[20] Traditionally there are 13 feasts in the ceremonial calendar. An observation of the crescent moon near the western horizon was correlated with the sun's horizon position when announcing ceremonial preparations and plantings. Each ceremony follows a new moon, and 'there is no moon not associated with a ceremony'.[21] Today the Hopi repeat the five standardized months of the winter ceremonial season in the succeeding summer agricultural season, 'like the right hand over the left hand'.[22] This stems from the dualism characteristic of Hopi thought where the crops being sown and harvested in the growing season have virtual and opposite reality in the underworld during the winter. The ceremonial schedule reflects this dualism by placing the major agricultural rituals in the winter, when the crops of the summer are being magically created in the underworld, but only brief recapitulations of these ceremonies in the summer. The dualism between winter and summer is also reflected in the nature of Hopi ceremonial life. From the winter solstice to just after the summer solstice, night dances featuring masked dancers called '*Kachinas*' are held inside the *kivas*, whereas in the second half of the lunar year, unmasked dancers perform in open plazas.[23] *Kivas* are subterranean chambers housing the ceremonial activities of the male members of matrilineal clans. *Kachinas* are believed to impersonate the beings of an invisible spirit world, particularly significant in rainmaking.[24] The Kachina Cult is believed to have arrived in the late 13th or early 14th century from the southern borders of the Chaco regional system, and may have replaced or integrated with earlier Chacoan ceremonies. The *Kachina* season lasts from the winter solstice ceremony, the '*Soyal*', until just after the summer solstice, the '*Niman*' ceremony in early July, when the *Kachinas* go to their spiritual home in the San Francisco Peaks, west of the Hopi villages by about 50 miles. Full moons corresponding with ceremonies are particularly important. Both Hopi and Zuni attempt to coordinate the public performances of the Winter Solstice ceremonies, *Soyal* or *Shalako* respectively, with the rise of the full moon.

Although the Hopi ceremonial year has seven named moons in its winter sequence corresponding to the *Kachina* feasts, only five are repeated in the summer sequence when non-*Kachina* ceremonials are performed. To these are added four or five other named moons (names vary according to village) which may overlap with the other moons when intercalation is necessary. In addition to the standard set of five winter lunar months, a sixth and seventh moons are added in the spring to reflect the planting season, bringing a total of twelve lunar months. A thirteenth month is added to intercalate the year with the seasons, but this varies according to village. Intercalation takes place approximately every three years. Alternative names for the standard moons of the summer sequence reflect the ceremonies enacted in these months. The names of the lunar months in Hopi order with their principle ceremonies are the following. Spellings follow the Hopi Dictionary.[25]

Winter Sequence: *'muya'*

1.) *Kyel'muya* – 'Initiates Moon'. Approx. November: *Wuwuchim*: 'New Fire'.
2.) *Kyaa'muya* – 'Dangerous Moon'. Approx. December: *Soyal*: 'Winter Solstice'. ......Arrival of Kachinas from the San Francisco Peaks
3.) *Paa'muya* – 'Play Moon', or 'Moisture Moon', Approx. January: Buffalo Dance.
4.) *Powa'muya* – 'Purification moon'. Approx. February: Bean Dance.
5.) *Oso'muya* – 'Cactus Blossom Moon'. Approx. March: *Palulukonti*: Horned Serpent drama.
6.) *Kwiya'muya* – 'Greasewood Fence Moon'. Approx. April: Early planting.
7.) *Hakiton'muya* – 'Waiting moon'. Approx. May: Plaza dances. (intercalary)

Summer Sequence: *'Tala'muya'*

8.) *Tala' Kyel'muya* – 'Planting Moon'. Approx June: 'Summer Solstice'.
9.) *Tala' Kyaa'muya* – 'Summer Home Going Moon'. Approx. July: *Niman* Ceremony: Departure of Kachinas to the San Francisco Peaks.
10.) *Tala' Pa'muya* – 'Moisture Moon', Approx. August: Snake Dance in odd years, and Flute Dance in even years.
11.) *Tala' Powa'muya* – 'Big Feast Moon'. Approx. September: *Mamzrau*, Women's Tableta Dance.
12.) *Tala' Isu'muya* – 'Harvest moon'. Approx. October: *Oa' qul* : Women's Basket Dance.
13.) *Angok'muya* – also 'Harvest Moon'. Approx. October. (intercalary)

Other named moons:

A.) *Uusu'muya* – 'Planting Moon'. Approx. June.
B.) *Niman'muya* – 'Homegoing Moon'. Approx. July.
C.) *Nashan'muya* – 'Big- Feast Moon'. Approx. September.
D.) *Tuhoosh'muya* – 'Basket Moon'. Approx. October.

Observation of New Moons

Although the solemnization of the calendar month was the responsibility of the priest presiding over the respective ceremony, all members of the community had a part in observing the new moon as a slender crescent in the west. The standpoint of observation was not a special shrine or structure, but the roof tops of the Pueblo. A. M. Stephens describes the ceremonial and secular observation of new moon in his entry for Thursday, Jan 19th, 1892:

> This evening the crescent of the new moon, Powa'muya, was visible for the first time; many of the women watched for it from the house tops. This evening elders smoke in somebody's house and set the time to begin the Powa'mu ceremony.[26]

New Moons were also reported according to their inclination to the horizon. Such terms as 'crescent first seen', 'moon vertical', 'crescent moon horizontal', were employed to characterize a lunar month. Phases after the first crescent were also named, such as 'moon half gone', and 'last of moon'.[27]

Ceremonial Authority of Calendar Priests

According to Elsie Clews Parsons, it was the responsibility of the chiefs of the major ceremonies to watch both sun and moon.

> The solstices are observed by the Sun chieftaincy on a sunrise horizon, which is also observed for a series of agricultural plantings; other ceremonial periods are determined either by solar or by lunar observation by the chief of the ceremony which is due. The sequence of the ceremonies is of course generally well known; it is formally noted, at least in some cases, by the chief of the concluding ceremony going to the chief of the ceremony next in order to say that it is his turn to watch the sun, that is, the sunrise, or to watch for the moon, the new moon.[28]

To this day, ranking among the Hopi and other Pueblo communities depends on membership in ceremonial societies, and the announcement of lunar months rested in the authority of those priests responsible for observing the sun and ceremonially making wands dressed with feathers, called '*pahos*', to be used as prayer offerings in local shrines. When the first crescent was observed the presiding chief for that month summoned other clan chiefs and together they deliberated whether to announce the forthcoming ceremony. These 'smoke talks' sanctified the coming ceremony and their purpose is to formalize the next eight or sixteen days of ritual activities. These culminate with a dance or puppet drama that will be attended by the community. In the case of the winter solstice ceremony it is intended to terminate on the full moon, but this is only the public part of the feast. At the ultimate close of the ceremony, the pahos are deposited in the cultic shrine a day or two after full moon.

In that authority for intercalation is given to the chief of the ceremony falling on these lunar periods, there is often rivalry with other chiefs as to the timing or absence of his decision. The rules for intercalation are somewhat loose, as they depend upon the authority of the calendar priest responsible for the seasonal ceremony. Even the practitioners get confused, and it is a common criticism that they have not done so correctly.[29] If the chief fails to intercalate as expected, he will be blamed for crop failures or other calamities that befall the community. Historically this has led to dissention, but shows that there was not a generally agreed upon calendar for intercalations, unlike the Gregorian calendar brought by the Franciscan Friars after the Spanish Entrada.

Intercalation

Intercalation is associated with periods of waiting observed by religious authorities, and is vital before the announcement of forthcoming ceremonies when adjustments in the lunar year are needed to be made. The information on intercalation is sometimes contradictory, and has been derived from separate villages, at different times and from different informants. The methods of intercalation are the following:

1. Additive: 3rd Mesa: village of Oravi: The springtime intercalary month, *Anguk'muya*, or 'after month' is inserted into a 12 month sequence after '*Kwia-muya*' (spelling,

Geertz) in April or May every three years.[30] Or, insert *Hakiton-muya* after *Kwyamuyan*. (1st Mesa: Walpi, Sichomove, Hano).[31] But other informants have found that the intercalary month is usually in autumn after *Tala'Powamuya*, and named '*Nosan'muya*' for the feasting after the harvest.[32]

2. Subtractive: intercalary month subtracted from a 13 month sequence: (3rd Mesa: Orabi, Hoteville): According to the Hopi dictionary intercalation is achieved by not counting one phase of the moon in reckoning the season: 'Sometimes when the moon gets ahead of the sun, they break up the months'. In other words when the lunar crescent is too late for the season, they skip a month.[33]

From these admittedly disparate sources, one pattern seems to emerge: intercalation seems to take place after an equinox, with observation of the following new moon, much like our practice of establishing Easter or Passover. After the Spanish King imposed a civil style government on the Pueblos in 1620 CE, the 12 month Gregorian calendar replaced the indigenous ceremonial calendars of all but the Western Pueblos of Zuni and Hopi, and the standard method of intercalation of lunar months, derived from the nineteen year Cycle of Meton was the norm. However there were certainly pre-conquest methods to bring the solar planting year into synchronization with the lunar ceremonial year. A calendar of thirteen months probably preceded the Colonial calendar, and according to Alexander Stephen, commenting in 1890, the Kachina feasts of the Hopi follow a 13 moon year as opposed to the 12 month secular year:

> The Katcina feasts are reckoned by a year of thirteen moons, consequentially, the New Year or Keli, of the katcina year only coincides with the secular Keli once in thirteen years, which period is called a Glad Year.[34]

Applying the Ethnographic Model to Prehistoric Lunar Observation

In Chaco Canyon, lat. 36°03', and in the Hopi mesas, lat. 36°15', the sunrise and sunset points at winter and summer solstice (on a level horizon) are very nearly 30° north and south of the east /west line of that latitude. This is convenient for archaeoastronomy, as the Hopi cardinal points are sixty degrees apart, like a hexagon. An Azimuth thirty degrees south of east, or 120°, will roughly correspond with winter solstice sunrise, whereas summer solstice sunset will take place at an azimuth of 300°. The Hopi celebrate the solar year annually during the *Soyal* ceremony near winter solstice, when the sun is ritually induced to turn back from its southern house by a warlike dance of shield bearers. *Soyal* concludes 16 days after the observance of the lunar crescent of *Kyaa'muya*, when prayer feathers, called '*pahos*' in the Hopi language, are deposited into the *kachina* shrine. Ideally, when the *kachinas* emerge from the *kivas* and dance in open plazas on the night of the public ceremony, *Soyal* full moon will rise on the winter solstice. But this ceremony rarely coincides with an actual winter solstice, because of the 11 day discrepancy between the 354 day lunar year, and the 365 day solar year. The Hopi see this as the sun moving either slow or fast according to the lunar year.[35] But as the solstitial points are firmly anchored to the local horizon yearly, unlike the moon which wanders over an 18.6 year cycle, the solar horizon calendar is always used as a cross-check on the lunar calendar.

Table 2 correlates observable lunar horizon phenomena with the Hopi directions on the horizon. Coincidentally, Stephen's observations took place during a major lunar standstill.

The Moon and Classic Chacoan Civilization

There are many rival interpretations of Chaco Civilization which try to account for its difference in scale and complexity in comparison to historic Pueblos. Because scarce resources and agricultural commodities actually had to be imported to the central place of Chaco Canyon, scholars have proposed that trade and ceremony, not agriculture, guided its development.[36] A simple subsistence model cannot explain the development, size, and centralized planning of the Great Houses of Chaco Canyon, which flourished for 100 years. And because there is a disparity between the number of rooms and the agricultural resources within Chaco Canyon itself, the population may have been custodial except when hosting festivals. To coordinate the festival calendar there should have been a commonly accepted means to choose dates and communicate over the vast regional system. A civil solar calendar may have been supplemented by a lunar ceremonial calendar, similar to the Hopi. But unlike the Chacoans, it appears the Hopi were more concerned with lunar phase than position when timing ceremonies. However there are many similarities between Hopi material culture and artifacts found archaeologically in Great Houses, for example ritual paraphernalia related to the plumed water serpent puppet drama such as bird effigies and the lightening frame.[37] A framework correlating the Hopi lunar calendar with the lunar standstill cycle over nineteen years is presented in table 3.

Summary and Conclusions

Thus far, no evidence has been presented by ethnographers that the Hopi or other historic Pueblos were concerned with the position of the moon on the horizon. But the well established lunar alignments at Chimney Rock and Mesa Verde are probably connected with anticipation periods that enabled the elite to organize community gatherings far in advance. These religious observances would have been particularly exciting during winter full moons at Major Northern Standstill. Because peak conditions for ceremonies vary with the 9 year oscillation between major and minor standstill, it must be assumed that ceremonies were multi-annual. In fact, there are great and small ceremonies in Hopi tradition, based on multiples of four years.[38] The New Year winter solstice ceremonial, such as Soyal in the Hopi villages, and Shalako at Zuni are also timed to coordinate with the full moon. The knowledge of the lunar excursions on the horizon would greatly assist ceremonial planning for these winter festivals. The traditional religious and ceremonial life of the Pueblo peoples has long been associated with the moon. But unlike scribal cultures, the tradition is passed orally and the canonical forms of calendar making are subject to several groups of authority, who may vary in their practice, especially in the methods of intercalating the solar-agricultural year with the lunar ceremonial year. Modern investigators of the prehistoric ancestors of the modern Pueblos have no reliable notation to reconstruct a definitive, or universal, method of calendar keeping. In the place of written procedures, calendar keeping seems to be embedded in the ceremonial life, and expressed through art, architecture and symbols able to integrate multiple communities under the

Ceremonial Month	Date	Lunar Phase	Azimuth (W)	Azimuth (E)
WINTER SEQUENCE:				
Kyel'muya	Oct. 22, 1892	First Crescent – west	245°	
	Nov. 3, 1892	Full Moon – east		88°
Kyaa'muya	Nov. 21, 1892	First Crescent-west	234°	
	Dec. 3, 1892	Full Moon – east		75°
Pa'muya	Dec. 20, 1892	First Crescent-west	235°	
	Jan. 1, 1893	Full Moon – east		55.5°
Powa'muya	Jan. 19, 1893	First Crescent – west	247°	
	Feb. 1, 1893	Full Moon – east		70°
Oso'muya	Feb. 17, 1893	First Crescent – west	260°	
	March 1, 1893	Full Moon – east		73°
Kwiya'muya	March 19, 1893	First Crescent – west	281° (#1)	
	March 31, 1893	Full Moon – east		93° (#2)
Hakiton'muya	April 17, 1893	First Crescent – west	294°	
	May 2, 1893	Full Moon – east		112°
SUMMER SEQUENCE:				
Tala'Kyel-muya	May 16, 1893	First Crescent – west	303°	
	May 30, 1893	Full Moon – east		124°
Tala'Kyaa'muya	June 15, 1893	First Crescent – west	304°	
	June 29, 1893	Full Moon – east		125°
Tala'Pa'muya	July 15, 1893	First Crescent – west	296°	
	July 29, 1893	Full Moon – east		119°
Tala'Powa'muya	August 13, 1893	First Crescent – west	227°	
	August 26, 1893	Full Moon – east		109°
Angok'Muya	Sept. 11, 1893	First Crescent – west	265° (#3)	
	Sept. 25, 1893	Full Moon – east		88° (#4)
Tala'Oso'Muya		FirstCrescent – west	246°	
	Oct. 24, 1893	Full Moon – east		77°

(#1) – Movement from south to north on western horizon; (#2) – movement from north to south on eastern horizon; (#3) – Movement from north to south on western horizon; (#4) – movement from south to north on eastern horizon

TABLE 2. Hopi Lunar Months observed by Alexander M. Stephen in the late 19th century (adapted from Ellis, 1975, and corrected by USNO Alt./Az. tables for Tuba City, Arizona).

Solar or lunar horizon position	Year in Lunar Standstill Cycle	Ceremonial Month	Ceremonial House	Azimuth
SUMMER SEQUENCE: Observations on the Western Horizon				
Full moonset: *Major Lunar Standstill: SS*	9 & 28 (2006 & 2025)	*Tala'Kyaa'muya*	"Unknown"	234°
Sun sets at Summer Solstice: June 21st	All years	*Tala'Kyaa'muya*	***Kwinini***	310°
Full moonset: *Minor Lunar Standstill: SS*	1 & 19 (1997 & 2015)	*Tala'Kyaa'muya*	"Unknown"	246° 247°
Full moonset: Major Lunar Standstill: AE	9 (2006)	*Ancock'muya* or *Nosan'muya*	"Unknown"	276.5°
Autumn Equinox Sun-set: Sept.22nd	All years	*Ancok'muya* or *Nosan'muya*	"Unknown"	270°
Full moonset: Minor Lunar Standstill: AE	1 & 19 (1997 & 2015)	*Ancok'muya* or *Nosan'muya*	"Unknown"	252° 256°
Full moonset: *Major Lunar Standstill: WS*	9 & 28 (2006 & 2025)	*Pa'muya*	"Unknown"	303°
Sun sets at Winter Solstice: Dec 21st	All years	*Pa'muya*	**Tevyuna**	240°
Full moonset: *Minor Lunar Standstill: WS*	1 & 19 (1997&2015)	*Pa'muya*	"Unknown"	294° 293°
WINTER SEQUENCE: Observations on the Eastern Horizon				
Full moonrise: Major Lunar Standstill: WS	9 & 28 (2006 & 2025)	*Pa'muya*	"Unknown"	55°
Sun rises at Winter Solstice	All years	*Pa'muya*	***Tatyuka***	120°
Full moonrise: Minor Lunar Standstill: WS	1 & 19 (1997 & 2015)	*Pa'muya*	"Unknown"	67° 68°
Full moonrise: Minor Lunar Standstill: VE	1 & 19 (1997 & 2015)	*Angktiwa-† Oso'muya*	"Unknown"	82° 82°
Sun Rise: Vernal Equinox	All years	*Angktiwa† Oso'muya*	"Unknown"	90°
Full moonrise: Major Lunar Standstill: VE	9 & 28 (2006 & 2025)	*Oso'muya*	"Unknown"	87°
Full moonrise: Minor Lunar Standstill: SS	1 & 19 (1997 & 2015)	*Tala'Kyaa'muya*	"Unknown"	114° 113.5°
Sun rises at Summer Solstice	All years	*Tala'Kyaa'muya*	***Hopoko***	60°
Full moonrise: Major Lunar Standstill: SS	9 & 28 (2006 & 2025)	*Tala'Kyaa'muya*	"Unknown"	127°

† Hopi for "waiting period"

TABLE 3. The Lunar Directions in the Chacoan World, following Hopi Ceremonial Order of the Sun (adapted from Altitude/Azimuth tables, USNO, for 1997–2025 CE) I have given dates that can be confirmed by documentary photography.

dramatic umbrella of periodic festivals, ultimately timed by direct observation of new lunar crescents within the structure of the solar year. However this does not preclude the possibility that the Ancestral Puebloans did not have a more elaborate lunar calendar. As yet no precise Hopi terminology for 'lunar houses' in the sky, or on the horizon, has been found by investigators, but the absence of such in the ethnographic record may simply be the consequence of never asking their informants. There might have been some sort of prehistoric method of lunar intercalation to blend 4-year ceremonies with solstitial full moons at the lunar standstills, combined with the established horizon directions of the sun. Predicting that a *Kya'muya* or *Tala'Kya'muya* full moon would be in lunar standstill position gave Chacoan calendar specialists the advantage of knowing here could not be a lunar eclipse during the all important culmination of the ceremony.[39] With the demise of Chaco as a central place, the wholesale migrations of the late 13th century, added by the arrival of the Kachina Cult, may have led to a calendrical reformulation that now focused only on lunar phases. Unlike the Chacoans, it appears the Hopi were more concerned with lunar phase than position when timing ceremonies. However there are still many unknowns. It would be surprising if lunar alignments were not important in the context of a lunar ceremonial calendar. Because Chaco archaeology has revealed material culture, and architectural prototypes related with historic Pueblo peoples, it is the subject of both retrospective ethnographic analogy, and forward projections into the historic present. These are still speculations which must be tested and demonstrated by many investigators and photographically confirmed, before we can confidently reconstruct the prehistoric lunar calendars of the Chaco Civilization. The intent of this paper is to provide a framework for the assessment of lunar alignments within the Chacoan context. But it must be realized that though the path of archaeoastronomy has yielded impressive alignments, this focus on the spatial realm minimizes the color, art, pageantry, and regalia still available for study as expressions of the Puebloan Cultures' romance with the moon.

Notes

1. An elaborate theory ascribing lunar directions to Chacoan Great Houses has been advanced by Anna Sofaer, researcher of the Fajada Butte 'Sun Dagger' phenomenon. In her book Chacoan Astronomy, An Ancient American Cosmology it is claimed that the base walls of major great houses are aligned with three classes of astronomical directions: 1) solar, 2) major lunar standstills, and 3) minor lunar standstills. The lunar directions have yet to be demonstrated in that the walls are not actually in position to view moonrise/set. Because minor standstill moon is within the horizon limits of the solar solstices, Great House orientations could just as likely be solar, and the argument that certain Great Houses are aligned with minor lunar standstill cannot be proven.
2. Ellis (1975), p. 64.
3. Jojola (1987), p. 93.
4. Lekson et al. (1988), p. 2.
5. Lekson et al. (1988), p. 10.
6. Cordell (1984), pp.6–7. The modern Pueblo languages within the Ancestral Pueblo Chacoan Regional System are divided geographically into eastern and western groups. To the west, the two principal languages are Hopi in Arizona, a Uto Aztecan language, and Zuni in far western New Mexico, linked to the Penutan language group of California. East of Zuni is Keresan, spoken in the Pueblos of Acoma and Laguna, with a differing dialect spoken in the Rio Grande villages of Zia, Santa Anna, Laguna, Santo Domingo, and Cochiti. In the Tanoan language group, Tewa is spoken in the villages of Santa Clara and Ok Owinge (San Juan), San Ildefonso and Nambe; Tiwa is spoken in Taos and Picuris north of Santa Fe, and

in Isleta, south of Albuquerque; and Towa is spoken at Jemez, east of Chaco Canyon. A fourth language group, Athabascan, is represented by the semi nomadic Navajo and Apache tribes who occupied the Chacoan territories shortly after the collapse of the system. Many of these modern peoples have oral traditions or legends about Chaco.

7. Fowler and Stein (1992), pp. 101–122.
8. McKenna and Toll (2001), p. 137.
9. McCluskey, (1992), pp. 38–39.
10. The Lunar Standstill Cycle derives from the moon's orbital inclination of 5.1 degrees to the earth's orbit's plane of revolution around the sun. Consequently the moon's declination to the celestial equator can be added or subtracted to the sun's, depending upon moon's position during its 27.3 day revolution about the earth. Because the precession of the moon's orbit wobbles about the earth in an 18.61 year cycle, this can vary from 23.5° + 5.1°= 28.6 degrees at maximum, to 23.5° – 5.1°= 18.4 degrees at minimum. Between these limits, the moon passes two midpoints when its orbit is parallel to the Earth's. As seen from Earth, the moon slows down to a 'standstill' every 14 days, or half of its orbital period, when it reaches the ends of its elliptical path as seen from Earth. The lunar synodic month of 29.5 days is out of step with the standstill cycle by about two days, therefore the 'standstill' falls behind the phases of the moon by about two days each month. Because of this phase shift, the moon will be visible as a new crescent on the western horizon in major southern standstill position near winter solstice, whereas the full moon 14 days later will be in major northern position.
11. Malville (2008), p. 99.
12. Judge (1990), p. 36, Malville and Malville (2001), pp. 339–340.
13. Malville (2008), p.129.
14. Malville (2008), p. 118.
15. Williamson et al. (1977), p. 203.
16. McCluskey (1993), pp. 38–39.
17. Hieb (1979), p. 577.
18. The 'kachina cult' is defined by masking, where celebrants of religious/ceremonial performances wear masks which disguise their secular identity and serve the purpose of identifying the wearer with a transcendent spirit world. The masks are not personal property, but correctly belong to the kiva or clan-house of their ceremonial association. Masked imagery does not appear in Ancestral Puebloan rock art, ceramics or wall paintings until the start of 14th century CE after the abandonment of the San Juan River communities, presumably because of drought in the late 13th century. With the arrival of the kachina cult there is a profound change in architecture and settlement patterns, which suggests there may be a discontinuity in ceremonial observances as well. The typical round kiva of the Chacoan horizon is replaced by a rectangular kiva, and in the new aggregated communities walled plazas become the site of communal festivities. However, contemporary Pueblo citizens often claim that their ceremonial life, even with kachina cult membership, is closely parallel to their ancestors' at the time of the Chacoan fluorescence. In fact, the original ceremonies may have their continuation in the un-masked festivals, such as Flute-Antelope dances during the summer. The arrival of the Kachina Cult is admirably documented by Adams (1991).
19. Parsons (1936), p. 159, Ellis (1982), p. 64.
20. Hopi Dictionary (1998), p. 264.
21. Ellis (1975), p. 65.
22. McCluskey (1982), p. 44, quoting Stephen (1894).
23. Frigout, (1979), p. 564.
24. Earl and Kenard (1971), p. 7.
25. Hopi Dictionary (1998), p. 833.
26. Parsons (1936), p. 136.
27. Ellis (1975), p. 64.
28. Parsons (1936), pp. XXXIX–XL.
29. Ellis (1975), p. 65.
30. Geertz (1987), p. 207.

31. Hopi Dictionary (1998), p. 833.
32. Frigout (1987), p. 567.
33. Hopi Dictionary (1998), p. 265.
34. Patterson (1994), p. 135.
35. McCluskey (1982), p. 52.
36. Judge (1990), pp. 35–36.
37. Vivian et al. (1978), pp.37–58.
38. Frigout (1979), pp. 567–568.
39. Malville and Judge (1993), pp. 4–5.

References

Adams, E. C., 1991, *The Origin and Development of the Pueblo Katsina Cult* (Tucson: University of Arizona Press).

Cordel, L. S., 1984, *Prehistory of the Southwest* (San Diego: Academic Press).

Earl, E., and Kennard, E. A., 1971, *Hopi Kachinas*, Second Edition, Revised. (New York: Museum of the North American Indian).

Ellis, F. H., 1975, "A Thousand Years of the Pueblo Sun-Moon-Star Calendar", in A. F. Aveni (ed.), *Archaeoastronomy in Pre-Columbian America* (Austin: University of Texas Press), 59–88.

Ford, C. D., 1931, "Hopi Agriculture and Land Ownership", *Journal of the Royal Anthropological Institute of Great Britain and Ireland* 61, 357–405.

Frigout, A., 1979, "Hopi Ceremonial Organization", in A. Ortiz (ed.), *Handbook of North American Indians, Vol. 9, Southwest* (Washington: Smithsonian Institution), 564–576.

Geertz, A W. and Lomatuway'ma, M., 1987, *Children of Cottonwood* (Lincoln: University of Nebraska Press).

Heib, L. A., 1979, "Hopi World View" in A. Ortiz (ed.), *Handbook of North American Indians, Vol. 9, Southwest* (Washington: Smithsonian Institution), 577–580.

Jojola, T. S., 1987, "The Sun Personified: Some Preliminary Observations Regarding Celestial Cosmology and the Ordering of Pueblo Indian Society", in J. B. Carlson and U. J. Judge (eds.), *Astronomy and Ceremony in the Prehistoric Southwest* (Albuquerque: Maxwell Museum of Anthropology).

Judge, W. J., 1990, "Resource Distribution and the Chaco Phenomenon", in J. M. Malville and G. Matlock (eds.), *The Chimney Rock Archaeological Symposium* (Fort Collins: US Dept. of Agriculture), 35–36.

Lekson, S. H., Windes, T. C., Stein, J. R., and Judge, W. J., 1988, "The Chaco Canyon Community", *Scientific American*, 259, No. 1, 100–109.

Lekson, S. H., 1999, *The Chaco Meridian, Centers of Political Power in the Ancient Southwest* (New York: AltaMira Press).

Malville, J. M., and Judge, J. W., 1993, "The Uses of Esoteric Astronomical Knowledge in the Chaco Regional System", Paper at the Fourth Oxford International Conference on Archaeoastronomy, Stara Zagora, Bulgaria

Malville, J. M. and Malville, N. J., 2001, "Pilgrimage and Periodic Festivals as Process of Social Integration in Chaco Canyon", *Kiva* 66, 327–344.

Malville, J. M., 2008, *Guide to Prehistoric Astronomy in the Southwest* (Boulder: Johnson Books).

McCluskey, S. C., 1993, "Space, time and the calendar in the traditional cultures of America", in C. Ruggles (ed.), *Archaeoastronomy in the 1990s* (Loughborough: Group D. Publications Ltd.), 33–44.

McKenna, P. J., and Toll, H. W., 1992, "Regional Paterns of Great House Development among the Totah Anazasi, New Mexico", in D. E. Doyel (ed.), *Anasazi Regional Organization and the Chaco System* (Albuquerque: Maxwell Museum or Anthropology), 133–146.

Patterson, A., 1994, *Hopi Pottery Symbols, Based on work by Alexander M. Stephen* (Boulder: Johnson Books).

Parsons, E. C. (ed.), 1936, *Hopi Journal of Alexander Stephen*, Columbia Contributions to Anthropology Number 23 (New York: Columbia University Press).

Sofaer, A., 2008, *Chaco Astronomy, An Ancient American Cosmology* (Santa Fe: Ocean Tree Books).

Vivian, R. G., Dodgen, D. N. and Hartman, G. H., 1978, *Wooden Ritual Artifacts from Chaco Canyon New Mexico, The Chetro Ketl Collection* (Tucson: University of Arizona Press).

Williamson, R. A., Fischer, H. J., and O'Flynn, D., 1977. "Anasazi Solar Observatories", in A. F. Aveni (ed.), *Native American Astronomy* (Austin: University of Texas Press), 203–218.

Adjusting Calculations to the Ideal in the Chinese and Japanese Calendars

Susan Tsumura

The Chinese calendar has a history of well over three thousand years, and given the extent of China, has probably been used by more people over a wider area than any other lunar calendar. Though the Gregorian calendar has been used officially since 1912, people are still familiar with the lunar calendar, and it is used for public and private festivals, including New Year.[1] As the Chinese calendar is lunisolar, it has to deal with some of the same issues as other lunisolar calendars, but its use of year-internal intercalary months and formal calendar procedures gives it some problems of its own.

The earliest Chinese dates we have are from the famous Shang (Yin) oracle bones dating from the late 2nd millennium BC. Not much is certain about how the calendar worked,[2] but some characteristics have continued down even to the present day. One is that months are named by numbers;[3] another is that days can be designated using a sexagenary cycle that continually repeats without regard to the starts of months, years, or anything else.[4]

The calendar developed over succeeding centuries, and versions of the early classical calendar were probably being used by around the middle of the 4th century BC, the "Warring States" period.[5]

Structure of the Classical Calendar

The first thing to note about the classical calendar is that it is calculated. Most of the official calendar procedures since 104 BC have been preserved. A procedure give constants such as the length of the solar year as well as instructions for calculating the new moons and solar terms for any given year. Although creating a procedure involves observation, using the procedure itself involves only mathematics.[6] The procedures were revised from time to time, and since 104 BC there have been almost fifty different official calendars.[7]

The calendar is lunisolar: that is, while the start of the month is determined by the phase of the moon, the start of the year is determined with reference to the solar year, i.e. the seasons.

Lunar Aspect

In the classical calendar, the new moon is the conjunction with the sun (*shuo* 朔), not the crescent moon.

The day is reckoned from midnight to midnight at the time of the capital city, and in principle, the first day of the month is that in which the new moon occurs.[8] For example,

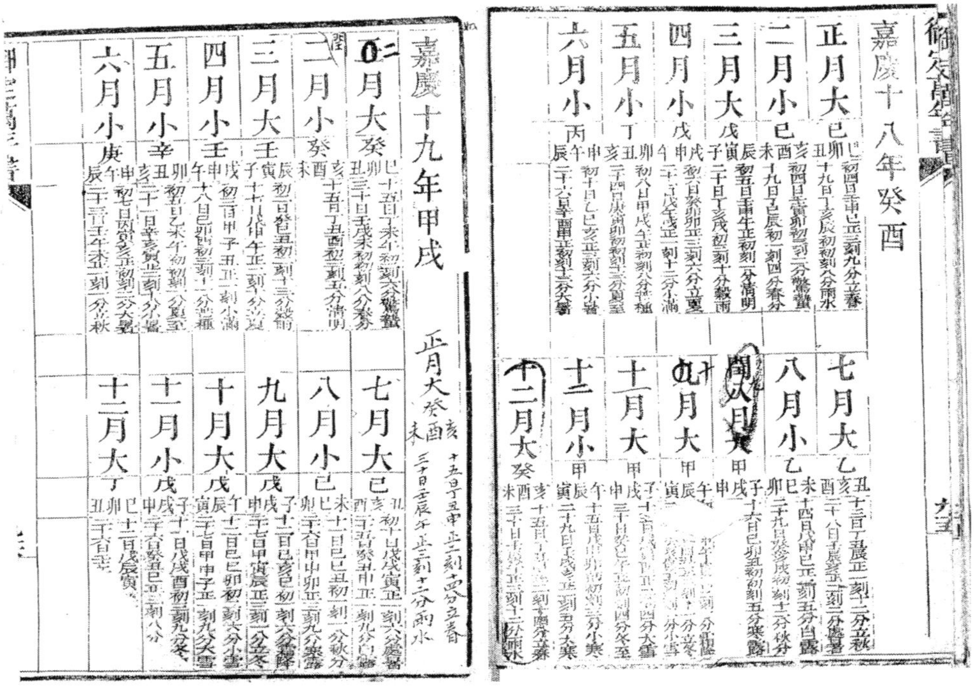

FIGURE 1. Calendar for Jiaqing 18–19 (1813–14) from Qintianjian (1799). In 1811 it was decided to change the intercalary month for those years in order that certain festivals should fall in their proper months. The owner of this copy made the necessary alterations with a brush. (See fig. 3) Used by permission of the Kyoto University Humanities Library.

February 1, 2010, the final day of the Living the Lunar Calendar Conference, was in the 12th month of the Chinese calendar. As the previous new moon was the afternoon of January 15 Beijing time, the 12th month started at the previous midnight, and February 1 was the 18th day of the 12th month. The next new moon (New Year's Day) was the morning of February 14th, so the 12th month ended February 13th; it was a 30-day month.

Of course, as in most lunar calendars, months are either 29 or 30 days long. No other lengths are allowed, even if month starts are artificially altered as we shall see later, and this rule seems to be absolute.

Solar Aspect: Intercalary Months

As a lunisolar calendar, the Chinese calendar has both 12- and 13-month years. References to a thirteenth month can be found in the Shang bone-oracles and in inscriptions for several centuries afterwards. However around the 7th century BC, intercalary months (*runyue* 閏月) start to appear during the year. [9]

Year-internal intercalary months are determined using a system of twenty-four solar terms (*jieqi* 節氣), points spread out evenly throughout the solar year that had been established to get a reckoning that was seasonally more accurate than the lunar calendar.[10]

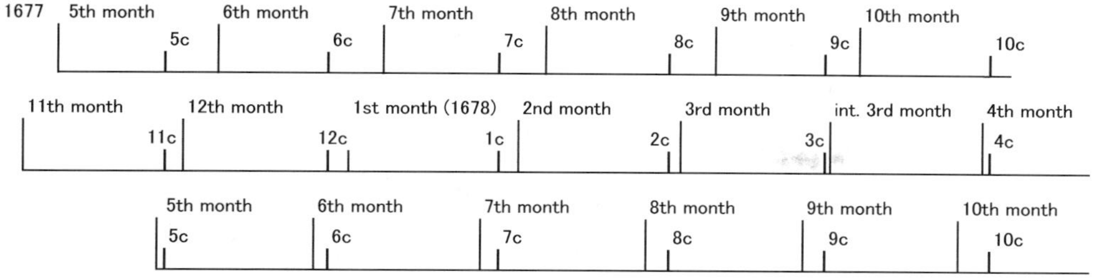

FIGURE 2. Intercalary months. The months are blocked off by thick lines; the central terms are marked by thin lines. Xc refers to the Xth-month central term. The central terms get later and later with respect to the month starts until there is a month with no central term, in this case the month after the 3rd month of 1678, which is made intercalary.

Twelve of these solar terms, including the solstices and equinoxes, are called central (or principal) solar terms (*zhongqi* 中氣).[11] These central terms are associated with months, and the designation of the month is determined by the central term that occurs in it.[12] For example, as one can see by figure 2, the first month is that in which the 1st-month central term (1c) occurs, and the tenth month is that in which the 10th-month central term (10c) occurs, and so forth, though the central terms can occur anywhere in the month from the first day to the last. Thus, using the central terms you have an ideal situation—twelve months tethered to twelve points distributed evenly throughout the solar year.

But of course, reality is not ideal and you need intercalary months. Since there are 12 central terms in the solar year, most central terms are 30 or 31 days apart, while the months are only 29 or 30 days long. Thus, central terms tend to occur later and later in the month, and every 33 months or so, there will be a month without a central term—such as the month following the 3rd month of 1678 in figure 2—and this month is designated the intercalary month of the previous month. In this case it is the intercalary 3rd month (*run sanyue* 閏三月), but intercalary months can occur after any month of the year.[13]

Which of the central terms is the first-month term has varied over the centuries from the central term before the winter solstice to the second central term after it, but at least since 104 BC, it has almost always been the second central term after the solstice. This means that in principle New Year's Day is the new moon closest to the traditional start of spring, which is half way between the winter solstice and the spring equinox. It also means that the important winter solstice is the 11th-month central term, a fact that will come up repeatedly in this paper.

Procedures

A general outline of the procedures for calculating the calendar is the following:[14] The procedure gives constants such as the year length, average month length, and a starting year.[15] First, by means of remainder mathematics, find the cyclic date and the time of day of the winter solstice (*donzhi* 冬至) and of the mean conjunction immediately preceding it. From those, calculate the cyclic dates and times of the mean conjunctions and mean solar terms for the following year.

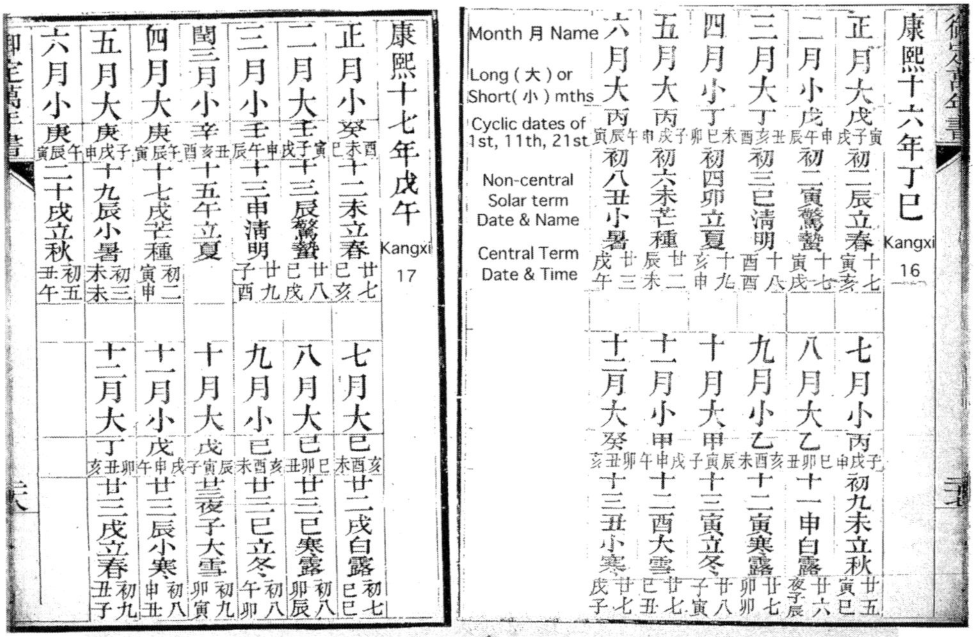

Figure 3. Calendar for Kangshi 16 and 17 (1677–78) in Qintianjian (1787). The months are listed from right to left. Kangshi 17 has an intercalary 3rd month 閏三月; note that it does not have a central term.

If the procedure uses the true conjunction, for each month adjust the time of the mean conjunction by calculating a solar and a lunar adjustment using tables and calculations. The solar adjustment is based on the position of the sun in its cycle from perihelion (= the winter solstice in most calendars) to perihelion, and the lunar adjustment is based on the position of the moon in its cycle from perigee to perigee. Add these two figures to the mean conjunction to get the time of the true conjunction.

Finally, from the cyclic dates of the conjunctions and central terms, determine the length of each month of the year and the position of any intercalary month.

Procedures also came to predict eclipses and the positions of the planets. Furthermore, divination matters such as auspicious and inauspicious days for various activities were usually included.

History of Calendar Changes

The above is the fundamental structure of the classical calendar, but details have changed throughout history.

Quarter-day Calendars

The earliest classical calendars were the quarter-day calendars (*shifen li* 四分曆), so called because their solar year was exactly 365 and one-quarter days long. Although they varied among themselves as to the exact time of the winter solstice or the new moon[16] or which month started the year, their principles were the same. They were very regular calendars. However, as the calendar procedures were changed to better reflect astronomical reality, irregularities appeared and some even caused trouble with rituals. In this paper I will briefly discuss three of these changes so as to give a clearer understanding of how the calendar worked.

Solar and Lunar Constants

However, first I will mention the most important constants of the procedures, the lengths of the average lunation and of the solar year.[17] With few exceptions, from AD 510 on, the average lunation was calculated as 29. 53059+ days. The solar year was more variable. The first procedure to change the quarter-day calendar, the 104 BC Taichu calendar 太初曆, put the length of the solar year at 365 385/1539 days . After that, they changed the length of solar year with each procedure, if even by a few seconds.

The length of the year was measured from winter solstice to winter solstice, and the moment of the solstice was traditionally determined by measuring the length of shadows cast by a gnomon (pole). When the great Shoushi calendar 授時曆 of 1281 was established, much work was put into making the huge gnomon and viewing its shadow.[18] The calendar used the year length 365.2425 days.

From Mean to True Conjunction

Now, the quarter-day calendar determined the start of the month using the mean conjunction, and one lunation was calculated as 29 499/940 (29.53085) days.[19] Therefore short and long months, that is 29- and 30-day months, alternated, except that every 15th or 17th month as calculations required there would be two long months in a row—the extra long month was not put off to a certain month of the year. The exact length of the lunation changed with the procedure, but the principle long remained the same. The irregularity of the lunar orbit was known by 52 BC, and from the AD 237 Jingchu calendar 景初曆 the true conjunction was calculated for predicting eclipses, but the start of the calendar month was still determined by the average conjunction.[20] However, during the preparations for the Yuanjia calendar 元嘉曆 of AD 445, it was proposed that months should start on the day of the true conjunction so that solar eclipses would always occur on the first day of the month. This proposal was rejected because it was known that with true conjunction there could be three long or two short months in a row, and while there were precedents for eclipses on the last or second day of the month, three long months in a row was too radical. At the start of the Tang period, the 619 Wuyin calendar 戊寅曆 finally used months based on the true conjunction, the makers accepting that there could be three long or short months in a row. However, in previous century the irregularity of the sun's movement around the ecliptic had been discovered, and this was used in computing the true conjunction for the Wuyin calendar. Thus twenty-six years later the new procedure calculated *four* long months in a row. This was terrible! They immediately returned to using mean conjunction. In 665, they adopted the Linde calendar 麟德曆, which did use true conjunction, but the calendars were changed every few decades, and they mostly managed

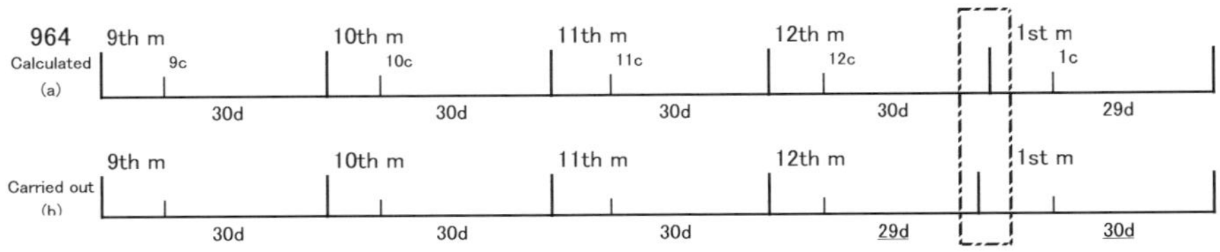

FIGURE 4. Avoiding 4 long months in a row in 964. 29d or 30d is the number of days in the month; Xc is the Xth central term. Resulting changes are underlined.

to avoid four long months until the 1281 Shoushi calendar, which placed astronomical accuracy above ideals.[21]

Now let us go to Japan. Unlike most other countries of East and South-east Asia, Japan never used the calendars calculated by the Chinese astronomical bureau, but rather adopted or adapted Chinese procedures at various times and did its own calculations. During the Tang period, the heyday of its borrowing from China, it successively adopted four Chinese calendar procedures, but at the end of the 9th century, official contact with China ceased, and Japan continued to use the 822 Xuanming (Jp. Senmyō) calendar 宣明暦, which it had adopted in 862, for over 800 years.[22]

Now, for the year 964 this procedure calculated that the 9th through the 12th months were all long months, making four long months in a row (figure 4a).[23] The Japanese knew the Chinese classics well, so naturally this sequence was unacceptable to them, as it had been to the Chinese three centuries before.[24] However, by now they could hardly go back to average conjunction, so they simply started the 1st month of 965 one day early, making it a long month of 30 days, and of course automatically making 12th month of 964 short, so there were only three long months in a row (figure 4b). This situation occurred another four times between 1018 and 1088, and each time either the first or the last month was made short.[25] Three centuries later in 1495 the procedure again calculated four long months, but by that time the Calendar Department had given up on changes to the calculations (except for the regular new-moon advancement, see note 8) and nothing was done.

These were simple changes, but the next ones I discuss get rather more complicated.

Breaking the Metonic Cycle

The Metonic cycle, the observation that 19 solar years are (almost) exactly equal to 235 months (i.e., 19 years x 12 regular months plus 7 intercalary months), played a major role in the development of lunisolar calendars. Though the evidence is much less secure in China than in Mesopotamia, it appears that in China, like in Mesopotamia, most 19-year cycles had close to 7 intercalary months at least from the end of the 8th century BC.[26] The quarter-day calendar and its successors formally incorporated the cycle (*zhang* 章) into their procedures: the mean month length and length of the solar year were kept in the ratio 235 months = 19 years. This meant that the intercalary months in the cycle were either constant or oscillated between two neighboring months. The winter solstice was the calendrical start of the solar year, and the calendar was constructed so that the moment of the solstice coincided exactly with the moment of the (mean) new moon every nineteen

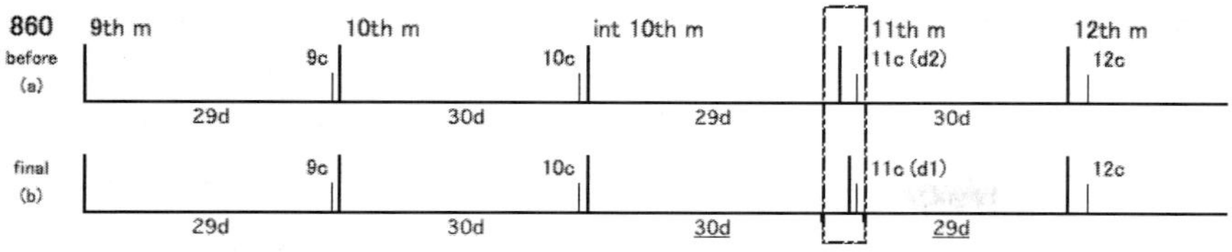

FIGURE 5. When the new moon is before the solstice.

years.[27] This day of the new-moon solstice was considered important as the beginning of the cycle, and had rituals associated with it.

However, the Metonic ratio is not completely accurate, and in the 5th and 6th centuries AD, it was given up so that a more accurate length could be used for the solar year.[28] This meant the new moon and the winter solstice might even occur on different days at the start of the cycle, and the later introduction of starting months on the true conjunction added yet another variable. But the cycle was so important that a commentary written soon after the break from the cycle stated that if necessary the starts of even several months should be altered to match the cycle.[29]

Now, during the Tang period (7th through 9th centuries) Japan was absorbing Chinese culture with all its ability.[30] In 784 the new moon occurred on the day of the winter solstice, so following the Chinese model, the court celebrated this felicitous day by celebrations and a rice-tax remission in the capital area, and made it one of the standard court rituals, the New-Moon Winter Solstice (*Sakutan Tōji* 朔旦冬至).[31] Fortunately for the Japanese Calendar Department, for several cycles, the new-moon solstice occurred regularly every 19 years. However, in 860, the new moon was calculated to occur the day before the winter solstice (the 11th central term), not on it (figure 5a). This was brought to the court's attention a week before the solstice, and there was a big discussion.[32] The calendrist explained that the new moon started before the solstice that year because of the "irregularity in the moon's speed... besides the fluctuation in the sun's movement, et cetera". As there was nothing in the procedures about changing calculations to bring about a felicitous day, he had left the new moon before the solstice, but the court had the authority to change the calendar. Chinese precedents were cited, including the Chinese commentary mentioned above. The emperor was assured that a change would not distort the following seasons or the lunar cycle, so it was agreed to start the 11th month one day later, and they now had the new-moon solstice (figure 5b).

The starts of the next three cycles were normal, but 936 had the opposite problem—the new moon was the day after the solstice, so the solstice (11c) was on the last day of the eleventh month instead of on the first (figure 6a). The Calendar Department realized it was the start of the cycle, but apparently could not figure out how to move the solstice to the beginning of the month, so they did not have the ritual that year.[33] However, 114 years later when the situation reoccurred, they were prepared! They moved the start of the intercalary 11th month one day earlier to the day of the winter solstice, making it the new-moon solstice (figure 6b–c). Since the winter solstice, i.e., the 11th central term, was now in a different month, the months changed names. The former 11th month now had

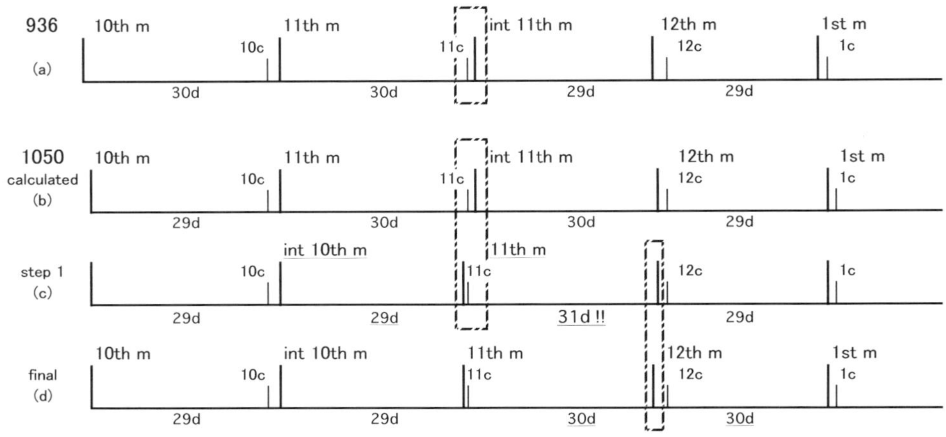

FIGURE 6. New moon after the solstice. In 936 they left it as is, but in 1050 they started the month earlier in the 11th and 12th months.

no central term and became the intercalary 10th month, while the former intercalary 11th month now had the winter solstice and so became the 11th month, and its first day was the new-moon solstice. However, the new 11th month now had 31 days, which was utterly impermissible. Therefore they started the 12th month one day earlier, so the 11th month had only 30 days, and they had their calendar (fig. 6d).

But, since the calendar was no longer calculated by the cycle, the new moon was getting later with respect to the solstice by an average of about 1 hour a cycle,[34] and the calendar had to be altered more and more often.[35] Most cycles it was sufficient to start the 11th month one day early, as in figure 6b to 6c, but in 1449, the lengths of as many as three other months also had to be adjusted.

But there were even more complex problems.[36] In 1202, the solstice was on the last day of the month (figure 7a), but if they moved the new moon earlier to the solstice, they would get 28-day and 31-day months (figure 7b), so they would have to move the beginnings of the intercalary 10th month and the 12th months also (figure 7c), and as a result, the 11th month through the 2nd month would be all long, i.e., there would be four long months in a row, which as we saw above was unacceptable. So instead, they simply moved the winter solstice later by an hour into the next month (figure 7d–e), so the month names came out right, and the lengths of the months were unchanged and presented no problem. However, moving solar terms was almost unprecedented, so when the same problem arose in 1316 (though this time they had to move the month start by two days), they moved the month start to the solstice (figure 8a–b), changed three more months to get 29- or 30-day months (figure 8c), then shortened the end of the 2nd month to avoid the four long months (figure 8d), the same technique they had used in 964 (figure 4). They did the same thing in 1335 and 1373.

In 1430 the new moon and solstice were again two days apart.

Finally they gave up. In 1487 the solstice was left at the end of the 11th month, and the new-moon-solstice ritual was not carried out. (As noted above, soon after this in 1495

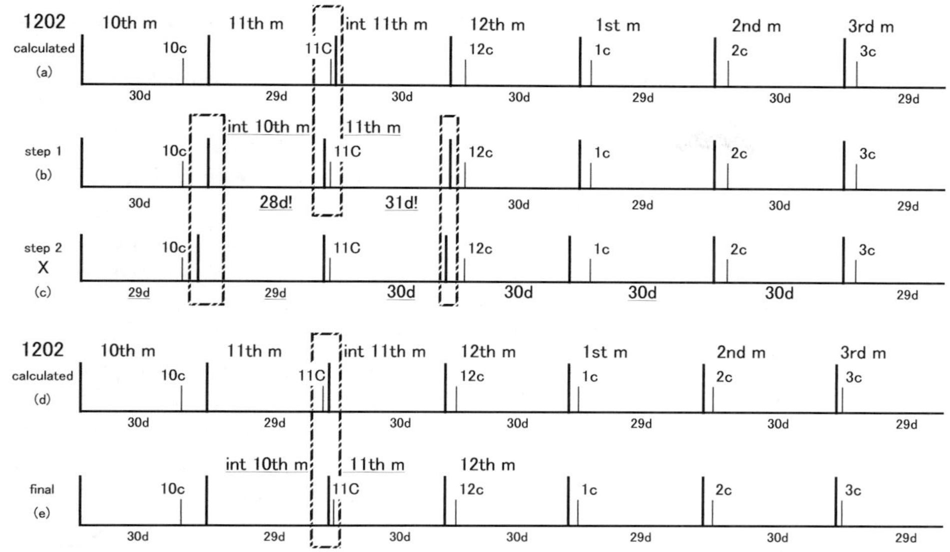

FIGURE 7. The new-moon solstice and the 4 long months problem. In 1202 if they moved the new moon to the solstice, they would end up with four long months, so they moved the solstice to the new moon instead.

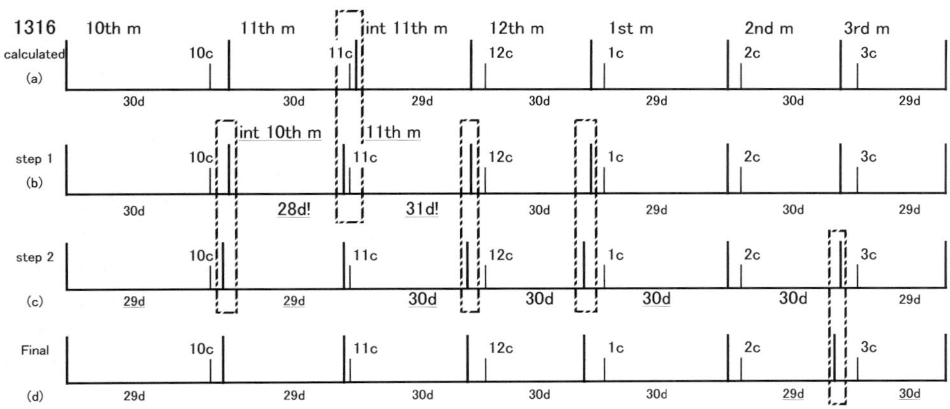

FIGURE 8. The new-moon solstice and 4 long months problem. In 1316 after moving the new moon to the solstice and making the necessary changes, there were 4 long months in a row (c), so the start of the 3rd month was altered also.

they did not make the change to avoid four long months either.) From then on until the Meiji Restoration (1868) with one exception, the festival was celebrated whenever the new moon fell naturally on the winter solstice, regardless of the 19-year cycle. The last time was in 1786.[37]

From Mean Solar Terms to True Solar Terms

The third change I will discuss is that from using mean solar terms (*pinqi* 平氣 or *hengqi* 恆氣) to using true solar terms (*dingqi* 定氣 or *shiqi* 實氣).

At the time the quarter-day calendars were developed it was assumed that the sun moved at an even speed around the ecliptic, and the solar terms were calculated by dividing the year into equal parts by time, starting with the moment of the winter solstice (mean solar terms). However, in the 6th century AD it was discovered that the sun's speed was uneven, so the summer solstice central term was not exact, and the equinox terms were almost two days off. This knowledge was used for computing astronomical phenomena as conjunctions and eclipses, but mean solar terms continued to be used for the solar terms published in the calendars and used for determining intercalary months. The rule for intercalary months was this: make months without central terms intercalary. With mean solar terms, the twelve central terms are approximately 30.44 days apart, that is 30 or 31 calendar days apart. As months have only 29 or 30 days, there can not be two central terms in one month, and so there is no problem with naming months by the central terms as the intercalary rule states.

However, in 1645, the beginning of the Qing dynasty, the western-influenced Shixian calendar 時憲曆 started using the present method of true solar terms, in which central terms occur when the sun has moved 30° along the ecliptic, like the western signs of the zodiac. This means central terms can be anywhere from 29 to 32 calendar days apart, and near the time of the earth's perihelion, now shortly after the winter solstice, there are years in the Metonic cycle when the new moons occur very close to the central terms for several months in a row, and the interrelationship can become complex (figure 9). Since some central terms are 29 days apart, it is even possible for two central terms to occur in one 30-day month. This presents a problem with naming months—if there are two central terms in the month, which term is the month named after? This problem is rare, but it is not just theoretical: the long-term Chinese calendars from the Qing period (1644–1912) include twelve periods with months that have two central terms, some of which are shown in figure 10.[38] In the figure, the first column shows months with letter names (the *a* month would be the winter-solstice month of the previous year; the *z* month, the winter-solstice month two years after the *a* month); the next column gives the central terms that occur in that month, and the shaded terms are those of the cardinal terms (solstices and equinoxes), the 5th, 8th, 11th, and 2nd central terms; the third column gives the name of the month, and the shaded months are the cardinal months, that is those associated with the cardinal terms, the 5th, 8th, 11th, and 2nd months.

The year 1678 is an ordinary intercalary year—the central terms and month names all agree, with the one month without a central term being an intercalary month. However, in 1813–14 month *l* contains both the 9th- and 10th-month terms, and *k* and *q* both have no central term. The period of 2033–34 even has two months with double central terms, and three months without a central term. How do you name months in cases like these?

Note that in such cases there are thirteen months from one 5th-month term to the next, so just one extra month in the period. This is because mathematically a month with two central terms must be both preceded and followed within a few months by a month without a central term, so there is one more non-central-term month than double-central-term month. Therefore if one wants the calendar to have all twelve months, one must

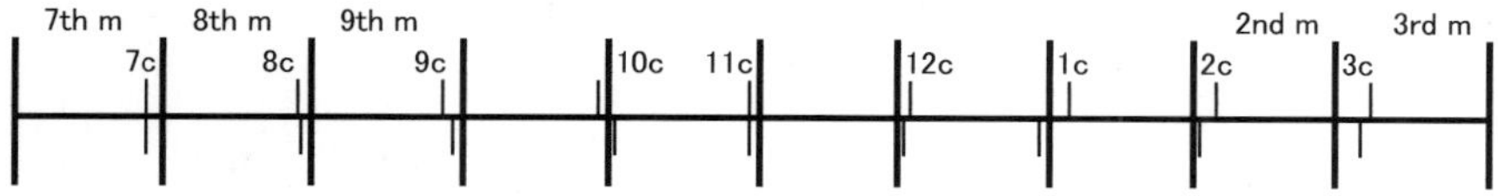

FIGURE 9. Mean central terms (thin lines above) and true central terms (thin lines below) for 1642-43. Note that the true solar terms stay within one day, before or after, of the month start (thick lines) from the 8th month to the 2nd month.

		term	month
1677–	g	5c	5m
	h	6c	6m
	i	7c	7m
	j	8c	8m
	k	9c	9m
	l	10c	10m
	m	11c	11m
	n	12c	12m
1678	o	1c	1m
	p	2c	2m
	q	3c	3m
	r	–	i
	s	4c	4m
	t	5c	5m

		term	month
1813–	g	5c	5m
1814	h	6c	6m
	i	7c	7m
	j	8c	8m
	k	–	
	l	9,10c	
	m	11c	
	n	12c	
	o	1c	
	p	2c	
	q	–	
	r	3c	3m
	s	4c	4m
	t	5c	5m

		term	month
1851–	g	5c	5m
1852	h	6c	6m
	i	7c	7m
	j	8c	8m
	k	–	
	l	9c	
	m	10c	
	n	11c	
	o	12,1c	
	p	2c	
	q	–	
	r	3c	3m
	s	4c	4m
	t	5c	5m

		term	month
2033–	g	5c	5m
2034	h	6c	6m
	i	7c	7m
	j	–	
	k	8c	
	l	9c	
	m	10,11c	
	n	–	
	o	12,1c	
	p	–	
	q	2c	2m
	r	3c	3m
	s	4c	4m
	t	5c	5m

FIGURE 10. The problem with true solar terms. How do you name the months here?

designate just one of the thirteen months as the intercalary month and number the other months in order, ignoring the central terms.[39] The question then is which one of the two or three non-central-term months will be the intercalary month.

Consciousness of this problem has waxed and waned, but there have been several methods for deciding the intercalary month for true solar terms.[40]

The Natural Rule

A double-central-term month is always preceded by non-central-term month. Therefore a calendrist who was used to the old rule for mean solar terms, would naturally make this first non-central-terms month the intercalary month as always, and continue numbering the following months in order. So when he came to a double-central-term month he would have to name the month by the first of the two central terms, then continue numbering in order, even when terms and months did not match, through the next non-central-term month, after which the central terms and month names would match again. In other words, the intercalary month is the first non-central-term month. I call this the natural rule. The N(atural) columns of figure 11 show how the months of various years would be numbered using this method.

		term	N&S
1677	g	5c	5m
	h	6c	6m
	i	7c	7m
	j	8c	8m
	k	9c	9m
	l	10c	10m
	m	11c	11m
	n	12c	12m
1678	o	1c	1m
	p	2c	2m
	q	3c	3m
	r	–	i
	s	4c	4m
	t	5c	5m

		term	N	S	(S)
1642	g	5c	5m	5m	5m
	h	6c	6m	6m	6m
	i	7c	7m	7m	7m
	j	8c	8m	8m	8m
	k	9c	9m	9m	9m
	l	–	i	10m	10m
	m	10,11c	10m	11m	11m
	n	–	11m	i	12m
	o	12,1c	12m	12m	1m
1643	p	–	1m	1m	i
	q	2c	2m	2m	2m
	r	3c	3m	3m	3m
	s	4c	4m	4m	4m
	t	5c	5m	5m	5m

		term	N&S	
1661	g	5c	5m	5m
	h	6c	6m	6m
	i	7c	7m	7m
	j	–	i	8m
	k	8c	8m	9m
	l	9c	9m	10m
	m	10c	10m	11m
	n	11,12c	11m	12m
	o	–	12m	i
1662	p	1c	1m	1m
	q	2c	2m	2m
	r	3c	3m	3m
	s	4c	4m	4m
	t	5c	5m	5m

		term	N	S
1813	g	5c	5m	5m
	h	6c	6m	6m
	i	7c	7m	7m
	j	8c	8m	8m
	k	–	i	9m
	l	9,10c	9m	10m
	m	11c	10m	11m
	n	12c	11m	12m
1814	o	1c	12m	1m
	p	2c	1m	2m
	q	–	2m	i
	r	3c	3m	3m
	s	4c	4m	4m
	t	5c	5m	5m

		term	N&S	
1832	g	5c	5m	5m
	h	6c	6m	6m
	i	7c	7m	7m
	j	8c	8m	8m
	k	9c	9m	9m
	l	–	i	10m
	m	10c	10m	11m
	n	11,12c	11m	12m
	o	1c	12m	1m
1833	p	–	1m	i
	q	2c	2m	2m
	r	3c	3m	3m
	s	4c	4m	4m
	t	5c	5m	5m

		term	N&S	
1851	g	5c	5m	5m
	h	6c	6m	6m
	i	7c	7m	7m
	j	8c	8m	8m
	k	–	i	9m
	l	9c	9m	10m
	m	10c	10m	11m
	n	11c	11m	12m
	o	12,1c	12m	1m
1852	p	2c	1m	2m
	q	–	2m	i
	r	3c	3m	3m
	s	4c	4m	4m
	t	5c	5m	5m

		term	N	S
1965	g	5c	5m	5m
	h	6c	6m	6m
	i	7c	7m	7m
	j	8c	8m	8m
	k	–	i	9m
	l	9,10c	9m	10m
	m	11c	10m	11m
	n	12c	11m	12m
1966	o	1c	12m	1m
	p	2c	1m	2m
	q	3c	2m	3m
	r	–	3m	i
	s	4c	4m	4m
	t	5c	5m	5m

		term	N	S	(S)
2033	g	5c	5m	5m	5m
	h	6c	6m	6m	6m
	i	7c	7m	7m	7m
	j	–	i	8m	8m
	k	8c	8m	9m	9m
	l	9c	9m	10m	10m
	m	10,11c	10m	11m	11m
	n	–	11m	i	12m
	o	12,1c	12m	12m	1m
2034	p	–	1m	1m	i
	q	2c	2m	2m	2m
	r	3c	3m	3m	3m
	s	4c	4m	4m	4m
	t	5c	5m	5m	5m

Figure 11. Calendars resulting from various placements of the intercalary month. N columns give the natural solution; S columns, the solution by the 1684 rule of the Shixian calendar. (S) follows the Shixin principle, but not the 1684 rule.

The Shixian Principle

Astronomically there is no trouble with the natural method. However, note that with the natural method in four years—1642,[41] 1813, 1965,[42] 2033—the winter solstice is in the 10th month. We have seen above how important the winter solstice was, and at times having the solstice in the 11th month was considered part of the Shixian calendar. I shall call this the Shixian principle. In just those cases where the natural rule would cause the solstice to be in the 10th month, a non-central-term month *after the solstice* is made intercalary instead.[43] The S(hixian) and (S) columns in figure 11 show how the months of various years would be numbered under this principle. But as the natural rule and the Shixian principle give the same result in most cases, it is only in the four years mentioned above that we can tell by looking at the calendar whether the Shixian principle was used or not.

The Shixian principle was incorporated into the Shixian calendar procedure of 1684, which contains a provision for determining the intercalary month. It gives a precise condition which insures that the solstice is in the same month under all circumstances:[44] if two winter solstices are thirteen months apart, the first month without a central term is designated intercalary.[45] (Because there are only 12 central terms, if there are thirteen months, at least one of the months does not have a central term.) The final clause goes beyond the

"11th-month solstice" requirement in specifying that it is the first non-central-term month after the solstice that is intercalary. This rule is possible and unambiguous in all cases. The provision only specifies which months are to be made intercalary, but if the winter solstices are twelve months apart, there of course can be no intercalary month, and months are numbered in order, without regard to central terms. If there is a non-central-term month among the twelve months, there is also a double-central-term month to cancel it out.[46] Of course in non-leap years or ordinary leap years like 1676–77 the rule gives the traditional result. This 1684 rule seems to have been forgotten sometime in the 18th century, however.[47]

The Fortunes of the Shixian Principle

The Shixian calendar was first calculated using a 1635 book on western astronomical calculation methods, the *Chongzhen Lishu* 崇禎曆書 (or the *Xiyang Xinfa Lishu* 西洋新法曆書) (Xu and Li (1635)). There was of course opposition to this "western calendar", some of which focused on its use of true solar terms, but except for a short interruption, the Shixian calendar was used officially until 1912, and its method of determining months and month names continues in the modern "Chinese lunar calendar".[48]

The *Chongzhen Lishu* does not discuss the problems truw solar terms cause for the naming of intercalary months.[49] However, a 200-year calendar was made around 1662[50] and there was a double-central-term month in 1669, so the problem was recognized and provision was made for it in the 1684 procedure.

But in the next three centuries consciousness of the problem came and went. The 1722 revision of the *Chongzhen Lishu*, the *Lixiang Kaucheng* 曆象考成, does not bring up the problem, and mentions the old mean-solar-term rule in passing.[51] It is perhaps not surprising that the long-term calendar made in 1727 did not use the Shixian principle—the winter solstice of 1813 is in the 10th month as with the natural method.[52] The 1741 long-term calendar calculated the calendars, or at least those for years after 1736, afresh, but it used the natural method for both 1642–43 and 1813–14; the intercalary months are in months *l* and *k* of figure 11 respectively.[53] The 1742 procedure, the *Lixiang Kaucheng Houbian* 曆象考成後編, apparently did not discuss the problem.[54]

However, by 1799 they had either remembered or recreated the Shixian principle, for the long-term calendar published that year was extended to 1995, and for 1965–66 the non-central-term month after the solstice was made intercalary (month *r*, not month *k*), so the solstice was in the 11th month. But, for the years before 1936 they reused the earlier printing blocks, just changing the year name when necessary, and 1642–43 and 1813–14 were left unchanged.[55] So in 1811 the Astronomical Bureau (*qintianjian* 欽天監) was shocked to find that in 1813 the solstice would be in the tenth month for the first time ever.[56] The Bureau clearly did not understand what was going on, as in an effort to discover the cause of the problem it looked at previous years with intercalary 8th months, not those with double-central-term months, but the calendar could not be left as it was, and the Bureau realized it could solve the problem by changing the intercalary month from the non-central-term month *k* to the next, month *q*. Now the calendars would conform to the Shixian calendar for the next two hundred years.[57] Figure 1 shows how one person marked the change with a brush in his copy of the 1799 long-term calendar—the names of the months from intercalary 8 (閏八月) to the 2nd month (二月) of the next year are changed, and the information about the solar terms is copied from the last month of 1813 to the head column of 1814. This corrected version of 1813–14 was the one actually used, and it appears in the long-term calendars of 1824 and 1851.

The long-term calendar of 1851 cut new printing blocks for 1965–66, recalculating the times of the solar terms and correcting a misprint in the earlier editions (6 long months in a row!); but it left the solstice in the 11th month. Furthermore it used the Shixian principle, not the natural rule, for 2033, making month *n* of figure 11 intercalary.

However, it seems that after that the Shixian principle was forgotten again. When the printing blocks were recut for the 1880 edition (or a previous edition), the solstice was put back in the 10th month of 1813.[58] Not only was this against the Shixian principle, it was unhistorical! In 1900 Xi and Wang carefully explained the calendrical issue involved in the change of 1813–14: the method used in 1727 (the natural method) was in accordance with calendar principals, but at the time of the change the Astronomical Bureau apparently gave the priority to the importance of the 11th month in calendrical theory. Though Xi and Wang praise the solution as matching up the central terms and months, they do not say the 1727 calendar was wrong.[59] Perhaps a low point in the understanding of the problem was the description of the calendar procedures written by an official of the Astronomical Bureau in the last days of the Qing dynasty: it just states "make non-central-term months intercalary".[60] (Of course, between 1870 and 1984 there were no double-central-term months in China, so he had never had to deal with them.) The 200-year calendar of 1959 did not include any years in which one could distinguish the natural and Shixian rules.[61]

However, the solution to the problem of intercalary months in the true-solar-term calendar has come back to its starting point. The long-term calendar of 1984 put out by the Chinese Purple Mountain Observatory has the winter solstice in the 11th month for 2033; Liu Baolin, who was involved in its publication, gives a rule for intercalary months that is essentially the same as that of the first rule of 1684.[62]

The Cardinal-Terms Rule

Let us go back to 1813–14. As mentioned above, in 1811 the Astronomical Bureau was concerned with the winter solstice, but that was not its only concern. Some of the most important court rituals of the Qing period were those on the days of the winter and summer solstices (*Jiao* 郊社), and those honoring Confucius in the months of the spring and autumn equinoxes (*Shangding* 上丁). However, as mentioned above, the 1799 long-term calendar used the natural rule for those years, so both the winter and spring festivals would be in the wrong months (see figure 11).[63]

The Shixian principle requires that the winter solstice always be in the 11th month, but its effect on the other cardinal terms is merely coincidental and is usually null. However, even though it was not part of the Shixian principle, Astronomical Bureau from its concern with the spring equinox and with the festivals, clearly also wanted all four cardinal terms be in the months named for them. It seems, though, that this condition was remembered even less that the Shixian principle. Though Xi and Wang say the Bureau probably was concerned with the winter solstice of 1813, they say nothing in particular about the spring equinox.[64]

In Japan, the Astronomy Office (*tenmon-kata* 天文方) used the Chinese *Lixiang Kaucheng Houbian* and western books on astronomy when creating the Kansei calendar 寛政暦 of 1798, but that calendar used mean solar terms as earlier Japanese calendars had

Japan 1851–52

	Natural term	Natural name	Shixian (1684) term	Shixian (1684) name	Cardinal Terms term	Cardinal Terms name
1851	5c	5m	5c	5m	5c	5m
	6c	6m	6c	6m	6c	6m
	7c	7m	7c	7m	7c	7m
	8c	8m	8c	8m	8c	8m
	9c	9m	9c	9m	9c	9m
	–	i	–	10m	–	10m
	10,11c	10m	10,11c	11m	10,11c	11m
	–	11m	–	i	–	12m
1852	12,1c	12m	12,1c	12m	12,1c	1m
	2c	1m	2c	1m	2c	2m
	–	2m	–	2m	–	i
	3c	3m	3c	3m	3c	3m
	4c	4m	4c	4m	4c	4m
	5c	5m	5c	5m	5c	5m

FIGURE 12. Three possibilities for the intercalary month for Japan in 1851-52. Japan used the cardinal-terms method.

done. However, around that time the authors of the calendar were asked how to deal with double central terms under a true-solar-term calendar, and one (naturally) suggested the natural rule—make the first non-central-term month intercalary.[65] Apparently the problem was not given thought, however, for even though the next calendar, the 1844 Tenpō calendar 天保暦 did use true solar terms, the procedure simply repeated the old rule that non-central-term months were intercalary.[66] However, a supplement to this procedure a few years later did explain the problem and established a rule.[67] The Astronomy Office looked at the annual Chinese calendars (the long-term calendars were apparently not available), and discovered that the natural method had been followed for 1775 and 1832 (for the latter see figure 11), which meant the winter solstice, the most important of the cardinal terms to the ancient sages, occurred in the 11th month. However in 1813, the second non-central term month was chosen, which made the winter solstice and spring equinox be in their own months. Thus, the four cardinal terms occurred in their corresponding months, in accordance with the intent of the ancient emperors who had determined the constellations for the four cardinal terms.[68] Accordingly, the rule given in the supplement is that if there are two central terms in a month, there will be a non-central-term month before and after it, and the intercalary month shall be chosen so that the two solstices and the two equinoxes will be in their corresponding months.[69]

Shortly afterwards, for 1851–52, the Tenpō procedure calculated two double-central-terms months, so there were three candidates for the intercalary month. Looking at figure 12, one can see that if the first non-central-term month was chosen (the natural method), the winter solstice would be in the wrong month, and if the second one was chosen (the first one after the solstice in accordance with the 1684 Shixian rule) the spring equinox would be wrong. In accordance with the cardinal-terms rule they chose the third non-central-term month as the intercalary month, and all four cardinal terms matched their months. (The Chinese Astronomical Bureau undoubtedly would have made the same choice if faced with this calendar in 1811.)

Anticipating criticism of the novelty of non-intercalary non-central-term months, the heir to the head of the Astronomy Office, Shibukawa Suketaka, launched a strong defense of the decision to use true solar terms: Calendrists are to follow the intentions of the sage-emperors, and match the calendar to the motions of the heavens. These ancients

Cardinal-terms Problems

		term	
1851	g	5c	5m
(China)		6c	6m
	i	7c	7m
	j	8c	8m
	k	–	i
	l	9c	9m
	m	10c	10m
	n	11c	11m
	o	12,1c	??
1852	p	2c	2m
	q	–	i
	r	3c	3m
	s	4c	4m
	t	5c	5m

		term	
2033	g	5c	5m
	h	6c	6m
	i	7c	7m
	j	–	i
	k	8c	8m
	l	9c	??
	m	10,11c	11m
	n	–	??
	o	12,1c	??
2034	p	–	??
	q	2c	2m
	r	3c	3m
	s	4c	4m
	t	5c	5m

Figure 13. In these years a good calendar using the cardinal-terms method is impossible.

established intercalary months in order to preserve the four seasons and determined the stars that mark the cardinal terms. If they reappeared now, they surely would approve of the current calendar. Heaven has its own rules, so as knowledge changes, calendars must also, as for example, the change from using the mean conjunction to using the true conjunction to determine months. Since now we understand the movement of the sun, it is in accordance with the intentions of the ancient sages that we match the calendar to the heavens and use real solar terms, even if we have to occasionally change the rule for intercalary months.[70] Shibukawa also pointed out that the winter solstice was the most important, and one should give preference to putting that in the 11th month.[71] However this caution was not in the supplement.

In 1873 the Gregorian calendar was adopted, but a lunisolar calendar, called the Kyūreki 旧暦 ("old calendar") has continued to be published; it follows the structure of the Tenpō calendar but uses modern astronomical methods to determine the new moons and solar terms. Hirayama's 1912 summary of the structure included the rule that the winter solstice be in the 11th month, the spring equinox be in the 2nd month, and similarly for summer and autumn.[72] This is essentially a restatement of the rule in the Tenpō calendar supplement. Almost anyone who discusses the intercalary months of the Tenpō or the Kyūreki calendar quotes Hirayama.[73] Accordingly, in 1965-66 the intercalary month was placed after the 3rd month of 1966, not after the 8th month of 1965, and the solstice was in the 11th month (see figure 11).

The Problem of the Cardinal-Terms Rule

However, it turns out that the cardinal-terms rule, being based on ideals and not astronomy, will not always work, because there are rare cases in which there is only one month between a solstice and equinox, while numbering requires at least two. Figure 13 shows such cases. In China, it was impossible to have both the winter solstice and the spring equinox in their proper months in 1852 as Shibukawa had realized (see note 68).[74] In Japan, the rule for the Kyūreki calendar will cause a problem in 2033, since it will be impossible both to have the autumn equinox in the 8th month and to have the winter solstice in the 11th month. There has been some discussion as to whether to give priority to the equinox or the solstice, i.e., the N or the S column of figure 11. There is no longer an official body deciding the old calendar in Japan, but opinion seems to be leaning towards having the winter solstice in the 11th month.[75] But which ever they choose, it will have to be accepted that here too, an ideal lunisolar calendar is unobtainable.

Notes

1. Japan adopted the Gregorian calendar suddenly in 1873. (In fact, there was so little preparation that the law stated there should be a leap day every four years! This was corrected in 1898, just in time for 1900; see Hirose (1978), pp. 96, 102.) The lunar calendar was long strong in rural areas, but with the increasing urbanization of the whole country, the solar calendar was used in almost all circumstances by the 1970's (Okada (1974), pp. 107–108, 123). Most people now have only the vaguest idea of the old calendar. Most festivals and commemorations, including New Year, were simply moved to the corresponding date in the new calendar, though some were moved to just one solar month later.
2. "The general picture that emerges [from the Shang oracles] is of a calendrical system that was evolving from observation and description toward one that was increasingly mathematical and prescriptive" (Keightley (1999), p. 251).
3. A multitude of other month names have used as well over the centuries, but numbers have always been the standard. However, though "first month" (*yiyue* 一月) is standard in the bone oracles (as in Shirakawa (1963), Plate 1 and p. 5), the term "proper month" (*zhengyue* 正月) also appeared (Keightley (1978), p. 114, n. 98), and it is this latter term that became the standard name of the month that starts the year. Nowadays in both Japan and China "first month" 一月 usually means the month of January of the western calendar. However, in this paper I use the English term "first month" to refer to the lunar first month.
4. The 60-cycle is represented by a pair of characters, the first from a cycle of ten "stem" characters, and the second from a cycle of twelve "branch" characters, which were much later given animal names. The 10- and 12-cycles change concurrently, so one gets (1,1) [1], (2,2) [2]....(10,10) [10], (1,11) [11], (2,12) [12], (3,1) [13], (4,2) [14],.....(9,11) [59], (10,12) [60]. Eventually the 60-cycle was used to indicate years as well, though nowadays often only the 12-cycle part is referred to. The 12-cycle can now also indicate direction, time, months, etc. Cyclic dates were often used instead of or along with the numbered days of the month, so we have dates such as "the 43rd cyclic day, which is in the 5th month" (*wuyue bingwu* 五月丙午). This had the advantage that the date was not dependent on the vagaries of the new moon, though dates in this form were still used centuries after the calendar was fully calculated. Cyclic dates are used by calendrists as a fixed grid for constructing calendars and by historians for reconstructing calendars from historical documents. The cyclic date of a day is also a basic factor in determining a large number of its divinatory properties.
5. Yabuuchi (1990), p. 6.
6. However, the modern Chinese and Japanese lunar calendars, called the "farmer's calendar" (*nongli* 農暦) and the "old calendar" (*kyūreki* 旧暦) respectively, get the solar terms and new moons from modern astronomical predictions.
7. I use the list with dates, year lengths, and average month lengths in Yabuuchi (1990), pp. 388–91. The year given for a procedure is the first year that it was used to calculate. Of course, some of the procedure changes were almost in name only, while some underwent major revisions though the name stayed the same.
8. In practice the month did not always start on the day of the calculated new moon, however. The first day could be moved for various reasons, some of which will be discussed below. The most frequent reason was "new-moon advancement" (*jinshuo* 進朔): in certain periods, when the new moon was "late" in the day, the start of the month was automatically put off till the next day; see Yabuuchi (1990), p. 97; Hiraoka (1966), p. 341; Uchida (1992), p. 497–498. During the period of the Xuanming (Jp. Senmyō) calendar 宣明暦, "late" was after 6:00 p.m.
9. Shima (1958), p. 505; Yabuuchi (1990), p. 278–279; Shinjō (1928), chart 春秋長暦図表 after p. 315.
10. Yabuuchi (1990), p. 276–277. They were probably originally determined by stars. The meaning of "spread out evenly" will be discussed in the section "From Mean Solar Terms to True Solar Terms."

11. The moments of the "true" central terms are the same as the moments the sun enters a new sign of the Western zodiac.
12. Note that the deciding factor is the *calendar* month that the central term is in, and the calendar month, like the day, starts at midnight. Thus, if the central term and the new moon occur on the same day, the central term is in the month of that new moon, and it does not matter which occurs first. (In Japan from 1771 to 1799 the conjunction was used as the standard instead of the calendar month, resulting in 3 cases where a calendar month with a central term was intercalary. However, this was criticized and finally abandoned; Shibukawa (1850), 巻4 (pp. 555–556 in reproduction); cf. Uchida (1973).
13. The traditional lunar calendar of India also tethers each lunar month to the solar year, but in a different way. There are both solar and lunar months. The solar month is determined by the sign of the (Indian) zodiac the sun is in. The moments the sun enters a new sign (*sakrāntis*), which mark the astronomical starts of the solar months, are spread out evenly and correspond to the Chinese central terms, though they are not the same place in the ecliptic. The lunar months are named by the solar month that they start in. If two lunar months start in the same solar month, the first one, the one without a *sakrānti*, is the intercalary (*adhika*) month, which means that an intercalary month has the same name as the following month. (Saha and Lahiri (1992), pp. 242, 247) On the other hand, a Chinese intercalary month has the same number as the preceding month. For more on the Indian calendar see fn. 39.
14. This is a generalization of the step-by-step description of the AD 822 Xuanming (Jp. Senmyō) calendar 宣明暦procedure in Uchida (1992), pp. 511–516. For detailed translations and discussions of some other calendar procedures, see Cullen (2002) for the AD 223 Qianxiang calendar 乾象暦, which contained the first thorough Chinese mathematical treatment of the motion of the moon, and Sivin (2009) for the Mongol-period 1281 Shoushi calendar 授時暦, which was the apex of Chinese calendrical studies.
15. Early calendars gave a theoretical starting year in the distant past in which the winter solstice, the new moon, and sometimes other factors as the moon's perigee and the "starting point" of planets, all occurred at midnight on cyclic day 1. The denominators of fractions of days also varied wildly from one procedure to the next. However, from AD 1281 on, the starting year was close to the year the calendar was made, and the cyclic dates and times of the winter solstice, etc. for that year were specified. Furthermore, decimal fractions were used.
16. Technically speaking, they varied by their starting year (see previous footnote).
17. See the tables in Yabuuchi (1990), pp. 388–393.
18. Yabuuchi (1990), pp. 289–290.
19. This figure comes directly from the year length and the Metonic ratio of year-length:month-length :: 235:19.
20. Yabuuchi (1990), pp. 37, 78.
21. Yabuuchi (1990), pp. 83–84, 87–88, 96–97, 144, 282; Momo (1966), p. 3; Hiraoka (1966).
22. The calendars in Japan were calculated by the imperial Calendar Department (*rekidō* 暦道), but during the middle ages, some major temple and shrines also produced regional calendars using the same Senmyō procedure, though with some different traditions and calculation results. However, in 1685 the shogunate adopted a version of the Chinese 1281 Shoushi calendar modified for 17th-century Japan, and used the opportunity to take control of the content of all calendars, including the regional calendars. For a discussion of the relationship between Japanese astronomy and calendar science and Chinese and western science see Nakayama (1969).
23. Some earlier procedures in Japan had calculated four long months, but it can not determined what was done in those cases (Uchida (1992), p. 496). The problem of four long months in the Senmyō calendar was thoroughly studied by Momo (1966); see also the tables for the individual years in Uchida (1992). Uchida gives the dates and times of the new moons and the solar terms as calculated by the relevant procedures in Japan for the years 445 to 1872 and indicates any deviations from the calculations found in the historical records.
24. For example, in 860 it had been noted that the calendar method allowed up to three long or three short months in a row; *Nihon Sandai Jitsuroku* 日本三代実録, 延喜2. intercalary 10.23 (vol. 1, p. 58). (Volume and page numbers for Japanese chronicles are taken from the Kurosaka (1972) series.)

25. This is clear from the cyclic dates recorded in diaries or chronicles such as the *Nihongi Ryaku* 日本紀略.
26. For Mesopotamian intercalation, see Britton (2007). For China, see Shinjō (1928), pp. 230–326, especially the chart 春秋長暦図表 after p. 315. According to Shinjō's reconstruction of the calendar of the *Spring and Autumn Annals* (722–479 BC), there were 7 leap years in most 19-year cycles. During the period, placement of the leap year became more and more regular, though even at the end of the period the placement of the intercalary month within the year was still rather irregular. It should be noted, however, that there have been many reconstructions of the *Annals*' calendar over the millennia.
27. Except in the Yuanjia calendar, which was constructed from the 1st-month term, not the solstice.
28. In the northern dynasties this started in AD 412; and in the southern dynasties, in 510 (Yabuuchi (1990), pp. 82, 84, 388–392).
29. The commentary was the *Shangshu Baishi* 尚書百釋, a commentary on the classic "Book of Documents" *Shangshu* 尚書 written during the southern Liang dynasty (AD 502–557), which was the first southern dynasty not to use the cycle. It is quoted in the *Nihon Sandai Jitsuroku* 日本三代実録, 延喜2.intercalary10.23 (vol. 1, p. 58).
30. The discussion of the Japanese new-moon solstice is based on the work of Momo Hiroyuki (1966), pp. 6–14, and (1974). Some of the data for the individual years in my discussion are taken from the yearly tables in Uchida (1992).
31. *Shoku Nihongi* 続日本紀, 延暦3.11.1 (vol.2, p. 502).
32. *Nihon Sandai Jitsuroku* 日本三代実録, 延喜2.intercalary10.23, 25; 11.1 (vol. 1, pp. 57–58). The month lengths for figure 5 are taken from this passage. I followed the Japanese rendition in Takeda and Imaizumi (1935), p. 104.
33. *Nihongi Ryaku* 日本紀略, 承平6.11.1–2 (vol. 3, p. 36).
34. That is, one hour a cycle by the Senmyō calendar, whose solar year was about 3 minutes too long. The modern figure is about 2 hours per cycle.
35. Similar adjustments were normally made to avoid the 8th month as the first intercalary month of the cycle or to keep the new-moon solstice from occurring on the 11th year of the cycle so they would not have to have the ceremony at the wrong time; see Momo (1974), pp. 94–100. (The cycle desperately wanted to have a "hiccup" of 11 years and reset.)
36. See Momo (1966), pp. 6–13.
37. In 1870 the new moon fell on the day of the solstice and there was discussion about whether the festival should be celebrated, but it was decided that the old ritual was not suited to the modern enlightened Meiji government (Momo (1974), p. 97).
38. The long-term calendars (*wannianshu* 萬年書; lit. "10,000-year book", see Zhang (1998), p. 486) Qintianjian (1787–1851) covered portions of the years from 1624 to 2050; the 1851 edition covered all. There were several later editions, but the only one I could find was that of 1880, apparently an extract of the one of 1876. For a discussion of the long-term calendars, including the years they were published, see Liu and Stephenson (1998a), pp. 34–35. The above calendars are all identical in the years they cover with respect to month names, except for 1813–14; in fact, my examination of them showed that the printing blocks from previous editions were normally reused. The years in the calendars that have double-central-term months are these: 1642–43, 1661, 1680, 1699, 1775, 1813, 1832, 1851, 1870, 1965, 1984, and 2033–34. Note that these are all in the same place in the 19-year cycle, except for the last, which has the 11-year shift.
39. Insisting on having all twelve non-intercalary months is not the only way of handing the matter. The same situation sometimes occurs in the Indian calendar (see n. 13), which has been using the equivalent of true central terms since about AD 1100. The calendar just follows its ordinary rules for naming: since a lunar month is named by the solar month it starts in, if no lunar month starts in a particular solar month, the lunar month associated with that solar month is missing (*kṣaya* "decayed") that year. For example, even if no lunar month starts during the 10th solar month, the lunar month that starts during the 11th solar month is still the 11th lunar month: there is no 10th lunar month; Saha and Lahiri (1992), pp. 248, 250, see also Yano (1989). This is probably the most logical solution. However, I cannot image that the Chinese even considered it.

40. I was made aware of the problem of the relationship between the solutions by Suchowan (2005). In particular he discusses 1813–14 in the long-term calendars Qintianjian (1787) and (1824) and in the 1811 report of the Astronomy Bureau and its connection with the Japanese calendar.
41. As true solar terms were not used until 1645, the calendar actually used in the years 1642–43 had no double central terms and so is not mentioned in discussions of the problem. However, the long-term calendars projected the Shixian calendar back to 1624, as can be seen by the solar-term dates, so the Shixian calendar for those years is still valid for investigating how double central terms were treated in the long-term calendars.
42. In modern Chinese calendars as Zhongguo Kexueyuan (1959) and (1984), the year 1965 does not have a double central term, though it does in modern calendars of Japan, which is one time zone to the east.
43. At present this is necessary only if a double-central-term month involves the 9th or 10th central terms. As the perihelion is now between the 11th and 12th central terms, it does not happen often, and this is the reason why the natural rule and the Shixian rule match in most cases.
44. The 1684 procedure is found in Zhao (1928), 卷47–49 =志22–24 = 時憲志3–5 (in vol. 7 of the 1928 edition). The intercalary rule is on 卷 48 =志23 = 時憲志4, fol. 7. It reads: "求閏月以前後兩年有冬至之月爲準中積十三月者以無中氣之月從前月置閏一歲中兩無者中氣者置在前無中氣之月爲閏." A supplementary procedure in 1724 states that intercalary months should be determined by the provision of 1684; Zhao (1928), 卷50 =志25 =時憲志6, fol.13.
45. In figure 11, if the winter solstice (11c) is in month *m*, there are thirteen months to the following solstice (month *z*) and the intercalary month is after the solstice; if it is in month *n*, it has been thirteen months since the preceding solstice (month *a*) and the intercalary month is before the solstice.
46. In figure 11, if the winter solstice (11c) is in month *n*, there are twelve months to the following solstice (month *z*); if it is in month *m*, it has been twelve months since the preceding solstice (month *a*). These (solar) years have no intercalary month.
47. The next even indirect reference I have found is in Yabuuchi (1969), p. 283, who quotes Zhao (1928) explicitly as the solution that was found to the problem of intercalary months with true solar terms. Awareness of it cannot have been very wide. The Astronomical Bureau in 1811, Shibukawa (1850), Xi and Wang (1900), Hirayama (1912), and Chen (1984) do not even hint at it when they discuss the problematic 1813-14 calendar. However, because the 1813–14 calendar was changed, all years have in fact conformed to the 1684 rule.
48. For the early history of the Shixian calendar and opposition to it see Zhang (1998), Huang (1998), and Zhao (1928), 卷45 =志20 = 時憲志1. One strident opponent of the calendar was Mei Wending 梅文鼎, who criticized it especially in his 1702 book *Lixue Yiwen* 曆學疑問 [Questioning Calendrical Studies].
49. Xu and Li (1635). It is possible I overlooked such a discussion, but Li gives no hint of this problem when he discusses intercalary months in his handbook to the *Chongzhen Lishu*, saying only that there could be a one-month difference in the placement of intercalary months under the mean- and true-solar-term calendars; Li (1847), 卷1, fols. 13-14, section 年月.
50. Zhang (1998), p. 486.
51. Qing (1898), 厤象考成上編卷4, fol. 90.
52. See Xi and Wang (1900), 卷3, fols. 25–26, which states which solar terms were in which months of 1813–14 in the 1727 calendar; they are the same as Qintianjian (1787). The calendar ignores the 1724 requirement that intercalary months should be set by the 1684 method (see note 44). I do not know whether the 1727 calendar included 1642–43.
53. Suchowan (2005) says the makers of Qintianjian (1787; note that the part that covered years before 1835 was composed in 1741) apparently were concerned that the winter-solstice festival should be held in the 11th month. However, the natural rule will cover all cases and explain the 10th-month solstices in 1642 and 1813. Suchowan's statement is probably based on his mistaken belief that the 10th month of 1794 had no central term.
54. The procedure is in Qing (1898). I could not find any discussion about the matter. The Japanese calendrists made use of the *Lixiang Kaucheng Houbian* when making the Kansei calendar 寛政暦 of 1798. For example, Asami and Yasuda (2000), pp. 570–574, are almost identical to Qing (1898), 厤象考成後編卷

4, fols. 14–35. The Japanese (mean-solar-term) intercalary month rule is given in the section just after this passage, but there is nothing corresponding to it in the *Lixiang Kaucheng Houbian*.

55. The pages for 1642–43 and 1813–14 are identical in the two editions, even to the shape of the characters, except for the year names of the latter.
56. The report is found in Liu J. (1959), 卷294:象緯1, 嘉慶16; vol. 3, pp. 10397–98. I would like to thank Sekiguchi Kayo for translating it as well as helping me with some other Chinese passages. This report will be discussed more below in the section on the cardinal-terms rule.
57. As the report is also concerned about the spring equinox, it is not clear whether the Bureau considered the Shixian calendar be concerned with just the winter solstice or with all four cardinal terms, or at least winter and spring (so Suchowan (2005)). I am inclined to think the former. The report says the calendar should be changed because the solstice is in the wrong month, and after that brings up the equinox. In Qintianjian (1799) except for 1813, the solstice is in the 11th month for all years after the start of the Shixian calendar, but though the spring equinoxes of both 1700 and 1852 are in the 1st month, there is no mention of that problem. Of course it could have been carelessness.
58. As Qintianjian (1880) is probably an extract of the 1876 long-term calendar, it is likely the change was made there, though it is possible that the change goes back to the 1862 calendar.
59. Xi and Wang (1900), vol. 3, fols. 25–26. The year 1900 had an intercalary 8th month, and this book was written to combat the idea that the Qing dynasty considered that month to be unlucky, and in particular the popular idea that the 1813–14 calendar had been changed to avoid it.
60. I would like to thank Prof. Liu Baolin for pointing this out to me. The description, the *Kezun Xiandu* 恪遵憲度 of Chen Xiling 陳希齡, is quoted in Chen Zungui (1984), vol. 3, pp. 1622, 1625 fn. 7; chapter 时宪书编造法 (in the 2006 edition, vol. 2, pp. 1161, 1163 fn. 6; note that Chen's 6-volume work of the same name published in Taipei is a different book). Chen Zungui (1984), vol. 3, 1625 fn. 7, says that the old rule given by Chen Xiling was inadequate and lists and explains the cases where non-central-term months were not made intercalary. However, he does not say clearly why, or even that, he has to treat 1813 differently from the others. In 1813 the month with the 10th and 11th central terms was "of course" the 11th month, but similarly in 1661 the month with the 1st central term was "of course" the first month. It seems that even at this late date understanding of the issue was not wide-spread.
61. Zhongguo Kexueyuan (1959). Also see n. 42.
62. Zhongguo Kexueyuan (1984), p. 194; Liu and Stephenson (1998a), p. 36; Liu and Stephenson (1998b). I am grateful to Prof. Liu for sending me a copy of his conference paper.
63. Liu, J. (1959), section 294:象緯1, 嘉慶16; vol. 3, pp. 10397–98.
64. Xi and Wang (1900), 卷3, fols. 25–26. Hirayama (1912), p. 510, states that a rule to deal with the problem of true solar terms, that is, always to keep each of the solstices and equinoxes in their traditional months, was first adopted in China during the Jiaqing period (1796–1820), a clear reference to 1813. He does not give a source, but it seems to be a standard Chinese work.
65. Shibukawa (1850), 卷4 (p. 558 in Asami and Yasuda (2005)).
66. The *Shinpō Rekisho* 新法暦書 [Record of the New Calendar Method], 卷2, section 推月離法, in Asami and Yasuda (2000), p. 726. Two more manuscript versions in the National Astronomical Observatory Library, the *Shinpō Rekisho* 新法暦書and the *Tenpō Rekisho* 天保暦書, are the same here. The exact wording was used in the previous Kansei calendar of 1798 (Asami and Yasuda (2000), p. 575), and goes back to the Jōkyō calendar 貞享暦 of 1685 (Asami and Yasuda (2000), p. 91).
67. The *Shinpō Rekihō Zokuhen* 新法暦法続編 of 1846, in Asami and Yasuda (2001), pp. 389–938. The discussion on intercalary months is in 卷4 二十四気 (pp. 453–454).
68. Shibukawa (1850), 卷4 (p. 558-60).
69. *Shinpō Rekihō Zokuhen*, in Asami and Yasuda (2001), p. 454.
70. Shibukawa (1850), 卷4 (p.556-61). He is also defending against the criticisms of Mei Wending, whose book was published in Japan in 1803 (see note 48).
71. He calculated the days of the new moons and the central terms of the Chinese calendar for 1851–1852—because of the time difference between Beijing and Kyoto they were not the same as the Japanese (see figures 11 and 12)—and predicted that the Chinese would put the winter solstice in the 11th month rather

than put the spring equinox in the 2nd month if they could not do both (Shibukawa (1850), 巻4 (pp. 560–61)).

72. Hirayama (1912), p. 511. At that time both the standard Gregorian calendar and Kyūreki lunar calendar were published by the Tokyo Astronomical Observatory (now the National Astronomical Observatory), so he undoubtedly got the information directly from the calendar makers there. A few years later he became head of the observatory.
73. For example, see Hirose (1978), p. 27; Uchida (1992), p. 543; Nishizawa (1994), p. 403.
74. China had not been able to have the spring equinox in the 2nd month in 1700 either.
75. Nishizawa (1994) presents the winter-solstice solution in the main section and the autumn-equinox solution in an appendix, preferring the former because the winter solstice is the point from which the solar year was traditionally calculated; he also considered the Chinese 1684 rule (p. 403). Others take a similar approach; one calendar had the equinox in the 8th month in its 1991 edition, but ten years later changed it to having the solstice in the 11th month. But, what will they do in some future millennium when the perihelion has changed and the choice is between the summer solstice and an equinox?

References

Asami Megumi 浅見恵 and Yasuda Ken 安田健 (eds.), 2000, *Nihon kagaku gijutsu kotenseki shiryō, tenmongaku hen* (1)日本科学技術古典籍資料, 天文学篇 1 [The Collected Historical Materials on Japanese Science and Technology/ The History of Japanese Astronomy (1)], Kinsei rekishi shiryō shūsei dai 3-ki 近世歴史資料集成第III期 [The Collected Historical Materials in Yedo Era (Third Series)], vol. 8 (Tokyo: Kagaku Shoin 科学書院). [This work contains reproductions of the official Japanese calendar procedure manuscripts.]

——, 2001, *Nihon kagaku gijutsu kotenseki shiryō, tenmongaku hen* (3)日本科学技術古典籍資料, 天文学篇 3, Kinsei rekishi shiryō shūsei dai 3-ki 近世歴史資料集成第III期, vol. 10 (Tokyo: Kagaku Shoin 科学書院).

——, 2005, *Nihon kagaku gijutsu kotenseki shiryō, tenmongaku hen* (5)日本科学技術古典籍資料, 天文学篇 5, Kinsei rekishi shiryō shūsei dai 4-ki 近世歴史資料集成第IV期, vol. 9 (Tokyo: Kagaku Shoin 科学書院).

Britton, J. P., 2007, "Calendars, Intercalations and Year-Lengths in Mesopotamian Astronomy," in J. M. Steele (ed.), *Calendar and Years: Astronomy and Time in the Ancient Near East* (Oxford: Oxbow Books), 115–132.

Chen Zungui 陈遵妫, 1984, *Zhongguo tianwenxue-shi* 中国天文学史 [History of Chinese Astronomy], 3 vols. (Shanghai : Shanghai Renmin 上海人民出版社).

Cullen, Christopher, 2002, "The First Complete Chinese Theory of the Moon: The Innovations of Liu Hong c. A. D. 200", *Journal for the History of Astronomy* 33, 21–39.

Hiraoka Takeo 平岡武夫, 1966, "Chōkei gannen no koyomi" 長慶元年の暦 [The Calendar of the First Year of the Changqing Era], *Tōhō Gakuhō* 東方学報 37, 341–344.

Hirayama Seiji 平山清次, 1912, "Taiin reki " 太陰暦 [The Lunar Calendar], *Nihon Hyakka Daijiten* 日本百科大辞典 [Japanese Encyclopedic Great Dictionary] (Tokyo: Sanseidō), vol. 6, 508–512. Also reprinted in Hirayama Seiji, *Rekihō oyobi jihō* 暦法及時法 [Calendars and Time] (Tokyo: Kōseisha 恒星社, 1938).

Hirose Hideo 広瀬秀雄, 1978, *Koyomi* 暦 [The Calendar], Nihonshi Shōhyakka 日本史小百科 [Little Encyclopedia of Japanese History Series] (Tokyo: Tōkyōdō).

Huang Yi-Long 黄一農 [1998], "East–West Cultural Confrontation and Compromise in Early Ch'ing China. A Case Study on Johann Adam Schall's Civil Calendars" (in Chinese), in R. Malek (ed.), *Western Learning and Christianity in China : The Contribution and Impact of Johann Adam Schall von Bell (1592–1666)*, Monumenta Serica monograph series 35 (China-Zentrum : Monumenta Serica Institute and Nettetal : Steyler Verlag), vol. 1, 430–473.

Keightley, D., 1978, *Sources of Shang History: The Oracle-Bone Inscriptions of Bronze Age China* (Berkeley: University of California Press).

——, 1999, "The Shang", in M. Loewe and E. L. Shaughnessy (eds.), *The Cambridge History of Ancient China: From the Origins of Civilization to 221 B.C.* (Cambridge: Cambridge University Press), 232–291.

Kurosaka Katsumi 黒坂勝美 (ed.), 1972, *Kokushi taikei* 国史大系 [Compendium of Japanese History] (Tokyo, Yoshikawa Kō Bunkan).

Li Tianjing 李天經 et al. (eds.), 1847, *Sūtei Rekisho Rekiin* 崇禎曆書曆引 [Guide to the "Chongzhen Calendrical Study"], Japanese edition ([Edo]: Watanabe ... 渡邊...藏版). (1855 reprint)

Liu Baolin 刘宝琳 and Stephenson, F. R., 1998a, "A Brief Contemporary History of the Chinese Calendar", *Orion* 56, no. 287, 33 –38.

——, and Stephenson, F. R., 1998b, "The Chinese Calendar and the Determination of Its Intercalary Months" (paper presented at the International Conference on Lunar Calendar Practices: A Common Heritage in Islamic, Chinese, Hindu and Other Civilizations, Universiti Sains Malaysi, Penang, Malaysia, November 24–25, 1998).

Liu Jinzao 劉錦藻 (ed.), 1959, *Huangchao xu-wenxian tongkao* 皇朝續文獻通考 [Continuation of the Collection of Documents of this Imperial Dynasty] (Taipei: Xinxing 新興書局). (Orig. pub. 1912; also known as 清朝續文獻通考.)

Momo Hiroyuki 桃裕行, 1966, "Shidai o sakeru koto" 四大を避けること [Avoiding Four Long Months], *Rekishi Chiri* 歴史地理 91:3, 1–18. Also repr. in vol. 7 of *Momo Hiroyuki chosakushū* 桃裕行著作集 (Tokyo: Shibunkaku, 1990), 252–290.

——, 1974, "Uruu-zuki to Sakutan Tōji (19 nen 7 jun e no nesshin)" 閏月と朔旦冬至（十九年七閏への熱心） [Intercalary Months and the New-moon Winter-solstice....], in Hirose Hideo 広瀬秀雄 (ed.), *Koyomi* 暦 [Calendars] (Tokyo: Daiyamondosha), 83–105. Also repr. in vol. 7 of *Momo Hiroyuki chosakushū* 桃裕行著作集 (Tokyo: Shibunkaku,1990).

Nakayama, Shigeru 中山茂, 1969, *A History of Japanese Astronomy: Chinese Background and Western Impact*, Harvard-Yenching Institute Monograph Series 18 (Cambridge, MA: Harvard University Press).

Nishizawa Yūsō 西澤宥綜 (ed.), 1994, *Rekijitsu taikan: Meiji kaireki 1873 nen~2100 nen* 暦日大鑑:明治改暦1873年~2100年 [Great Compendium of the Calendar, from the Meiji Calendar Change of 1873 until 2100] (Tokyo: Shinjinbutsu Ōraisha).

Okada Yoshirō 岡田芳郎, 1974, "Meiji no kaireki" 明治の改暦 [The Meiji Calendar Change], in Hirose Hideo 広瀬秀雄 (ed.), *Koyomi* 暦 [Calendars] (Tokyo: Daiyamondosha), 107–123.

Qing Yunlu 清允祿 et al. (eds.), 1898, *Yuzhi lixiang kaucheng shangbian 16 juan xiabian10 juan houbian 10 juan* 御製厤象考成上編十六卷下編十卷後編十卷 [Studies of Calendar Phenomena, Parts 1 and 2 and Supplement (Total 36 volumes)] (Fuqiang Zhaijing 富強齋敬印).

Qintianjian 欽天監 [Astronomical Bureau] [1787], *Yuding wannianshu fu yuzuanlidai sanyuanjiazi biannian—Qianlong-Daoguang kan* 御定萬年書附御纂歴代三元甲子編年—乾隆道光刊 [Official long-term calendar...Qianlong and Daoguang editions], vols. 1 and 2. [The Qianlong-era edition, covering 1624–1935.]

—— [1799], *Yuding wannianshu bufenjuan yuzuanlidai sanyuanjiazi biannian—Jiaqing zhong kanben* 御定萬年書不分卷 御纂歴代三元甲子編年 嘉慶中 刊本 [Official long-term calendar...Jiaqing edition]. [The Jiaqing-era edition, covering 1624–1995.]

—— [1824], *Yuding wannianshu fu yuzuanlidai sanyuanjiazi biannian—Qianlong-Daoguang kan* 御定萬年書附御纂歴代三元甲子編—乾隆道光刊 [Official long-term calendar... Qianlong and Daoguang editions], vols. 3 and 4. [The Daoguang-era edition, covering 1624–2020].

—— [1851], *Yuding wannianshu bufenjuan yuzuanlidai sanyuanjiazi biannian—Xianfeng zhong kanben* 御定萬年書不分卷 御纂歴代三元甲子編年 咸豐中 刊本 [Official long-term calendar... Xianfeng edition]. [The Xianfeng-era edition, covering 1624–2050.]

——, 1880, *Yuding wannianshu fu yuzuanlidai sanyuanjiazi biannian—Guangxu kan* 御定萬年書附御纂歴代三元甲子編年—光緒刊 [Official long-term calendar... Guangxu edition] (Saoyeshan Fangzangban 掃葉山房藏板). [A Guangxu-era edition, covering 1744–1933.]

Saha, M. N., and Lahiri, N. C., 1992, *History of the Calendar in Different Countries Through the Ages* (New Delhi: Council of Scientific & Industrial Research). [Originally published 1955.]

Shibukawa Suketaka 渋川佑賢 [1850?], *Seigaku suchi* 星学須知 [Essentials of Astronomy]. [One manuscript of it is reproduced in Asami and Yasuda (2005).]

Shima Kunio 島邦男, 1958, *Inkyo bokuji kenkyū* 殷墟卜辞研究 [Studies of the Oracles from the Yin Ruins] (Hirosaki, Japan: Chūgoku-gaku Kenkyūkai 中国学研究会). [1975 reprint.]

Shinjō Shinzō 新城新蔵,1928, *Tōyō tenmongaku-shi kenkyū* 東洋天文学史研究 [Studies in the History of East Asian Astronomy] (Tokyo: Kōbundō 弘文堂書房).

Shirakawa Shizuka 白川静, 1963, *In kōkotsu bunshū* 殷・甲骨文集 [Collection of the Yin Oracle-bone Inscriptions] (Tokyo: Nigensha 二玄社).

Sivin, N., 2009, *Granting the Seasons : The Chinese Astronomical Reform of 1280, with a Study of its Many Dimensions and a Translation of its Records* (with the research collaboration of Kiyosi Yabuuti [藪内清]and Shigeru Nakayama [中山茂]) (New York: Springer).

Suchowan [(?) Suga Takashi 須賀隆], 2005, "Tenpō-reki chijun-hō no nazo" 天保暦置閏法の謎 [The puzzle of the placement of intercalary months in the Tenpō calendar], http://hosi.org/u/ wiki.cgi? Calendar/When/Exe/暦説明/本編/日本/補足/天保暦置閏法の謎.

Takeda Yūkichi 武田祐吉 and Imaizumi Tadayoshi 今泉忠義 (eds.), 1935, *Kokubun Rikkoku-shi* 国文六国史 [The "Six Histories" in Japanese], vol. 8, *Sandai Jitsuroku Jō* 三代実録上 (Tokyo: Ōokayama 大岡山書店).

Uchida Masao 内田正男, 1973, "Hōreki rekihō chijun no gimon" 宝暦暦法置閏の疑問 [Questioning the Placement of Intercalary Months in the Hōreki Calendar], *Tenmon Geppō* 天文月報 66:3, 67–68.

—— (ed.), 1992, *Nihon rekijitsu genten* 日本暦日原典 [Sourcebook for the Japanese Calendar], 4th ed. (Tokyo: Yuzankaku).

Xi Zhi 襲穉 and Wang Xiqi 王錫祺, 1900, *Runbayue kao* 閏八月攷 [Concerning the Intercalary Eighth Month] (Xiaofanhuzhai 小方壷齋印).

Xu Guangqi 徐光啓 and Li Tianjing 李天經, [1635], *Chongzhen lishu* 崇禎暦書 [The Chongzhen Calendrical Study]. [Edo-period manuscript copy in the Japan National Diet Library.]

Yabuuchi Kiyoshi 藪内清, 1969, *Chūgoku no tenmon rekihō* 中国の天文暦法 [Chinese Astronomical Calendrical Procedures] (Tokyo: Heibonsha).

——, 1990, *Chūgoku no tenmon rekihō, kaitei zōho* 中国の天文暦法　改訂増補 [Chinese Astronomical Calendrical Procedures, revised and expanded] (Tokyo: Heibonsha).

Yano Michio 矢野道雄, 1989, "Kṣayamāsa (ketsugetsu) ni tsuite" Kṣayamāsa (欠月)について [On the kṣayamāsa (missing month)], *Indo Shisōshi Kenkyū* インド思想史研究 6 (1989), 119–126.

Zhang Dawei 张大卫 [1998], "The 'Calendar Case' in the Early Qing Dynasty Re-Examined," in R. Malek (ed.), *Western Learning and Christianity in China: The Contribution and Impact of Johann Adam Schall von Bell (1592-1666)*, Monumenta serica monograph series 35 (China-Zentrum : Monumenta Serica Institute and Nettetal : Steyler Verlag), vol. 1, 475–495.

Zhao Erxun 趙爾巽 et al. (eds.) [1928], *Qingshigao* 清史稿 [Draft History of the Qing Dynasty] (Beijing: Qingshi-guan 清史館).

Zhongguo Kexueyuan, Zijinshan Tianwentai 中国科学院紫金山天文台 [Chinese Academy of Science, Purple Mountain Observatory] (ed.), 1959, *Erbai nian libiao: 1821-2020 nian* 二百年历表 : 1821–2020 年 [200-Year Calendar: 1821–2020] (Beijing: Kexue 科学出版社).

——, 1984, *Xinbian wannianli: 1840-2050 nian* 新编万年历: 1840~2050年 [New Long-Term Calendar: 1840–2050] (Beijing: Kexuepuji 科学普及出版社).

Living with a Lunar Calendar in Mesopotamia and China

John M. Steele

True lunar calendars which base the beginning of the month on the observation of a particular phase of the moon (generally the first sighting of the new moon crescent) are subject to an inherent uncertainty in the length of each month caused by unpredictable variations in atmospheric conditions, cloud cover, and the eyesight of the observer. These factors prevent members of a community who use a lunar calendar knowing in advance the precise number of days between an event in one month and a planned event in the next month. In addition, different observers may first see the new moon crescent on different nights (sometimes because of false sighting of an expected new moon, a phenomenon well attested among historical and modern observers from many cultures),[1] leading to the possibility of two individuals using calendars which are out of sync by one or more days. For a true lunar calendar to operate effectively within a community, therefore, it is necessary for there to be an agreed 'observation' of the new moon crescent, either arising from a consensus of the members of that community or, more commonly, an observation made by an individual or small group charged with the authority to proclaim the beginning of a new month.[2] But although this process leads to an 'official' calendar, used by all members of a particular community, it still does not solve the problem of lack of foreknowledge of the length of any particular month. Furthermore, whilst it is possible to agree on when an observation of the moon has defined the beginning of a new month within a small community—for example a town or, conceivably, a city—communicating that information to a larger community—a country, for example, or even an empire—is nearly impossible and so it is difficulty to ensure that everyone in that large community has a calendar which is exactly in sync with everyone else's.

In addition to the uncertainty over when lunar months are deemed to have begun, a second problem arises where communities choose to use not just a simple lunar calendar, but a luni-solar calendar in which the months are kept in line with the seasons. The average length of the lunar month is close to 29½ days, making twelve lunar months equal about 354 days. The solar year, however, is a little less than 365¼ days in length, about 11¼ days longer than twelve lunar months. Many cultures have solved this problem by adding an extra month into certain years. Choosing when to add this extra month, however, is complicated because three 11¼ days total 33¾ days, which is a little more than three days greater than the length of one lunar month. This means that the extra month needs to be added on average slightly more often than once every three years: occasionally it will need to be added after only two years. The decision of when to insert these extra 'intercalary' months adds another level of uncertainty to a lunar calendar.

The two problems just outlined—uncertainty over when months begin and when to intercalate—are not only astronomical problems but impact on many aspects of daily life in societies which use a lunar calendar. For example, how does one calculate interest on a loan taken out on the fifteenth of one month to be repaid on the fifth of next month if one does not know whether there will be nineteen or twenty days between these two dates? If taxes are to be collected on the first day of the new year, how do you budget if you do not know in advance whether the twelfth month of the current year is the last month of this year or whether there will be an additional intercalary month before the new year begins? If a ritual is to be performed on a particular day, how do you ensure that every subject of an empire performs the ritual on the same day if there is uncertainty in when the month begins? And, more practically, how can one prepare cultic foods and other perishable items for a new month ritual or festival if one does not know whether the month will begin tomorrow or the day after? One can think of many more examples where calendrical uncertainly impacts on religious, civic and personal life in societies which rely on an observational lunar calendar. Unsurprisingly, therefore, many societies have sought a way to reduce uncertainty in their calendar through the use of astronomy to calculate in advance whether, for example, a year will contain an intercalary month or the new moon crescent will be seen on the thirtieth or the thirty-first day of a month. Astronomical calculation can be used to approximate observation, providing a 'best guess' of what the calendar will be that can subsequently be adjusted once an observation has been made, it may serve as a guide to observation, reducing the risk of human error, or it may partially or fully replace observation.

In this paper I discuss how the lunar calendar impacted on life in Mesopotamia during the first millennium BC and China from the third century BC to the end of the Imperial period. These two case studies will provide evidence for a wider discussion of the role of calculation in the operation of lunar calendars and the impact this has on the role of the calendar in daily life.

Mesopotamia

The calendar used by the people of Mesopotamia from at least the late fourth millennium BC to the beginning of the first millennium AD was a luni-solar calendar in which the beginning of the month was dependent upon the visibility of the new moon crescent and intercalary months were added in certain years.[3] During the first millennium, and quite probably earlier as well, the beginning of the month was determined as follows: At the beginning of the thirtieth day of a month a watch was kept for the new moon crescent (the Babylonian day began at sunset). If the lunar crescent was seen (or, by the last few centuries BC, if it was calculated to be visible) then the thirtieth day which had just begun was 'turned back' (*turru*) and this day became the first day of the new month. If the lunar crescent was not seen, the day which had just begun was 'completed' (*kunnu*) as the thirtieth day of the current month and the new month would begin the following evening.[4] It was not necessary for the new moon crescent to be seen on the thirty-first day for the new month to begin—it if had not been seen on the thirtieth day then the new month started by default on the next day. This ensured that months always had either twenty-nine or thirty days, never twenty-eight or thirty-one or more (which will occur in a true lunar calendar because of bad weather preventing the observation of a theoretically visible new moon crescent).

Alongside the civil luni-solar calendar, at least two simplified calendars were used in Mesopotamia. An early 'administrative' calendar in which each month contained thirty days (and therefore a year contained either 360 days in a normal year or 390 days in a year with an intercalary month) was used in bureaucratic contexts in the late third and parts of the second millennium BC.[5] A further simplification of the calendar in which the year was assumed to always have twelve months and each month thirty days, making a year of 360 days, is attested in a variety of second and first millennium contexts. This 'schematic' calendar, as I shall refer to it, is described in the fifth tablet of the creation epic *Enūma Eliš*, referred to in Old Babylonian prayers, and is used in certain astronomical and astrological texts from the second and the first millennium BC.[5] It should be stressed that both the administrative and the schematic calendars were used alongside the civil calendar in certain restricted contexts. Neither calendar replaced the luni-solar civil calendar used in everyday life.

As I have discussed in the introduction, calendars impact upon many aspects of daily life. In the following I briefly outline five examples of how the lives of people in Mesopotamia during the first millennium BC were affected by the calendar.

(i) Administration

Taxes, loans, wages and rations all have to be paid on agreed dates. As mentioned above, in the late third and early second millennium BC, calculations of rations etc were usually made using the administrative calendar in which the length of each month was artificially set at thirty days. It is not known whether similar practices were followed in the first millennium BC, or whether the true length of the civil month of twenty-nine or thirty days was taken into account. In general it seems that during this period wages were paid either for the outcome of a specified piece of work, regardless of the length of time it took (comparable to paying an author a set amount for a two-page article rather than by the hour, for example), or for long-term continuous work such as building work a salary was paid monthly or occasionally yearly.[7] Promissory notes generally state the interest to be paid on loans monthly, although occasional yearly interest rates are given.[8] A variety of interest rates are found in different texts,[9] but the most common rate is one shekel per mina per month. In a normal twelve-month year, this is equivalent to twelve shekels per mina or twenty percent (assuming standard interest), a convenient figure because of the metrology. Unfortunately, insufficient research has been undertaken to answer the question of how interest was assessed in years that contained an intercalary month. In particular, it would be interesting to know whether there are any cases of promissory notes which state a twelve shekel per mina annual interest rate in years where there are thirteen months.

(ii) Good and bad luck days

Hemerological texts list for every day of the year whether that day is favourable or unfavourable for a variety of activities including marriage, the signing of business contracts, the beginning of a legal case, and the signing of treaties.[10] The texts usually provide entries for thirty days of each month of the year, including intercalary months, but were clearly intended to be used with the civil luni-solar calendar—if a month only had twenty-nine days, then the entry in the hemerology could simply be ignored; similarly entries for an intercalary month could be ignored if the year did not contain an intercalary month.

We have ample evidence from the Neo-Assyrian period of how seriously the recommendations from the hermerologies were taken. For example, Essarhaddon's chief scribe Issar-šumu-ereš wrote to the king shortly after his accession to the throne advising him on the dates on which emissaries should sign the oath of allegiance, saying that the hemerologies indicate that it is bad to swear oaths on the fifteenth day, but favourable to do so on the sixteenth.[11] Livingstone has shown that the use of the hemerologies was widespread outside of the palace as well. Analysing the dates of preserved contracts and reports of extispicies, Livingstone found a clear correlation between dates on which they are most frequently attested and the favourable days for those activities given in the hemerologies. [12]

In order to use a hemerology it is necessary to know the date in the civil calendar. The favourable or unfavourable character of a day depended, for example, on whether that day was the thirtieth of one month or the first of the next month. Clearly this could cause difficulties in preparing for the signing of contracts or treatises at the beginning of the month if the length of the month was not known in advance.

(iii) Medicine

Several medical texts specify the day of the month on which certain remedies should be prepared or administered to a sick patient. In addition, a small number of texts correlate the ingredients used to make a medical remedy with the day of the year. These texts are among a group of texts, generally referred to as '*kalendertexte*', which use a numerical scheme to relate dates in the year with positions in the sky.[13] The dates and positions are then associated with various cultic objects including stones, plants, temples, cities, or the ingredients used to make a medical remedy. Two examples of this last type of *kalendertexte* are known from the collection of tablets belonging to a *mašmaššu* priest named Iqīšâ who lived in the city of Uruk during the late fourth century BC.[14] Each tablet covers a single month (SpTU III 104 concerns Month IV and SpTU III 105 concerns Month VIII) gives in four columns the name of the month, the day of the month from one to thirty, the sign of the zodiac, and the position within the zodiacal sign in degrees. Accompanying each entry is a statement of the ingredients to be used in the remedy to heal the patient.[15] For example, the first line of one of Iqīšâ's *kalendertexte* reads: 'Month IV 1 Aries 7 Sheep-blood, sheep-fat, and sheep-hair, you anoint'. In each line, the animal from which the remedy is to be made is linked with the sign of the zodiac given by the numerical scheme (in the present example Aries, which was traditionally the 'Hired Man' is identified as the 'Sheep', an identification known from several late texts).[16]

Texts such as the two *kalendertexte* belonging to Iqīšâ provide evidence for the use of calendrical data in the practice of medicine. The numerical scheme which underlies the *kalendertexte* is based upon the schematic calendar of twelve 30-day months, but, as with the hemerologies, in practice the *kalendertexte* were probably equated with the luni-solar civil calendar and unneeded day thirties simply ignored. In any case, it is clear that knowing the first day of the month was of importance in using the *kalendertexte* in the treatment of patients. An error of one day would mean that the wrong zodiacal sign would be given by the numerical scheme, and therefore the wrong ingredient used to produce the remedy.

(iv) Celestial divination

The correspondence between the Neo-Assyrian kings Esarhaddon and Assurbanipal and scholars employed to provide advice based upon the interpretation of ominous events provides ample evidence for the importance of celestial divination in the politics of Neo-Assyrian Mesopotamia.[17] The scholars performed a number of roles including the observation of celestial (and terrestrial) phenomena, the interpretation of these observations through their knowledge of the scholarly traditions of divination, and providing advice to the king (ostensibly, but not always in practice, based upon their observation and interpretation of ominous phenomena). Among the preserved correspondence sent by scholars to the king the most frequent subject of discussion was the observation of (or sometimes the failure to observe) the new moon crescent.[18] Typically, these discussions focus upon two issues: (i) the ominous interpretation of the new moon based upon the length of the previous month (generally speaking twenty-nine days is a bad omen, thirty days is a good omen), the appearance of the moon when it is first visible (for example, its colour, brightness, the shape of its horns), the presence of a halo, its position among the constellations, etc., and (ii) the establishment of the beginning of the month for the civil calendar. The importance of the new moon crescent in celestial divination may also be seen in the fact that lunar omens make up the first twenty-two of the seventy tablets of the standard celestial divination series *Enūma Anu Enlil*. Of these twenty-two, the first fourteen form their own subseries with the title IGI.DU$_8$.A.ME *šá* 30 'visibilities of the moon', and most of these omens deal with the appearance of the moon at first visibility.[19]

(v) Rituals and other cultic activities

Major public festivals such as the *Akītu* or 'New Year' festival were part of a detailed calendar of cultic activities in Mesopotamia.[20] The Akītu festival provided an ideological link between the king, the people and the gods, legitimizing the king's rule.[21] It was therefore important that all of the people celebrated the festival at the same time, necessitating advance planning for the first day of the new year. Other cultic activities such as the dressing and procession of the statues of gods were also strictly regulated in the cultic calendar. Calendrical uncertainty could clearly impact upon the preparation for and the proper performance of these rituals, as we can see from several examples of letters discussing the problem. For example, Marduk-šakin-šumi, the king's chief exorcist, wrote to the king during the sixth month of 671 BC to ask when a festival should take place on account of the king's decision to proclaim an intercalary sixth month:

> [Concerni]ng the intercalation [of] the year [about which the k]ing said as follows: "Let us add an intercalary Elul (VI)!" – the matter is (now) settled. [May the kin]g, my lord, live forever on account of that! [The king, my lo]rd, knows that Bel is dressed (for the festival) [on the 7]th of Tishri (VII); on the 8th day the gate (of the temple) is kept open, and the procession of Bel sets out as the month Nisa[n (I). The cerem]onies of the city of Der are conducted in the same way. [In fa]ct, [the king], should (now) decide what t[o d]o (with these ceremonies) [and send word] (about it).[22]

Should the ceremony scheduled for Month VII now take place in the following month (i.e., Month VI_2, which is the seventh month of the year), or should they wait until Month VII (now the eighth month of the year)? The king's answer was that the ceremony should wait until Month VII. But by the time his reply reached Babylon, the ceremony (which takes place over several days) had already started. The king's agent in Babylon, Mar-Issar, told the king that on learning of the king's decision, the ceremony had been suspended, and would resume the following month:

> As to what the king, my lord, wrote to me: "The month Elul (VI) is intercalary; do not perform the ceremonies this month" – Ammu-salam entered Babylon on the evening of the 6th day; the god Nabû had come before him, on the 3rd. The gate was kept open before Bel and Nabû on the 4th, the 5th and the 6th, and sacrifices were performed. When I saw the king my lord's sealed order, I issued the order: the rest of the ceremonies of Elul (VI) will be performed in the coming month, as the king, my lord, wrote to me.[23]

This demonstrates the very real practical problems for the maintenance of proper cultic activity by the uncertainty caused by lack of foreknowledge of when intercalation would take place.

The uncertainty of whether a month will have twenty-nine or thirty days also caused problems for the advance preparation for rituals, especially those that took place on the first day of the month. From the Neo-Babylonian period we have letters from temple workers writing to more important cult centres to ask about whether the month just ending had twenty-nine or thirty days and therefore whether the current day was, for example, the fourth or the fifth of the month.[24]

The examples described above show that the calendar played an important role in many aspects of daily life for the people of ancient Mesopotamia. Uncertainty surrounding the length of the month and in whether there would be an intercalary month this year could result in deviations from the proper time for cultic activities, the undertaking of a business venture on a 'bad-luck' day, or even the proscription of the wrong medical remedy. It has been argued by David Brown that a regulated calendar was '*insignificant* to the general population',[25] but the examples just discussed point instead to the importance of the calendar in everyday life, and the problems caused by calendrical uncertainty. While it may be true that a well-regulated calendar, astronomically speaking, was irrelevant to most people, a calendar that was regulated to the extent that it was commonly agreed upon and preferably known in advance, was clearly of benefit to a broad range of the population.

Given the importance of the calendar in civic, cultic and everyday life, therefore, it is not surprising that as soon as we have evidence for real interest in predictive astronomy in Mesopotamia (around the eighth century BC), we find attempts to use astronomy to provide ways of removing these uncertainties from the calendar. The correspondence of the Neo-Assyrian scholars includes several letters that contain discussions of attempts to predict in advance the length of the month.[26] Other letters concern whether intercalation is going to be necessary in the current or coming year.[27] The basis for these discussions may not be astronomically accurate, although in truth we know little about how such calculations were made during the Neo-Assyrian period, but clearly some methods were being used to try to provide regularity and advance knowledge of the calendar. Ultimate responsibility for the calendar itself, however, remained with the king, who was still able to make the final decision as to whether to intercalate and when the month began, and occasionally, no doubt, he made decisions for short-term benefit rather than on the basis of any advice from scribes knowledgeable in astronomy.

During the Neo-Babylonian period, the Babylonian king similarly retained the right to declare intercalations. For example, in letter from king Nabonidus (ruled 556–536 BC) to a certain Kurbanni-Marduk, the king announced that he had added an intercalary month XII in the current year. Similar letters announcing intercalations during the reigns of the Persian kings Cyrus (ruled 536–530 BC) and Cambyses (ruled 530–522 BC) come from the officials of the Esagila temple;[29] it is not certain whether this reflects a shift in responsibility for the decision on when to intercalate from the king to the temple or whether they simply acted a conduit for the kings instructions. Around this time, however, we find a significant change in intercalation practices in Babylonia: an increasing regularity of intercalation repeating in either an eight- or a nineteen-year cycle, with only occasional unexpected or delayed intercalations, often during periods of political instability.[30] Beginning in the early part of the fifth century BC a regular 19-year cycle of intercalation was used continuously until the end of the cuneiform record. Clearly, at some point during the early Persian period, the king transferred the authority for deciding when to intercalate to the astronomers. I have argued elsewhere that the transfer of power over intercalation from the king to his astronomical officials meant that the king lost the discretionary ability to implement intercalations which might provide him with short-term benefits (for the example, by omitting an intercalation, the king would receive tribute a month sooner than if the intercalation took place), but in return the country, including its government, received the long-term benefits of calendrical stability and predictability.[31] But handing over power for intercalation to his officials also may have meant an ideological break with centuries of tradition and it is tempting to ask whether this was only possible because the country was under foreign rule. A 'Babylonian' king may not have felt able to relinquish the power over the calendar in the way that a Persian king of Babylonia could.

A second change in operation of the Babylonian calendar took place over the Neo-Babylonian and Persian periods: an increasing interest in predicting the length of the month in advance. By the beginning of the sixth century BC methods for predicting the day of the beginning of the month using past observations of the 'lunar six' had been developed. The lunar six are six measurements of the time interval between the rising and setting of the sun and moon made on the evening of the moon's first visibility, on the morning and evenings around full moon, and on the morning of last visibility of the moon.[32] Babylonian astronomers developed simple but highly accurate methods for predicting in advance the values of future 'lunar six' using observations made eighteen and eighteen and a half years earlier. Several later tables describe these procedures, most notably an almost completely preserved tablet from the Seleucid period,[33] but records of lunar six observations and predictions from the late seventh and sixth centuries BC indicate that the methods were already known by about 600 BC.[34] The tablets which describe the procedure for calculating the lunar six also set out methods for predicting the length of the month based upon the calculated lunar six values.[35] Studies of preserved records of the length of the month indicate that during the Seleucid period (and quite probably earlier), most, probably all, month lengths were predicted in advance using these methods, rather than by observation.[36]

The last seven centuries therefore witnessed a transformation from a luni-solar calendar based primarily on observation, regulated at the discretion of the king, to a calculated lunisolar calendar, regulated by astronomical scholars. This transition brought advantages to Babylonian society in the form of increased certainty for administration, the preparation of rituals and the harmonization of cultic and civil life throughout the empire, but at a cost (for the king) of a reduction in his ability to manipulate the calendar for his own purposes and, perhaps, a lessoning of his prestige as the traditional focus of calendrical authority. Evidently, this was a price that was considered worth paying for the benefits of calendrical certainty.

China

The Chinese word *li* 曆 is usually translated as 'calendar', but encompasses a range of distinct concepts that are not all implied by the English word 'calendar'. Nathan Sivin has pointed to four different meanings for the word *li*: i) the practice of computing the times and celestial locations of certain astronomical phenomena in the past or future, or what is commonly called 'mathematical astronomy'; ii) a set of astronomical procedures for producing an ephemeris of predicted astronomical and calendrical events for a coming year; iii) a computational treatise describing a system of mathematical astronomy; and iv) the ephemerides produced by a system of mathematical astronomy and published in an almanac.[37] This last sense may be closest to what we consider a 'calendar' in English, but even here there are significant differences between the 'calendars' found in almanacs and, for example, a wall calendar or a desk diary. Almanacs contain not only a listing of the days of each month of the year but also predictions of astronomical events such as eclipses. The almanacs also provide a substantial body of hemerological data by which readers can decide on the appropriate day to undertake various activities.[38]

The yearly publication of the almanac was an important ideological and ceremonial event in Imperial China. The elaborateness of the ceremony accompanying the publication of the almanac can be seen from a detailed account by the missionary Peter Hoang of the ceremony as it was practiced in the first years of the twentieth century by the last rulers of the Qing 清 dynasty:

> Every year, on the 1st of the 2nd month, the Board of Mathematics presents to the Emperor three copies of the Annual Calendar for the following year, namely in Chinese, in Manchou and in Mongolian. Approbation being given, it is engraved and printed. Then on the 1st of the 4th month, two printed copies in Chinese are sent to the *Fan-t'ai* (Treasurer) of each province, that of *Chih-li* excepted; one of which, stamped with the seal of the Board of Mathematics, is to be preserved in the archives of the Treasury, while the other is used for engraving and printing for public use in the province.
>
> On the 1st day of the 10th month, early in the morning, the Board of Mathematics goes to offer Calendars to the Imperial court. The copies destined to the Emperor and Empresses are borne upon a sedan-like stand painted with figures of dragons (*Lung-t'ing*), those for the Princes, the Ministers and officers of the court being carried on eight similar stands decorated with silk ornaments (*Ts'ai-t'ing*). They are accompanied by the officers of the Board with numerous attendants and the Imperial band of music. On arriving at the first entrance of the palace, the Calendars for the Emperor are placed upon an ornamented stand, those for other persons being put upon two other stand on each side. The copies for the Emperor and his family and not stamped with the seal of the Board of

Mathematics, while the others are. The middle stand is taken into the palace, where the officers of the Board make three genuflections, each followed by three prostrations, after which the Calendars and handed to the eunuchs who present them to the Emperor, the Empress-mother, the Empress and other persons of the seraglio, two copies being given to each, viz. one in Chinese and one in Manchou. The master of ceremonies then proceeds to the entrance of the palace where the two other stands were left, and where the Princes, the Ministers with the civil and military mandarins, both Manchous and Mongols all in robes of state are in attendance. The master of ceremonies reads the Imperial decree of publication of the Calendars, namely: "The Emperor presents you all with the Annual Calendar of the year, and promulgates it throughout the Empire", which proclamation is heard kneeling. Then follow the three genuflections and nine prostrations, after which all receive the Calendar on their knees, the Princes two copies, one in Chinese and one in Manchou, the ministers and other officers only one, each in his own language. Lastly the Corean envoy, who must attend every year on that day, is presented kneeling with one hundred Chinese copies, to take home with him.

In the provinces, the *Fan-t'ai* (Treasurer), after getting some printed copies of the Calendar stamped with a special seal, also on the 1st of the 10th month, sends them on a sedan-like stand to the Viceroy or Governor, accompanies by the mandarin called *Li-wen-t'ing*, who is intrusted with the printing of the Calendar. The Viceroy or Governor receives them to the sound of music and of three cannon shots. The Calendars being set upon a stand between two tapers in the tribunal, the Viceroy or Governor, in robes of state, approaches the stand, and turning towards that quarter where Peking is situated, makes three genuflections and nine prostrations, after which ceremony he reverently receives the Calendars. The Treasurer sends the Calendar to all the civil and military Mandarins, all of whom, except those of inferior degree, receive it with the same forms. Any copies left are sold to the people. The reprinting of the Calendar is forbidden under a penalty (except in Fu-chien and Kuang-tong where it is tolerated). If therefore any copy is found without seal or with a false one, its author is sought after and punished. Falsification of the Calendar is punished with death; whoever reprints the Annual Calendar is liable to 100 blows and two months cangue.[39]

Hoang's account of the official promulgation of the calendar describes a ceremony that has a history stretching back over two thousand years: a remarkably similar description of the calendar presentation ritual is found in the *Jinshu* 晉書, the official history of the Jin 晉 dynasty compiled in the mid-seventh century AD, and there are discussions of a similar monthly ceremony from the Eastern Han 漢 dynasty (ruled AD 25–220) in the *Houhanshu* 後漢書, the official history of the Eastern Han compiled in the first part of the fifth century AD.[40] This ceremony of 'granting the seasons' was an important duty that the Chinese emperor needed to fulfill every year in order to demonstrate and maintain the legitimacy of his rule. Even in periods of non-native Chinese rule, such as the Mongol Yuan 元 dynasty (AD 1264–1367) or the Manchu Qing 清 dynasty (AD 1644–1911), emperors continued the tradition of performing the calendar promulgation ceremony.[41]

The Chinese emperor's role in the calendar promulgation ceremony was part of his duty to act as a link between heaven and earth.[42] The emperor ruled under the mandate of heaven (*tianming* 天命), a belief that rule was granted to him by heaven.[43] But this mandate could also be removed at any time if heaven was displeased by the emperor's rule. In order to retain the mandate, the emperor had to correctly perform the rituals of state to maintain the harmony of the cosmos. Issuing the annual almanacs and performing rituals on the correct dates given in these almanacs was part of this process. Thus the calendar in China not only had a practical role but was also essential to dynastic legitimacy. When a new dynasty came to power one of its first acts had to be to produce the calendar. The *Shiji* 史記 (compiled in the second century BC) explains:

> When a new dynasty rose by accepting the Heavenly Ordinance, at first it had to be prudent. It had to obey the will of Heaven by renewing the basis of all things: the calendar and the colour ... In founding a new dynasty, the Emperor should not depend upon former institutions.[44]

Many reforms of the calendar (*li*, in its sense of a whole system of mathematical astronomy) in China can be traced to this political need for change, and do not represent significant improvements over earlier calendars.

The belief in the mandate of heaven had a further consequence for the calendar in China. One way in which it could be claimed that an emperor's rule was loosing the mandate of heaven was in an increase in the number of observed portents.[45] According to the *Chunqiu fanlu* 春秋繁露 (second century BC):

> The source of all portents and anomalies lies in the faults that exist within the state. When faults have just begun to germinate, Heaven sends forth fearful portents to warn and inform the ruler of these faults. ... If one examines these portents and anomalies carefully, one will observe Heaven's intent. Heaven's intent desires certain things and does not desire other things. As for those things that Heaven desires and does not desire, if one examines oneself, one will surely find such warnings within oneself. If one observes affairs around oneself, one will surely find verification (of these warnings) in the state. Thus Heaven's intent is manifested in these portents and anomalies.[46]

This idea was developed into a political philosophy by the Western Han scholar Dong Zhongshu 董仲舒 in which astronomical and other portents were seen as a commentary on imperial rule.[47] As it applied to astronomical events, in broad terms unexpected astronomical events were of great importance as portents whereas astronomical events which had been predicted in advance were of little significance. The calendar, through its prediction of eclipses and other astronomical phenomena which were published in the almanacs, served as a way of regulating the heavens, reducing the number of potential portents. The calendar (in its broadest sense), therefore, was of great political significance.

The calendar (in the sense of a reckoning of days) that was used in China from the last few centuries BC until the end of the Imperial period was a calculated luni-solar calendar. Months began on the day the sun and moon were calculated to be in conjunction, an event that cannot be directly observed. Years contained twelve months unless intercalation was deemed necessary in which case an extra month was inserted into the year. Intercalation was also determined through calculation, usually by the use of a simple rule: beginning with the winter solstice the solar year was divided into twenty-four equal parts called *qi* 氣. The odd numbered *qi*, known as *zhongqi* 中氣, mark points in the year separated by about 30½ days. Since the mean synodic month is only about 29½ days there will occasionally occur a month which does not contain one of the *zhongqi*. When this happens, that month is designated as an intercalary month.[48] In addition a sixty-day cycle, the *ganzhi* 干支 cycle, was used to keep track of days and years.[49]

As I have discussed above, the Chinese luni-solar calendar was part of a bigger calendrical system that was geared towards the maintenance of cosmic harmony through the calculation of astronomical phenomena and the regulation of time.[50] As part of this astronomical system, the luni-solar calendar only made sense as a calendar based upon calculation, not on observation. Furthermore, cosmic harmony demanded that the calendar was the same throughout the country, something that could not be achieved with an observational lunar calendar.

Some Reflections

In general, when we think about lunar calendars, we tend to think of them being based upon the observation of the moon, with all the uncertainties that this implies. But in both China and Mesopotamia (at least at some periods) we find lunar calendars that are governed by calculation and therefore do not suffer from the same problems as an observed lunar calendar. The same is true of the late rabbinic Jewish calendar after the fourth century AD,[51] and at least some calendars used in the ancient Greek world.[52] But other calendars, such as the Islamic calendar, remained strictly based upon observation. It is interesting to ask why some cultures retain observational lunar calendars and other cultures adopt calculated calendars. Part of the answer may lie in the large amount of territory covered by China and the various empires that ruled Mesopotamia during the first millennium BC, but this cannot be the only reason. Other cultures which cover similarly large areas retained lunar calendars based strictly upon observation, the Islamic calendar being the obvious example. Instead we must look to the various ideological and practical reasons which lie behind the decision to adopt a calculated calendar in any particular culture.[53]

Although both China and Mesopotamia came to rely upon calculation for the operation of the calendar, there are differences in the way a calculated calendar was implemented in each culture. In China, a reliance on calculation to fix the calendar appears already in the earliest calendars that we know of in any detail (c. 300–200 BC). These early calculated luni-solar calendars rely only upon a value for the mean synodic month and knowledge of the 19-year cycle equating 235 lunar months with 19 solar years for the calculation of the beginning of each month and the determination of intercalations.[54] The calculation of both these aspects of the calendars was made without any allowance being made for the variable velocity of the sun and moon (solar and lunar anomaly). The first calendar to incorporate lunar and solar anomaly into the calculation of the beginning of the month was the *Lindeli* 麟德曆 designed by Li Chunfeng 李淳風 and adopted in AD 665; solar anomaly was ignored, however, in the calculation of the *zhongqi* used in determining intercalations until the *Shixianli* 時憲曆 compiled by Jesuit astronomers and adopted in AD 1644.[55] By contrast, in Mesopotamia the transition to a calculated calendar only took place after astronomical understanding had achieved a level where the resulting calendar was as accurate as the observed calendar was under ideal observing conditions (for example, perfect weather).[56] The reason may be that in China, the main motivation for calculating the calendar was ideological, in particular as one aspect of the emperor's role in maintaining cosmic harmony and as a demonstration of his continued possession of the mandate of heaven, whereas in Mesopotamia the transition to a calculated calendar was driven by practical needs, in particular through the simplification of administration and bureaucracy, the desire for certainty in the preparation for rituals and in ensuring all subjects of the empire maintained the same calendar. I should stress that in both China and Mesopotamia there were certainly other factors in the decision to rely upon a calculated calendar—some practical ones in China, some ideological ones in Mesopotamia, and some that fall outside of these reasons in both cultures—but these two general categories may provide useful models for understanding the motivations behind different types of lunar calendars in other cultures.

Notes

1. For modern examples, see Doggett and Schaefer (1994); for a historical example, see Steele (2008), p. 86.
2. See, for example, the case of the Rabbinic Jewish calendar discussed by Stern in this volume.
3. On the structure of the Mesopotamian calendar, see for example Britton (2007), Cohen (1993), Steele (2007), and Steele (2011b). It is uncertain whether intercalation was practiced in the Assyrian calendar of the second millennium BC; for contrasting views see Weidner (1928–29), Koch (1989), Reade (2000), Veenhof (2000), and Bloch's paper in the present volume.
4. Beaulieu (1993), Britton (2007).
5. Englund (1988), Brack-Bernsen (2007).
6. The 360-day calendar is sometimes referred to as the 'ideal' calendar in the scholarly literature because it apparently represented the perfect state of the universe at creation. In Steele (2011b) I have raised the question of whether the 'schematic' 360-day calendar found in astronomical texts was considered to be the same as the 'ideal' calendar, but for the present purposes it suffices to treat them as one.
7. Jursa (2010), pp. 660–680.
8. Wunsch (2002), pp. 234–238.
9. Jursa (2010), pp. 660–680.
10. For examples of Neo-Assyrian hemerologies, see Labat (1939) and Casaburi (2003). Many examples of Late Babylonian hemerologies remain unpublished.
11. Parpola (1993), no. 6.
12. Livingstone (1993), Livingstone (1997).
13. Brack-Bernsen and Steele (2004).
14. On Iqīšâ and his tablet collection, see Hunger (1971), Oelsner (2000), Robson (2008), pp. 227–240.
15. The texts are edited in von Weiher (1988), nos. 104 and 105.
16. Steele (2006), Steele (2011a). I have argued in Steele (2011a), pp. 337–338 that the ingredients given in these texts are not to be read literally (it can hardly have been practical to make remedies out of a lion), but should be understood as secret names ('dreckapotheke'). This has interesting consequences for how we understand the dreckapotheke tradition as the association between the 'secret' names and the real ingredients has its own meaning when the choice of ingredient is determined by the association of the secret name with an astronomical phenomena. I explore this issue further in a forthcoming paper.
17. Jean (2010). The correspondence is edited by Hunger (1992) and Parpola (1993).
18. Casaburi (2000–01).
19. Verderame (2002).
20. Landsberger (1915), Cohen (1993), Robbins (1996), Linssen (2004).
21. Bidmead (2002).
22. Parpola (1993), no. 253.
23. Parpola (1993), no. 357.
24. Beaulieu (1993).
25. Brown (2000), p. 195 (his italics).
26. Beaulieu (1993), Brown (2000), pp. 198–199.
27. Brown (2000), pp. 196–197.
28. YOS 3 115, see Parpola (1983), p. 504.
29. Parker and Dubberstein (1956), pp. 1–2, Parpola (1983), pp. 504–505, Kleber (2008), pp. 267–268.
30. Britton (2007).
31. Steele (2011b).

32. The term 'lunar six' is due to Sachs (1948), who gives a detailed description of the six intervals.
33. Brack-Bernsen and Hunger (2002).
34. Huber and Steele (2007), Huber and Britton (2007).
35. Brack-Bernsen (2002).
36. Steele (2007). For a more cautious conclusion, which argues that some month lengths were predicted and some were based upon observation, see Stern (2008).
37. Sivin (2011), pp. 41–42.
38. Sivin (2011), p. 42, Smith (1992), Chang (1940). For an English paraphrase of one traditional Chinese almanac, see Palmer (1986).
39. Hoang (1904), pp. 4–6. I have retained Hoang's Wade-Giles transcriptions of Chinese in the quotation.
40. Sivin (2011), p. 40.
41. Sivin (2011).
42. Loewe (1994), pp. 121–141.
43. On the 'mandate of heaven', see, for example, Fung (1953), pp. 71–73, Yabuuti (1973), Bodde (1991), p. 242, and Loewe (2004), pp. 421–456.
44. Translation taken from Yabuuti (1974), p. 53.
45. Eberhard (1957), Loewe (1994), pp. 121–141.
46. Translation taken from de Bary (1960), vol. 1, p. 164.
47. Loewe (1994), pp. 138–139.
48. The account I have given here is somewhat simplified. For further details, see, for example, Yabuuti (1963), Steele (2000), pp. 170–175, or the paper by Tsumura in the present volume.
49. On the early history of the *ganzhi* cycle, see, for example, Smith (2011).
50. See also Sivin (1969), p. 7.
51. See Stern (2001), p. 139, and the paper by Schiffman in the present volume.
52. For example the calendar found on the Antikythera Mechanism; see Freeth, Jones, Steele and Bitsakis (2008).
53. The papers by Ben-Dov and Feldman in the present volume discuss similar ideas in the context of Jewish calendars.
54. Sivin (1969), Sivin (2011).
55. Yabuuti (1963), pp. 460–470.
56. This level of understanding was reached with the discovery of the so-called 'Goal-Year' methods; see Brack-Bernsen (2002).

References

Beaulieu, P.-A., 1993, "The Impact of Month-lengths on the Neo-Babylonian Cultic Calendar", *Zeitschrift für Assyriologie* 83, 66–87.

Bidmead, J., 2002, *The Akītu Festival: Religious Continuity and Royal Legitimation in Mesopotamia* (Piscataway, New Jersey: Gorgias Press).

Bodde, D., 1991, *Chinese Thought, Society, and Science: The Intellectual and Social Background of Science and Technology in Pre-Modern China* (Honolulu: University of Hawaii Press).

Brack-Bernsen, L., 2002, "Predictions of Lunar Phenomena in Babylonian Astronomy", in J. M. Steele and A. Imhausen (eds.), *Under One Sky: Astronomy and Mathematics in the Ancient Near East* (Münster: Ugarit Verlag), 5–19.

——, 2007, "The 360-Day Year in Mesopotamia", in J. M. Steele (ed.), *Calendars and Years: Astronomy and Time in the Ancient Near East* (Oxford: Oxbow Books), 83–100.

——, and Hunger, H., 2002, "TU11: A Collection of Rules for the Prediction of Month Lengths", *SCIAMVS* 3, 3–90.

——, and Steele, J. M., 2004, "Babylonian Mathemagics: Two Mathematical Astronomical-Astrological Texts", in C. Burnett, J. P. Hogendijk, K. Plofker & M. Yano (eds.), *Studies in the History of the Exact Sciences in Honour of David Pingree* (Leiden: Brill), 95–125.

Britton, J. P., 2007, "Calendars, Intercalations and Year-Lengths in Mesopotamian Astronomy", in J. M. Steele (ed.), *Calendars and Years: Astronomy and Time in the Ancient Near East* (Oxford: Oxbow Books), 115–132.

Brown, D., 2000, *Mesopotamian Planetary Astronomy-Astrology* (Groningen: Styx).

Casaburi, M. C., 2000–01, "Calendrical Lunar Phenomena and their Interpretation According to Neo-Assyrian Sources: A Preliminary Survey", *Annali Instituto Universitario Orientale* 60–61, 9–57.

——, 2003, *Ūmē Ṭābūti "I Giorni Favorevoli"* (Padova: S.a.r.g.o.n. Editrice e Libreria).

Chang Hsueh-yen, 1940, *The Lunar Calendar as a Social Control Mechanism in Chinese Rural Life* (PhD thesis, Cornell University).

Cohen, M. E., 1993, *The Cultic Calendars of the Ancient Near East* (Bethesda, Maryland: CDL Press).

De Bary, T., 1960, *Sources of Chinese Tradition* (New York: Columbia University Press).

Doggett, L. E., and Schaefer, B. E., 1994, "Lunar Crescent Visibility", *Icarus* 107, 307–314.

Eberhard, W., 1957, "The Political Function of Astronomy and Astronomers in Han China", in J. K. Fairbank (ed.), *Chinese Thought and Institutions* (Chicago: University of Chicago Press), 33–70.

Englund, R. K., 1988, "Administrative Timekeeping in Ancient Mesopotamia", *Journal of the Economic and Social History of the Orient* 31, 121–185.

Freeth, T., Jones, A., Steele, J. M., and Bitsakis, Y., 2008, "Calendars with Olympiad Display and Eclipse Prediction on the Antikythera Mechanism", *Nature* 454, 614–617

Fung Yu-lan, 1953, *A History of Chinese Philosophy. Volume II: The Period of Classical Learning* (Princeton: Princeton University Press).

Hoang, P., 1904, *A Notice of the Chinese Calendar and a Concordance with the European Calendar* (Zi-ka-wei: Catholic Mission Press).

Huber, P. J., and Britton, J. P., 2007, "A Lunar Six Text from 591 B.C.", *Wiener Zeitschrift für die Kunde des Morgenlandes* 97, 213–218.

——, and Steele, J. M., 2007, "Babylonian Lunar Six Tablets", *SCIAMVS* 8, 3–36.

Hunger, H., 1971, "Die Tafeln des Iqīša", *Welt des Orients* 6, 163–165.

——, 1992, Astrological Reports to Assyrian Kings (Helsinki: Helsinki University Press).

Jean, C, 2010, "Divination and Oracles at the Neo-Assyrian Palace: The Importance of Signs in Royal Ideology", in A. Annus (ed.), *Divination and Interpetation of Signs in the Ancient World* (Chicago: The Oriental Institute), 267–275.

Jursa, M., 2010, *Aspects of the Economic History of Babylonia in the First Millennium BC* (Münster: Ugarit-Verlag).

Kleber, K., 2008, *Tempel und Palast. Die Beziehungen zwischen dem König und dem Eanna-Tempel im spätbabylonischen Uruk* (Münster: Ugarit-Verlag).

Koch, J., 1989, *Neue Untersuchungen zur Topographie des babylonischen Fixsternhimmels* (Weisbaden: Harassowitz).

Labat, R., 1939, *Hémérologies et Ménologies d'Assur* (Paris: Libraireie d'Amérique et d'orient adrien-maisonneuve).

Landsberger, B., 1915, *Der Kultische Kalender der Babylonier und Assyrer* (Leipzig: Hinriscs'sche Buchhandlung)

Linssen, M. J. H., 2004, *The Cults of Uruk and Babylon: The Temple Ritual Texts as Evidence for Hellenistic Cult Practice* (Leiden: Brill–Styx).

Livingstone, A., 1993, "The Case of the Hemerologies: Official Cult, Learned Formulation and Popular Practice", in Eiko Matsushima (ed.), *Official Cult and Popular Religion in the Ancient Near East* (Heidelberg: Universitätsverlag C. Winter), 97–113.

——, 1997, "New Dimensions in the Study of Assyrian Religion", in S. Parpola and R. M. Whiting (eds.), *Assyria 1995* (Helsinki: The Neo-Assyrian Text Corpus Project), 165–177.

Loewe, M., 1994, *Divination, Mythology and Monarchy in Han China* (Cambridge: Cambridge University Press).

——, 2004, *The Men Who Governed Han China* (Leiden: Brill).

Oelsner, J., 2000, "Von Iqīšâ und einigen anderen spätgeborenen Babyloniern", in S. Graziani (ed.), *Studi su vicino Oriente antico dedicati alla memoria di Luigi Cagni* (Naples: Instituto Universitario Orientale), 797–813.

Palmer, M., 1986, *T'ung Shu: The Ancient Chinese Almanac* (Boston: Shambhala).

Parker, R. A., and Dubberstein, W. H., 1956, *Babylonian Chronology 626 B.C. – A. D. 75* (Providence: Brown University Press).

Parpola, S., 1993, *Letters from Assyrian and Babylonian Scholars* (Helsinki: Helsinki University Press).

Reade, J., 2000, "Absolute Dates and Assyrian Calendars", *Akkadica* 119–120, 151–153.

Robbins, E., 1996, "Tabular Sacrifice Records and the Cultic Calendar of Neo-Babylonian Uruk", *Journal of Cuneiform Studies* 48, 61–87.

Robson, E., 2008, *Mathematics in Ancient Iraq* (Princeton: Princeton University Press).

Sivin, N., 1969, "Cosmos and Computation in Early Chinese Mathematical Astronomy", *T'oung Pao* 55, 1–73.

——, 2011, "Mathematical Astronomy and the Chinese Calendar", in J. M. Steele (ed.), *Calendars and Years II: Astronomy and Time in the Ancient and Medieval World* (Oxford: Oxbow Books), 39–51.

Smith, A., 2011, "The Chinese Sexagenary Cycle and the Ritual Origins of the Calendar", in J. M. Steele (eds.), *Calendars and Years II: Astronomy and Time in the Ancient and Medieval World* (Oxford: Oxbow Books), 1–37.

Smith, R. J., 1992, *Chinese Almanacs* (Hong Kong: Oxford University Press).

Steele, J. M., 2000, *Observations and Predictions of Eclipse Times by Early Astronomers* (Dordrecht: Kluwer).

——, 2006, "Greek Influence on Babylonian Astronomy?", *Mediterranaen Archaeology and Archaeometry* 6, 153–160.

——, 2007, "The Length of the Month in Mesopotamian Calendars of the First Millennium BC", in J. M. Steele (ed.), *Calendars and Years: Astronomy and Time in the Ancient Near East* (Oxford: Oxbow Books), 133–148.

——, 2008, *A Brief Introduction to Astronomy in the Middle East* (London: Saqi Books).

——, 2011a, "Astronomy and Culture in Late Babylonian Uruk", in C. L. N. Ruggles (ed.), *Archaeoastronomy and Ethnoastronomy: Building Bridges Between Cultures* (Cambridge: Cambridge University Press, 2011), 331–341.

——, 2011b, "Making Sense of Time: Observational and Theoretical Calendars", in K. Radner and E. Robson (eds.), *The Oxford Handbook of Cuneiform Culture* (Oxford: Oxford University Press, 2011), 470–485.

Stern, S., 2001, *Calendar and Community: A History of the Jewish Calendar 2nd Century BCE – 10th Century CE* (Oxford: Oxford University Press).

Veenhof, K. R., 2000, "Old Assyrian Chronology", *Akkadica* 119–120, 137–150.

Verderame, L., 2002a, "Enūma Anu Enlil Tablets 1–13", in J. M. Steele and A. Imhausen (eds.), *Under One Sky: Astronomy and Mathematics in the Ancient Near East* (Münster: Ugarit-Verlag), 447–457.

Weidner, E. F., 1928–29, "Der altassyrische Kalender", *Archiv für Orientforschung* 5, 184–185.

von Weiher, E., 1988, *Spätbabylonische Texte aus Uruk.* Teil III (Berlin: Mann).

Wunsch, C., 2002, "Debt, Interest, Pledge and Forfeiture in the Neo-Babylonian and Early Achaemenid Period: The Evidence from Private Archives", in M. Hudson and M. van de Mieroop (eds.), *Debt and Economic Renewal in the Ancient Near East* (Bethesda, Maryland: CDL Press), 221–253.

Yabuuti Kiyoshi, 1963, "Astronomical Tables in China from the Han to the T'ang Dynasties", in *Chūgoku Chūsei Kagaku Gijutsushi No Kenkyū* (Tokyo), 445–492.

——, 1973, "Chinese Astronomy: Development and Limiting Factors", in Shigeru Nakayama and N. Sivin (eds.), *Chinese Science: Explorations of an Ancient Tradition* (Cambridge, MA: The MIT Press), 91–103.

——, 1974, "The Calendar Reforms in the Han Dynasties and Ideas in their Background", *Archives Internationales d'Histoire des Sciences* 24, 51–65.